Advanced Sensors for Real-Time Monitoring Applications

Advanced Sensors for Real-Time Monitoring Applications

Editors

Olga Korostynska
Alex Mason

MDPI • Basel • Beijing • Wuhan • Barcelona • Belgrade • Manchester • Tokyo • Cluj • Tianjin

Editors
Olga Korostynska
Oslo Metropolitan University
Norway

Alex Mason
Norwegian University of Life Sciences
Norway

Editorial Office
MDPI
St. Alban-Anlage 66
4052 Basel, Switzerland

This is a reprint of articles from the Special Issue published online in the open access journal *Sensors* (ISSN 1424-8220) (available at: https://www.mdpi.com/journal/sensors/special_issues/Sensors_Real-Time_Monitoring_Applications).

For citation purposes, cite each article independently as indicated on the article page online and as indicated below:

LastName, A.A.; LastName, B.B.; LastName, C.C. Article Title. *Journal Name* **Year**, *Volume Number*, Page Range.

ISBN 978-3-0365-0426-1 (Hbk)
ISBN 978-3-0365-0427-8 (PDF)

© 2021 by the authors. Articles in this book are Open Access and distributed under the Creative Commons Attribution (CC BY) license, which allows users to download, copy and build upon published articles, as long as the author and publisher are properly credited, which ensures maximum dissemination and a wider impact of our publications.

The book as a whole is distributed by MDPI under the terms and conditions of the Creative Commons license CC BY-NC-ND.

Contents

About the Editors vii

Preface to "Advanced Sensors for Real-Time Monitoring Applications" ix

Irina Yaroshenko, Dmitry Kirsanov, Monika Marjanovic, Peter A. Lieberzeit, Olga Korostynska, Alex Mason, Ilaria Frau and Andrey Legin
Real-Time Water Quality Monitoring with Chemical Sensors
Reprinted from: *Sensors* 2020, 20, 3432, doi:10.3390/s20123432 1

Thi Thi Zin, Pann Thinzar Seint, Pyke Tin, Yoichiro Horii and Ikuo Kobayashi
Body Condition Score Estimation Based on Regression Analysis Using a 3D Camera
Reprinted from: *Sensors* 2020, 20, 3705, doi:10.3390/s20133705 25

Ritesh Misra, Hoda Jalali, Samuel J. Dickerson and Piervincenzo Rizzo
Wireless Module for Nondestructive Testing/Structural Health Monitoring Applications Based on Solitary Waves
Reprinted from: *Sensors* 2020, 20, 3016, doi:10.3390/s20113016 35

Daehyeon Yim, Won Hyuk Lee, Johanna Inhyang Kim, Kangryul Kim, Dong Hyun Ahn, Young-Hyo Lim, Seok Hyun Cho and Sung Ho Cho
Quantified Activity Measurement for Medical Use in Movement Disorders through IR-UWB Radar Sensor
Reprinted from: *Sensors* 2019, 19, 688, doi:10.3390/s19030688 53

Ning Chen, Jieji Zheng and Dapeng Fan
Pre-Pressure Optimization for Ultrasonic Motors Based on Multi-Sensor Fusion
Reprinted from: *Sensors* 2020, 20, 2096, doi:10.3390/s20072096 67

Se-beom Oh, Yong-moo Cheong, Dong-jin Kim and Kyung-mo Kim
On-Line Monitoring of Pipe Wall Thinning by a High Temperature Ultrasonic Waveguide System at the Flow Accelerated Corrosion Proof Facility
Reprinted from: *Sensors* 2019, 19, 1762, doi:10.3390/s19081762 83

David Gillett and Alan Marchiori
A Low-Cost Continuous Turbidity Monitor
Reprinted from: *Sensors* 2019, 19, 3039, doi:10.3390/s19143039 93

Nessrine Zemni, Fethi Bouksila, Magnus Persson, Fairouz Slama, Ronny Berndtsson and Rachida Bouhlila
Laboratory Calibration and Field Validation of Soil Water Content and Salinity Measurements Using the 5TE Sensor
Reprinted from: *Sensors* 2019, 19, 5272, doi:10.3390/s19235272 111

Wen Sha, Jiangtao Li, Wubing Xiao, Pengpeng Ling and Cuiping Lu
Quantitative Analysis of Elements in Fertilizer Using Laser-Induced Breakdown Spectroscopy Coupled with Support Vector Regression Model
Reprinted from: *Sensors* 2019, 19, 3277, doi:10.3390/s19153277 129

Baohua Zhang, Pengpeng Ling, Wen Sha, Yongcheng Jiang and Zhifeng Cui
Univariate and Multivariate Analysis of Phosphorus Element in Fertilizers Using Laser-Induced Breakdown Spectroscopy
Reprinted from: *Sensors* 2019, 19, 1727, doi:10.3390/s19071727 143

Joost Laurus Dinant Nelis, Laszlo Bura, Yunfeng Zhao, Konstantin M. Burkin,
Karen Rafferty, Christopher T. Elliott and Katrina Campbell
The Efficiency of Color Space Channels to Quantify Color and Color Intensity Change in
Liquids, pH Strips, and Lateral Flow Assays with Smartphones
Reprinted from: *Sensors* **2019**, *19*, 5104, doi:10.3390/s19235104 153

Chin-Chi Cheng, Yen-Hsiang Tseng and Shih-Chang Huang
An Innovative Ultrasonic Apparatus and Technology for Diagnosis of Freeze-Drying Process
Reprinted from: *Sensors* **2019**, *19*, 2181, doi:10.3390/s19092181 173

Lu Peng, Genqiang Jing, Zhu Luo, Xin Yuan, Yixu Wang and Bing Zhang
Temperature and Strain Correlation of Bridge Parallel Structure Based on Vibrating Wire
Strain Sensor
Reprinted from: *Sensors* **2020**, *20*, 658, doi:10.3390/s20030658 187

Xiangming Liu and Valéri L. Markine
Train Hunting Related Fast Degradation of a Railway Crossing—Condition Monitoring and
Numerical Verification
Reprinted from: *Sensors* **2020**, *20*, 2278, doi:10.3390/s20082278 203

X. Liu and V. L. Markine
Correlation Analysis and Verification of Railway Crossing Condition Monitoring
Reprinted from: *Sensors* **2019**, *19*, 4175, doi:10.3390/s19194175 223

Qiuming Nan, Sheng Li, Yiqiang Yao, Zhengying Li, Honghai Wang, Lixing Wang and
Lizhi Sun
A Novel Monitoring Approach for Train Tracking and Incursion Detection in Underground
Structures Based on Ultra-Weak FBG Sensing Array
Reprinted from: *Sensors* **2019**, *19*, 2666, doi:10.3390/s19122666 245

Lin Li, Hongli Hu, Yong Qin and Kaihao Tang
Digital Approach to Rotational Speed Measurement Using an Electrostatic Sensor
Reprinted from: *Sensors* **2019**, *19*, 2540, doi:10.3390/s19112540 261

Zhinong Jiang, Yuehua Lai, Jinjie Zhang, Haipeng Zhao and Zhiwei Mao
Multi-Factor Operating Condition Recognition Using 1D Convolutional Long
Short-Term Network
Reprinted from: *Sensors* **2019**, *19*, 5488, doi:10.3390/s19245488 281

Ran Jia, Biao Ma, Changsong Zheng, Xin Ba, Liyong Wang, Qiu Du and Kai Wang
Comprehensive Improvement of the Sensitivity and Detectability of a Large-Aperture
Electromagnetic Wear Particle Detector
Reprinted from: *Sensors* **2019**, *19*, 3162, doi:10.3390/s19143162 299

Andrzej Felski, Krzysztof Jaskólski, Karolina Zwolak and Paweł Piskur
Analysis of Satellite Compass Error's Spectrum
Reprinted from: *Sensors* **2020**, *20*, 4067, doi:10.3390/s20154067 319

About the Editors

Olga Korostynska has a B.Eng. (1998) and M.Sc. (2000) in Biomedical Engineering from National Technical University of Ukraine (KPI); Ph.D. (2003) in Electronics and Computer Engineering from the University of Limerick, Limerick, Ireland; and LLB (2011) from the University of Limerick, Limerick, Ireland. Currently, she is a Professor in Biomedical Engineering at Oslo Metropolitan University (OsloMet), Oslo, Norway. Before that, she was a Senior Lecturer in Advanced Sensor Technologies at the Liverpool John Moores University, Liverpool, UK. She was an EU Postdoctoral Research Fellow developing electromagnetic wave sensors for real-time water quality monitoring, as well as a Postdoctoral Researcher at the University of Limerick, working on a number of projects, including projects funded by IRCSET, EI, and EU FP7. She also was a Lecturer in Physics at Dublin Institute of Technology, Dublin, Ireland. Olga has co-authored a book, 15 book chapters, 4 UK patents, and over 200 scientific papers in peer-reviewed journals and conference proceedings.

Alex Mason has a B.Sc. (Hons), 2005, in Computer and Multimedia Systems from the University of Liverpool and Ph.D., 2008, in Wireless Sensor Networks and their Industrial Applications from Liverpool John Moores University. He led a sensor research team in Liverpool for several years as Reader in Sensor Technologies, before moving to Norway to take a dual industry/academic role. Today he is a Project Engineer at Animalia AS, Oslo, as well as Research Professor at the Norwegian University of Life Sciences (NMBU), Ås. Alex has a strong track record of over 200 publications, including several patents. He also leads a team working on food automation topics at NMBU, and amongst other activities co-ordinates the H2020 project RoBUTCHER, where sensing and robotics are key themes.

Preface to "Advanced Sensors for Real-Time Monitoring Applications"

It is impossible to imagine the modern world without sensors, or without real-time information about almost everything—from local temperature to material composition and health parameters. We sense, measure, and process data and act accordingly all the time. In fact, real-time monitoring and information is becoming the key to a successful business, an assistant in life-saving decisions that healthcare professionals make, a facility to optimize value-chains in manufacturing, and a tool in research that could revolutionize the future. To ensure that sensors address the rapidly developing needs of various areas of our lives and activities, scientists, researchers, manufacturers, and end-users have established an efficient dialogue so that the newest technological achievements in all aspects of real-time sensing can be implemented for the benefit of the wider community. This book documents some of the results of such a dialogue and reports on advances in sensors and sensor systems for existing and emerging real-time monitoring applications.

Olga Korostynska, Alex Mason
Editors

Review

Real-Time Water Quality Monitoring with Chemical Sensors

Irina Yaroshenko [1], Dmitry Kirsanov [1,*], Monika Marjanovic [2], Peter A. Lieberzeit [2], Olga Korostynska [3,4], Alex Mason [4,5,6], Ilaria Frau [6] and Andrey Legin [1]

[1] Institute of Chemistry, St. Petersburg State University, Mendeleev Center, Universitetskaya nab. 7/9, 199034 St. Petersburg, Russia; irina.s.yaroshenko@gmail.com (I.Y.); andrey.legin@gmail.com (A.L.)
[2] Faculty for Chemistry, Department of Physical Chemistry, University of Vienna, Waehringer Strasse 42, 1090 Vienna, Austria; Monika.Marjanovic@univie.ac.at (M.M.); Peter.Lieberzeit@univie.ac.at (P.A.L.)
[3] Faculty of Technology, Art and Design, Department of Mechanical, Electronic and Chemical Engineering, Oslo Metropolitan University, 0166 Oslo, Norway; olga.korostynska@oslomet.no
[4] Faculty of Science and Technology, Norwegian University of Life Sciences, 1432 Ås, Norway; alex.mason@nmbu.no
[5] Animalia AS, Norwegian Meat and Poultry Research Centre, P.O. Box 396, 0513 Økern, Oslo, Norway
[6] Faculty of Engineering and Technology, Liverpool John Moores University, Liverpool L3 3AF, UK; i.frau@2016.ljmu.ac.uk
* Correspondence: d.kirsanov@gmail.com; Tel.: +7-921-333-1246

Received: 25 May 2020; Accepted: 14 June 2020; Published: 17 June 2020

Abstract: Water quality is one of the most critical indicators of environmental pollution and it affects all of us. Water contamination can be accidental or intentional and the consequences are drastic unless the appropriate measures are adopted on the spot. This review provides a critical assessment of the applicability of various technologies for real-time water quality monitoring, focusing on those that have been reportedly tested in real-life scenarios. Specifically, the performance of sensors based on molecularly imprinted polymers is evaluated in detail, also giving insights into their principle of operation, stability in real on-site applications and mass production options. Such characteristics as sensing range and limit of detection are given for the most promising systems, that were verified outside of laboratory conditions. Then, novel trends of using microwave spectroscopy and chemical materials integration for achieving a higher sensitivity to and selectivity of pollutants in water are described.

Keywords: water quality; real-time monitoring; multisensor system; molecularly imprinted polymers; functionalised coating; microwave spectroscopy

1. Introduction

Water is one of the major natural resources for people. In 2012 it was declared that a safe water supply for every person is a crucially important task worldwide [1]. There are special water sustainability guides issued by the World Health Organization and regulated water quality standards [2]. The United Nations Sustainable Development Goals are the blueprint to achieving a better and more sustainable future for all-goal six specifically aims to ensure clean and accessible water. This, in turn, requires adequate water quality monitoring solutions specific to the situation. For example, summer 2019 was marked by catastrophic events in Norway, when more than 2000 people became sick, with more than 60 being hospitalized and 2 people dying as a result of an outbreak of *Campylobacter* and *Escherichia Coli* (*E. Coli*) that arose in the drinking water in Askøy, on the west coast of Norway. It is even more remarkable that this occurred in a country which has the status of being one of the countries with the highest quality of water in the world. The exact origin of the bacterial contamination that has caused this is still not confirmed, but the fact that there is a need for real-time monitoring of all drinking water reservoirs everywhere is undisputed. Therefore, this review examines various technologies that could meet these demands.

The conventional approach to qualitative water analysis assumes application of various chemical, physical and microbiological methods [3]. Most such methods demand specialized laboratories equipped with expensive and sophisticated scientific devices. Furthermore, highly qualified personnel are needed to operate such devices and special efforts and manpower must be spent for representative water sampling. More effective water quality control methods must be developed. Such methods should be fast, low-cost, with minimum automatic sampling and, ultimately, provide real-time results.

2. Current Situation with Online Water Analysis

A comprehensive description of the current situation with online water analysis can be found in [4]. Mobile chemical analysis stations were used in this work to monitor different water parameters. The systems were deployed in specially produced trailers (Figure 1) that were towed to the river banks.

Figure 1. Mobile station for water quality monitoring and a sample of typical output of this station. Reprinted from Science of the Total Environment, Vol. 651, Angelika M. Meyer, Christina Klein, Elisabeth Fünfrocken, Ralf Kautenburger, Horst P. Beck, Real-time monitoring of water quality to identify pollution pathways in small and middle scale rivers, Pages 2323–2333, Copyright (2019), with permission from Elsevier.

The sensors, based on various principles, measured temperature, total phosphorus, pH and ammonium ions (by standard electrochemical sensors), dissolved oxygen, conductivity, nitrate ions and total organic carbon (by optical sensors). The measurements were performed by different stations in different locations. The measurement accuracy was within 10% for most measured parameters over 5 years of experiments in 35 locations along 25 small- and medium-size rivers. Such stations may likely improve our understanding of pollution types and pathways depending on water basins, seasonal factors and anthropogenic load. However, such stations cannot be considered as a practical instrument for wide-scale water quality monitoring due to their high cost, need for maintenance and significant power supply requirements.

Chemical sensors are attractive instruments for water quality analysis. The electrochemical or optical properties of such sensors may depend on the concentration of analytes in the water. Such sensors are already widely applied to the analysis of natural and potable water [5].

The growth of publication numbers in the field is shown in Figure 2.

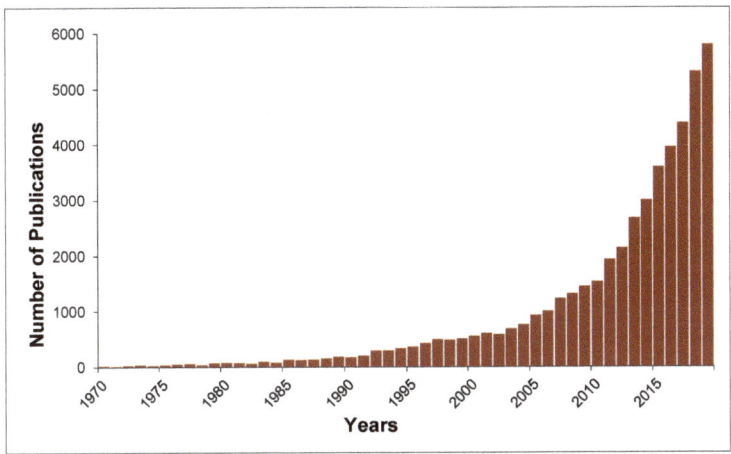

Figure 2. Number of publications published on the topic of the review in the last 50 years. Search keywords: "real-time water quality sensors". Scopus (October 2019).

3. Water Quality Monitoring Systems

The first sensor really suitable for water monitoring was the glass pH electrode, which appeared, in the present shape, along with a pH meter, around 1930. Since then, pH is a primary parameter of most water monitoring devices.

It is obvious, however, that multiple water parameters must be evaluated to responsibly judge its quality and multisensor systems should be applied for such purposes.

There have been multiple attempts to develop multisensor systems that could be applied for water quality control, e.g., [6,7]. However, the first efforts mostly dealt with laboratory water analysis rather than being applied in real-time, online mode.

For instance, a voltammetric sensor array with four electrodes (Au, Pt, Ir and Rh) served for multisite water quality monitoring at a water treatment plant [8]. The aqueous samples were taken at nine filtration steps, as well as before and after the complete procedure of water purification. The voltammetric data were processed by principal component analysis (PCA), revealing pronounced difference between raw, rapidly filtered and clean water. However, it was observed that the samples collected after treatment by several slow filters were close to the rapidly filtered water samples on the PCA scores graph. This can potentially be explained by the low efficiency of these filters. Therefore, it was concluded that a multisensor system approach is suitable for continuous control of water quality at treatment facilities, indication of the possible malfunctioning units and for checking the water status after maintenance. Significant influence of sensor drift and the necessity to compensate this drift was pointed out.

The system developed in [9] was designed to measure pH, temperature, dissolved oxygen, conductivity, redox potential and turbidity. This set of parameters is the most common one in water quality assessment since the sensors for these parameters can run in continuous mode. The whole set of sensors was mounted on aluminum oxide. All sensors were united into a single PVC body and their outputs were collected by the data acquisition system, which could also perform remote data transmission. The work suggests that the body might be dipped into water or even built into water flow. The device also included a set of electric valves and pumps for sampling, cleaning and calibration. The authors proposed that such a portable system can be suitable for water quality monitoring from different sources.

Chinese authors published a paper where a multisensor system was applied for the determination of several elements such as iron, chromium, manganese, arsenic, zinc, cadmium, lead and copper [10].

The device comprised three analytical detection systems: a multiple light-addressable potentiometric sensor (MLAPS) based on a thin chalcogenide film for simultaneous detection of Fe(III) and Cr(VI) and two groups of electrodes for detection of other elements using anodic and cathodic stripping voltammetry. The following detection limits were obtained: Zn—60 µg/L, Cd—1 µg/L, Pb—2 µg/L, Cu—8 µg/L, Mn—60 µg/L, As—30 µg/L, Fe—280 µg/L and Cr—26 µg/L. The authors recommended their method to determine metals simultaneously in seawater and wastewater; however, the possibility of application of this device for online analysis was unclear.

Potable water quality is of primary interest to people. Such type of water was studied in [11], using two sensor stations. The first one was used for detecting free chlorine with a precision of 0.5% and limit of detection (LoD) of 0.02 mg/L, as well as total chloride by colorimetric method with precision of 5% and LoD 0.035 mg/L. The second station used was a multisensor for detection of pH, redox potential, dissolved oxygen, turbidity and conductivity. Eleven different contaminants were injected into the flow of the studied liquid, namely pesticides, herbicides, alkaloids, E. coli, mercury chloride and potassium ferricyanide. It was demonstrated that the set of sensors produces a response for each type of contaminants. Unfortunately, the work does not report any data about the precision of such systems during long-term application.

Another device for online water analysis was suggested in [12]. Fourteen buoys were installed in a freshwater lake; each of them was equipped with three ion-selective electrodes detecting the concentration of ammonium ions along with nitrate and chloride. Wireless connection between buoys could be implemented using Global System for Mobile Communications (GSM) and General Packet Radio Services (GPRS) protocols. The data was accumulated in a single place. The data was accessible via the internet allowing real-time control of the system performance (Figure 3).

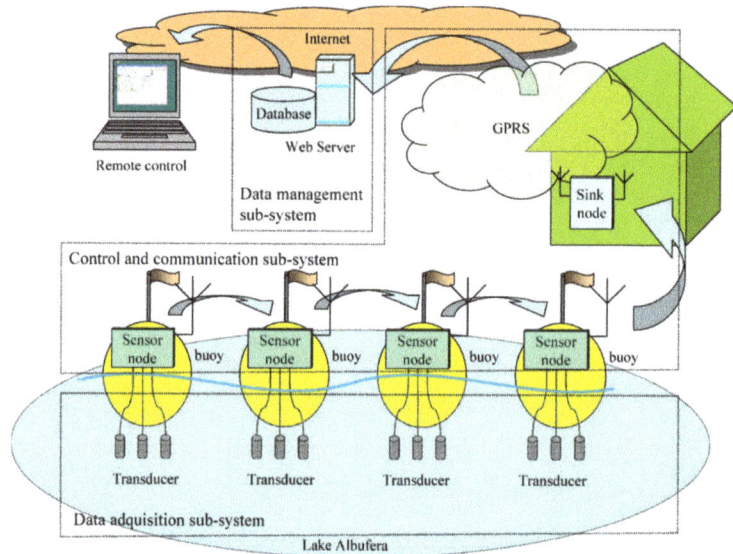

Figure 3. System for water quality monitoring. Reprinted from Talanta, Vol. 80, J.V. Capella, A. Bonastre, R. Ors, M. Perisa, A Wireless Sensor Network approach for distributed in-line chemical analysis of water, Pages 1789–1798, Copyright (2010), with permission from Elsevier.

The authors reported daily drift of sensor readings of about 1.5 mV. Since the measurements were performed for only 7 days, the accumulated drift was not that significant. However, the influence of drift can be critical over longer periods of time.

Two multisensor systems were suggested in [13] for environmental monitoring of various contaminants. One was dealing with contents of ammonium, potassium, and sodium in a river with low anthropogenic load. The second system was installed in river water in a populated region and was designed for detection of heavy metals such as copper, lead, zinc and cadmium. The systems were also equipped with radio transmission devices. The system was tested just for 8 h, which is obviously too short a period for any serious conclusions about such a technology.

Wider research was performed in [14], which was conducted over a period of 12 months. A sensor array comprising eight conducting polymer sensors for gas phase analysis was used to detect abrupt changes in the wastewater quality. Free gas emanating from bubbled liquid in the flow cell with constant temperature was delivered to the sensor chamber for analysis. The results of field tests at the water treatment plants using automatic systems produced water quality profiles and displayed the possibility of determining both random and model contaminants. This approach showed high sensitivity and flexibility and low dependence on long-term drift, daily oscillations, temperature and humidity. It must be noted, however, that the described experiments were carried out not at a real water treatment plant, but in a pilot system. Thus, the diversity of its performance may not be representative for real-world conditions. Besides, the idea to follow water quality via headspace analysis is obviously limited: it is impossible to follow contaminants which are not volatile enough.

4. Application of Biosensors and Optical Sensors for Water Quality Assessment

Biosensors were also used for water quality control, though quite a few of them were applied for online flow analysis. Pesticides were the main target of biosensors.

A system of biosensors capable of determining dichlorvos and methylparaoxon in the water was suggested in [15]. The systems consisted of three amperometric biosensors based on various AChE (acetylcholinesterase) enzymes. These enzymes were immobilised in a polymeric matrix onto the surface of screen-printing electrodes. The enzymes solutions were deposited over the electrode surface and irradiated by light, inducing photo polymerisation of the azide groups in the molecules. Such a sensor array was built into a flow system permitting automatic analysis. Bottled and river water was studied. The concentration of pesticides was detected in the ranges 10^{-4}–0.1 µM for dichlorvos and 0.001–2.5 µM for methylparaoxon. Solely spiked samples were considered; therefore it is necessary to further verify performance of such a system in online mode.

Another work [16], reports on using Pt electrodes instead of screen-printed ones and self-made carbon paste was applied as a sensing layer. The paper implies that such a procedure may improve sensitivity of the substrate for some of the immobilized enzymes. The total number of biosensors in the array was eight. The ultimate aim of this research was not a quantification of pollutants but a global evaluation of water quality. It is doubtful though, if such a quality can be precisely determined by biosensors, which are highly selective to the main substance and would exhibit low cross-sensitivity to many other analytes present in the natural water.

One more attempt to evaluate global water toxicity by biosensors is described in [17]. The online toxicity monitoring system employed sulfur oxidizing bacteria (SOB) and consisted of three reactors. No toxicity changes in the natural flow water were observed over a period of six months. When the flow was spiked with diluted pig farm waste, the activity of sulfur oxidizing bacteria decreased by 90% in 1 h. The addition of 30 µg/L of nitrite ions or 2 µg/L of dichromate ions resulted in full degradation of sulfur oxidizing bacteria activity in 2 h. Thus, the sensitivity of the system to both inorganic and organic pollutants was demonstrated. It must be noted that one or two hours is a rather long period of time for detecting acute contaminations; functionality of the system could be regained only by introducing a new portion of bacteria, which significantly impairs real-time, online application of such device.

Optical sensors were also recently applied for water quality analysis, however, these are mostly discrete sensors, though tuned sometimes for integral parameters such as water color, turbidity or even COD and BOD. Discrete sensors were used to determine chlorophyll in the seawater on the basis of its fluorescence [18], for evaluation of water opacity and color evolution by LED [19], for analysis of water turbidity and color in online mode [20] as well as for determination of heavy metal ions [21].

5. Biomimetic Approaches for Sensing Water Quality

5.1. Chemical Sensors for Sensing in "Real-Life Environments"

Biosensors reveal exceptional selectivity and often sensitivity, but usually are limited in terms of ruggedness and technical applicability in non-physiological conditions. One way to overcome this is to implement bioanalogous selectivity into systems that are able to withstand harsh and non-physiological conditions, so-called biomimetic systems [22]. Molecularly imprinted polymers (MIPs) are a promising example of such synthetic materials [23], since they are robust due to their highly cross-linked nature. Furthermore, they come at much lower costs than natural materials and provide longer storage and use periods. MIPs can also be produced for molecules that cannot be detected by natural receptors [24].

MIPs are generally synthesized by co-polymerization of functional and cross-linking monomers in the presence of a template (see Figure 4). Initially, a complex forms between functional monomers and the template through weak, noncovalent interactions (mainly hydrogen bonds, Van-der-Waals or π-π interactions), followed by polymerisation with cross-linking monomers to form a rigid, three-dimensional polymeric network. Removal of the template leads to recognition sites (cavities) within the polymer that are complementary to the target molecule in size, shape, and chemical functionality and are suitable to selectively rebind the analyte [25].

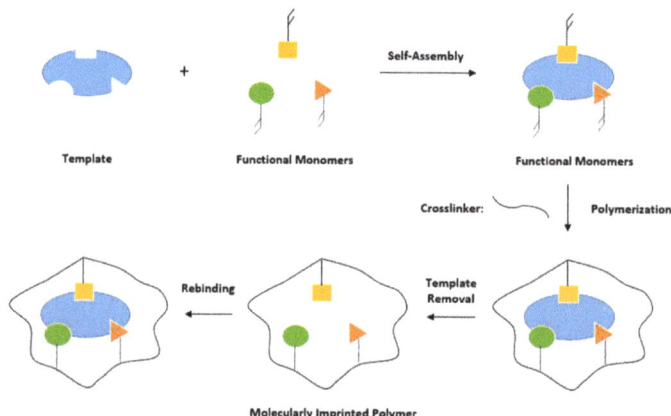

Figure 4. Schematic overview of molecular imprinting.

Except for MIPs, target recognition can also be obtained using other strategies. Aptamers, for example, are single-stranded RNA or DNA oligonucleotides, whose tertiary structure selectively binds their target molecules [26]. Another option is whole-cell-based sensors, which were also already applied to real wastewater samples [27,28]. In this case, mammalian cells were used for detecting harmful and toxic compounds, because their closeness physiology is close to that of humans. Although this strategy may not be regarded biomimetic in the strict sense of the word, it provides direct information about the overall toxicity of samples rather than detecting or quantifying one specific substance. When applying this method, unknown or new chemicals and pollutants may be detected. Kubisch et al. used rat myoblast cells in combination with a commercially available multiparametric readout system to measure impedance (morphological integrity) and two metabolic parameters—acidification and respiration—to investigate the overall toxicity and bioavailability of substances in water samples [27]. This was achieved by using three different types of electrodes on a single chip surface: impedance, pH and oxygen (CLARK) electrodes. After testing different test compounds, including metal ions and neurotoxins, the system was exposed to real wastewater samples. It responded to different contaminants and was indeed suitable for monitoring unknown, harmful compounds in water. Similarly, the group of C. Guijarro applied rat liver cells in a whole-cell-based sensor system to monitor environmental contaminants, including an insecticide and a flame retardant, in water samples using the same analyzing system [28].

The aforementioned advantages of MIPs make them an attractive tool for different applications, such as solid phase extraction, drug targeting, development of sensors for various types of analytes, and environmental monitoring. Although the number of publications concerning MIP-based sensors is rising, only a small amount is actually applied to real-life environments or complex matrices. This part of the review will provide an overview of chemical sensors that were tested in (real-life) water samples with a special focus on receptor layers based on MIPs.

5.2. MIP-Based Sensors for Water Analyses

In 2018, Ayankojo et al. introduced a sensor system capable of detecting pharmaceutical pollution in aqueous solutions [29]. They chose amoxicillin as the model analyte and implemented a hybrid MIP, consisting of organic and inorganic components, on the gold surface of a surface plasmon resonance (SPR) transducer. The hybrid MIP film was synthesized by applying the sol-gel technique and using methacrylamide as organic monomer and vinyltrimethoxysilane as inorganic coupling agent to form a stable and rigid polymeric network. Sol-gels have a highly porous structure and recognition sites are usually formed in a more ordered way. This results in enhanced sensitivity and faster sensor response times. Rebinding experiments of the amoxicillin MIP in phosphate-buffered saline (PBS) and tap water revealed an imprinting factor of 16 compared to the nonimprinted polymer (NIP) and a limit of detection LoD = 73 pM. Furthermore, the MIP responded almost exclusively to its target analyte thus exhibiting utmost specificity. In the same year, the group of Cardoso also developed a sensor for detecting chloramphenicol, an antibiotic used in fish farms [30] (see Figure 5). The corresponding MIPs were electro-polymerized on screen-printed carbon electrodes.

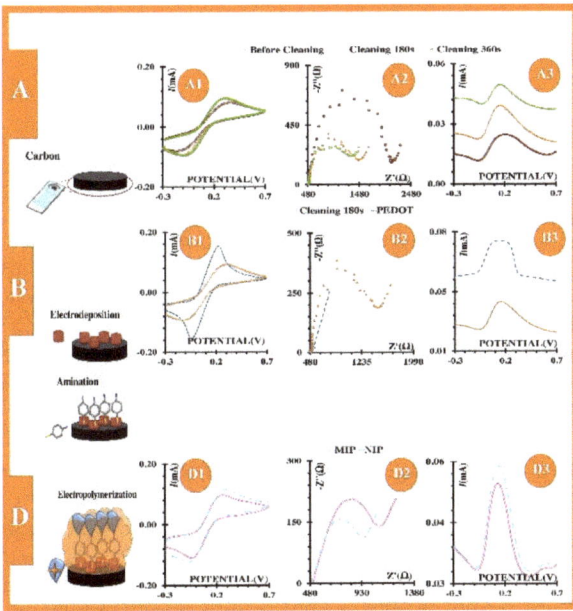

Figure 5. Construction principle and setup of chloramphenicol sensor. Reproduced from [30] with permission © Elsevier B.V. 2018.

Impedance and square wave voltammetry (SVW) in both electrolyte solution and water from a fish tank served to investigate the performance of the recognition element. In case of impedance measurements in electrolyte solution, sensor characteristics were linear in a concentration range from 1 nM to 100 µM, achieving an LoD = 0.260 nM; SVW yielded similar characteristics and an LoD = 0.653 nM. In real-life samples—water from a fish tank—sensors responded linearly down to 1 nM and achieved an LoD of 0.54 nM and 0.029 nM for impedance and SVW measurements, respectively. These results suggest that there is no significant impact on sensor behavior when switching from standard solutions to real water samples leading to reproducible and sensitive sensor characteristics over five orders of magnitude down to 1 nM.

The real-life feasibility of MIP-based sensor systems were also demonstrated in case of detection of faecal contamination of seawater samples [31]. MIP nanoparticles were fabricated for sensing *Enterococcus faecalis* (*E. faecalis*) serving as faecal indicator to assess the water quality. Such MIP nanoparticles have the advantage of a higher surface-to-volume ratio, which means that the resulting cavities or binding sites are easier to access by target analytes [5]. *E. faecalis*-imprinted nanoparticles demonstrated good SPR sensor performance in aqueous and real seawater samples:

As can be seen from Figure 6, changes in refractive index were linear in a concentration range from 2×10^4–1×10^8 CFU/mL covering four orders of magnitude with a limit of detection of 1.05×10^2 CFU/mL. Selectivity studies with structurally similar bacteria revealed higher affinity of MIP nanoparticles towards the imprinted analyte compared to the other competitors. Selectivity coefficients for *E. coli*, *Staphylococcus aureus* and *Bacillus subtilis* were as follows: 1.38, 1.25 and 1.37.

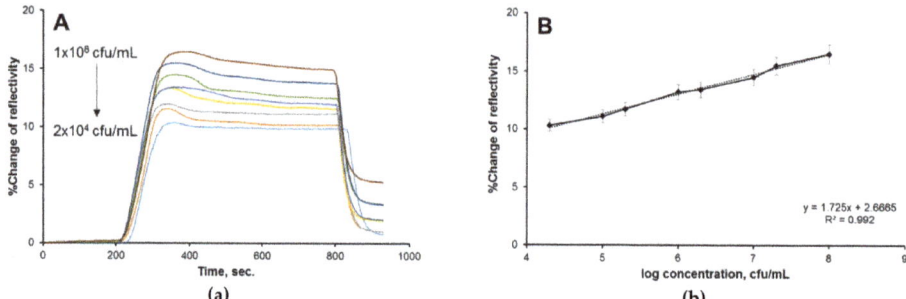

Figure 6. Sensor responses for faecal indicators, showing the (**a**) % change of reflectivity by time and (**b**) its linear correlation with the concentration. Reproduced from [31] with permission © Elsevier B.V. 2019.

Khadem et al. fabricated an electrochemical sensor for detecting diazinon, an insecticide, based on a modified carbon paste electrode combined with MIPs and multi-wall carbon nanotubes (MWCNTs) [32]. Using the latter modifier improves conductivity, whereas MIPs offer the necessary sensitivity towards the template molecule. After optimizing electrode composition, the method was first validated in aqueous standard solutions. SVW measurements revealed that the MIP showed much higher affinity to the analyte than the reference, the nonimprinted polymer; the system achieved linear performance in the concentration range from 5×10^{-10} to 1×10^{-6} mol/L with a calculated LoD = 1.3×10^{-10} mol/L. Furthermore, it was considerably more selective to the analyte than to other tested substances (ions and other pesticides). To investigate the applicability of the system to real biological and water samples, different amounts of diazinon were spiked to urine, tap and river water. In all these cases the sensors detected the target analyte with high recovery rates (>92%). This work demonstrates the use of MIP-based sensors in real-life samples and environments without the need of special sample pretreatment or preconcentration steps.

Another example for pesticide detection is presented in the work of Sroysee et al. [33]. They developed an MIP-based quartz crystal microbalance (QCM) sensor for quantification of carbofuran

(CBF) and profenofos (PFF). For that purpose, an in-house-developed dual-electrode system was used, where one electrode pair served as reference with the upper electrode being coated with the NIP. Doing so offers the advantage of measuring MIP and NIP simultaneously under the same conditions. Applying the bulk imprinting method, MIPs for PFF were based on polyurethanes whereas CBF MIPs were synthesized using acrylic monomers. Frequency measurements of MIP- and NIP-coated QCMs are shown in Figure 7.

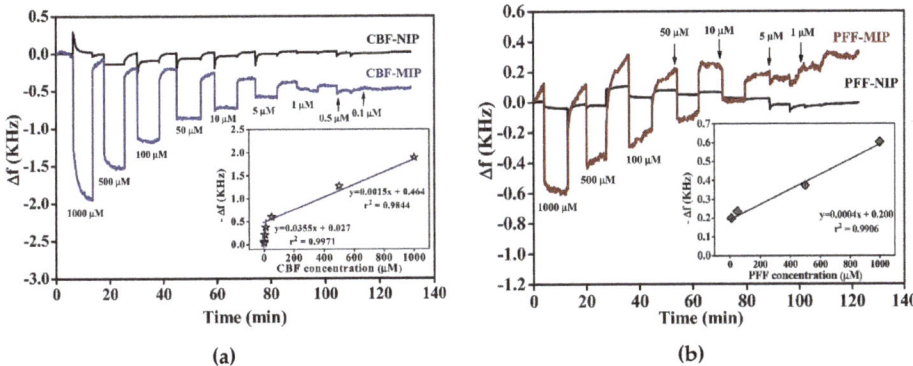

Figure 7. Frequency measurements of MIP- and NIP-coated QCMs for detection of (**a**) CBF and (**b**) PFF at different analyte concentrations. Reproduced from [33] Creative Commons License CC BY-NC-ND 4.0.

One can clearly see that both CBF and PFF MIPs led to linear sensor responses between 0.5–1000 µM and 5–1000 µM for CBF and PFF, respectively, whereas the frequency signal of the NIP stayed more or less constant.

Polycyclic aromatic hydrocarbons (PAH) are organic compounds which consist of at least two condensed aromatic rings. They are released into the environment through incomplete combustion of organic materials and considered to be mutagenic and carcinogenic. They usually occur in mixtures and their concentrations in air, water and sediments can be very low. Therefore, detection systems for PAH analysis need to be sensitive and selective. In particular, fluorescent sensors based on MIPs have gained in popularity due to their advantageous properties, such as high specificity, sensitivity and reversibility. Having a linear concentration dependency and low LoDs, those sensors seem to be quite promising for rapid detection of PAHs in aqueous solutions [34].

Sensors for the detection of nutrient components have been developed as well. For example, Warwick et al. reported a detection system based on MIPs combined with conductometric transducer for monitoring phosphates in environmental water samples [35]. Previous studies demonstrated that N-allylthiourea was the appropriate monomer for phosphate recognition [36]. The thiourea-based MIP was first optimized in terms of the optimal cross-linking monomer and ideal ratio of functional monomer to template (phenylphosphonic acid). Of all cross-linking monomers that were tested, ethylene glycol dimethacrylate (EGDMA) had the highest capacity of retaining phosphate as well as a monomer to template ratio of 2:1. After optimization, MIP membranes were integrated into the conductometric measuring cell. Selectivity tests in laboratory samples revealed no cross-talk to other ions, nitrate and sulfate. Both types of samples—standard and real-life ones—led to a linear increase in conductance with increased phosphate concentrations. In wastewater samples spiked with different amounts of potassium phosphate, the system allowed for LoD and LoQ values of 0.16 mg/L and 0.66 mg/L, respectively. The maximum acceptable amount of phosphate in wastewater is 1–2 mg/mL. The implemented sensor system with a linear range from 0.66 to 8 mg/mL therefore seems very promising for detecting small amounts of phosphate in environmental samples.

5.3. MIP-Based Sensors for On-Site Applications

For a pollutant sensor to be applicable on-site in real-life environments, it needs to be selective, reusable, robust and able to withstand harsh conditions. In 2015, Lenain et al. [37] reported a sensor that met all those criteria: it consists of spherical MIP beads deposited onto the electrodes of a capacitive transducer via electro-polymerization. Corresponding MIP beads for detecting metergoline—a model analyte for small molecules such as insecticides and pharmaceuticals—were synthesized through emulsion polymerization. Different concentrations of metergoline in PBS buffer were measured with the difference in capacitance between MIP and NIP representing the specific binding of the analyte as depicted in Figure 8. As shown in the figure, capacitance decreased with increased concentrations (10–50 µM). Furthermore, the system was able to regenerate itself without adding regeneration buffer, demonstrating reusability of the sensor.

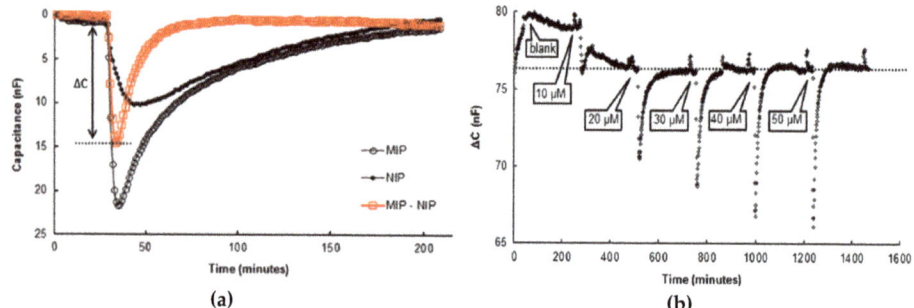

Figure 8. (a) Capacitance (nF) vs. time (min) of MIP, NIP and the differential signal ΔC (MIP-NIP). (b) Differential signal (ΔC in nF)) for different concentrations of metergoline vs. time (min). Reproduced with permission from [37] © Elsevier. MIP: molecularly imprinted polymers, NIP: nonimprinted polymers.

The sensor was also able to withstand harsh environments and achieved both a low LoD (1 µM) and low cross-selectivity. All these results suggest its suitability for monitoring pollutions originating from substances like pesticides or antibiotics in water samples (rivers, seawater) on-site.

Another example of a method suitable to the monitoring of contaminants in water *in situ* was introduced by Cennamo et al. [38]. It consists of an SPR sensor with an integrated plastic optical fiber (POF) combined with MIPs for detecting the model analyte perfluorobutanesulfonic acid (PFBS). With an LoD of 1 ppb, an interface software and the ability to connect to the internet directly, the SPR-POF-MIP technique is inherently suitable to detect small concentrations of different toxic or harmful compounds in real water samples *in situ*. Other advantageous features such as its reduced size, robustness and remote sensing abilities are further key factors for industrial applications of MIP-based sensor systems.

5.4. Mass Production of MIP-Based Sensor Systems

To prove that MIP-based sensors are inherently suitable for mass manufacturing, Aikio et al. developed a low-cost and robust optical sensor platform based on integrated Young interferometer sensor chips, where waveguides were fabricated on top of a carrier foil via roll-to-roll manufacturing techniques [39]. For chemical sensing of melamine, MIPs were used as recognition materials, whereas for biosensing of multiple biomolecules, sensor chips were functionalized with antibodies. In case of melamine sensing, the change in phase depended on the analyte concentration: Sensor responses increased with higher concentrations. Furthermore, the reference (NIP) led to much lower phase changes compared to sensor responses of the MIP. However, injection of higher concentrations (>0.5 g/L) led to saturation effects. For multianalyte biosensing, the sensor chip was functionalised with antibodies for C-reactive protein (CRP) and human chorionic gonadotropin (hCG) via inkjet printing. The results indicated that the Young interferometer bearing a specific antibody indeed selectively detected its corresponding protein. This work demonstrated the use of large-scale production techniques to develop a cost-efficient and rugged sensor system.

6. Functionalised Electromagnetic Wave Sensors

6.1. Microwave Spectroscopy and Water Analysis

Using electromagnetic (EM) waves at microwave frequencies for sensing purposes is an active research approach with potential for commercialization. This novel sensing approach has several advantages, including noninvasiveness, nondestructiveness, immediate response when the EM waves are in contact with a material under test, low-cost and power. Microwave spectroscopy provides the opportunity to guarantee continuous monitoring of water resources and intercept unexpected changes in water quality [40].

During the last 3 decades, microwave spectroscopy for liquid sensing has been investigated [41]. A water sample is placed in direct contact through a sensing structure and measured in real-time using an EM source (such as a vector network analyzer). The EM field interacts with the sample under test in a unique manner, depending on the polarization of water molecules and other compounds in the water samples, which produces a specific reflected or transmitted signal. The spectral response at specific frequencies depends on the conductivity and permittivity of the material under test [42]. Considering the variability of the sensing structures, the most successful experiments for detecting water quality were obtained using resonant cavities and planar sensors, due to the practicability of measuring a liquid sample.

Cavities resonate when the wavelength of the excitation within the cavity coincides with the cavity's dimension [43]. They enable noncontact, real-time measurements, as liquid samples in plastic or glass containers with known dimensions and properties can be inserted into the cavity. Several experiments have shown the resonant cavity ability to detect the presence and concentration of various materials. Specifically, cylindrical cavities were used to determine water hardness [43], nitrates [44], silver materials [45] and mixtures such as NaCl and $KMnO_4$ [46]. A rectangular resonant cavity was developed and tested for measuring pork-loin drip loss for meat production industry applications [47]. Another rectangular cavity was designed and tested for monitoring water quality, specifically the presence and concentration (>10 mg/L) of sulphides and nitrates [48].

During the last few years, several planar microwave sensors with different conformations have been developed and tested for differentiate compositions of water for both qualitative and quantitative concentration measurements [49–53]. Between planar sensing structures, Korostynska et al. [49] confirmed the action of a novel planar sensor with a sensing element consisted of interdigitated electrodes (IDE, also defined as interdigital by other researchers) metal patterns (silver, gold and/or copper) for water analysis observing changes in the microwave part of the EM spectrum analysing 20 µL of deionized water (DIW), KCl, NaCl and MnCl at various concentrations (Figure 9a,b). Then, Mason et al. [50], Moejes et al. [51] and Frau et al. [52] demonstrated the ability to detect respectively Lincomycin and Tylosin antibiotics, *Tetraselmis suecica* and lead ions (Pb^{2+}) using gold (Au) eight-pair IDE sensors.

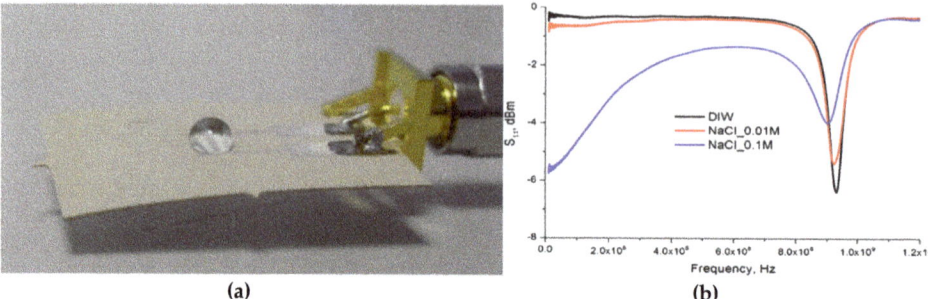

Figure 9. (a) Microwave flexible sensor with 20 µL of water sample placed the silver IDEs and (b) its output sensing response comparing deionized water (DIW), and 0.01 and 0.1 M of NaCl. Reproduced with permission from [49] © 2020 IOP Publishing Ltd.

6.2. Progresses and Challenges in Microwave Spectroscopy

Microwave spectroscopy is an attractive option for detecting changes in materials in a noninvasive manner, at low cost with the option of portability and rapid measurements. This strategy, however, suffers from a deficiency of specificity, related to low sensitivity (ΔdB related with small changes in material changes) and selectivity (diverse spectral response for similar pollutants) [48,53]. Some of the disadvantages are also related to the capability to detect minor changes in the water sample which are not related to the changes in the target analyte, such as temperature and density [54].

There has been increasing research and development on understanding and improving the sensing performance of microwave spectroscopy for a deeper analysis of specific pollutants and small concentration changes related to them. Also, changes in the shape pattern of the sensing structure are not able to improve the performance of required sensitivity and selectivity of pollutants [55]. The bigger problem remains the detection of more than two pollutants at low concentrations.

Novel strategies are being adopted to improve sensitivity and selectivity using microwave spectroscopy, but no one has yet demonstrated the feasibility of distinguishing low concentrations of similar substances in water. Amirian et al. [56] simulated the feasibility of distinguishing between pure liquid materials, such as ethanol, ammonia, benzene and pentene using a novel sensor design and mathematical approach, reaching higher sensitivity and noise reduction. Harnsoongnoen et al. [57] demonstrated the discrimination of organic and inorganic materials using planar sensors and principal component analysis (PCA). Magnitude spectra at 2.3–2.6 GHz were able to measure specific concentrations of sucrose, glucose, NaCl and $CaCl_2$ citric acid between others, generating linear and nonlinear prediction models, correlating the transmission coefficient (S_{21}) and R^2. PCR method was used to divide samples into two groups using the S_{21} magnitude: sugars and organic acids (blue oval) and salts (red oval) in Figure 10.

The following year Harnsoongnoen et al. [58] proposed a novel approach to discriminate between phosphorus and nitrate using the transmission coefficient and the ratio between the resonance frequency and the frequency bandwidth at the magnitude of 10 dB. This proposed method offers high sensitivity for both nitrate and phosphates, but not yet the required specificity.

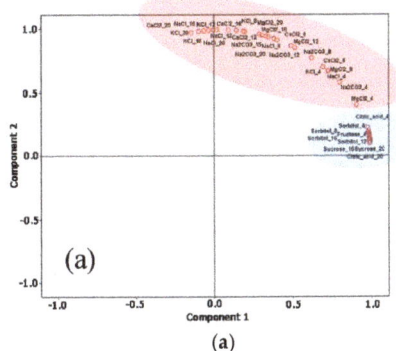
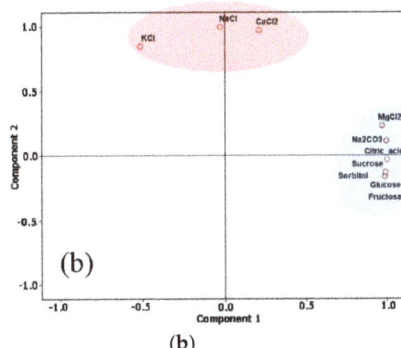

Figure 10. Discrimination between sugars and organic acid (blue oval) and salts (red ovals) using principal component analysis (PCA). (**a**): magnitudes of the transmission coefficient (S21) between 2.3 and 2.6 GHz; (**b**): amplitude of S21 at the resonance frequency. Reproduced with permission from [57] © Elsevier.

Other researchers are using machine learning features for selecting and distinguishing a target material [44] The developed model has been able to estimate the presence of nitrate in deionized water above the threshold, but it has not been able to quantify the precise concentration. Mason et al. [49] adopted a combined sensor approach using microwave analysis, combined with optical and impedance measurements for a more selective and sensitive determination of antibiotics, tylosin and lincomycin at different frequencies (at 8.7 and 1.8 GHz respectively) reaching high sensitivity (Figure 11a,b). Specifically, the selected planar microwave sensors were able to detect 0.20 μg/L of lincomycin and 0.25 μg/L of tylosin, a common concentration found in both surface and groundwater.

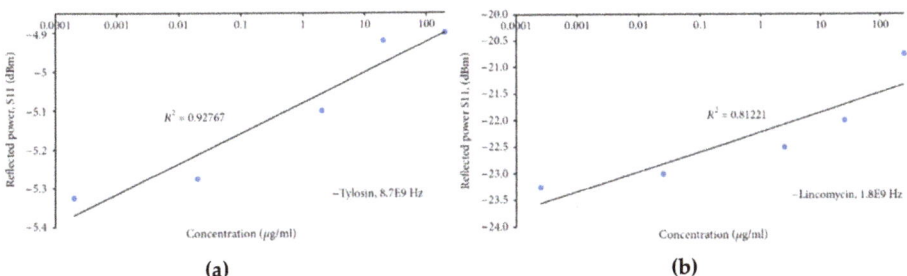

Figure 11. (**a**) Dependence of the S_{11}-transmitted signal on tylosin and lincomycin at respectively, 8.7 GHz and (**b**) 1.8 GHz. Copyright © [50] under the Creative Commons Attribution License.

6.3. Microwave and Materials Integration

Further improvement in sensitivity and selectivity are essential steps for water quality sensing, especially in complex mixtures [59]. A recent attractive approach is the integration of sensing materials onto the sensing structure, which has been experimented with the use of electrochemical impedance spectroscopy (EIS) and IDE sensors [60] using diverse coating thicknesses and for microwave gas sensing [61]. Planar sensors are an attractive option for the implementation of materials such as thin and thick films or microfluidic structures [53].

The synergy between microwave sensing technology and chemical materials provides interesting advantages in the field of quality monitoring for adapting this method to a specific purpose and it is consequently a promising area of research and development. The integration of specific materials in the form of thin and thick films onto planar sensors has been recently recognized as a novel and attractive approach for reaching higher selectivity, sensitivity and specificity for a selected material under tests using microwave and impedance spectroscopy [59,62–64]. The principle is based on two processes: the sensing process, where the target analyte interacts, via physical or chemical interaction, with the material on the sensing structure, and the transduction process, where the interaction between EM waves, material and sample generate a particular signal [61]. A vector network analyzer is used as a microwave source and can be configured with one or two ports. One-port configuration (S_{11} measurement) measures the reflection coefficient of a material under test, which depends on how much the incident wave propagates through, or is reflected by the sample. Two-port configuration (S_{21} measurement) allows for measuring the transmission coefficient, which depends on how much EM power propagates from one port (port 1) through the sample and is received at the second port (port 2). This configuration allows the determination of both transmitted and reflected signals. S-parameters vary with frequency. By functionalizing planar sensors with certain sensitive materials using screen-printing technology, which are defined functionalized electromagnetic sensors (f-EM) sensors, it is possible to obtain the desired sensitivity and/or selectivity to one or more specific analyte in water obtaining an immediate specific response as the sample is placed onto the microwave sensor (uncoated or f-EM sensor) [65] (Figure 12). Accordingly, such work has the foundations for developing new methods, based on EM sensors and functional chemical materials, capable of determining specific chemicals in water, both qualitatively and quantitatively [65]. The sensing response as S_{11} can be determined by direct contact between the material and the analyte under test, which changes the permittivity of each component, and the consequent overall complex permittivity changes. The improvement can be associated with the increase of material thickness as well as the composition itself [66].

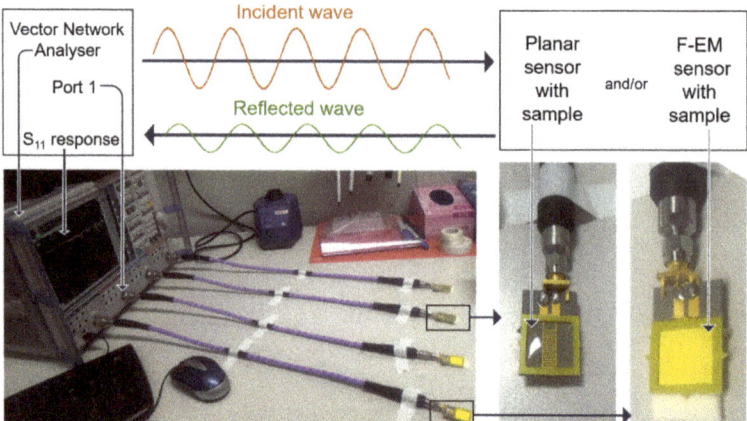

Figure 12. Scheme of the microwave interaction generated from a vector network analyser (VNA) which interact with a planar sensor (uncoated or functionalized with a coating, defined f-EM sensor) and a water sample placed onto it which generate a specific reflected signal (S_{11}) at selected frequencies. Reproduced from [65], Creative Commons Attribution 4.0 International License.

6.4. F-EM Sensors for Toxic Metals Analysis

Among the toxic elements, copper (Cu) and zinc (Zn) belong to the most common contaminants associated with mine wastes, which pollute water resources. Progress has been made in the last decade in developing chemo-sensors using mostly optical and electrochemical techniques. These are able to recognize specific metal ions using synthetic, natural and biological receptors [67], zeolites, inorganic oxides [68], organic polymers, biological materials [69], carbon-based materials [70] and hybrid ion-exchangers [71]. The interaction between the material and metal ions is the base for accredited optical and electrochemical sensing systems for detecting small concentrations in water

Currently, no certified method can guarantee real-time monitoring of toxic metals in water. Microwave sensing technology is promising for facing this challenge, although new strategies must be developed for obtaining more specific response. The integration of certain sensitive materials onto the planar sensing structure can be used to obtain the desired sensitivity and/or selectivity to specific analytes in water. Among these functional chemical compounds, inorganic oxide compositions are considered advantageous owing to their strong adsorption and rapid electron transfer kinetics [68,72]. Inorganic materials have attracted considerable attention owing to their low cost, compatibility and strong adsorption of toxic metal ions [69]. For instance, zinc oxide (ZnO) nanoparticles are well-known for strongly adsorbing Cu and Pb ions [73].

Frau et al. [74] laid the foundations for developing new methods, based on EM sensors and functional chemical materials, capable of determining in real-time metal content in mining-impacted waters, both qualitatively and quantitatively. Specifically, integrating planar electromagnetic wave sensors operating at microwave frequencies with bespoke thick film coatings was proven feasible for monitoring of environmental pollution in water with metals caused by mining [74]. A recently developed f-EM sensor based on L-CyChBCZ (acronym for a mixture based on l-cysteine, chitosan and bismuth zinc cobalt oxide) was tested by probing a polluted water sample spiked with Cu and Zn solutions using the standard addition method. The S_{11} response has shown linear correlation with Cu and Zn concentration at three resonant frequencies. Especially at 0.91–1.00 GHz (peak 2) the f-EM sensor shows an improvement for Cu detection with an improvement in sensitivity, higher Q-factor and low LoD compared with an uncoated (UNC) sensor, shown in bold in Table 1 [74]. The sensor was able to detect Cu concentration with a limit of detection (LoD) of 0.036, just above the environmental quality standards for freshwater (28–34 µg/L) Responses for Cu and Zn were then compared by analysing microwave spectral responses using a Lorentzian peak fitting function and investigating multi-peaks (peaks 0–6) and multi-peaks' parameters (peak center, xc, FWHM, w, area, A, and height, H, of the peaks) for specific discrimination between these two similar toxic metals [74]. It is useful to compare additional parameters for determining the selectivity, as it was demonstrated by Harnsoongnoen et al. [57], to distinguish between sugars and salts. However, more work is required for reaching higher discrimination between similar contaminants.

Table 1. Comparison of statistical features between uncoated and coated sensors for a water sample collected in a mining area spiked with Cu using the standard addition method. Reproduced from [74], Creative Commons Attribution 4.0 International License. LoD: limit of detection.

	R^2		CV (dB)		Sensitivity (ΔdB/mg/L)		LoD (mg/L)		Q-Factor	
	UNC	f-EM	UNC	f-EM	UNC	f-EM	UNC	f-EM	UNC	f-EM
Peak 0	0.970	0.928	0.20	0.25	0.362	0.222	0.194	0.379	/	/
Peak 1	0.963	0.981	0.02	0.03	0.354	0.260	0.146	0.409	2.60	6.57
Peak 2	0.888	**0.983**	0.01	0.02	0.824	**1.651**	0.083	**0.036**	30.71	135.48

The microwave sensor functionalized with a 60-µm-thick β-Bi$_2$O$_3$-based film was developed specifically for the detection of Zn in water [66]. It showed an improved performance compared with the uncoated sensor and repeatedly detected the changes of Zn concentrations in water at 0–100 ppm levels with a linear response (Figure 13). Globally, Zn concentrations in mine water can be greater than 500 ppm, with typical concentrations ranging from 0.1 to 10 ppm. Thus, the proposed system can be adapted as a sensing platform for monitoring Zn in water in abandoned mining areas that would be able to detect unexpected events of pollution and to clarify metal dynamics. As recently reported by Vélez et al. [75], the f-EM sensor based on β-Bi$_2$O$_3$ is the one that exhibits the best performance (but a limited dynamic range) comparing glucose and NaCl, between others. Between the evaluated sensors, this sensor was the one that presented the highest sensitivity and the best resolution.

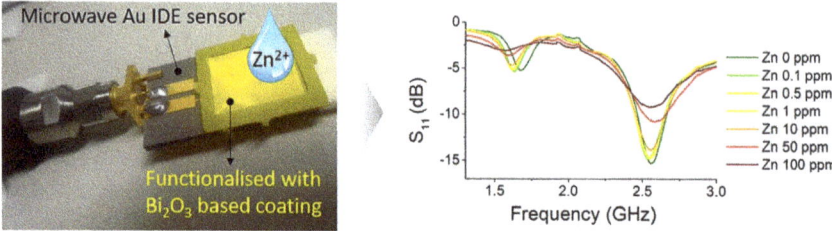

Figure 13. Microwave sensor functionalized with Bi_2O_3-based coating for the detection of Zn in water (**left**), and the dependence of the reflected signal (S_{11}) response on Zn concentration (**right**). Reproduced with permission from [66] © Elsevier.

6.5. F-EM Sensors for On-Site and In Situ Applications

The sensing system developed in [74] was tested on real water samples from two abandoned mining areas in Wales, UK, and evaluated the possibility of detecting both small and big differences between mining-impacted waters by comparing peak parameters in multiple peaks (Figure 14). Between the four samples, PM (a drainage adit from Parys Mountain mining district) was the most polluted (with 9.3 mg/L of Cu and 10.5 mg/L of Zn) compared to the other three samples (FA, MR and NC, from Wemyss mine) (with <0.001 mg/L of Cu and 2.9–9.3 mg/L of Zn). The sensor was able to distinguish between the two groups of samples in all the peaks, and at peaks 2–6 between the four samples, with both resonant frequency and amplitude shifts. Multiple peak characterization can give more information about the water composition. The sensor was able to perform a real-time identification of more- and less-polluted samples with high repeatability (with a coefficient of variation <0.05 dB), and evaluate changes in water composition. However, interferences were noticed, and more work is necessary for achieving higher selectivity.

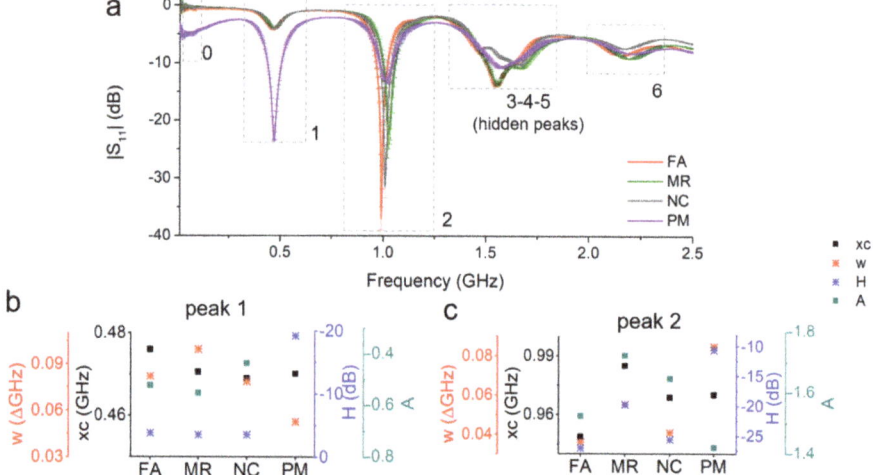

Figure 14. (**a**) Microwave spectral output for four polluted mining-impacted water samples collected in Wales (UK) analysed with an f-EM sensor based on L-CyCHBCZ coating and (**b**) peak parameters (w, xc, H and A) comparison for peaks 1 and (**c**) 2. Reproduced from [74], Creative Commons Attribution 4.0 International License.

Therefore, f-EM sensors can be considered a feasible option for real-time monitoring of water quality for a broad range of pollutants. Choosing the right coating or sensor functionalization allows for adapting the system to selected pollutants in a given scenario to ensure sensor response with high selectivity and sensitivity.

The sensing structure was adapted for directly probing the water, for then providing an *in situ* monitoring. Despite many recent technological advances and positive results obtained using this novel sensing system, significant work remains to be accomplished before a reliable smart sensor for water quality monitoring is achievable. Real mine water is more complex and characterized by high levels of dissolved metals and sulphate and, frequently, low pH. Though, there are obviously several challenges that must be overcome before this technology can ensure that measurements correctly identify distinct contaminants, and qualifying and quantifying the interferences caused by complex water matrices and similar pollutants.

7. New Trends in Water Quality Monitoring

The most interesting and recent efforts of online water quality monitoring rely on multisensor systems based on electrochemical methods along with advances in data processing and automatization of measurements.

A miniature device was reported in [76], where the authors combined a pH meter and a conductometer for evaluating drinking water quality. The system was tested over a period of 30 days in water streams of different speeds. It was shown that the device was working stably under these conditions. Though pH and conductivity are important water parameters, they are obviously not enough for comprehensive evaluation of water quality.

The demand for simplicity of the analysis is gaining momentum in the recent years. A combination of paper-based sensors and a smartphone application is described as an analytical instrument for water quality monitoring in [77]. The paper sensors generate colorimetric signals depending on the content of certain analytes and the cell phone captures such signals and compares them to that from a clean control sample. The smartphone would also transmit the results to the special site mapping the water quality. The schematic of this system is shown in Figure 15.

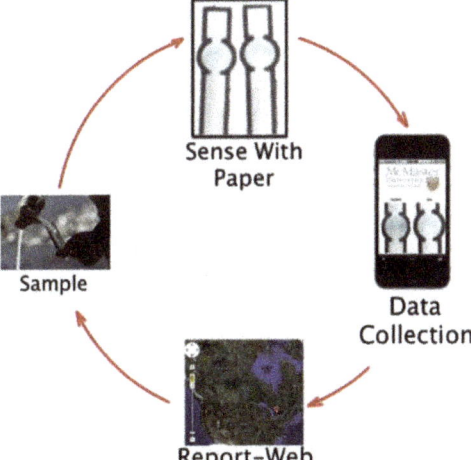

Figure 15. Concept of water quality evaluation using paper-based colorimetric sensors and a smartphone. Reprinted from Water Research, Vol. 70, Clemence Sicard, Chad Glen, Brandon Aubie, Dan Wallace, Sana Jahanshahi-Anbuhi, Kevin Pennings, Glen T. Daigger, Robert Pelton, John D. Brennan, Carlos D.M. Filipe, Tools for water quality monitoring and mapping using paper-based sensors and cell phones, Pages 360–369, Copyright (2015), with permission from Elsevier.

It was demonstrated that the system may quantify some organophosphorous pesticides like paraoxon and malathion in the natural water at the level 10^{-8}–10^{-6} mol/L for both analytes. It is not clear however if such a device would perform correctly and precisely if multiple interfering species are present.

The successful application of a multisensor array for continuous online monitoring of processed water quality at an aeration plant was reported in [78]. The responses of 23 sensors of the array were continuously registered every seven seconds over a period of 26 days. The sensors were located in a special container with short direct connection to outlet water line of the aeration plant. The results of the multisensor were available immediately in a real-time mode. The use of topological data analysis (TDA) allowed for exploration of a very large dataset (295,828 measurements) accumulated through the whole period of continuous measurements. The achieved precision of analysis was suitable to monitor possible alarm events. No significant changes in water quality were observed over the experimental period. TDA helped visualizing some sharp changes related to sensor cleaning procedures and electrical power shortages.

Figure 16 shows a multisensor system (MSS) applied to evaluate integral and discrete parameters of wastewater at two urban water treatment plants around St. Petersburg (Russia) [79]. A closely similar system was used for evaluation of the water quality from the Ganga river and city ponds in Kolkata (India). Good correlations (R^2 = 0.85 for cross-validation) of MSS readings with COD values produced by laboratory chemical analysis were observed for all locations. This proves once again the applicability of MSS for real-time water quality analysis. A number of traditional water analytical parameters, such as ammonium and nitrate nitrogen and phosphorous were also determined with precision around 25% using the same MSS. All integral and discrete characteristics were calculated on the basis of the same set of measurements with MSS, without using any additional laboratory procedures, materials or qualified staff.

Figure 16. The sensor array after long-term online measurements at an aeration water plant. The sensors were cleaned by intensive washing. In spite of significant contamination, the sensors were stable for at least two months [79].

Along with traditional water quality analysis by analytical devices, a totally different approach has been developed. The global water toxicity (safety) can be also evaluated with the help of living creatures—from single-cell microorganisms to fishes, crustacea and mollusks. As a result, an ISO standard appeared in 1989, which suggested fresh water algae such as *Scenedesmus subspicatus* and *Selenastrum capricornutum* as a test. The decay of growth and reproduction of these bioassays is occurring in the contaminated water and can be used as its quality measure [80]. Another widespread method is based on the application of *Vibrio fischeri* bacteria exhibiting variations of luminescence dependent on water pollution [81]. Unfortunately, such methods have also been burdened by specific drawbacks such as the necessity to strictly maintain the livestock of biotests and realistic analysis times of at least 15–30 min (and up to few days) depending on bioassay applied. Therefore, these approaches can hardly be treated as online ones.

It must be noted that multisensor devices can also perform like biotests. The toxicity of polluted water samples was evaluated in [82] by a standard bioassay method and a potentiometric multisensor system comprising 23 cross-sensitive electrodes. Real wastewater samples from different regions in Catalonia (Spain) in addition to a set of model aqueous solutions of hazardous substances (54 samples in total) were used for the measurements. The obtained data set was treated by several regression algorithms; the results of the bioassay tests, expressed as EC50 (the concentration of sample causing a 50% luminescence reduction), were taken as Y-variable. The regression models were validated by full cross-validation and randomized test set selection. It was demonstrated that the proposed system was able to evaluate integral water toxicity with the errors of EC50 prediction from 20% to 25%. The suggested sensor array can be implemented in online mode, unlike bioassay techniques, which makes it a beneficial tool in industrial water quality monitoring.

8. Conclusions

The global need for developing novel platforms for real-time monitoring of various water pollutants is well-recognized. This paper provides a critical assessment of recent achievements in real-time water quality monitoring with chemical sensors in particular. The focus is given to those systems that were reportedly tested online, with real water samples and their feasibility for long-term use is considered. The review shows that there are still many obstacles for having one sensing approach that would satisfy different situations. The most successful systems based on chemical sensing or its combination with other methods rely on specificity of a coating material that is capable of accurate detection of certain water pollutants, with molecularly imprinted polymers providing an increased flexibility for the designing of those systems.

Author Contributions: Conceptualization, A.L. and D.K.; writing—original draft preparation, I.Y.; writing—review and editing, A.L., I.F., A.M., O.K., M.M., P.A.L., and D.K. All authors have read and agreed to the published version of the manuscript.

Funding: D.K. and A.L.: acknowledge the funding from the RSF project #18-19-00151. I.Y acknowledges partial financial support of this work from the RFBR project #18-53-80010.

Conflicts of Interest: The authors declare no conflict of interest.

References

1. The United Nations World Water Development Report 4: Managing Water under Uncertainty and Risk, Executive Summary. Available online: https://unesdoc.unesco.org/ark:/48223/pf0000217175 (accessed on 19 October 2019).
2. Environment. Available online: http://ec.europa.eu/environment/water/water-drink/legislation_en.html (accessed on 19 October 2019).
3. Methods Approved to Analyze Drinking Water Samples to Ensure Compliance with Regulations. Available online: https://www.epa.gov/dwanalyticalmethods (accessed on 19 October 2019).
4. Meyer, A.M.; Klein, C.; Fünfrocken, E.; Kautenburger, R.; Beck, H.P. Real-time monitoring of water quality to identify pollution pathways in small and middle scale rivers. *Sci. Total. Environ.* **2019**, *651*, 2323–2333. [CrossRef]
5. Lvova, L.; Di Natale, C.; Paolesse, R. Chemical Sensors for Water Potability Assessment. In *Bottled Packaged Water*; Grumezescu, A., Holban, A.M., Eds.; Elsevier Science: Amsterdam, The Netherlands, 2019; Volume 7, pp. 177–208.
6. Legin, A.V.; Bychkov, E.A.; Seleznev, B.L.; Vlasov, Y.G. Development and analytical evaluation of a multisensor system for water quality monitoring. *Sens. Actuators. B Chem.* **1995**, *27*, 377–379. [CrossRef]

7. Rudnitskaya, A.; Ehlert, A.; Legin, A.; Vlasov, Y.G.; Büttgenbach, S. Multisensor system on the basis of an array of non-specific chemical sensors and artificial neural networks for determination of inorganic pollutants in a model groundwater. *Talanta* **2001**, *55*, 425–431. [CrossRef]
8. Krantz-Rülcker, C.; Stenberg, M.; Winquist, F.; Lundström, I. Electronic tongues for environmental monitoring based on sensor arrays and pattern recognition: A review. *Anal. Chim. Acta* **2001**, *426*, 217–226. [CrossRef]
9. Martınez-Manez, R.; Soto, J.; Garcıa-Breijo, E.; Gil, L.; Ibanez, J.; Gadea, E. A multisensor in thick-film technology for water quality control. *Sens. Actuators. A Phys.* **2005**, *120*, 589–595. [CrossRef]
10. Men, H.; Zou, S.; Li, Y.; Wang, Y.; Ye, X.; Wang, P. A novel electronic tongue combined MLAPS with stripping voltammetry for environmental detection. *Sens. Actuators B Chem.* **2005**, *110*, 350–357. [CrossRef]
11. Yang, Y.J.; Haught, R.C.; Goodrich, J.A. Real-time contaminant detection and classification in a drinking water pipe using conventional water quality sensors: Techniques and experimental results. *J. Env. Manag.* **2009**, *90*, 2494–2506. [CrossRef] [PubMed]
12. Capella, J.V.; Bonastre, A.; Ors, R.; Peris, M. A Wireless Sensor Network approach for distributed in-line chemical analysis of water. *Talanta* **2010**, *80*, 1789–1798. [CrossRef] [PubMed]
13. Mimendia, A.; Gutierrez, J.M.; Leija, L.; Hernandez, P.R.; Favari, L.; Munoz, R.; del Valle, M. A review of the use of the potentiometric electronic tongue in the monitoring of environmental systems. *Environ. Modell. Softw.* **2010**, *25*, 1023–1030. [CrossRef]
14. Bourgeois, W.; Gardey, G.; Servieres, M.; Stuetz, R.M. A chemical sensor array based system for protecting wastewater treatment plants. *Sens. Actuators. B Chem.* **2003**, *91*, 109–116. [CrossRef]
15. Valdes-Ramirez, G.; Gutierrez, M.; del Valle, M.; Ramirez-Silva, M.T.; Fournier, D.; Marty, J.-L. Automated resolution of dichlorvos and methylparaoxon pesticide mixtures employing a Flow Injection system with an inhibition electronic tongue. *Biosens. Bioelectron.* **2009**, *24*, 1103–1108. [CrossRef] [PubMed]
16. Czolkos, I.; Dock, E.; Tønning, E.; Christensen, J.; Winther-Nielsen, M.; Carlsson, C.; Mojzíková, R.; Skládal, P.; Wollenberger, U.; Nørgaard, L.; et al. Prediction of wastewater quality using amperometric bioelectronic tongues. *Biosens. Bioelectron.* **2016**, *75*, 375–382. [CrossRef] [PubMed]
17. Hassan, S.H.A.; Gurung, A.; Kang, W.-G.; Shin, B.-S.; Rahimnejad, M.; Jeon, B.-H.; Kim, J.R.; Oh, S.-E. Real-time monitoring of water quality of stream water using sulfur oxidizing bacteria as bio-indicator. *Chemosphere* **2019**, *223*, 58–63. [CrossRef] [PubMed]
18. Attivissimo, F.; Carducci, C.G.C.; Lanzolla, A.M.L.; Massaro, A.; Vadrucci, M.R. A portable optical sensor for sea quality monitoring. *IEEE Sens. J.* **2015**, *15*, 146–153. [CrossRef]
19. Murphy, K.; Heery, B.; Sullivan, T.; Zhang, D.; Paludetti, L.; Lau, K.T.; Diamond, D.; Costa, E.; O'Connor, N.; Regan, F. A low-cost autonomous optical sensor for water quality monitoring. *Talanta* **2015**, *132*, 520–527. [CrossRef]
20. Skouteris, G.; Webb, D.P.; Shin, K.L.F.; Rahimifard, S. Assessment of the capability of an optical sensor for in-line real time wastewater quality analysis in food manufacturing. *Water. Resour. Ind.* **2018**, *20*, 75–81. [CrossRef]
21. Vaughan, A.A.; Narayanaswamy, R. Optical fibre reflectance sensors for the detection of heavy metal ions based on immobilised Br-PADAP. *Sens. Actuators. B Chem.* **1998**, *51*, 368–376. [CrossRef]
22. Lieberzeit, P.A.; Dickert, F.L. Sensor technology and its application in environmental analysis. *Anal. Bioanal. Chem.* **2007**, *387*, 237–247. [CrossRef]
23. Haupt, K.; Mosbach, K. Molecularly imprinted polymers and their use in biomimetic sensors. *Chem. Rev.* **2000**, *100*, 2495–2504. [CrossRef]
24. Latif, U.; Qian, J.; Can, S.; Dickert, F.L. Biomimetic receptors for bioanalyte detection by quartz crystal microbalances—From molecules to cells. *Sensors* **2014**, *14*, 23419–23438. [CrossRef]
25. Wackerlig, J.; Lieberzeit, P.A. Molecularly imprinted polymer nanoparticles in chemical sensing - Synthesis, characterisation and application. *Sens. Actuators B Chem.* **2015**, *207*, 144–157. [CrossRef]
26. Zhang, Y.; Lai, B.S.; Juhas, M. Recent advances in aptamer discovery and applications. *Molecular* **2019**, *24*, 941. [CrossRef] [PubMed]
27. Kubisch, R.; Bohrn, U.; Fleischer, M.; Stütz, E. Cell-based sensor system using L6 cells for broad band continuous pollutant monitoring in aquatic environments. *Sensors* **2012**, *12*, 3370–3393. [CrossRef] [PubMed]
28. Guijarro, C.; Fuchs, K.; Bohrn, U.; Stütz, E.; Wölfl, S. Simultaneous detection of multiple bioactive pollutants using a multiparametric biochip for water quality monitoring. *Biosens. Bioelectron.* **2015**, *72*, 71–79. [CrossRef] [PubMed]

29. Ayankojo, A.G.; Reut, J.; Öpik, A.; Furchner, A.; Syritski, V. Hybrid molecularly imprinted polymer for amoxicillin detection. *Biosens. Bioelectron.* **2018**, *118*, 102–107. [CrossRef] [PubMed]
30. Cardoso, A.R.; Tavares, A.P.M.; Sales, M.G.F. In-situ generated molecularly imprinted material for chloramphenicol electrochemical sensing in waters down to the nanomolar level. *Sens Actuators B Chem.* **2018**, *256*, 420–428. [CrossRef]
31. Erdem, Ö.; Saylan, Y.; Cihangir, N.; Denizli, A. Molecularly imprinted nanoparticles based plasmonic sensors for real-time Enterococcus faecalis detection. *Biosens. Bioelectron.* **2019**, *126*, 608–614. [CrossRef]
32. Khadem, M.; Faridbod, F.; Norouzi, P.; Rahimi Foroushani, A.; Ganjali, M.R.; Shahtaheri, S.J.; Yarahmadi, R. Modification of Carbon Paste Electrode Based on Molecularly Imprinted Polymer for Electrochemical Determination of Diazinon in Biological and Environmental Samples. *Electroanalysis* **2017**, *29*, 708–715. [CrossRef]
33. Sroysee, W.; Chunta, S.; Amatatongchai, M.; Lieberzeit, P.A. Molecularly imprinted polymers to detect profenofos and carbofuran selectively with QCM sensors. *Phys. Med.* **2019**, *7*, 100016. [CrossRef]
34. Nsibande, S.A.; Montaseri, H.; Forbes, P.B.C. Advances in the application of nanomaterial-based sensors for detection of polycyclic aromatic hydrocarbons in aquatic systems. *Trends Anal. Chem.* **2019**, *115*, 52–69. [CrossRef]
35. Warwick, C.; Guerreiro, A.; Gomez-Caballero, A.; Wood, E.; Kitson, J.; Robinson, J.; Soares, A. Conductance based sensing and analysis of soluble phosphates in wastewater. *Biosens. Bioelectron.* **2014**, *52*, 173–179. [CrossRef] [PubMed]
36. Warwick, C.; Guerreiro, A.; Wood, E.; Kitson, J.; Robinson, J.; Soares, A. A molecular imprinted polymer based sensor for measuring phosphate in wastewater samples. *Water Sci. Technol.* **2014**, *69*, 48–54. [CrossRef] [PubMed]
37. Lenain, P.; De Saeger, S.; Mattiasson, B.; Hedström, M. Affinity sensor based on immobilised molecular imprinted synthetic recognition elements. *Biosens. Bioelectron.* **2015**, *69*, 34–39. [CrossRef] [PubMed]
38. Cennamo, N.; Arcadio, F.; Perri, C.; Zeni, L.; Sequeira, F.; Bilro, L.; Nogueira, R.; D'Agostino, G.; Porto, G.; Biasiolo, A. Water monitoring in smart cities exploiting plastic optical fibers and molecularly imprinted polymers. In Proceedings of the 2019 IEEE International Symposium on Measurements & Networking (M&N), Catania, Italy, 8–10 July 2019.
39. Aikio, S.; Zeilinger, M.; Hiltunen, J.; Hakalahti, L.; Hiitola-Keinänen, J.; Hiltunen, M.; Kontturi, V.; Siitonen, S.; Puustinen, J.; Lieberzeit, P.; et al. Disposable (bio)chemical integrated optical waveguide sensors implemented on roll-to-roll produced platforms. *RSC Adv.* **2016**, *6*, 50414–50422. [CrossRef]
40. Mohammadi, S.; Nadaraja, A.V.; Roberts, D.J.; Zarifi, M.H. Real-time and hazard-free water quality monitoring based on microwave planar resonator sensor. *Sens. Actuators A Phys.* **2020**, *303*, 111663. [CrossRef]
41. Zhang, K.; Amineh, R.K.; Dong, Z.; Nadler, D. Microwave sensing of water quality. *IEEE Access* **2019**, *7*, 69481–69493. [CrossRef]
42. Korostynska, O.; Mason, A.; Al-Shamma'a, A.I. Flexible microwave sensors for real-time analysis of water contaminants. *J. Electromagn. Waves Appl.* **2013**, *27*, 2075–2089. [CrossRef]
43. Teng, K.H.; Shaw, A.; Ateeq, M.; Al-Shamma'a, A.; Wylie, S.; Kazi, S.N.; Chew, B.T.; Kot, P. Design and implementation of a non-invasive real-time microwave sensor for assessing water hardness in heat exchangers. *J. Electromagn. Waves Appl.* **2018**, *32*, 797–811. [CrossRef]
44. Cashman, S.; Korostynska, O.; Shaw, A.; Lisboa, P.; Conroy, L. Detecting the presence and concentration of nitrate in water using microwave spectroscopy. *IEEE Sens. J.* **2017**, *17*, 4092–4099. [CrossRef]
45. Ateeq, M.; Shaw, A.; Garrett, R.; Dickson, P. A proof of concept study on utilising a non-invasive microwave analysis technique to characterise silver based materials in aqueous solution. *Sens. Imaging* **2017**, *18*, 13. [CrossRef]
46. Kapilevich, B.; Litvak, B. Microwave Sensor for Accurate Measurements of Water Solution Concentrations. In Proceedings of the 2007 Asia-Pacific Microwave Conference, Bangkok, Thailand, 11–14 December 2007.
47. Mason, A.; Abdullah, B.; Muradov, M.; Korostynska, O.; Al-Shamma'a, A.; Bjarnadottir, S.G.; Lunde, K.; Alveike, O. Theoretical basis and application for measuring pork loin drip loss using microwave spectroscopy. *Sensors* **2016**, *16*, 182. [CrossRef] [PubMed]
48. Gennarelli, G.; Soldovieri, F. A non-specific microwave sensor for water quality monitoring. *Int. Water Technol. J.* **2013**, *3*, 70–77.

49. Korostynska, O.; Ortoneda-Pedrola, M.; Mason, A.; Al-Shamma'A, A.I. Flexible electromagnetic wave sensor operating at GHz frequencies for instantaneous concentration measurements of NaCl, KCl, $MnCl_2$ and CuCl solutions. *Meas. Sci. Technol.* **2014**, *25*, 065105. [CrossRef]
50. Mason, A.; Soprani, M.; Korostynska, O.; Amirthalingam, A.; Cullen, J.; Muradov, M.; Carmona, E.N.; Sberveglieri, G.; Sberveglieri, V.; Al-Shamma'a, A. Real-time microwave, dielectric, and optical sensing of lincomycin and tylosin antibiotics in water: Sensor fusion for environmental safety. *J. Sens.* **2018**, *2018*, 7976105. [CrossRef]
51. Moejes, K.; Sherif, R.; Dürr, S.; Conlan, S.; Mason, A.; Korostynska, O. Real-time monitoring of tetraselmis suecica in a saline environment as means of early water pollution detection. *Toxics* **2018**, *6*, 57. [CrossRef]
52. Frau, I.; Korostynska, O.; Mason, A.; Byrne, P. Comparison of electromagnetic wave sensors with optical and low-frequency spectroscopy methods for real-time monitoring of lead concentrations in mine water. *Mine Water Environ.* **2018**, *37*, 617–624. [CrossRef]
53. Zarifi, M.H.; Daneshmand, M. Liquid sensing in aquatic environment using high quality planar microwave resonator. *Sens. Actuators B Chem.* **2016**, *225*, 517–521. [CrossRef]
54. Al-Kizwini, M.A.; Wylie, S.R.; Al-Khafaji, D.A.; Al-Shamma'a, A.I. The monitoring of the two phase flow-annular flow type regime using microwave sensor technique. *Measurement* **2013**, *46*, 45–51. [CrossRef]
55. Salim, A.; Lim, S. Review of recent metamaterial microfluidic sensors. *Sensors* **2018**, *18*, 232. [CrossRef]
56. Amirian, M.; Karimi, G.; Wiltshire, B.D.; Zarifi, M.H. Differential narrow bandpass microstrip filter design for material and liquid purity interrogation. *IEEE Sens. J.* **2019**, *19*, 10545–10553. [CrossRef]
57. Harnsoongnoen, S.; Wanthong, A.; Charoen-In, U.; Siritaratiwat, A. Planar microwave sensor for detection and discrimination of aqueous organic and inorganic solutions. *Sens. Actuators B Chem.* **2018**, *271*, 300–305. [CrossRef]
58. Harnsoongnoen, S.; Wanthong, A.; Charoen-In, U.; Siritaratiwat, A. Microwave sensor for nitrate and phosphate concentration sensing. *IEEE Sens. J.* **2019**, *19*, 2950–2955. [CrossRef]
59. Zarifi, M.H.; Farsinezhad, S.; Abdolrazzaghi, M.; Daneshmand, M.; Shankar, K. Selective microwave sensors exploiting the interaction of analytes with trap states in tio2 nanotube arrays. *Nanoscale* **2016**, *8*, 7466–7473. [CrossRef] [PubMed]
60. Afsarimanesh, N.; Mukhopadhyay, S.C.; Kruger, M. Performance assessment of interdigital sensor for varied coating thicknesses to detect ctx-i. *IEEE Sens J.* **2018**, *18*, 3924–3931. [CrossRef]
61. Li, F.; Zheng, Y.; Hua, C.; Jian, J. Gas sensing by microwave transduction: Review of progress and challenges. *Front. Mater.* **2019**, *6*. [CrossRef]
62. Azmi, A.; Azman, A.A.; Kaman, K.K.; Ibrahim, S.; Mukhopadhyay, S.C.; Nawawi, S.W.; Yunus, M.A.M. Performance of coating materials on planar electromagnetic sensing array to detect water contamination. *IEEE Sens. J.* **2017**, *17*, 5244–5251. [CrossRef]
63. Ebrahimi, A.; Withayachumnankul, W.; Al-Sarawi, S.; Abbott, D. High-sensitivity metamaterial-inspired sensor for microfluidic dielectric characterisation. *IEEE Sens. J.* **2014**, *14*, 1345–1351. [CrossRef]
64. Chen, T.; Li, S.; Sun, H. Metamaterials application in sensing. *Sensors* **2012**, *12*, 2742. [CrossRef]
65. Frau, I.; Wylie, S.; Byrne, P.; Cullen, J.; Korostynska, O.; Mason, A. New sensing system based on electromagnetic waves and functionalised em sensors for continuous monitoring of zn in freshwater. In Proceedings of the 11th ICARD | IMWA | MWD Conference—"Risk to Opportunity", Pretoria, South Africa, 10–14 September 2018.
66. Frau, I.; Wylie, S.; Byrne, P.; Cullen, J.; Korostynska, O.; Mason, A. Detection of zn in water using novel functionalised planar microwave sensors. *Mater. Sci. Eng. B* **2019**, *247*, 114382. [CrossRef]
67. Aragay, G.; Pons, J.; Merkoci, A. Recent trends in macro-, micro-, and nanomaterial-based tools and strategies for heavy-metal detection. *Chem. Rev.* **2011**, *111*, 3433–3458. [CrossRef]
68. Sen Gupta, S.; Bhattacharyya, K.G. Kinetics of adsorption of metal ions on inorganic materials: A review. *Adv. Colloid Interface Sci.* **2011**, *162*, 39–58. [CrossRef] [PubMed]
69. Cui, L.; Wu, J.; Ju, H. Electrochemical sensing of heavy metal ions with inorganic, organic and bio-materials. *Biosens. Bioelectron.* **2015**, *63*, 276–286. [CrossRef] [PubMed]
70. Wanekaya, A.K. Applications of nanoscale carbon-based materials in heavy metal sensing and detection. *Analysis* **2011**, *136*, 4383–4391. [CrossRef] [PubMed]
71. Chatterjee, P.K.; SenGupta, A.K. Toxic metal sensing through novel use of hybrid inorganic and polymeric ion-exchangers. *Solvent Extr. Ion Exch.* **2011**, *29*, 398–420. [CrossRef]

72. Gumpu, M.B.; Sethuraman, S.; Krishnan, U.M.; Rayappan, J.B.B. A review on detection of heavy metal ions in water—An electrochemical approach. *Sens. Actuators B Chem.* **2015**, *213*, 515–533. [CrossRef]
73. Bhatia, M.; Satish Babu, R.; Sonawane, S.H.; Gogate, P.R.; Girdhar, A.; Reddy, E.R.; Pola, M. Application of nanoadsorbents for removal of lead from water. *Int. J. Environ. Sci. Technol.* **2017**, *14*, 1135–1154. [CrossRef]
74. Frau, I.; Wylie, S.R.; Byrne, P.; Cullen, J.D.; Korostynska, O.; Mason, A. Functionalised microwave sensors for real-time monitoring of copper and zinc concentration in mining-impacted water. *Int. J. Environ. Sci. Technol.* **2019**, *17*, 1861–1876. [CrossRef]
75. Vélez, P.; Muñoz-Enano, J.; Gil, M.; Mata-Contreras, J.; Martín, F. Differential microfluidic sensors based on dumbbell-shaped defect ground structures in microstrip technology: Analysis, optimisation, and applications. *Sensors* **2019**, *19*, 3189. [CrossRef]
76. Banna, M.H.; Najjaran, H.; Sadiq, R.; Imran, S.A.; Rodriguez, M.J.; Hoorfar, M. Miniaturized water quality monitoring pH and conductivity sensors. *Sens. Actuators. B Chem.* **2014**, *193*, 434–441. [CrossRef]
77. Sicard, C.; Glen, C.; Aubie, B.; Wallace, D.; Jahanshahi-Anbuhi, S.; Pennings, K.; Daigger, G.T.; Pelton, R.; Brennan, J.D.; Filipe, C.D.M. Tools for water quality monitoring and mapping using paper-based sensors and cell phones. *Water Res.* **2015**, *70*, 360–369. [CrossRef]
78. Belikova, V.; Panchuk, V.; Legin, E.; Melenteva, A.; Kirsanov, D.; Legin, A. Continuous monitoring of water quality at aeration plant with potentiometric sensor array. *Sens. Actuators. B Chem.* **2019**, *282*, 854–860. [CrossRef]
79. Legin, E.; Zadorozhnaya, O.; Khaydukova, M.; Kirsanov, D.; Rybakin, V.; Zagrebin, A.; Ignatyeva, N.; Ashina, J.; Sarkar, S.; Mukherjee, S.; et al. Rapid Evaluation of Integral Quality and Safety of Surface and Wastewater s by a Multisensor System (Electronic Tongue). *Sensors* **2019**, *19*, 2019. [CrossRef] [PubMed]
80. International Organization for Standardization. *Water Quality—Fresh Water Algal Growth Inhibition Test with Scenedesmus Subspicatus and Selenastrum Capricornutum, ISO 8692*; International Organization for Standardization: Geneva, Switzerland, 1989; p. 12.
81. International Organization for Standardization. *Water Quality—Determination of The Inhibitory Effect of Water Samples on The Light Emission of Vibrio Fischeri (Luminescent bacteria test). ISO 11348-3*; International Organization for Standardization: Geneva, Switzerland, 1998; p. 13.
82. Zadorozhnaya, O.; Kirsanov, D.; Buzhinsky, I.; Tsarev, F.; Abramova, N.; Bratov, A.; Muñoz, F.J.; Ribó, J.; Bori, J.; Riva, M.C.; et al. Water pollution monitoring by an artificial sensory system performing in terms of Vibrio fischeri bacteria. *Sens. Actuators. B Chem.* **2015**, *207*, 1069–1075. [CrossRef]

© 2020 by the authors. Licensee MDPI, Basel, Switzerland. This article is an open access article distributed under the terms and conditions of the Creative Commons Attribution (CC BY) license (http://creativecommons.org/licenses/by/4.0/).

Letter

Body Condition Score Estimation Based on Regression Analysis Using a 3D Camera

Thi Thi Zin [1],*, Pann Thinzar Seint [1], Pyke Tin [1], Yoichiro Horii [2] and Ikuo Kobayashi [3]

[1] Graduate School of Engineering, University of Miyazaki, 1 Chome-1 Gakuenkibanadainishi, Miyazaki 889-2192, Japan; pantenzasein.t9@cc.miyazaki-u.ac.jp (P.T.S.); pyketin11@gmail.com (P.T.)
[2] Center for Animal Disease Control, University of Miyazaki, 1 Chome-1 Gakuenkibanadainishi, Miyazaki 889-2192, Japan; horii@cc.miyazaki-u.ac.jp
[3] Field Science Center, Faculty of Agriculture, University of Miyazaki, 1 Chome-1 Gakuenkibanadainishi, Miyazaki 889-2192, Japan; ikuokob@cc.miyazaki-u.ac.jp
* Correspondence: thithi@cc.miyazaki-u.ac.jp

Received: 29 May 2020; Accepted: 29 June 2020; Published: 2 July 2020

Abstract: The Body Condition Score (BCS) for cows indicates their energy reserves, the scoring for which ranges from very thin to overweight. These measurements are especially useful during calving, as well as early lactation. Achieving a correct BCS helps avoid calving difficulties, losses and other health problems. Although BCS can be rated by experts, it is time-consuming and often inconsistent when performed by different experts. Therefore, the aim of our system is to develop a computerized system to reduce inconsistencies and to provide a time-saving solution. In our proposed system, the automatic body condition scoring system is introduced by using a 3D camera, image processing techniques and regression models. The experimental data were collected on a rotary parlor milking station on a large-scale dairy farm in Japan. The system includes an application platform for automatic image selection as a primary step, which was developed for smart monitoring of individual cows on large-scale farms. Moreover, two analytical models are proposed in two regions of interest (ROI) by extracting 3D surface roughness parameters. By applying the extracted parameters in mathematical equations, the BCS is automatically evaluated based on measurements of model accuracy, with one of the two models achieving a mean absolute percentage error ($MAPE$) of 3.9%, and a mean absolute error (MAE) of 0.13.

Keywords: body condition score; 3D surface roughness parameters; rotary parlor; 3D camera; regression analysis

1. Introduction

Tracking body condition scores (BCS), and using them to avoid rapid fluctuations in body weight during the production cycle, has a positive impact on decision-making in dairy farm management, and makes economic sense. It is also useful for improving milk production, health, and reproduction (pregnancy rate) throughout the production cycle. The resulting improved monitoring provides an opportunity to fine-tune nutrition, and healthcare more generally. Although various methods are available for evaluating body condition, many producers use the BCS system, which ranks cattle using an arbitrary scale, and does not rely on body weight [1]. The BCS is assigned by scoring the amount of fat that is observed on several skeletal parts of the cow. Various scoring systems are used to arrive at the BCS, which are used to assign a number as the score. As the system most commonly used, the BCS ranges from 1 to 5, in increments of 0.5 or 0.25. Very thin cows are given a BCS of '1', and very obese cows are rated as '5'. The intermediate stages of BCS can be characterized as thin, ideal and obese. A 'very thin' cow has prominent hips and spine. The hips and spine of a 'thin' cow are easily felt without pressure, and those of an 'ideal' cow can be felt with firm pressure. A 'very obese' cow is heavily covered by fat. BCS '3' is considered ideal. Cows can then be managed and fed according to the requirements for attaining an optimal BCS.

Research results indicate that optimizing BCS can positively influence the health and productivity of dairy cows. In addition, a rapid decrease in the BCS after calving closely correlates with metabolic disorders and other problems [2]. Current interpretations of available evidence indicate that metabolic disorders affect the immune system of dairy cows during the critical transition from calving [3]. BCS decreases during the approximate 100-day period from calving through early lactation to peak milk, and then increases through dry-off. Generally, maintaining an optimal BCS is needed to avoid extremes of too fat or too lean [4]. For a proper evaluation of BCS, the observer must be familiar with skeletal structures and fat reserves, as described in [5]. In the measurement of BCS using vision-based technology, tailhead and loin areas are of primary concern. Many researchers have evaluated BCS by manually checking off significant anatomical points on digital 2D images. Hook angles and tailhead depressions are formulated to estimate BCS, using a technique introduced in 2008 by Bewley et al. [6]. In this technique, the skeletal checkpoints associated with anatomical structures are used in the assessment of BCS. However, automating the identification of these checkpoints with 2D images is difficult. A new perspective for measuring fat levels in cows is proposed using ultrasonography in [7]. It shows that the larger the BCS, the more the increase in the fat reserves. In recent years, single 3D camera and multiple 3D cameras with multiple viewpoints have been introduced to evaluate the body condition score by using machine vision technology. In our system, we introduce the BCS automation system by using a single 3D camera that is mounted above the rotary parlor.

2. Related Work

To rate BCS, a technique was introduced by Edmonson et al. in [8], which consists of manually assessing the amount of body fat around the tailhead, as well as by palpation of the tailhead (the depression beneath the tail), and the pelvis (hook and pin bones). In an automated system, an image analysis technique was introduced to derive relevant characteristics from anatomical points, and from intensities or depth values in regions of interest and cow contours. By using a low-cost 3D camera, an automatic body condition scoring system was developed by implementing an image-processing technique and regression algorithms [9]. In this system, fourteen features correlated with BCS were used (such as age, weight, and height), including some features that were derived from video images, and automatically derived from farm records. The accuracy of the entire system was 0.26 of mean absolute error (MAE). In [10], an automated BCS rating system was introduced, which assesses scores from 1.5 to 4.5 by extracting multiple features related to body condition from three viewpoints. In this system, body images are recorded using 3-dimensional cameras positioned above, behind, and to the right. Anatomical landmarks are automatically identified, and then bony prominences and surface depressions are quantified to evaluate BCS and provide the result.

In our own previous work, we proposed a noninvasive method for automatically evaluating BCS [11]. This method starts with a 3D image, from which two analytical models are created, one using the root-mean-square deviation (RMSD), and the other using the convex hull volume parameter. This method resulted in a standard error of 0.35 using RMSD, and 0.19 using convex hull volume. We also noticed that convex hull volume has a strong correlation with BCS. The benefits of continuously monitoring BCS are intuitive to most dairy producers, nutritionists, and others involved in dairy farming. A few dairy farms have incorporated such monitoring as part of their management strategy, as described in [12]. In our previous paper of body condition indicators described in [13], we noted a strong link between BCS and parameters such as convex hull volume and mean height, with BCS ratings between 3.5 and 3.75. That system also introduced variations in BCS trends during the calving and lactation intervals using values for monthly mean height. In [14], a low-cost monitoring system was proposed for unobtrusively and regularly monitoring BCS, lameness, and weight using 3D imaging technology. In the paper described in [14], a new approach for assessing BCS based on a rolling ball algorithm was validated by achieving repeatability within ±0.25 BCS. Our approach included automatic image selection steps for each cow in the parlor that was targeted for a smart application of continuous monitoring. The approach also featured a newly developed BCS estimation model using two region of interests (ROI) visible from above. Finally, this approach also involves extracting 3D surface roughness features, and generalizing two linear regression models to estimate BCS by applying the proposed parameters.

3. Data Collection and Preprocessing

In our proposed system, a 3D camera is mounted 3.4 m above a rotary parlor. Data collection is done at a large-scale dairy farm in Oita Prefecture, Japan. The position of the 3D camera and an illustration of cows in the parlor are seen in Figure 1a,b. The 3D camera generates a resolution of 132 × 176 pixels in X, Y, and Z directions. X and Y are the x and y coordinates of the image and Z is the distance information (z or D_0) of the image.

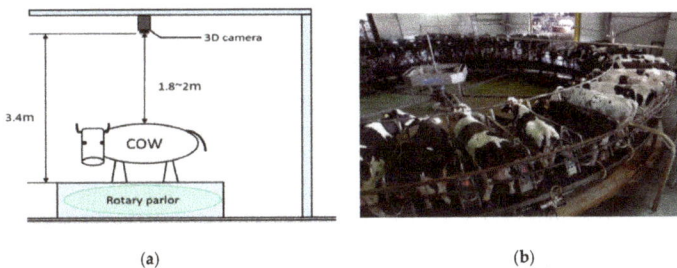

Figure 1. (a) Position of 3D camera; (b) image of cows in rotary parlor.

The proposed system uses a single 3D camera and generates data in csv format (comma-separated values). Each line of the csv data has 23,232 values for distance information, which is preprocessed into image dimensions of 132 × 176 pixels. The transformed image has a maximum of three cows. An original image obtained using this camera is shown in Figure 2a. The original image shows the distance data (D_0) from the camera center to the image plane. To obtain real-world data for the distance from the ground, the difference between D_0 and the camera height (3.4-D_0) m is calculated. The distance range between 1.21 m and 2.1 m is considered to be the cow region, as shown in Figure 2b. Conversely, the background region is automatically removed by the extraction of the cow region. Each of the cows has their related ID number, using radio frequency identification (RFID). Therefore, we only extract the middle cow image as the desired ID number on the rotary parlor, as shown in Figure 3a,b. The sided images of other ID numbers are conversely removed. In Figure 3a, ROI 1 and ROI 2 are the two regions of interest used for BCS estimation, for which details are discussed in Section 4.

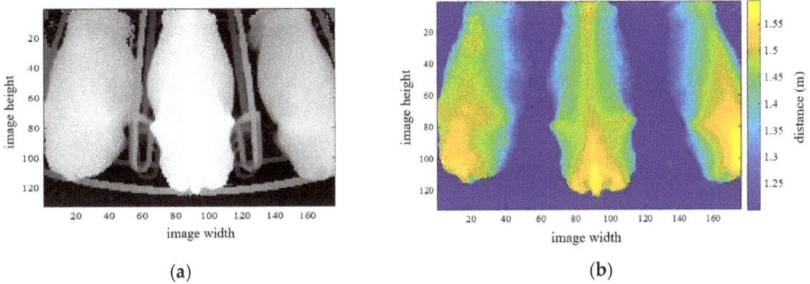

Figure 2. Cow region extraction from 3D camera. (a) Original image in rotary parlor from 3D camera; (b) Cow region extraction by distance information.

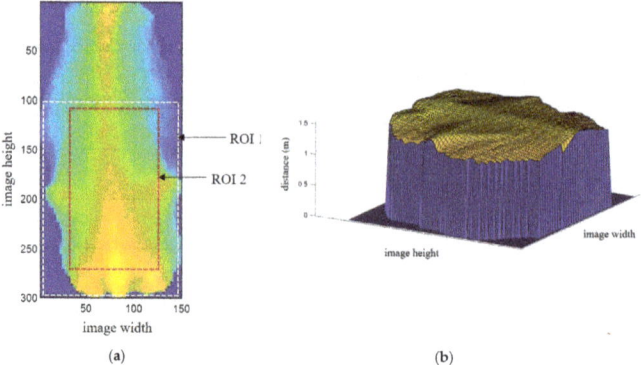

Figure 3. The processed cow image. (**a**) Distance image of cow (color expresses the distance); (**b**) The cow image in 3D space.

4. Automatic Image Selection Process by Filtering

Sometimes, the 3D camera returns distorted images. These bad images are removed using geometric and hole areas. A sample of images discarded due to distortion and touching between cows is shown in Figure 4a. The selected cow images are grouped by ID number. From all of the cow images recorded, filtering ensures that only one good position of cow image for the same ID number on the same day is selected by using the symmetricity parameter, though the camera generally captures three images per day in the milking parlor. Filtering is performed by a comparison of symmetricity. The filtering for cow ID "LA982123529378694" is shown in Figure 4b. This cow has camera capture times (at 04:57 a.m., 13:22 p.m., and 21:21 p.m. on 4th November 2019), and the selected filtered image is shown by a red marker. When the left and right sides of the image are symmetric, the difference between the two areas is nearly zero. The image with the least difference in value is selected as the most symmetric image, filtered from the images collected every day.

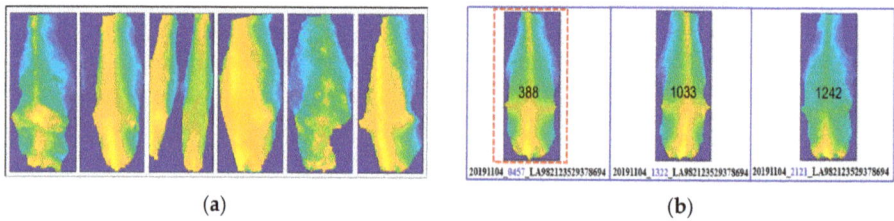

Figure 4. Sample discarded and filtered images. (**a**) Sample of discarded images; (**b**) Image selection by symmetricity.

A detailed workflow for the automatic image selection process by filtering is shown in Figure 5. After making the cow extraction with the filtering step, each of the selected or filtered images is stored by its ID group in the database for further implementation of the smart system. In our system, approximately 20,000 images were automatically discarded by geometric area as bad images, and over 140,000 selected images were recorded from August 2018 to February 2020.

The proposed work is performed on Windows 10, an Intel ® Core ™ i7-7700 CPU, @ 3.6 GHz. The processing time for the cow selection process from each set of csv data is approximately 0.2 s. BCS is a good management tool for developing nutrition and care programs for specific situations. This is the first step in improving the use of BCS. Follow-on steps include developing a BCS monitoring program for each individual cow, determining the BCS at calving, and then monitoring changes in BCS during lactation. Optimal scores can be devised for each cow at each stage of the production cycle, i.e., the optimal score for calving is 3.25 and the optimal score at the start of breeding is 3, and so on. Therefore, we launched this study to automate an accurate assessment of BCS in the next step of BCS modeling.

This section includes a discussion of automatic collecting of cow images from a 3D camera. The remaining sections of the paper include a discussion of collecting cow images and their use in building models in Section 4, experimental use and performance evaluation of the two analytical models in Section 5, and a presentation of conclusions, as well as prospective work, in Section 6.

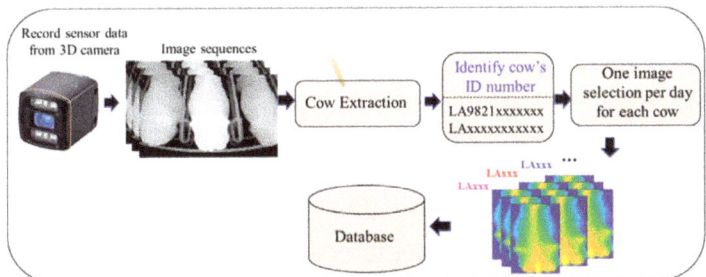

Figure 5. Detailed workflow for automatic image selection.

5. Proposed BCS Modeling

BCS is generally evaluated by experts who have been trained in its use. Though this conventional method is time- and labor-intensive, automation can ease the burden. Using manual measurements for BCS as a baseline or for referencing a model, a reliable automated system can be established. To confirm the performance of the proposed system, we used images of cows with manually measured BCS values in the range of 2.5 to 4. These cow images were collected from two different farms: (1) the Sumiyoshi Livestock Science Station, Field Science Center, University of Miyazaki, and (2) a large-scale dairy farm in Oita Prefecture, Japan. Two experts performed these initial manual measurements. Although the possible BCS scores range from 1 to 5 for very thin to very obese cows, respectively, BCS values between 2.5 and 4 are the most frequently seen. Our BCS dataset is shown in Table 1. In order to evaluate BCS values, two analytical models ($M1$ and $M2$) are proposed for the automated system. In total, 52 cows were used in the experiment, 32 for training, and an additional 20 for testing. $M1$ was applied to ROI 1, and $M2$ was applied to ROI 2, as seen in Figure 3a. The learning parameters were extracted for the two regions of interest (ROI 1 and ROI 2) using the concept of 3D roughness texture, which can be seen in surface texture analysis ASME-B46.1 (American Society of Mechanical Engineers 2002).

Table 1. BCS dataset taken by experts.

Body Condition Score (BCS)	2.5	2.75	3	3.25	3.5	3.75	4
No. of Cows	1	1	6	24	14	5	1

5.1. BCS Estimation Model 1

In the 3D image of a cow's backbone, BCS estimation model 1 was used for ROI 1, which is two-thirds of the whole cow body starting from the tailhead, as seen in Figure 3a. Variations in the amount of fat reserves or in the energy balance are apparent in that region. Moreover, the more that body fat covers the bones, the larger the BCS. Visually, we can clearly differentiate thin cows from fat ones by this coverage by fat. Therefore, roughness parameters are extracted to determine the BCS. Figure 6 shows images composed of cross-sectional slices in ROI 1 for each BCS value between 2.5 and 4 in increments of 0.25. In this proposed region, the following parameters were extracted:

i. Arithmetic mean height (A_1);

ii. Convex hull volume (A_2);
iii. Difference between convex hull volume and 3D volume (A_3);
iv. Difference between peak height and valley depth in fifteen maximums and minimums for all profiles (A_4).

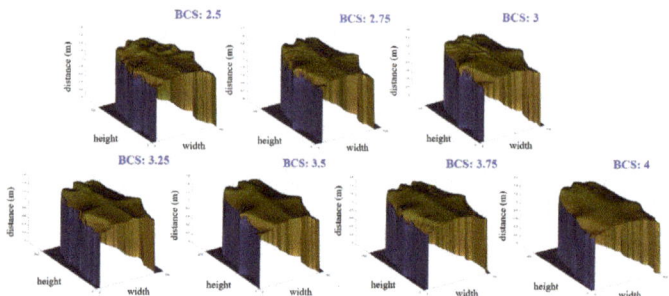

Figure 6. Images composed of cross-sectional slices for ROI 1 used in BCS estimation model 1.

Arithmetic mean height is calculated by following Equation (1):

$$\text{Arithmetic mean height} = \bar{z} = \frac{1}{N} \sum_{i=1}^{N} z_i \qquad (1)$$

where z is the height or distance parameter on the roughness profile, and N is the total number of all height values in ROI 1.

Convex hull volume is calculated by [15]:

$$\text{Convex hull volume} = \frac{1}{3} \times \sum_{F} (height \times area\ of\ face) \qquad (2)$$

where F represents the faces of the polyhedron.

The difference between peak height and valley depth in fifteen maximums and minimums is calculated by using Equation (3):

$$\text{Difference between peak and valley points} = \frac{1}{15} \sum_{i=1}^{15} z_i - \frac{1}{15} \sum_{j=1}^{15} z_j \qquad (3)$$

By using A_1, A_2, A_3, and A_4 features, a stepwise linear regression model ($M1$) is generated by the following Wilkinson notation:

$$M1 \sim 1 + A_3 + A_1 * A_2 + A_2 * A_4 + A_1{}^2 + A_4{}^2 \qquad (4)$$

where $M1$ is the BCS obtained by proposed method 1.

5.2. BCS Estimation Model 2

To build BCS estimation model 2, ROI 2 in Figure 3a was used, for which an image composed of cross-sectional slices, is shown in Figure 7. In this region, the following features are extracted:

i. Arithmetic mean height (B_1);
ii. Difference between peak height and valley depth (B_2).

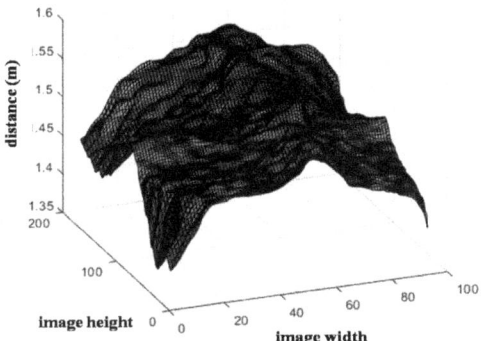

Figure 7. Image composed of cross-sectional slices for ROI 2 used in BCS estimation model 2.

The stepwise linear regression model (M2) was generated using two learning parameters (B_1 and B_2). The output of the second proposed BCS model is defined using Wilkinson notation, as seen in Equation (5).

$$M2 \sim 1 + B_1 * B_2 + B_2{}^2 \tag{5}$$

where $M2$ is the BCS obtained by proposed method 2.

Stepwise linear regression is a semi-automated process used for building a model. It can be generated as a way of adding or removing predictor parameters. Parameters or features to be added or removed are picked up from statistics on the test of estimated coefficients used to reach the target output.

In the selected cow image, the BCS estimation model is established after the parameter extraction. The processing time to obtain the BCS output is about 0.13 s. The results for training and testing BCS measurements for the two proposed models are seen in Table 2.

Table 2. Comparison of BCS obtained manually by experts, and BCS obtained by proposed models 1 and 2.

Training with Two Proposed Methods by Regression Analysis				Testing with Two Proposed Methods by Regression Analysis			
Cow No.	BCS by Experts	BCS by Proposed Method (Model 1)	BCS by Proposed Method (Model 2)	Cow No.	BCS by Experts	BCS by Proposed Method (Model 1)	BCS by Proposed Method (Model 2)
1	3	3.39	3.32	1	3	3.29	3.27
2	3.5	3.31	3.35	2	3.25	3.34	3.35
3	3.25	3.35	3.64	3	3.25	3.14	3.29
4	3.25	3.28	3.56	4	3.25	3.39	3.28
5	3.25	3.29	3.30	5	3.25	2.79	3.36
6	3.25	3.54	3.29	6	3.5	3.47	3.53
7	3	3.21	3.12	7	3.25	3.40	3.24
8	3.25	3.20	3.07	8	3.5	3.77	3.36
9	3	3.09	3.34	9	3.5	3.40	3.43
10	3.25	3.10	3.29	10	3.5	3.37	3.31
11	3.5	3.52	3.42	11	3.25	3.28	3.25
12	3.5	3.18	3.06	12	3.5	3.34	3.18
13	3.5	3.59	3.28	13	3.25	3.44	3.28
14	3.25	3.10	3.17	14	3.5	3.27	3.50
15	3.25	3.48	3.21	15	3.75	3.59	3.20
16	3.5	3.38	3.51	16	3.75	3.52	3.53
17	3.25	3.05	3.12	17	3.25	3.25	3.21
18	3.75	3.73	3.53	18	3	3.04	3.16
19	3.25	3.29	3.30	19	3.25	3.41	3.43
20	3.25	3.20	3.40	20	3.25	3.08	3.38
21	3.25	3.40	3.26				
22	3.5	3.45	3.32				
23	3.5	3.48	3.18				
24	3.25	3.30	3.37				
25	3.75	3.67	3.67				
26	3.25	3.11	3.25				
27	2.5	2.62	2.89				
28	3.5	3.32	3.27				
29	2.75	2.88	3.21				
30	3.75	3.41	3.41				
31	4	3.81	3.61				
32	3	3.28	3.26				

6. Performance Evaluation

For automating BCS estimation, we proposed two analytical models, each used for a 3D image of the top view of the region of interest (ROI). The analytical parameters were extracted from pixel depth values. The training models were built using data collected on 32 cows, including one cow for each of the BCS scores of 2.5, 2.75, and 4, four cows for BCS 3, fourteen cows for BCS 3.25, eight cows for BCS 3.5, and three cows for BCS 3.75. In this experiment, training data were collected between BCS 2.5 and 4. A total of 20 cows were used to determine the accuracy of the models, including two cows for BCS 3, ten cows for BCS 3.25, six cows for BCS 3.5, and two cows for BCS 3.75. Manual assessments of BCS were performed by experts, and these assessments were compared with the results of using the proposed models. The measurable parameters were mean absolute error (MAE), and mean absolute percentage error ($MAPE$). Performance evaluations for models 1 and 2 are shown in Table 3. According to test results, the mean absolute error percentage for model 2 ($M2$) was less than that for model 1 ($M1$); i.e., small error values indicate good predictive capability. Calculations for MAE and $MAPE$ were performed using Equations (6) and (7):

$$MAPE = \frac{100\%}{n} \Sigma \left| \frac{y - y_i}{y} \right| \qquad (6)$$

$$MAE = \frac{1}{n} \Sigma |y - y_i| \qquad (7)$$

where n = the number of cows tested, y = BCS by experts, y_i = BCS by the proposed method.

Table 3. Performance evaluation for models 1 and 2.

BCS Model	Training		Testing	
	MAE	MAPE	MAE	MAPE
Model 1 ($M1$)	0.14	4.31%	0.15	4.64%
Model 2 ($M2$)	0.19	5.89%	0.13	3.87%

7. Discussion and Conclusions

On large-scale dairy farms, management using manual labor is impractical for numerous reasons. Therefore, automated systems of farm management have been a big focus of dairy farmers. Automated milking robots have recently been introduced to replace manual labor, which is time-intensive and requires one-on-one attention to each cow. The nutritional status of each cow can affect milk production. As one of the most important tools for evaluating nutritional status, BCS is a frequent topic of research. Accurately assessing BCS and regularly monitoring BCS trends have become critical requirements. Our belief is that more and more dairy producers will implement automated BCS evaluation as an important part of smart dairy farming.

The initial step in the proposed system involves automatically selecting images for individual cows. We have tested this process by collecting images in various groups to implement in the advance application. The processing time for this proposed work is acceptable for large-scale data sources. In the next step, automated BCS estimation models are introduced by combining proposed parameter extraction and the two stepwise regression analysis to evaluate BCS. Since our dataset contains only one cow each for BCS of 2.5, 2.75, and 4, we could not test for these BCS in our testing process although they are used in the training for having a good estimation model. Our proposed regression models are tested on 20 cows of BCS (3, 3.25, 3.5, and 3.75). The first model obtains minimum and maximum errors of 0.0025 and 0.45, and the second model obtains minimum and maximum errors of 0.0024 and 0.55, respectively. To measure the proposed model accuracies, the mean absolute parameter (MAE) is used and the MAE for the first and second models was 0.15 and 0.13, respectively. Although the proposed stepwise regression worked reasonably well for developing useful models in this existing data, it cannot always work well for all new data. Therefore, our future work entails collecting more fruitful data of various BCS for the purpose of discerning variations in our proposed automated BCS evaluation process.

Author Contributions: The major portion of work presented in this paper was carried out by the first author T.T.Z. and P.T.S., supported the experimental work; P.T. and Y.H., provided valuable advice on statistical analysis and animal behavior theory respectively; I.K., cooperated on setting up video cameras in the farm. The first two authors analyzed the experimental results and discussed the preparation and revision of the paper. All authors read and approved the final manuscript.

Funding: This work is supported in part by SCOPE: Strategic Information and Communications R&D Promotion Program (Grant No. 172310006).

Acknowledgments: We would like to give our sincere thanks to Kodai ABE for providing cow image data.

Conflicts of Interest: The authors declare no conflict of interest.

References

1. Kellogg, W. Body Condition Scoring with Dairy Cattle. Available online: https://www.uaex.edu/publications/pdf/FSA-4008.pdf (accessed on 2 April 2020).
2. Gearhart, M.A.; Curtis, C.R.; Erb, H.N.; Smith, R.D.; Snien, C.J.; Chase, L.E.; Cooper, M.D. Relationship of changes in condition score to cow health in Holsteins. *J. Dairy Sci.* **1990**, *73*, 3132–3140. [CrossRef]
3. Roche, J.R.; Meier, S.; Heiser, A.; Mitchell, M.D.; Walker, C.G.; Crookenden, M.A.; Riboni, M.V.; Loor, J.J.; Kay, J.K. Effects of precalving body condition score and prepartum feeding level on production, reproduction, and health parameters in pasture-based transition dairy cows. *J. Dairy Sci.* **2015**, *98*, 7164–7182. [CrossRef] [PubMed]
4. Heinrichs, A.J. *Body-Condition Scoring as a Tool for Dairy Herd Management*; Cooperative Extension, College of Agriculture, Pennsylvania State University: University Park, PA, USA, 1980; Volume 363.
5. Rossi, J.; Wilson, T.W. Body Condition Scoring Beef Cows. Available online: https://secure.caes.uga.edu/extension/publications/files/pdf/B%201308_3.PDF (accessed on 23 April 2020).
6. Bewley, J.M.; Peacock, A.M.; Lewis, O.; Boyce, R.E.; Roberts, D.J.; Coffey, M.P.; Kenyon, S.J.; Schutz, M.M. Potential for Estimation of Body Condition Scores in Dairy Cattle from Digital Images. *J. Dairy Sci.* **2008**, *91*, 3439–3453. [CrossRef] [PubMed]
7. Alapati, A.; Kapa, S.R.; Jeepalyam, S.; Rangappa, S.M.P.; Yemireddy, K.R. Development of the body condition score system in Murrah buffaloes: Validation through ultrasonic assessment of body fat reserves. *J. Dairy Sci.* **2010**, *11*, 1–8. [CrossRef] [PubMed]
8. Edmonson, A.J.; Lean, I.J.; Weaver, L.D.; Farver, T.; Webster, G. A body condition scoring chart for Holstein dairy cows. *J. Dairy Sci.* **1989**, *72*, 68–78. [CrossRef]
9. Spoliansky, R.; Edan, Y.; Parmet, Y.; Halachmi, I. Development of automatic body condition scoring using a low-cost 3-dimensional Kinect camera. *J. Dairy Sci.* **2016**, *99*, 7714–7725. [CrossRef] [PubMed]
10. Song, X.; Bokkers, E.A.M.; van Mourik, S.; Koerkamp, P.G.; van der Tol, P.P.J. Automated body condition scoring of dairy cows using 3-dimensional feature extraction from multiple body regions. *J. Dairy Sci.* **2019**, *102*, 4294–4308. [CrossRef] [PubMed]
11. Imamura, S.; Zin, T.T.; Kobayashi, I.; Horii, Y. Automatic evaluation of cow's body-condition-score using 3D camera. In Proceedings of the 2017 IEEE 6th Global Conference on Consumer Electronics (GCCE), Nagoya, Japan, 24–27 October 2017; pp. 1–2.
12. Krukowski, M. Automatic Determination of Body Condition Score of Dairy Cows from 3D Images. Available online: https://pdfs.semanticscholar.org/a9e1/bddb0fdc862859b90d03e20b34d4cfdf4b93.pdf (accessed on 23 April 2020).
13. Zin, T.T.; Seint, P.T.; Tin, P.; Horii, Y. The Body Condition Score Indicators for Dairy Cows Using 3D Camera. In Proceedings of the International Workshop on Frontiers of Computer Vision (IW-FCV), Ibusuki, Japan, 20–22 February 2020; pp. 1–8.
14. Hansen, M.F.; Smith, M.L.; Smith, L.N.; Jabbar, K.A.; Forbes, D. Automated monitoring of dairy cow body condition, mobility and weight using a single 3D video capture device. *Comput. Ind.* **2018**, *98*, 14–22. [CrossRef] [PubMed]
15. Polyhedron. Available online: https://en.wikipedia.org/wiki/Polyhedron (accessed on 20 April 2020).

© 2020 by the authors. Licensee MDPI, Basel, Switzerland. This article is an open access article distributed under the terms and conditions of the Creative Commons Attribution (CC BY) license (http://creativecommons.org/licenses/by/4.0/).

Article

Wireless Module for Nondestructive Testing/Structural Health Monitoring Applications Based on Solitary Waves

Ritesh Misra [1], Hoda Jalali [2], Samuel J. Dickerson [1] and Piervincenzo Rizzo [2,*]

[1] Mixed-Signal Multi-Domain Systems Laboratory, Department of Electrical and Computer Engineering, University of Pittsburgh, 3700 O'Hara Street, 1206 Benedum Hall, Pittsburgh, PA 15261, USA; rim39@pitt.edu (R.M.); dickerson@pitt.edu (S.J.D.)
[2] Laboratory for Nondestructive Evaluation and Structural Health Monitoring Studies, Department of Civil and Environmental Engineering, University of Pittsburgh, 3700 O'Hara Street, 729 Benedum Hall, Pittsburgh, PA 15261, USA; HOJ14@pitt.edu
* Correspondence: pir3@pitt.edu; Tel.: +1-412-624-9575; Fax: +1-412-624-0135

Received: 12 May 2020; Accepted: 24 May 2020; Published: 26 May 2020

Abstract: In recent years, there has been an increasing interest in the use of highly nonlinear solitary waves (HNSWs) for nondestructive evaluation and structural health monitoring applications. HNSWs are mechanical waves that can form and travel in highly nonlinear systems, such as granular particles in Hertzian contact. The easiest setup consists of a built-in transducer in drypoint contact with the structure or material to be inspected/monitored. The transducer is made of a monoperiodic array of spherical particles that enables the excitation and detection of the solitary waves. The transducer is wired to a data acquisition system that controls the functionality of the transducer and stores the time series for post-processing. In this paper, the design and testing of a wireless unit that enables the remote control of a transducer without the need to connect it to sophisticated test equipment are presented. Comparative tests and analyses between the measurements obtained with the newly designed wireless unit and the conventional wired configuration are provided. The results are corroborated by an analytical model that predicts the dynamic interaction between solitary waves and materials with different modulus. The advantages and limitations of the proposed wireless platform are given along with some suggestions for future developments.

Keywords: highly nonlinear solitary waves; nondestructive evaluation; wireless sensing; Bluetooth technology

1. Introduction

Highly nonlinear solitary waves (HNSWs) are mechanical waves that can form and travel in highly nonlinear systems such as mono-periodic arrays of elastically interacting spherical particles, sometimes indicated as granular crystals [1–18]. One of the key features of HNSWs is that properties like duration, amplitude, and speed can be tuned without electronic equipment by simply adding static precompression on the array or varying the particles' material and geometry. This tunability makes the use of HNSWs appealing in some engineering applications such as nondestructive testing (NDE), structural health monitoring (SHM) [19–22] or acoustics [23–28].

The scheme of an HNSW-based NDE/SHM is shown in Figure 1. One end of an array of identical spheres is in dry contact with the structure to be inspected/monitored. The top particle of the chain is lifted and released to create a mechanical impact that triggers the formation of a solitary wave, hereinafter referred to as the incident solitary wave (ISW). When the mass of the striker is equal to the mass of the other particles composing the chain, a single pulse is generated. The ISW propagates through the array, is detected by a sensing system embedded in the chain, and then reaches the surface of the structure to be inspected/monitored. At the interface between the chain and the structure, most of the acoustic energy carried by the ISW is reflected back, giving rise to one or two reflected solitary waves, typically referred to as the primary and the secondary reflected solitary waves (PSW and SSW). These reflected waves are then detected by the same sensing system embedded in the array.

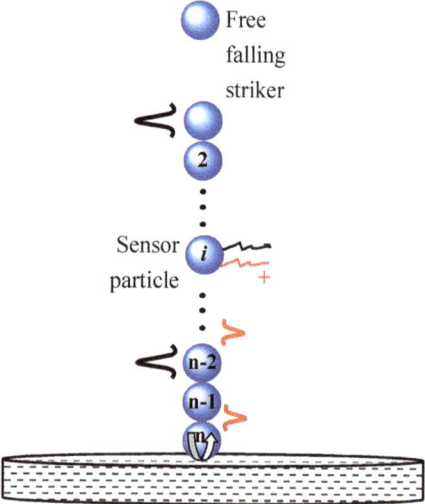

Figure 1. General scheme of nondestructive evaluation/structural health monitoring paradigm using highly nonlinear with different modulus. The advantages and limitations of the proposed wireless platform are given along with some suggestions for future developments.

Many researchers proved that the amplitude and the arrival time of the reflected solitary waves depend on key properties of the adjacent structure/material [19–22,29–38]. For example, Yang et al. studied numerically, analytically, and experimentally the reflection of HSNWs at the interface of a large thin plate, and found that the amplitude and the arrival time of the reflected waves are affected by the plate thickness, the particles size, and the boundary conditions at a critical distance from the plate edges [39]. Kim et al. indicated experimentally and numerically that solitary waves can be used to detect delamination in carbon fiber reinforced polymer plates [31]. A numerical study on the interaction of HNSWs with composite beams showed that solitary waves are helpful to evaluate the directional elastic parameters of the composites [40]. Others reported that the method is effective at detecting subsurface voids [38], assessing the quality of adhesive joints [22,30], composites [31,34,40,41], orthopedic and dental implants [32,42], and at measuring internal pressure [43–46] and axial stress [47,48].

In all the studies cited above, the array of particles is coaxially wired to electronic equipment that controlled the striker and digitized the time-waveforms for post-processing analysis. Although this is acceptable for periodic inspections, i.e., NDE applications, it may be detrimental when continuous monitoring, i.e., SHM protocol, is preferred or necessary. SHM systems, however, can become expensive in large structures or high-dense sensor systems due in part to cabling networks. SHM using wireless sensors can overcome the limitations of traditional wired methods with many attractive advantages such as wireless communication, onboard computing, battery power, ease of installation, and so on. Platforms for SHM containing wireless smart sensing (WSS) have been developed during the past years. WSSs are devices that have a sensor, microprocessor, radio frequency transceiver, memory, and power source integrated into one small size unit and are characterized by their capabilities of sensing, computation, data transmission, and storage, all achieved by a single device. WSS represents an attractive alternative to their wired counterparts because of the lower cost achieved by removing the need for cables (including cost labor reduction), and by the widespread production of micro-electro-mechanical sensors. The wireless communication capability allows flexible network topology and hence enables a decentralized monitoring scheme, which adds robustness to the SHM system compared with the centralized approach in wired systems [49]. For these reasons, in the last two decades there has been a great deal of research and development of wireless sensors for SHM applications that were summarized in a few excellent reviews [50–53]. However, as the NDE/SHM method based on HNSW has been developed only recently, there is no wireless sensing technology available in support of nonlinear solitary wave propagation and detection.

To fill this gap, a wireless sensor system to support the generation and detection of HNSWs for NDE or SHM applications was designed, assembled, and tested. The system is not a sensor per se, but rather an autonomous data acquisition node to which a traditional HNSW transducer, designed and developed in our lab, can be attached. The new device can be best viewed as a platform in which mobile computing and wireless communication elements converge with the transducer. This capability is particularly advantageous in the context of SHM, in which several HNSW transducers can be deployed on the structure of interest to perform structure interrogation and remote communication to a remote repository. Other advantages of this newly designed system are cost reduction in the installation phase, autonomous data processing, denser sensing capability in large structural systems, just to mention a few.

The paper is organized as follows: Section 2 provides an overview of the designed wireless platform and is divided into sub-sections that describe the single components. Section 3 presents the experimental results in which the wireless platform is compared to the performance of conventional sensing with the use of a data acquisition system. Section 4 supplements the experimental finding with the outcomes of a numerical analysis that models the dynamic interaction of the solitary waves with different materials. Lastly, Section 5 provides some concluding analyses and remarks about the study presented here.

2. Wireless HNSW Sensor System Overview

The new wireless sensor network is comprised by three components, the HNSW transducer, a custom printed circuit board (PCB), and a mobile computing device. A schematic of the overall platform is shown in Figure 2.

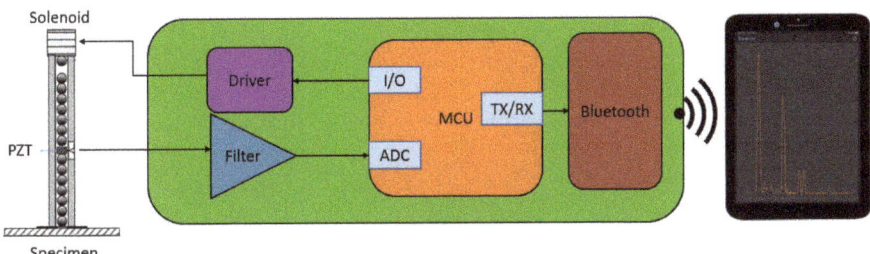

Figure 2. Schematics of the overall HNSW wireless sensor node. The "specimen" in the figure representsany material/structure to be monitored using the solitary waves.

2.1. HNSW Transducer: Conventional Design

The HNSW transducer, hereinafter simply referred to as the *transducer*, contains eight 19.05 mm spheres, a commercial electromagnet (Uxcell 12 V DC) able to lift and release the striker, a sensor, and a frame. All the particles except the striker were made of a non-ferromagnetic material. The current flowing through the solenoid generates a magnetic field strong enough to lift the first particle of the array, whereas flow interruption causes the striker to impact the chain and generate the ISW. The sensing system consists of a lead zirconate titanate ($Pb[Zr_xTi_{1-x}]O_3$) wafer transducer (PZT), embedded between two 6.05 mm thick, 19.05 mm diameter disks. Kapton tape insulated the PZT from the metal. In the conventional (wired) configuration, the transducer is connected to and driven by a National Instruments PXI running in LabVIEW using a graphical user interface created ad hoc for the experiments. The hardware generates the signal to open/close the electrical current to feed the solenoid, while the user interface allows for the selection of the repetition rate of the striker (by controlling the output signal), the number of measurements, the sampling frequency of the digitized waveforms, and the storage of the waveforms for post-processing analysis. The *transducers* used in this study were also successfully used elsewhere [19,43–45,54] and were, therefore, the starting point to develop the proposed wireless platform. However, it is emphasized here that the proposed platform can be adopted and adapted to a chain of any size, length, and particle materials.

2.2. Circuit Board

In the proposed wireless sensor node, the role of the data acquisition system used in the conventional configuration is replaced by a PCB, which is the centerpiece of the schematics of Figure 2. The PCB includes the driver to provide the current for the solenoid, a filter to remove white noise and provide anti-aliasing, and a protocol for the remote communication to and from mobile devices.

Figure 3 shows a photo of the PCB with color-coded sections corresponding to the individual components. The driver allows the microcontroller (MCU) to deliver the DC current to the solenoid with an input/output (I/O) pin on-demand (GPIO). Although the term "GPIO" suggests that both an output and an input are necessary to interface with the driver, the latter is designed such that the movement of the striker onto the array can be controlled through a single digital output. The waves, sensed by the PZT, are filtered by an analog filter and subsequently sampled by the analog to digital converter (ADC) within the MCU. The MCU then sends the data samples to an integrated circuit (IC) enabled for Bluetooth Low Energy (BLE) communication using the Universal Asynchronous Receiver/Transmitter (UART) protocol. The protocol allows for the transmission of the data to any mobile device capable of BLE communication.

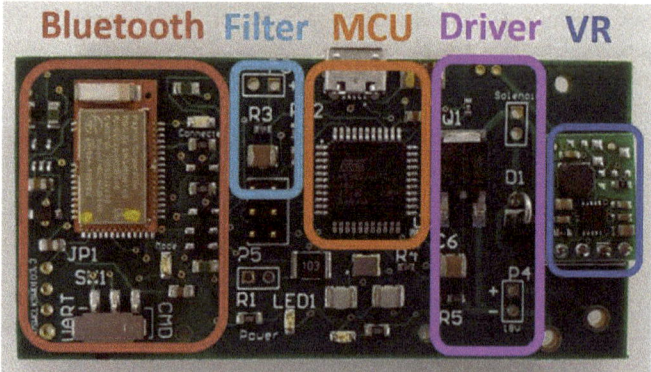

Figure 3. Photo of the final printed circuit board (PCB), which includes a Bluetooth transceiver, filter, microcontroller (MCU), transducer driver, and a voltage regulator (VR). The board is 76.2 mm wide and 36.8 mm high.

The final PCB (76.2 × 36.8 mm^2) was designed to optimize space and therefore enhance portability. The MCU was an ATMega32u4 with 32 kB of flash memory for storing embedded programs, 2 kB of SRAM for storing measurement data, all of the peripherals required to induce and measure the signal, and libraries that allowed for easy communication with the Bluefruit LE module. The MCU has also its own USB controller, making local data collection possible without adding an IC to do FTDI to UART conversion. Overall, the created design fares better than the conventional design because the new system is more compact, is dedicated to the specific application with HNSWs, and can be driven by smart devices such as tablets or even smartphones.

2.3. Actuation: Wireless Configuration

In the wired configuration, the DC current used to drive the electromagnet is delivered by a power supply external to or embedded to a data acquisition system. In the proposed wireless node, the DC current is provided by the PCB. Following the protocol implemented in the PXI in previous studies [19,43–45,54], the solenoid is energized for 250 ms, an interval sufficiently long to lift the striker until it touches the electromagnet before falling freely onto the array. The energy necessary to deliver the current necessary to operate the electromagnet is significant with respect to the other electronic components and is directly proportional to the weight and the falling height of the striker. To supply the necessary energy, a custom power source for the solenoid and driver circuit was chosen to allow the control of the striker while maintaining portability. A 14.8 V rechargeable LiPo battery with a maximum discharge rate of 47 A and a capacity of 1050 mAh was chosen. For the 250 ms interval, 675 mA is consumed. In future, the duration and/or the energy consumption can be reduced by shortening the falling height of the striker, by making the striker lighter (in order to be able to use smaller solenoids), or by minimizing the friction between the striker and the inner wall of the guide.

Under this design, the battery can continuously drive the striker for 93 minutes, which means that the battery can theoretically power 22,390 = (93 × 60 × 4) impacts. This value was obtained by multiplying the capacity of the battery by the inverse of the current consumption. The number of strikes can be also increased using a battery with higher capacity within the limit of size and weight constraints necessary to make the whole node practical. In the present study, the 1050 mAh battery was chosen because it was sufficient to complete a full round of experiments (see Section 3).

Figure 4 shows the control circuit for the actuation. A 1N4003 diode was added in parallel to the solenoid; this flyback diode prevents voltage spike that results from turning off the solenoid from damaging the metal–oxide–semiconductor field-effect transistor (MOSFET), which would, in turn, shorten the life of the overall system [55]. The MOSFET itself acts as an open circuit if the GPIO pin is off, and acts as a closed circuit if the GPIO pin is on, allowing for the control of the current through the solenoid via software. The MOSFET NTD3055-150 from ON Semiconductor was chosen. According to the manufacturer, this specific MOSFET is designed for low voltage, high-speed switching applications in power supplies, converters and power motor controls and bridge circuits, and dissipates 28.8 W continuously, which provides a threefold factor of safety compared to the 9.45W (14 V × 0.675 A) required to lift the striker. The calculation for the required wattage was based on experiments done with a power supply that justified the choice of a 14.8 V battery in the first place. An RC circuit at the gate of the transistor provides a slight delay between turning the GPIO pin off in software and the moment at which the magnetic ball on top of the chain drops. The specific resistor and capacitor values used are not critical, what is most important is that the time constant of the RC circuit is three times larger than the minimum delay that the MCU can produce. This prevents the worst-case scenario of the MCU sampling the ADC after the incident waves passed the sensor disk. The delay that the RC circuit introduces safeguards against mechanical adjustments to the transducer that can reduce the amount of time it takes for the striker to fall. The RC circuit consisted of a 10 kΩ resistor and a 33 nF capacitor, resulting in a time constant of 333 µs.

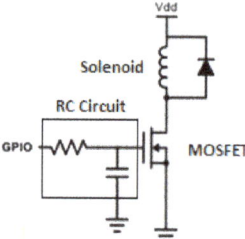

Figure 4. Schematic of the control circuit used to drive the direct current through the solenoid. V_{dd} represents the 14.8 V power source while the coil directly below V_{dd} represents the electromagnet used to lift the topmost ball.

2.4. Filter Design

A passive low-pass filter was added to remove white noise and provide anti-aliasing. The cutoff frequency was determined by examining the frequency spectrum of solitary waves recorded conventionally at a sampling rate of 2 MHz by placing a *transducer* above a 12.7 mm thick steel plate. Figure 5 shows one of the time waveforms and the corresponding Fourier transform associated with this control test. The 99% bandwidth of this signal (the frequency range across which 99% of the power in the signal is contained) is 13.93 kHz. Our initial approach was to place the cutoff frequency somewhere between 50 and 100 kHz so as to eliminate white noise and also have at least a 4x factor of safety from losing any important information. This would also have served to provide anti-aliasing. However, setting the cutoff frequency to a point within that range would introduce a significant amount of phase lag in the relevant frequencies, which could distort the measurements of the arrival time of the pulses of interest. The phase response of a filter is the radiant phase shift added to the phase of each sinusoidal component of the input signal, and the term "phase lag" is used when these phase shifts cause delays in time. In the case that the primary wave and secondary wave have the exact same frequency content, both waves would experience the same amount of delay. However, if the reflected waves are different in terms of the frequency content, a narrow low-pass filter may cause artifacts. Based on the above, a cutoff frequency of 2.4 MHz was chosen in order to minimize the phase lag across the frequencies with relevant information.

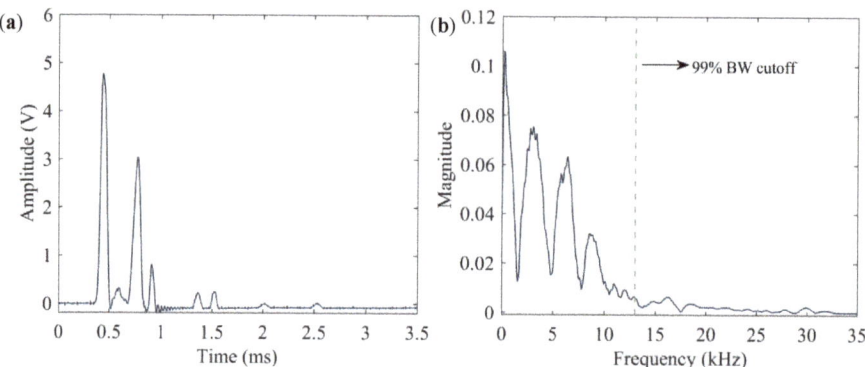

Figure 5. (a) Time waveform and (b) corresponding Fast-Fourier transform (FFT) measured during a control test to design a low-pass filter.

Figure 6 compares two cutoff frequencies, namely 100 kHz and 2.4 MHz, and their effects on the phase response of the low pass filter. The phase lag in the 10–100 kHz range associated with the 2.4 MHz cutoff is essentially flat. At 50 kHz, the phase lag is 0.2 degrees which corresponds to a 0.01

μs delay, well below the sampling period typically used in HNSW tests. However, for the cutoff at 100 kHz, the phase lag at 50 kHz is equal to 4.8 degrees, which yields a 0.26 μs delay. A delay of 0.26 μs is negligible at a sampling period of 13.3 μs, which is the sampling period of the current design. However, as discussed within Section 6, we plan on increasing the sampling frequency in future iterations of this circuit board to at least 1 MHz. In that context, the usage of a lower cutoff frequency could distort the HNSW measurements. The components of the filter's final design have values equal to 2 Ω and 33 nF, resulting in a cutoff frequency of 2.411 MHz. It is noted here that the design of the filter can be modified based upon the specific application of the transducers as the characteristics of the particles in the array and/or the properties of the structure to be monitored, modify the duration of the incident and reflected waves.

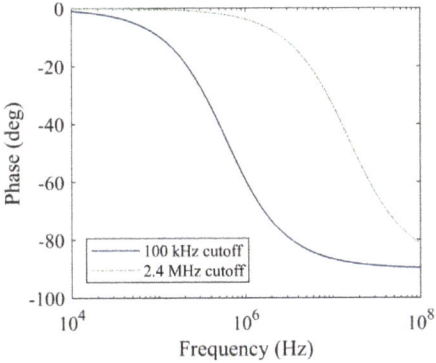

Figure 6. Phase lag as a function of frequency, induced by low pass filters with cutoffs at 100 kHz and 2.4 MHz.

2.5. Wireless Communication

Bluetooth technology was chosen as the communication protocol between the newly proposed wireless device and laptops, tablets, or phones. For short-distance communication, Bluetooth technology is appealing owing to its compatibility with the majority of smartphones, tablets, and laptops. Additionally, Bluetooth communication does not rely on any external network. The Bluetooth LE UART module we used relies on the general-purpose, ultra-low power System-on-Chip nrF51822 to provide wireless communication with any BLE-compatible device. The term "System-on-Chip" means that the nrF51822 is a complete computer system within a single chip that can act independently from the MCU. This could be essential for improvements to future iterations of the PCB as discussed in the Discussion and Conclusions section. The nrF51822 has the ability to choose between UART and SPI communication with external devices, and sleep modes for power preservation. According to the manufacturer, the range of the module is approximately 60 m in an indoor environment, assuming a lack of obstacles. The module was flashed with the firmware provided by the manufacturer [56].

2.6. Mobile Application

An app (Figure 7) was designed to communicate with the *transducer* via the PCB from a smart device. The app was adapted from a general app framework provided by the manufacturer [57]. The custom adaptation added a data streaming mode capable of compartmenting the data it received from different runs into separate graphs. These plots can also be exported as data files for further processing. The data streaming was designed to work with the messaging protocol programmed into the MCU, as discussed later in Section 2.7. First, the app provides a list of Bluetooth devices within the vicinity. After the user selects the appropriate device (Figure 7a), the "Data Stream" menu option allows the user to remotely drive the striker and collect data from the embedded sensor disk (Figure 7b). As shown in Figure 7c, selecting "Data Stream" prompts the user to select the number of strikes and the length (data points) of the signal. Once the PCB receives the command, it actuates the *transducer*, collects samples of the time waveform from the ADC, sends the data to the mobile device, and iterates the process as many times as the number of strikes chosen by the user. During the process, the waveforms are displayed in real-time on the smart device (Figure 7d). The app and the PCB together make a self-contained system that only requires a basic knowledge of smart mobile devices to operate.

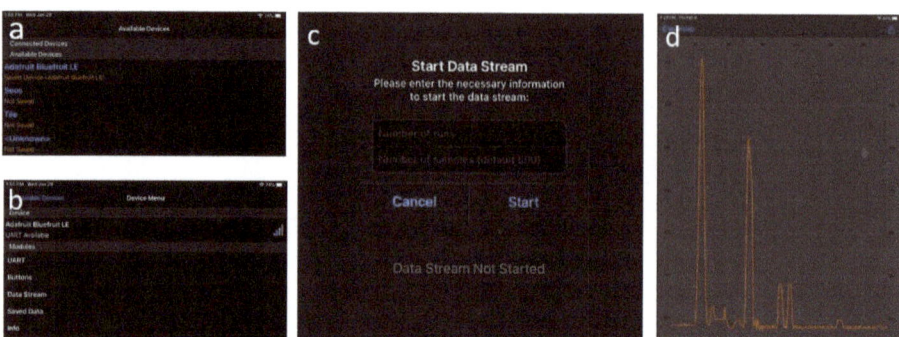

Figure 7. Screenshots of the custom application for mobile devices such as tablets and phones. (**a**) Panel to select the appropriate module from a list of devices. (**b**) Panel to select a data stream from a device menu. (**c**) Panel to enable the user to select the desired number of samples and runs. (**d**) Display of a single measurement from a sheet of aluminum plotted on the mobile device.

2.7. Implementation

The ADC within the AtMega32u4 was used to digitize the signals detected by the embedded sensor disk. The ADC clock was set equal to 1 MHz, as setting the clock to a higher frequency would reduce the resolution for this particular device. A single conversion takes 13 clock cycles, and the clock frequency was set to 16 MHz, so the highest sampling frequency we could theoretically achieve was 1 (MHz)/13 = 77 kHz. The ADC uses a sample-hold capacitor, which is first charged by the signal and then closed-off from the input signal so that the voltage of the signal at that time can be indirectly read through the voltage on the capacitor at that moment. A 5 V power supply for the ATMega32u4 and the Bluetooth module was generated with a 3.7 V single-cell LiPo and a Pololu 5 V Step-Up Voltage Regulator U1V11F5. The U1V11F5 can handle input voltages in a range of 1 to 5.5 V, so it is robust to small voltage drops caused by the discharging of the single-cell LiPo. The PCB follows a protocol for collecting data and sending it to a mobile app. After the first time the PCB is turned on, it waits for a mobile device to connect to it. After a device has connected, the PCB turns the solenoid on and off again, starts a timer, and then collects ADC samples until the ADC reading passes a certain threshold. This allows it to learn the timing between it dropping the ball and observing a HNSW. It then allows the app to send it the desired number of samples and runs and then executes the appropriate number of runs while recording the desired number of samples in time for each run.

Figure 8 shows the final prototype used in the experiments presented in the next section. The PCB is connected to both batteries it needs for all of its functionalities and is in a state where it could be connected to any Bluetooth-compatible mobile device.

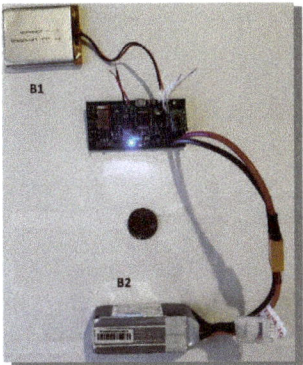

Figure 8. Photo of the final PCB with two batteries, B1 and B2. The 3.7 V battery (B1) powers the microcontroller and Bluetooth module. The 14.8 V battery (B2) powers the solenoid.

3. Experimental Setup and Procedure

To compare the performance of the proposed wireless module to a conventional configuration, two identical *transducers* like those illustrated in Section 2.1 were used. Figure 9 shows part of the setup. *Transducer* T1 was wired to the PCB whereas *transducer* T2 was cabled to a National Instruments —PXI. The latter controlled an external DC power supply to energize the electromagnet.

Figure 9. Photo of the two identical *transducers* used in this study. *Transducer* T1 was wired to the PCB. *Transducer* T2 was wired to a National Instruments PXI.

Three sets of experiments were conducted. In the first two, both *transducers* were placed above a steel thick plate and then above a foam board. The latter was laying above an optical table. Then, one *transducer* only was placed above a granite block and connected to the wireless node first and then to the PXI. The steel specimen was $304.8 \times 152.4 \times 76.2$ mm^3, the foam board was $604.5 \times 457.2 \times 6.35$ mm^3, and the granite block was $152.4 \times 152.4 \times 304.8$ mm^3. For each experiment, 100 impacts were triggered in order to collect a statistically significant amount of data. For the steel plate, one transducer at the time was placed at the center of the 304.8×152.4 mm^2 area. For the foam test, the two *transducers* were placed symmetrically, 101.6 mm away from the center of the board and 228.6 mm from the sides in order to secure identical boundary conditions. The three materials were chosen as representatives of different Young moduli. For the steel plate test, the sampling frequency was equal to 61 kHz and then raised to 75 kHz for the other two tests for both the wireless node and the PXI. Based on the discussion of Section 2.4, when the sampling frequency was 61 kHz, the filter used a 51 Ω resistor and a 330 nF capacitor, whereas when the sampling frequency was 75 kHz, a 2 Ω resistor and a 330 nF capacitor were used.

4. Experimental Results

4.1. Steel Plate: Control Test

The first round of experiments was conducted on the thick steel plate and served as a control test to troubleshoot any possible problems associated with the wireless module. Figure 10 shows one of the time waveforms collected by *transducer* T1. The first peak is the ISW whereas the second peak around 550 µsec is the PSW. Both the ISW and the PSW were tailed by small humps. The origin of these small pulses is described in Section 5. It is anticipated here that these pulses are not detrimental to the successful implementation of NDE/SHM strategies with HNSWs. As conventionally done in HNSW-based NDE/SHM, the time-of-flight (ToF) of the PSW was calculated, as the difference between the arrival time of the PSW with respect to the ISW at the sensing disk. This wave feature is significant for NDE-based or SHM-based applications, as it has been demonstrated that the ToF is sensitive to the presence of damage and/or variation in local stiffness, etc.

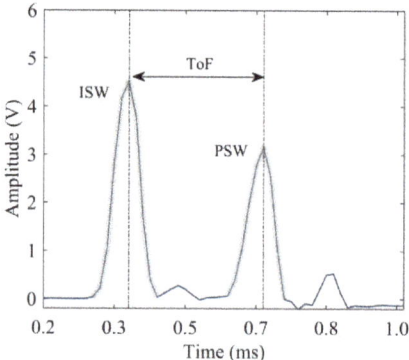

Figure 10. Steel plate test. Example of one of the hundred-time waveforms recorded during the experiment.

To compare the results from both the wired and the wireless systems, the time-series obtained by averaging the one hundred waveforms from both (wired and wireless) systems are displayed in Figure 11a. As the trigger for the two systems was different, the graphs in Figure 11a were shifted horizontally in order to overlap the arrival time of the ISW. Figure 11a proves that the two time-waveforms are remarkably similar to each other, in terms of pulse amplitude and corresponding ToF. The slight differences can be attributable to manual *transducer*-to-*transducer* differences associated with fabrication and assembly. To identify differences at each measurement, Figure 11b shows the number of occurrences for a given ToF for both setups. Three values, namely 295, 311, and 328 µs were measured. It is noted here that the width of each bar is equal to the sampling period, which, for the steel plate test, was equal to 16.4 µs.

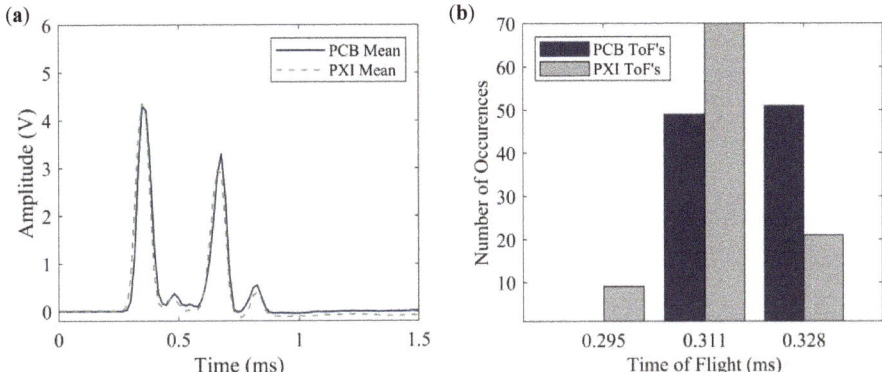

Figure 11. Steel plate test. (**a**) Time-series obtained by averaging the 100-time waveforms measured with the sensor disks. (**b**) Number of occurrences of a given ToF. For each color, the total number of occurrences is 100.

4.2. Foam Specimen

Similar to Figures 11 and 12 presents the results associated with the foam board that was probed with both *transducers* acting sequentially. For this material, the amplitude of the ISW measured with the PXI is larger than the corresponding amplitude measured with the wireless platform. This trend confirms what seen in Figure 11a but for the latter case the difference was smaller. The origin of such difference is discussed in Section 4.3. Nonetheless, it is important to emphasize that for NDE/SHM applications, the trends of interests are those associated with variations from baseline data when material degradation of damage occurs. As such any *transducer* to *transducer* difference in the amplitude of the ISW is not detrimental to prove the main hypothesis of the present research, i.e., that the wireless platform can be used for HNSW measurements in lieu of sophisticated electronics. Another distinctive feature of the wireless platform is that it was able to detect a twin peak associated with the dynamic interaction of the waves with the foam. The origin of these twin-peaks, originally identified and detailed in [58] and elaborated further in [43], was found in several HNSW-based NDE related to soft materials including rubber [44]. For the sake of completeness, the dynamic interaction among the particles that induce the "twin-peaks" is discussed in Section 5.

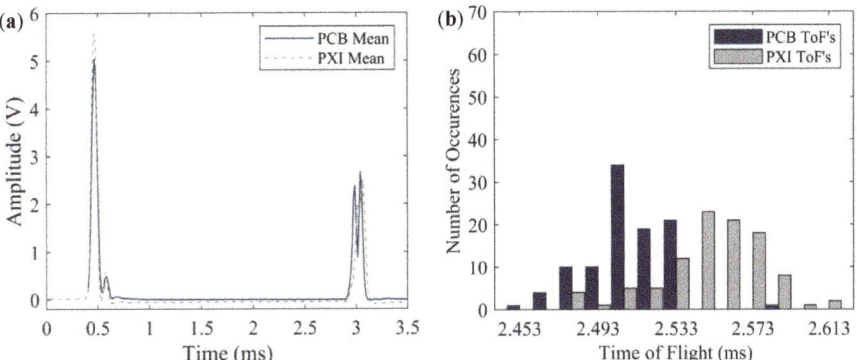

Figure 12. Foam board test. (**a**) time-series obtained by averaging the 100-time waveforms measured with the sensor disks. (**b**) Number of occurrences of a given ToF. For each color, the total number of occurrences is 100.

Figure 12b shows that the number of occurrences of the ToF is more spread with respect to the steel test. This is attributable to the soft nature of the foam specimen. Additionally, the occurrences of the ToF measured with the wireless platform are skewed to the left, i.e., the PSW traveling along *transducer* T1 seems faster than in *transducer* T2. While the origin of such skew is not clear, and more insights could have been gained by inverting the terminals of the *transducers*, it is believed that what was observed is more related to the dynamic interaction of the waves with the foam rather than a variability associated with the *transducers*. In the future, the experiments will be repeated by connecting the terminals of the solenoid and the sensor disk to the PCB first and then to the PXI.

4.3. Granite Specimen

To investigate the origin of the small discrepancies observed for the foam board, one transducer only was used above the granite block and connected to the PCB first and then to the PXI. The results are presented in Figure 13. The time-waveforms (Figure 13a) confirms the trend observed for the foam (Figure 12a). The amplitude of the ISW measured with the PXI is about 20% higher than the corresponding amplitude measured with the wireless node. As such, it can be concluded that differences in the impedance between the two devices result in differences in the amplitude of the detected incident wave. Figure 13b shows the number of occurrences associated with the granite test. The distribution of the ToF is in between the foam and the steel specimen. This confirms that the scattering of the results (i.e., the number of histograms plotted in each figure) is related to Young's modulus of the material. Contrary to what is observed in Figure 12b or Figure 13b shows that the ToF measured with the PCB is higher than the ToF measured with the PXI.

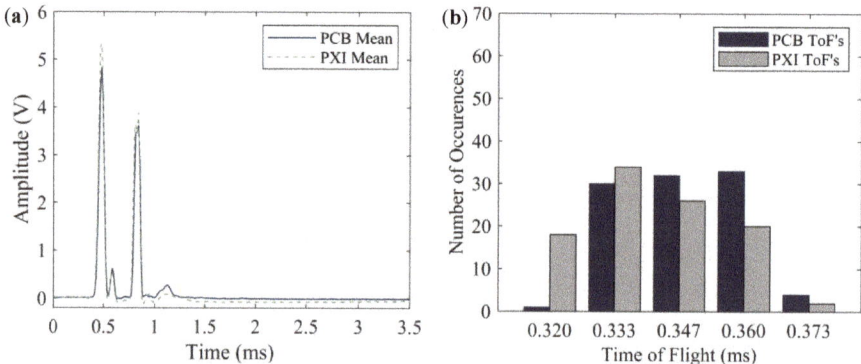

Figure 13. Granite test. (**a**) time-series obtained by averaging the 100-time waveforms measured with the sensor disks. (**b**) Number of occurrences of a given ToF. For each color, the total number of occurrences is 100.

5. Numerical Results

To interpret the experimental results presented in Section 4, an analytical study was performed about the dynamic interaction of solitary waves with the test specimens, using a model that simulates the propagation of solitary waves along the granular array using a series of point masses interacting via nonlinear Hertzian contact forces (Figure 14). Furthermore, the test specimen in contact with the granular chain was modeled as a linear elastic media using a finite element model. The two models were integrated at the interaction point between the granular chain and the test specimen based on the Hertzian contact law. Owing to the scope of this paper, the numerical formulation and the analytical equations used for the simulation are not presented here and interested readers are referred to [27,43–48,54] in order to gain more insight on the subject.

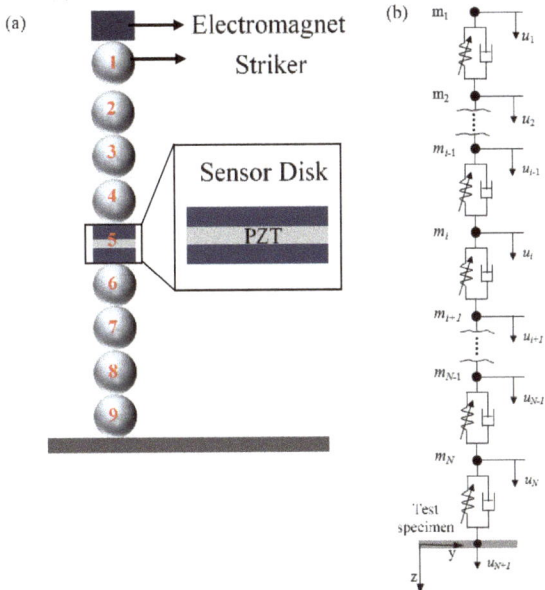

Figure 14. Modeling the interaction between the granular chain and the test specimen. (**a**) Schematics of the HNSW transducer with particle site number. (**b**) Schematic of the mono-periodic array as a series of point masses interacting via nonlinear Hertzian contact (In this figure, N = 9).

For the models, the properties listed in Table 1 were used for the three materials, namely steel, granite, and foam considered in this study.

Table 1. Mechanical properties, specimen thickness, and the predicted ToFs of the steel, granite, and foam specimens in the numerical study.

Material	Material Properties			Specimen Thickness (mm)	Numerical Time of Flight (ms)
	Density (kg/m^3)	Modulus of Elasticity (GPa)	Poisson's Ratio		
Steel	7800	200	0.3	6.35	0.333
Granite	2700	70	0.25	304.8	0.359
Foam	60	0.03	0.3	6.35	2.494

Figure 15 shows the amplitude of the dynamic force measured at the sensor disk (site #5 in Figure 14a). The waveforms represent the average interaction force between the sensor particle and the adjacent particles (sites #4 and #6). As observed in the experiments (Figure 11a, Figure 12a or Figure 13a), Figure 15 shows that the ISW (highlighted in light blue background) is trailed by a small pulse emphasized with red circles. This pulse is the result of a rebound between the sensor disk and particle #6, caused by the local increase of contact stiffness between the disk and the sphere. The higher (with respect to the contact between two spheres) contact stiffness at the sensor particle results in the increase of its maximum particle velocity. As the ISW passes through the sensor and arrives at bead #6, the disk and the particle #6 separate. This yields a momentary zero interaction force at the sensor particle (Figure 15). Then, owing to its higher momentum, the sensor particle strikes particle #6 again, yielding to the small hump circled in Figure 15. The same mechanism causes the small pulses trailing the PSW in the steel and granite tests.

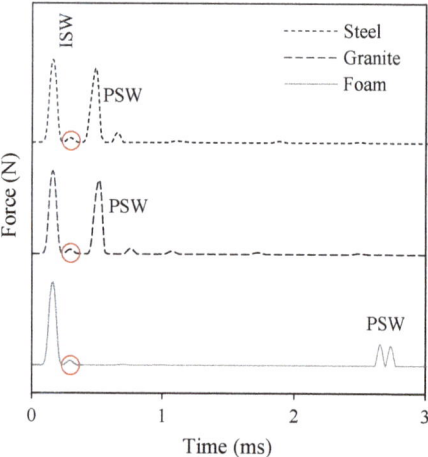

Figure 15. Solitary wave force profiles obtained from the numerical modeling of the interaction of solitary waves and steel, granite, and foam specimens.

When the *transducer* was used above the foam board, both the experimental (Figure 12a) and the numerical (Figure 15) results show a twin PSW. Owing to its soft nature, the acoustic energy carried by the ISW causes a relatively large deformation chain/foam interface. This deformation induces separation between particles that are much larger than the separation that occurs when the chain is above stiff materials. When the reflected wave propagates back to the chain, the separation between bead #6 and the sensor particle and the separation of the sensor particle and bead #4 is large enough that two distinct impacts, i.e., peaks, are seen. The first peak is therefore the upward impact of particle #6 onto the sensor disk, and the second peak is the upward impact of the disk onto particle #4.

Figure 15 also confirms what was observed experimentally (Figure 11b, Figure 12b, or Figure 13b) about the ToF that is inversely proportional to the stiffness of the interface material. The results of the numerical study are summarized in Table 1. The numerical ToF predictions are in very good agreement with the experimental results displayed in Figure 11b, Figure 12b or Figure 13b. At most, the discrepancy between the numerical and the experimental values is within 2–3 data samples. As the thin foam board was placed on a rigid optical table, it is believed that the mechanical properties of foam only material were not representative of the real test object. As such, Young's modulus foam listed in Table 1 was calibrated to match the numerical and the experimental results.

6. Discussion and Conclusions

In statistics, the Pearson's correlation coefficient is generally used to measure the linear correlation between two variables. The coefficient can be comprised between +1 and −1, where 1 is the total positive linear correlation, 0 is no linear correlation, and −1 is the total negative linear correlation. Following past studies [59,60], Pearson's correlation coefficient was used in the work presented in this article to measure the similarity between the waveforms collected with the PXI and the wireless platform. The values of the coefficients for the steel, foam, and granite were 0.975, 0.878, and 0.957, respectively. Although this observation needs data from more materials and a greater number of tests in order to be definitive, the trend of the coefficients seems to be related to Young's modulus of the test specimen. In general, a coefficient greater than 0.8 is considered to be strong evidence of a linear relationship [61]. In this context, it means that it is very likely that the mean waveforms from the PXI and PCB are simply scaled versions of each other. The Pearson coefficients confirm the research hypothesis that the proposed wireless module can replace the wired configuration to perform NDE/SHM using HNSWs.

The quantitative results relative to the ToF are summarized in Table 2, which presents the average value of the ToF, and the corresponding standard deviation and coefficient of variation (CoV) for all the experiments conducted in this study. Additionally, the difference in percent between the numerical and the PCB-measured ToF is reported. Overall, it can be said that the experimental results are in very good agreement with the numerical values and the difference is within the margin associated with the sampling period of the instrumentation, i.e., related to the accuracy provided by the sampling frequency adopted in the experiments. By observing the standard deviations, they are within 1–2 data samples. In addition, the coefficient of variation is quite small proving the repeatability of the experiments.

Table 2. Summary of the experimental and numerical results relative to the time-of-flight of the primary reflected solitary wave. The CoV represents the coefficient of variation, i.e., the ration of the standard deviation to the mean value.

Parameter	Material		
	Steel	Granite	Foam
ToF numerical (μsec)	333	359	2494
ToF experim.: PCB mean (μsec)	320	347	2510
ToF experim.: PCB Standard deviation (μsec)	8.20	12.1	20.5
ToF experim.: PCB CoV (%)	2.56	3.47	0.82
ToF experim.: PXI mean (μsec)	313	341	2551
ToF experim.: PXI Standard deviation (μsec)	8.76	14.2	27.1
ToF experim.: PXI CoV (%)	2.79	4.16	1.06
Numerical/PCB difference (%)	3.90	3.34	0.64

In conclusion, in the study presented in this article, a wireless node was designed, assembled, and tested to support the generation, propagation, and detection of highly nonlinear solitary waves. The aim was to develop the components of a custom system that allows for structural health monitoring applications based on solitary waves. The development of the wireless sensor node included the design of a filter, a driver, and a mobile application. To prove the feasibility of the first prototype, three different interface materials were tested using two HNSW transducers. One transducer was conventionally wired to a National Instruments PXI running under LabView and one transducer was connected to a wireless node and driven with a mobile App using a tablet. The experimental results showed an excellent agreement with respect to each other and with respect to an analytical model that described the dynamic interaction between the waves and the material in dry contact with the chain. This is proof of the reliability of our wireless platform.

In the future, the wireless system needs to be improved over a few fronts. The sampling frequency should be increased to at least 1 or 2 MHz in order to reduce the sampling period and to make the system more sensitive to variations within the material being probed. The power necessary to drive the solenoid needs to be reduced in order to increase the battery life and allow the system to last longer before calling upon battery replacement. The development of a power solar cell to harvest energy would allow the wireless node to operate indefinitely. The latter would also be a suitable solution to power low voltage electronics. With the current design, the 3.7 V, 2500 mAh battery can power the low voltage electronics for 14 h. Besides harvesting energy with a solar module, another remedy could be adding both a separate smaller battery to power the Bluetooth module while it sleeps in a low power mode and hardware to allow the Bluetooth module to turn on the rest of the board once it receives a wake-up signal.

Author Contributions: Conceptualization, P.R. and S.J.D.; methodology, P.R. and S.J.D.; software, R.M. and H.J.; validation, R.M. and H.J.; formal analysis, R.M. and H.J.; investigation, R.M. and H.J.; resources, P.R.; data curation, R.M. and H.J.; writing—original draft preparation, R.M. and H.J.; writing—review and editing, P.R. and S.J.D.;

visualization, P.R.; supervision, P.R. and S.J.D.; project administration, P.R.; funding acquisition, P.R. All authors have read and agreed to the published version of the manuscript.

Funding: This research was funded by the U.S. National Science Foundation, grant number 1809932. The second author acknowledges the 2019 Fellowship Research Award from the American Society for Nondestructive Testing. The authors acknowledge the contribution of Samuel Peterson in the initial design of the mobile application.

Conflicts of Interest: The authors declare no conflict of interest.

References

1. Nesterenko, V. Propagation of nonlinear compression pulses in granular media. *J. Appl. Mech. Tech. Phys.* **1984**, *24*, 733–743. [CrossRef]
2. Nesterenko, V. *Dynamics of Heterogeneous Materials*; Springer Science & Business Media: New York, NY, USA, 2013.
3. Coste, C.; Falcon, E.; Fauve, S. Solitary waves in a chain of beads under Hertz contact. *Phys. Rev. E* **1997**, *56*, 6104. [CrossRef]
4. Coste, C.; Gilles, B. On the validity of Hertz contact law for granular material acoustics. *Eur. Phys. J. B* **1999**, *7*, 155–168. [CrossRef]
5. Manciu, M.; Sen, S.; Hurd, A.J. The propagation and backscattering of soliton-like pulses in a chain of quartz beads and related problems.(I). Propagation. *Physica A* **1999**, *274*, 588–606. [CrossRef]
6. Manciu, M.; Sen, S.; Hurd, A.J. The propagation and backscattering of soliton-like pulses in a chain of quartz beads and related problems.(II). Backscattering. *Physica A* **1999**, *274*, 607–618. [CrossRef]
7. Daraio, C.; Nesterenko, V.; Herbold, E.; Jin, S. Strongly nonlinear waves in a chain of Teflon beads. *Phys. Rev. E* **2005**, *72*, 016603. [CrossRef] [PubMed]
8. Daraio, C.; Nesterenko, V. Strongly nonlinear wave dynamics in a chain of polymer coated beads. *Phys. Rev. E* **2006**, *73*, 026612. [CrossRef]
9. Daraio, C.; Nesterenko, V.; Herbold, E.; Jin, S. Tunability of solitary wave properties in one-dimensional strongly nonlinear phononic crystals. *Phys. Rev. E* **2006**, *73*, 026610. [CrossRef]
10. Porter, M.A.; Daraio, C.; Szelengowicz, I.; Herbold, E.B.; Kevrekidis, P. Highly nonlinear solitary waves in heterogeneous periodic granular media. *Physica D* **2009**, *238*, 666–676. [CrossRef]
11. Nesterenko, V.; Daraio, C.; Herbold, E.; Jin, S. Anomalous wave reflection at the interface of two strongly nonlinear granular media. *Phys. Rev. Lett.* **2005**, *95*, 158702. [CrossRef]
12. Khatri, D.; Daraio, C.; Rizzo, P. Coupling of highly nonlinear waves with linear elastic media. In Proceedings of the Sensors and Smart Structures Technologies for Civil, Mechanical, and Aerospace Systems 2009, San Diego, CA, USA, 9–12 March 2009; p. 72920P.
13. Hong, J.; Ji, J.-Y.; Kim, H. Power laws in nonlinear granular chain under gravity. *Phys. Rev. Lett.* **1999**, *82*, 3058. [CrossRef]
14. Hong, J. Universal power-law decay of the impulse energy in granular protectors. *Phys. Rev. Lett.* **2005**, *94*, 108001. [CrossRef] [PubMed]
15. Job, S.; Melo, F.; Sokolow, A.; Sen, S. How Hertzian solitary waves interact with boundaries in a 1D granular medium. *Phys. Rev. Lett.* **2005**, *94*, 178002. [CrossRef] [PubMed]
16. Vergara, L. Scattering of solitary waves from interfaces in granular media. *Phys. Rev. Lett.* **2005**, *95*, 108002. [CrossRef] [PubMed]
17. Fraternali, F.; Porter, M.A.; Daraio, C. Optimal design of composite granular protectors. *Mech. Adv. Matter. Struct.* **2009**, *17*, 1–19. [CrossRef]
18. Ni, X.; Rizzo, P.; Daraio, C. Laser-based excitation of nonlinear solitary waves in a chain of particles. *Phys. Rev. E* **2011**, *84*, 026601. [CrossRef] [PubMed]
19. Zheng, B.; Rizzo, P.; Nasrollahi, A. Outlier analysis of nonlinear solitary waves for health monitoring applications. *Struct. Health Monit.* **2019**, 1475921719876089. [CrossRef]
20. Cai, L.; Rizzo, P.; Al-Nazer, L. On the coupling mechanism between nonlinear solitary waves and slender beams. *Int. J. Solids Struct.* **2013**, *50*, 4173–4183. [CrossRef]
21. Rizzo, P.; Nasrollahi, A.; Deng, W.; Vandenbossche, J.M. Detecting the presence of high water-to-cement ratio in concrete surfaces using highly nonlinear solitary waves. *Appl. Sci.* **2016**, *6*, 104. [CrossRef]

22. Ni, X.; Rizzo, P. Use of highly nonlinear solitary waves in nondestructive testing. *Mater. Eval.* **2012**, *70*, 561–569.
23. Spadoni, A.; Daraio, C. Generation and control of sound bullets with a nonlinear acoustic lens. *Proc. Natl. Acad. Sci. USA* **2010**, *107*, 7230–7234. [CrossRef] [PubMed]
24. Carretero-González, R.; Khatri, D.; Porter, M.A.; Kevrekidis, P.; Daraio, C. Dissipative solitary waves in granular crystals. *Phys. Rev. Lett.* **2009**, *102*, 024102. [CrossRef] [PubMed]
25. Daraio, C.; Ngo, D.; Nesterenko, V.; Fraternali, F. Highly nonlinear pulse splitting and recombination in a two-dimensional granular network. *Phys. Rev. E* **2010**, *82*, 036603. [CrossRef] [PubMed]
26. Karanjgaokar, N.; Kocharyan, H. Influence of Lateral Constraints on Wave Propagation in Finite Granular Crystals. *J. Appl. Mech.* **2020**, *87*, 071011.
27. Jalali, H.; Rizzo, P.; Nasrollahi, A. Asymmetric propagation of low-frequency acoustic waves in a granular chain using asymmetric intruders. *J. Appl. Phys.* **2019**, *126*, 075116. [CrossRef]
28. Porter, M.A.; Daraio, C.; Herbold, E.B.; Szelengowicz, I.; Kevrekidis, P. Highly nonlinear solitary waves in periodic dimer granular chains. *Phys. Rev. E* **2008**, *77*, 015601. [CrossRef]
29. Yang, J.; Silvestro, C.; Khatri, D.; De Nardo, L.; Daraio, C. Interaction of highly nonlinear solitary waves with linear elastic media. *Phys. Rev. E* **2011**, *83*, 046606. [CrossRef]
30. Ni, X.; Rizzo, P. Highly Nonlinear Solitary Waves for the Inspection of Adhesive Joints. *Exp. Mech.* **2012**, *52*, 1493–1501. [CrossRef]
31. Kim, E.; Restuccia, F.; Yang, J.; Daraio, C. Solitary wave-based delamination detection in composite plates using a combined granular crystal sensor and actuator. *Smart Mater. Struct.* **2015**, *24*, 125004. [CrossRef]
32. Yang, J.; Silvestro, C.; Sangiorgio, S.N.; Borkowski, S.L.; Ebramzadeh, E.; De Nardo, L.; Daraio, C. Nondestructive evaluation of orthopaedic implant stability in THA using highly nonlinear solitary waves. *Smart Mater. Struct.* **2011**, *21*, 012002. [CrossRef]
33. Yang, J.; Sangiorgio, S.N.; Borkowski, S.L.; Silvestro, C.; De Nardo, L.; Daraio, C.; Ebramzadeh, E. Site-specific quantification of bone quality using highly nonlinear solitary waves. *J. Biomech. Eng.* **2012**, *134*, 101001. [CrossRef] [PubMed]
34. Yang, J.; Restuccia, F.; Daraio, C. Highly nonlinear granular crystal sensor and actuator for delamination detection in composite structures. In *International Workshop on Structural Health Monitoring, Stanford, CA*; 2011; pp. 123–134.
35. Nasrollahi, A.; Deng, W.; Rizzo, P.; Vuotto, A.; Vandenbossche, J.M. Nondestructive testing of concrete using highly nonlinear solitary waves. *Nondestruct. Test. Eval.* **2017**, *32*, 381–399. [CrossRef]
36. Li, F.; Zhao, L.; Tian, Z.; Yu, L.; Yang, J. Visualization of solitary waves via laser Doppler vibrometry for heavy impurity identification in a granular chain. *Smart Mater. Struct.* **2013**, *22*, 035016. [CrossRef]
37. Shelke, A.; Uddin, A.; Yang, J. Impact identification in sandwich structures using solitary wave-supporting granular crystal sensors. *AIAA J.* **2014**, *52*, 2283–2290. [CrossRef]
38. Schiffer, A.; Alkhaja, A.; Yang, J.; Esfahani, E.; Kim, T.-Y. Interaction of highly nonlinear solitary waves with elastic solids containing a spherical void. *Int. J. Solids Struct.* **2017**, *118*, 204–212. [CrossRef]
39. Yang, J.; Khatri, D.; Anzel, P.; Daraio, C. Interaction of highly nonlinear solitary waves with thin plates. *Int. J. Solids Struct.* **2012**, *49*, 1463–1471. [CrossRef]
40. Schiffer, A.; Kim, T.-Y. Modelling of the interaction between nonlinear solitary waves and composite beams. *Int. J. Mech. Sci.* **2019**, *151*, 181–191. [CrossRef]
41. Singhal, T.; Kim, E.; Kim, T.-Y.; Yang, J. Weak bond detection in composites using highly nonlinear solitary waves. *Smart Mater Struct.* **2017**, *26*, 055011. [CrossRef]
42. Berhanu, B.; Rizzo, P.; Ochs, M. Highly nonlinear solitary waves for the assessment of dental implant mobility. *J. Appl. Mech.* **2013**, *80*, 0110281–0110288. [CrossRef]
43. Nasrollahi, A.; Rizzo, P.; Orak, M.S. Numerical and experimental study on the dynamic interaction between highly nonlinear solitary waves and pressurized balls. *J Appl. Mech.* **2018**, *85*, 031007. [CrossRef]
44. Nasrollahi, A.; Lucht, R.; Rizzo, P. Solitary Waves to Assess the Internal Pressure and the Rubber Degradation of Tennis Balls. *Exp. Mech.* **2019**, *59*, 65–77. [CrossRef]
45. Nasrollahi, A.; Orak, M.S.; James, A.; Weighardt, L.; Rizzo, P. A Nondestructive Evaluation Approach to Characterize Tennis Balls. *ASME J. Nondestruct. Eval.* **2019**, *2*, 011004. [CrossRef]
46. Nasrollahi, A.; Rizzo, P. Modeling a new dynamic approach to measure intraocular pressure with solitary waves. *J. Mech. Behav. Biomed.* **2020**, *103*, 103534. [CrossRef]

47. Nasrollahi, A.; Rizzo, P. Axial stress determination using highly nonlinear solitary waves. *J. Acoust. Soc. Am.* **2018**, *144*, 2201–2212. [CrossRef]
48. Nasrollahi, A.; Rizzo, P. Numerical Analysis and Experimental Validation of an Nondestructive Evaluation Method to Measure Stress in Rails. *ASME J. Nondestruct. Eval.* **2019**, *2*, 031002. [CrossRef]
49. Li, J.; Mechitov, K.A.; Kim, R.E.; Spencer, B.F., Jr. Efficient time synchronization for structural health monitoring using wireless smart sensor networks. *Struct. Contorl. Health Monit.* **2016**, *23*, 470–486. [CrossRef]
50. Lynch, J.P.; Loh, K.J. A summary review of wireless sensors and sensor networks for structural health monitoring. *Shock. Vib.* **2006**, *38*, 91–130. [CrossRef]
51. Noel, A.B.; Abdaoui, A.; Elfouly, T.; Ahmed, M.H.; Badawy, A.; Shehata, M.S. Structural health monitoring using wireless sensor networks: A comprehensive survey. *IEEE Commun. Surv. Tutor.* **2017**, *19*, 1403–1423. [CrossRef]
52. Aygün, B.; Gungor, V.C. Wireless sensor networks for structure health monitoring: Recent advances and future research directions. *Sens. Rev.* **2011**, *31*, 261–276. [CrossRef]
53. Abdulkarem, M.; Samsudin, K.; Rokhani, F.Z.; A Rasid, M.F. Wireless sensor network for structural health monitoring: A contemporary review of technologies, challenges, and future direction. *Struct. Health Monit.* **2019**. [CrossRef]
54. Jalali, H.; Rizzo, P. Highly nonlinear solitary waves for the detection of localized corrosion. *Smart Mater. Struct.* **2020**. accepted for publication.
55. Easwaran, S.N.; Weigel, R. 1.3 A,-2V Tolerant Solenoid Drivers for Pedestrian Protection in Active Hood Lift Systems. In Proceedings of the 2018 IEEE International Symposium on Circuits and Systems (ISCAS), Florence, Italy, 27–30 May 2018.
56. Thach, H.; Townsend, K. Adafruit_BluefruitLE_Firmware, GitHub repository. Available online: https://github.com/adafruit/Adafruit_BluefruitLE_Firmware (accessed on 1 May 2020).
57. García, A. Bluefruit LE Connect v2, GitHub repository. Available online: https://github.com/adafruit/Bluefruit_LE_Connect_v2 (accessed on 1 May 2020).
58. Herbold, E.B. Optimization of the Dynamic Behavior of Strongly Nonlinear Heterogeneous Materials. Ph.D. Thesis, University of California, San Diego, CA, USA, 2008.
59. Rebonatto, M.T.; Schmitz, M.A.; Spalding, L.E.S. Methods of comparison and similarity scoring for electrical current waveforms. In Proceedings of the 12th Iberian Conference on Information Systems and Technologies (CISTI), Lisbon, Portugal, 14–17 June 2017; pp. 1–6.
60. Di Lena, P.; Margara, L. Optimal global alignment of signals by maximization of Pearson correlation. *Inform. Process. Lett.* **2010**, *110*, 679–686. [CrossRef]
61. Akoglu, H. User's guide to correlation coefficients. *Turk. J. Emerg. Med.* **2018**, *18*, 91–93. [CrossRef] [PubMed]

© 2020 by the authors. Licensee MDPI, Basel, Switzerland. This article is an open access article distributed under the terms and conditions of the Creative Commons Attribution (CC BY) license (http://creativecommons.org/licenses/by/4.0/).

Article

Quantified Activity Measurement for Medical Use in Movement Disorders through IR-UWB Radar Sensor [†]

Daehyeon Yim [1,‡], Won Hyuk Lee [1,‡], Johanna Inhyang Kim [2,7], Kangryul Kim [2], Dong Hyun Ahn [3,7], Young-Hyo Lim [4], Seok Hyun Cho [5], Hyun-Kyung Park [6,7,*] and Sung Ho Cho [1,*]

1. Department of Electronics and Computer Engineering, Hanyang University, 222 Wangsimini-ro, Seongdong-gu, Seoul 04763, Korea; ldh166@hanyang.ac.kr (D.Y.); lwh9886@gmail.com (W.H.L.)
2. Department of Psychiatry, Hanyang University Medical Center, 222-1 Wangsimni-ro, Seongdong-gu, Seoul 04763, Korea; iambabyvox@hanmail.net (J.I.K.); kangryul@naver.com (K.K)
3. Department of Psychiatry, Hanyang University College of Medicine, 222 Wangsimini-ro, Seongdong-gu, Seoul 04763, Korea; ahndh@hanyang.ac.kr
4. Division of Cardiology, Department of Internal medicine, Hanyang University College of Medicine, 222 Wangsimni-ro, Seongdong-gu, Seoul 04763, Korea; mdoim@hanyang.ac.kr
5. Department of Otorhinolaryngology-Head and Neck Surgery, Hanyang University College of Medicine, 222 Wangsimni-ro, Seongdong-gu, Seoul 04763, Korea; shcho@hanyang.ac.kr
6. Department of Pediatrics, Hanyang University College of Medicine, 222 Wangsimini-ro, Seongdong-gu, Seoul 04763, Korea
7. Hanyang Inclusive Clinic for Developmental Disorders, Hanyang University Medical Center, 222-1 Wangsimni-ro, Seongdong-gu, Seoul 04763, Korea
* Correspondence: neopark@hanyang.ac.kr (H.-K.P.); dragon@hanyang.ac.kr (S.H.C.); Tel.: +82-2-2220-0390 (H.-K.P.); +82-2-2290-8397 (S.H.C.)
† This paper is an extended version of our paper published in Lee, W.H., Cho, S.H., Park, H.K., Cho, S.H., Lim, Y.H., Kim, K.R. Movement Measurement of Attention-Deficit/Hyperactivity Disorder (ADHD) Patients Using IR-UWB Radar Sensor. In Proceedings of the 2018 International Conference on Network Infrastructure and Digital Content (IC-NIDC), 2018.
‡ These authors contributed equally to this work.

Received: 31 December 2018; Accepted: 4 February 2019; Published: 8 February 2019

Abstract: Movement disorders, such as Parkinson's disease, dystonia, tic disorder, and attention-deficit/hyperactivity disorder (ADHD) are clinical syndromes with either an excess of movement or a paucity of voluntary and involuntary movements. As the assessment of most movement disorders depends on subjective rating scales and clinical observations, the objective quantification of activity remains a challenging area. The purpose of our study was to verify whether an impulse radio ultra-wideband (IR-UWB) radar sensor technique is useful for an objective measurement of activity. Thus, we proposed an activity measurement algorithm and quantitative activity indicators for clinical assistance, based on IR-UWB radar sensors. The received signals of the sensor are sufficiently sensitive to measure heart rate, and multiple sensors can be used together to track the positions of people. To measure activity using these two features, we divided movement into two categories. For verification, we divided these into several scenarios, depending on the amount of activity, and compared with an actigraphy sensor to confirm the clinical feasibility of the proposed indicators. The experimental environment is similar to the environment of the comprehensive attention test (CAT), but with the inclusion of the IR-UWB radar. The experiment was carried out, according to a predefined scenario. Experiments demonstrate that the proposed indicators can measure movement quantitatively, and can be used as a quantified index to clinically record and compare patient activity. Therefore, this study suggests the possibility of clinical application of radar sensors for standardized diagnosis.

Keywords: IR-UWB radar sensor; movement disorder; hyperactivity; actigraphy

1. Introduction

Movement disorders, such as Parkinson's disease, dystonia, tic/Tourette's disorder, and attention-deficit/hyperactivity disorder (ADHD), are clinical syndromes with either an excess of movement or a paucity of voluntary and involuntary movements. The assessment of many movement disorders has heavily relied on clinical observation and rating scales, which are inherently subjective, and results vary according to the informant [1]. There is an increasing need for tools that objectively evaluate the level of activity. We focused on ADHD, among various movement disorders, to explore the possibility of a new evaluation method. ADHD is a common neurodevelopmental disorder characterized by inattention, impulsivity, and hyperactivity [2]. In contrast to the research on objective measurements of inattention, such as the continuous performance test (CPT), the assessment of hyperactivity in clinical settings is based on subjective reports from caregivers and from the observations of clinicians [3]. As hyperactivity is influenced by environmental factors and cognitive demands, discrepancy regarding the description of hyperactivity often occurs, thereby making the diagnosis of ADHD challenging [4].

Studies have been conducted, using infrared cameras (QbTest; Qb Tech, Stockholm, Sweden), 3D cameras (Microsoft Kinect, Redmond, US), or actigraphy (ActiGraph, Florida, US), to measure the objective level of activity in young people having ADHD [5]. However, these sensors have not been applied widely in clinical settings due to several limitations. A recent study reported that the QbTest is insufficient as a diagnostic test for ADHD, as it is unable to differentiate ADHD from other neurodevelopmental disorders [6]. Examination methods using infrared cameras, such as QbTest, are not perfectly non-contact, and measure the patient's concentration, but do not measure activity [7]. It can be difficult to judge exact whole-body motions, as these methods reflect only the movement of a specific part of the body. In the case of a depth camera or a 3D camera, the angle of view is limited to about 60 degrees, the performance varies depending on the indoor lighting environment, and the maximum measurable distance is as short as several meters [8]. Actigraphy has been the most commonly used device in measuring hyperactivity in ADHD [9]. Its primary use is measuring sleep and wakefulness, but it can also measure the movement of the subject in the x, y, and z axes, through an acceleration sensor [10]. It is not only possible to measure the amount of activity by obtaining the number of steps and the vector magnitude with this acceleration data, but the position can be estimated (even though the error is cumulative). However, as the device is worn on a certain part of the body, such as ankles and wrists, activity measurement does not reflect the movements of the whole body. The device is attached to the skin, and it may cause inconvenience for the user. Currently, actigraphy is considered to be useful in monitoring motor activity during treatment, but there is little evidence supporting the use of actigraphy in the diagnosis or as a screening tool for ADHD [11].

Impulse radio ultra-wideband (IR-UWB) radar sensors are capable of detecting objects without interference from other sensors through the use of ultra-wideband frequencies. Despite sending and receiving signals with very low power to comply with Federal Communications Commission (FCC) standards, they have enough range and resolution to observe the indoor environment. The primary advantages for clinical application of an IR-UWB radar sensor are its very low power and high spatial resolution. IR-UWB radar signals typically have a high resolution, so it can be used to detect the fine motion of objects [12]. Moreover, it is harmless to the human body and enables the diagnosis of the subject by a non-contact method, causing no inconvenience for the patient. The radar sensor is in a sustainable form and has no contact or requirement for the patient. As it has excellent penetrability, it can be installed on the wall invisibly, and so it is able to observe the target without attracting any attention from the target. Due to these characteristics, the measurement and quantification of activity in clinical movement disorders using an IR-UWB radar sensor is very promising. The IR-UWB radar sensor is capable of detecting not only large movements of the human body but also small movements, such as breathing. Recently, communications, localization, positioning, and tracking using the IR-UWB radar sensor have been studied. Most applications can be performed simultaneously using the same hardware [13–15].

The purpose of this study was to calculate the objective quantity of movement by using four radar sensors to find the position of the subject, and to calculate the amount of body movement in a testing room, during an attention task called the comprehensive attention test (CAT), which is a computerized CPT widely used for ADHD patients [16]. Through this study, we will quantify the movement of subjects and present new indicators that can potentially be applied in the measurement of activity in movement disorders, such as ADHD. All of the different radar functions have one thing in common: The information is based on human movement. Therefore, movement information was obtained from radar signals and two types of movements were defined which could be measured by radar, based on changes in position. In regard to spatial movement (which refers to the movement of the subject accompanied with position change), the degree of movement can be calculated by replacing the position change amount with a vector by tracking. In regards to sedentary movement (which refers to the movement of the subject accompanied by little or no position change), the degree of movement was calculated by continuously measuring the amount of change in the magnitude of the reflected signal from the target.

The following sections introduce the signal model of the IR-UWB radar and the basic concept of the algorithm for the activity measurement. Then, a detailed description of the algorithm, based on the tracking and signal magnitude, is presented. Finally, after introducing the experimental environment and methods, experimental results are presented and analyzed.

2. Problem Statement

2.1. Signal Model and Basic Signal Processing

The impulse signal $s[k]$, emitted by the radar to observe the target area, is delayed and scaled while being reflected from the surrounding environment. The received signal of the radar is generated by the reflected $s[k]$ through N_{path} paths from the surrounding environment. The signal received by the i-th radar can be represented by the sampled signal $x_i[k]$, including the environment noise $\mathcal{N}[k]$, as follows

$$x_i[k] = \sum_{m=1}^{N_{path}} a_{m,i} s[k - \tau_{m,i}] + \mathcal{N}[k]. \tag{1}$$

The sampled time index k can be called a distance index, and is represented by a natural number from 0 to L_{signal}, which is the distance index of the maximum observable distance. When $s[k]$ is reflected on the m-th path of the i-th radar, $a_{m,i}$ and $\tau_{m,i}$ are the scale values and delays, respectively [15].

In an indoor environment, there are a lot of objects and walls and so there are various signals received, in addition to people. Reflected signals from the background are called clutter signals, and usually have a large and constant magnitude. Removing the clutter signals is necessary to observe only the reflected signal from the target, and requires a detection algorithm to detect people while excluding noise. Hence, we can only obtain signals for the target in the indoor environment. The basic signal processing procedure for detecting people is shown in Figure 1.

Background removal algorithms are used frequently in indoor environments to observe only the desired targets. A signal $y_i[k]$ with background removed from the received signal $x_i[k]$ can be obtained. The purpose of the initialization phase is to create a threshold. This threshold should reflect the characteristics of the experimental environment, so the initialization should proceed without any humans in the observation area of the IR-UWB radar. The environment-adapted threshold $T_i[k]$ is, then, used for detection in the real-time process.

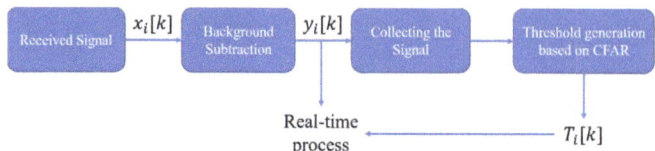

Figure 1. Basic signal processing.

Background subtraction is a technique for separating the foreground from the background, where walls or static objects correspond with the background, while the observed target corresponds with the foreground [17]. With this algorithm, background signals can be removed, and only the signal components of a moving target can be detected. The background clutter signal $C_{i,n}[k]$ is continually updated from the previous clutter signal $C_{i,n-1}[k]$ and $x_{i,n}[k]$, where n is the sequence number of the received signal in each radar, and $C_{i,n}[k]$ is subtracted from the radar signal $x_{i,n}[k]$ to obtain the background subtraction signal $y_{i,n}[k]$, which is expressed as

$$y_{i,n}[k] = x_{i,n}[k] - C_{i,n}[k],$$
$$C_{i,n}[k] = \alpha C_{i,n-1}[k] + (1-\alpha)x_{i,n}[k]. \quad (2)$$

To more accurately detect the signal of the target, the distance between the subject and the radar can be calculated from the background subtraction signal using the Constant False Alarm Rate (CFAR) algorithm [18]. Generally, it is common to detect using the cell-averaging (CA-CFAR) method with a certain window size in one-frame data received from the radar. However, the $y_i[k]$ collected by observing the environment without a target for a certain duration is represented as $\mathbf{Y}_i[k] = [y_{i,0}[k], y_{i,1}[k], y_{i,2}[k], \cdots, y_{i,N_c}[k]]^T$, and is used for threshold value-generation, based on the CFAR method, where N_c is the number of collected $y_i[k]$. This allows the probability of false alarms to be set to a certain value by comparing the received signal from the target with the threshold level $T_i[k]$, expressed as

$$T_i[k] = \beta \sigma_i[k] + \mu_i[k], \quad (3)$$

where the subscript i points to the i-th radar, β is a parameter to adjust the false alarm rate, and $\mu_i[k]$ and $\sigma_i[k]$ are the mean and standard deviation of $\mathbf{Y}_i[k]$, respectively. When the background signal is removed, only the target signal and the noise remain, and the $y_i[k]$, applied with the background subtraction algorithm, can be expressed as

$$y_i[k] = \hat{r}_i[k] + \mathcal{N}_i[k], \quad (4)$$

where $\hat{r}_i[k]$ is the target signal estimated by the clutter removal in i-th radar and $\mathcal{N}_i[k]$ is the noise [15]. Therefore, if there is no target, such as when collecting a signal for a threshold, $y_i[k]$ only has noise. To detect the target separately from the noise, we can obtain the mean and variance of $\mathbf{Y}_i[k]$, and create a threshold as shown in Equation (3).

2.2. Basic Concept of Activity Measurement

Generally, because the reflection coefficient of the electromagnetic wave to the target does not change, there is no change in $x_i[k]$ if there is no movement. Conversely, when there is movement, the value of $x_i[k]$ changes because some paths differ from those of the previous environment. Additionally, the distance measurement to the target is calculated as k, where $x_i[k]$ is largely changed. The distance resolution of the UWB radar has units of a few millimeters, which can detect very small changes within the range of the radar. Therefore, it is impossible for a radar to miss even the very small movements of a human, and the amount of activity of the target can be measured by the change of the magnitude of the radar signal and the moving distance information. However, because the radar can measure only one-dimensional distance data, it is limited to observing the target with one radar. Thus, multiple radars were used to measure the position of the target while simultaneously measuring the change in the magnitude of the signal, which was represented as activity.

In this paper, human movement is divided into two types: A type with a change in position, and a type with no change in position (such as sitting). The former is defined as spatial movement, and the latter is defined as sedentary movement. In the past, these two movements have been measured in different ways. One way is to measure the target's motion intensity (e.g., actigraphy sensor [19]), and the other is to track the target [5]. These two modes of motion are independent of each other, and have different characteristics. Spatial movements are observable movements from a macroscopic point of view, and sedentary movements are observable movements from a microscopic point of view. If the position of the target does not change, the measured distance of the radar does not change significantly, so the positioning information will not reflect sedentary movements well [20]. Conversely, if there is a change the position, the positioning information may reflect this movement, but it is difficult for the received signal magnitude to reflect the movement state, such as the position change or movement speed. Therefore, because it is difficult to confirm the amount of activity in all cases with one measurement method, an algorithm is proposed in this paper that can numerically compare the amount of activity through two indicators.

2.3. Experiment Scenario

We designed several scenarios to check our proposed indicators. The criteria for dividing the scenario first broadly, based upon whether there is any spatial movement, divides scenarios into two groups. These groups are then divided into several scenarios, depending on the degree of movement. This study is not intended to recognize specific actions, because it is aimed at measuring movement by projecting human motion using one-dimensional data. Therefore, each scenario was designed to include random behavior with minimal limitations. The list of scenarios is as follows:

1. When a person sits and concentrates on one thing;
2. When a person has a relatively small motion in a sitting position;
3. When a person has a relatively large motion in a sitting position;
4. When a person walks slowly in the room in a narrow radius;
5. When a person walks slowly in the room in a large radius;
6. When a person walks quickly in the room in a narrow radius;
7. When a person walks quickly in the room in a large radius.

The scenarios were designed to account for situations including movement during the test, and were only for reproducing other test environments. The proposed indicators have no particular dependency on CAT. Scenario 1 is a situation in which the target is focused on the test and does not move. In this scenario, the subject should minimize any actions other than the restricted, small movements required for the test. Scenarios 2 and 3 assumed that the target was seated for CAT, but cared about other things. Scenario 2 is a situation in which the limbs and the head move while the torso is fixed (such as looking around or touching something else). Scenario 3 is a situation in which the entire body moves in a sitting position, such as sitting with the chair tilted back. Scenarios 4–7 are four scenarios created using two opposing features. Scenarios 4 and 6 include walking near the center of the room, while Scenarios 5 and 7 include roaming the entire room. While Scenarios 4 and 5 are relatively slow walking scenarios, Scenarios 6 and 7 are relatively fast walking scenarios.

The greater the torso movement, the greater the sedentary movement index. This is because the torso takes up most of the human body. Thus, for our scenarios, the size of the sedentary movement index can be expected to decrease in order of Scenarios 3, 2, and 1. In Scenarios 4–7, it was expected that walking around a wide area or moving quickly would be observed as a larger movement than walking around a narrow area or moving slowly.

3. Algorithm for Measuring Activity

3.1. Measuring the Sedentary Movement

If there is a person with any movement, the corresponding $y_i[k]$ deviates greatly from the probability characteristics of the vacant state, so a person can be easily detected by comparing $y_i[k]$ with the threshold $T_i[k]$. If there is no person at the position of the k-th sample for i-th radar, $\hat{r}_i[k]$ is estimated to be zero in Equation (4), and $y_i[k]$ is not helpful in measuring activity. Previously, movement was simply represented as the sum of the differences in signal amplitude [21,22]. In this case, however, even when the difference between two consecutive frames is obtained, noise cannot be reduced. In cases where the motion is small, the amplitude of the target signal can be reduced. Therefore, to make only the signals from the target into activity indicators, we only used the samples for k where $y_i[k]$ exceeds the threshold $T_i[k]$, which can be expressed as

$$E_i[n] = \sum_{k=0}^{L_{signal}} g_{i,n}[k], \quad g_{i,n}[k] = \begin{cases} |y_{i,n}[k] - y_{i,n-1}[k]|^2 & if \quad y_{i,n}[k] > T_i[k] \\ 0 & if \quad y_{i,n}[k] \leq T_i[k] \end{cases}. \tag{5}$$

Of course, the received signal of a radar differs greatly, according to the distance to the target, and so, for a single radar, the degree of movement will vary greatly depending on the position. However, to compensate for this difference, it is necessary to consider not only the attenuation compensation along the distance, but also the compensation according to the antenna pattern in three dimensions. Further studies are needed to consider the relationship between position dependent signal attenuation and the clutter cancelling signal. We used the median of the data obtained by installing four radars at each corner of the four directions to apply the minimum compensation. If the target is too close to (or far from) any radar, it will be measured as too large (or too small), so the maximum and minimum values of E_i are excluded. Therefore, the proposed indicator to observe the sedentary movement can be expressed as

$$M_{sedentary}[n] = \text{Median}\left(E_0[n], E_1[n], E_2[n], \cdots, E_{N_r}[n]\right), \tag{6}$$

where N_r is the number of radars for measurement and $\text{Median}(\cdot)$ is the function that returns the median value of the input value.

3.2. Measuring the Spatial Movement

There are not many people who move at a constant speed or only perform one action. Human behavior varies, and human movement is closely related to many forces; examples include friction forces and reaction forces from the earth. Due to these forces, the human movement state is constantly changing. However, for people who are not moving, the only movement is due to breathing. This means that there is almost no change in force. Because the force that moves a target is represented by the acceleration of the target, a person with active motion will have a greater acceleration than a person with slight motion. For this reason, it is possible to measure the amount of activity of an object through actigraphy [9]. As we cannot mathematically model the random movements of a person, we can use numerical differentiation to obtain the acceleration value from the data measured. Although the accuracy of the calculated acceleration may be low, we do not need the exact value [23].

The process of obtaining the acceleration begins by obtaining the distance value from each radar signal $y_i[k]$, obtained in Section 2.1. Methods for distance measurement in radar are already well known [24]. When the target exists in the observation region, multipath causes the signal magnitude to change at the distance index behind the target signal [15]. This magnitude change can be sufficiently above the threshold. Hence, in the signal $y_i[k]$, the shortest distance from the radar to the target can be obtained by using the minimum value of k satisfying $y_i[k] > T_i[k]$. Using the sampling frequency of the radar to convert from k to the actual distance unit, calculated as $d_i = c/f_s \times k$, d_i, gives the distance from the i-th radar to the target measured.

The position of the target can be obtained by using the obtained d_i and the least-squares (LS) method. To apply LS, it is necessary to change the equation to be more simple. The circle equation, with radius d_i centered on the location of the i-th radar, (x_i, y_i, z_i) is represented by

$$(x - x_i)^2 + (y - y_i)^2 + (z - z_i)^2 = d_i^2. \tag{7}$$

Equation (7) for the l-th radar and the m-th radar can be rearranged, as Equation (8), to convert the quadratic equation into a linear equation:

$$2x(x_m - x_l) + 2y(y_m - y_l) + 2z(z_m - z_l) = d_l^2 - d_m^2 - x_l^2 + x_m^2 - y_l^2 + y_m^2 - z_l^2 + z_m^2. \tag{8}$$

To obtain the solution in the LS scheme, we can convert Equation (8) to matrix equation form, $Ax = b$, where A and b can be expressed as

$$A = 2 \begin{bmatrix} x_1 - x_0 & y_1 - y_0 & z_1 - z_0 \\ x_2 - x_1 & y_1 - y_0 & z_1 - z_0 \\ \cdots & \cdots & \cdots \\ x_{N_r} - x_{N_r-1} & y_{N_r} - y_{N_r-1} & z_{N_r} - z_{N_r-1} \end{bmatrix}, \quad b = \begin{bmatrix} C_1 \\ C_2 \\ \cdots \\ C_{N_r} \end{bmatrix}. \tag{9}$$

C_i replaces the right side of Equation (8) as $d_i^2 - d_{i-1}^2 - x_i^2 + x_{i-1}^2 - y_i^2 + y_{i-1}^2 - z_i^2 + z_{i-1}^2$. As the solution of LS is well known as the right side of Equation (10), we can obtain the position $p = [x_t, y_t, z_t]^T$ of the target:

$$p = (A^T A)^{-1} A^T b. \tag{10}$$

Position data can be obtained in real-time using the positioning method mentioned above. The position data for time n can be represented as $p[n] = [x_t[n], y_t[n], z_t[n]]^T$, where $x_t[n]$, $y_t[n]$, and $z_t[n]$ are the three-dimensional coordinates of the target. Using numerical differentiation, the velocity and acceleration of the target can be expressed as:

$$\begin{aligned} v[n] &= (p[n] - p[n-1])/t_r \\ a[n] &= (v[n] - v[n-1])/t_r. \end{aligned} \tag{11}$$

The observation period t_r of the radar is the sampling period for the target position data. The initial value can be specified by $v[0] = v[1] = a[0] = 0$. However, $a[n]$ will be closer to the acceleration at $n-1$, and not exactly at n. If the t_r is sufficiently small, there will be little time difference between n and $n-1$, and delay by one sample will not have a large impact. Therefore, even if the acceleration is not accurate, the acceleration and speed are calculated with Equation (11) to maintain real-time processing. As a result, the activity indicator for the spatial movement can be represented as:

$$M_{spatial}[n] = \beta \cdot a[n] = \gamma(p[n] - 2p[n-1] + p[n-2]). \tag{12}$$

As the amount of activity is not mathematically defined, the goal is not to create an accurate mathematical model in this paper. In other words, our goal is not to prove the exact relationship between acceleration and activity, but rather to suggest an indicator for objectively comparing activity. Therefore, we modeled the amount of activity and acceleration as a linearly proportional relationship.

4. Experiment Results

The XK300-MVI (Xandar Kardian, Toronto, ON, Canada) radar was used to verify the above algorithm. The X4M03 can select various center frequencies, from 7.29 GHz to 8.748 GHz, by adjusting various parameters according to local regulations. In these experiments, we selected the parameter with a center frequency of 8.748 GHz and a bandwidth of 1.5 GHz as −10 dB concept. The radiation power of the radar was 68.85 µW. The radar receiver can sample at 23.328 GS/s. The four radars were installed in each ceiling corner of the experimental room, as shown in Figure 2; which was 2.4 m in width, 3.0 m in length, and 2.4 m in height. In the indoor space, a distance error may occur due to the volume of a person; the radars were installed radially to minimize this error. The signal from the radars can be disturbed by movement of the arms or legs of the target. Hence, in order to reduce the effect of limbs as much as possible, radars were installed on the ceiling to observe the target. All of these radars were connected to the PC through the USB interface. The signal frames received from the radar were converted into digital values and transmitted to the PC by USB. These data were processed in MATLAB 2018b using the signal processing algorithm. The operating system of the PC was Windows 10. The frame-per-second (FPS) value of the signal received by the radar was 30.

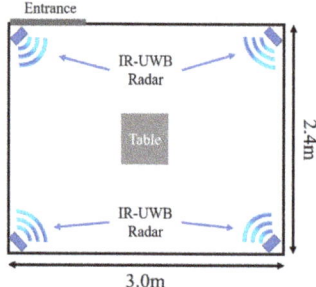 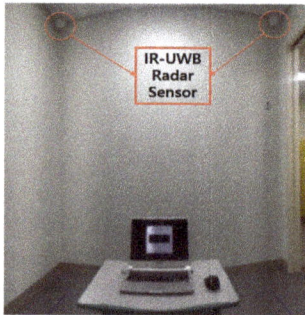

Figure 2. Experimental environment.

A table and a laptop were placed in the middle of the room for the experiment. In Scenario 1, the tester focused on the notebook. Scenarios 2 and 3 were also measured for sedentary movement situations, sitting in front of the table. From the center table, a space of approximately 1.2 m by 1.2 m was designated as a narrow area for Scenarios 4 and 6, and the entire room area was designated as a wide area for Scenarios 5 and 7. All of the experimenters performed scenarios consecutively, and they acted in a condition for about three minutes per scenario. Additionally, the empty room was used to generate the threshold value by measuring data for about three minutes.

The actigraphy data was measured, for comparison with the proposed IR-UWB radar base. Actigraphy sensors wGT3X-BT (ActiGraph, Florida, US) were worn on the right wrist and right ankle. In each actigraphy sensor, there is an accelerometer for measuring the movement of the body part. With a dedicated-license software called ActiLife (Actigraph, Florida, US), vector magnitude values were extracted to the PC at a sampling rate of 1 s. The data of the actigraphy sensors were scaled to compare the trends.

4.1. Experiment Results for Each Scenario

During the experiment, it was possible to check the data in real time. However, the distribution of results is more useful to characterize each scenario. Five researchers participated in the experiment, and were assigned three minutes per scenario. The results of applying the algorithm for sedentary movement are shown in Figure 3. The experimental results of $M_{sedentary}$ for Scenarios 1–3 are shown, with a significant difference, in Figure 3a. The experimental results show that $M_{sedentary}$ values increased in the order of Scenarios 1–3. Individual behaviors may vary in the same scenario, so the outcome varied slightly for each person. However, the trends in the results were all similar. As the scenario progressed from 1 to 3, the value of $M_{sedentary}$ increased, which indicates that the results of each scenario were consistent and relevant. Conversely, in Scenarios 4 to 7, the results for spatial movement showed no significant difference in histogram and real-time measurement, and the values tended to be similar. This is effective for distinguishing the degree of sedentary movement with the algorithm used in Section 3.1, but it is insufficient for judging the degree of spatial movement around the room.

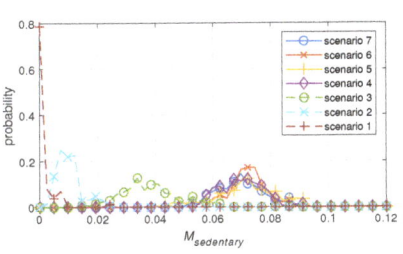
(a) Histograms of $M_{sedentary}$ for each scenario

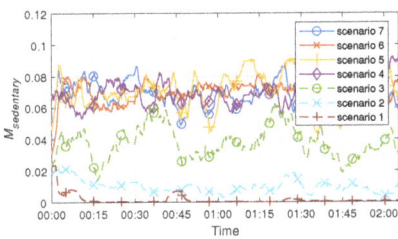
(b) Real-time measurements of $M_{sedentary}$ for each scenario

Figure 3. Experimental results of the sedentary movement index, $M_{sedentary}$, for each scenario.

In Section 3.2, the algorithm to find the position and acceleration of the subject was applied to the seven scenarios. As a result, the algorithm was made based on changes in the target's position, so Scenarios 1 to 3 were smaller than Scenarios 4 to 7, and were not well distinguished. As the motions corresponding to Scenarios 4 to 7 consisted of traveling around the inside of the experimental room, the velocity wes not constant, and the acceleration varied instantaneously. Therefore, the variation of the $M_{spatial}$ value for Scenarios 4 to 7 was larger than that of Scenarios 1 to 3. Additionally, Scenarios 5 and 7 consisted of traveling around the lab within a large radius, and are generally larger than Scenarios 4 and 6, which consisted of traveling within a small radius, as can be identified in Figure 4. Scenario 7, which involved moving quickly within a large radius, had the highest value in most real-time measurement areas. Each scenario can be distinguished by the $M_{spatial}$ value derived from the algorithm used, and it was shown that the degree of movement can be presented by measuring the spatial movement using the suggested indicator.

The mean values of each scenario obtained from the sedentary movement indicator, $M_{sedentary}$, and spatial movement indicator, $M_{spatial}$, are shown in Tables 1 and 2. It is difficult to compare the values of $M_{sedentary}$ and $M_{spatial}$ by themselves, because the algorithms used were different, and there are no units.

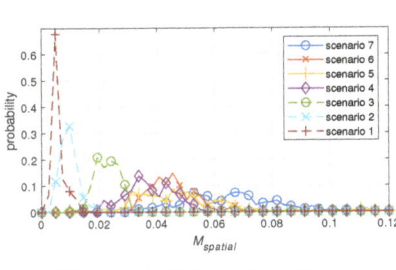
(a) Histograms of $M_{spatial}$ for each scenario

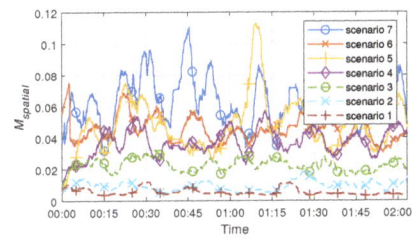
(b) Real-time measurements of $M_{spatial}$ for each scenario

Figure 4. Experiment results of the spatial movement index, $M_{spatial}$, for each scenario.

Table 1. Numerical results are shown for each target. The results for the $M_{sedentary}$ indicator are summarized (no unit).

Scenario	Mean of $M_{sedentary}$					
	A	B	C	D	E	Total
1	0.16	0.16	0.14	0.30	0.19	0.19
2	1.01	1.77	0.76	0.73	0.85	1.02
3	3.81	4.27	2.30	2.20	2.01	2.92
4	6.99	5.10	2.78	5.72	4.74	5.07
5	7.11	5.04	4.64	5.11	4.43	5.27
6	7.15	6.60	3.06	5.04	7.02	5.77
7	6.94	5.39	4.57	3.85	6.25	5.40

Table 2. Numerical results are shown for each target. The results for the $M_{spatial}$ indicator are summarized (no unit).

Scenario	Mean of $M_{spatial}$					
	A	B	C	D	E	Total
1	0.55	0.68	0.56	0.54	0.47	0.56
2	0.95	1.25	1.20	1.72	1.56	1.34
3	2.28	2.65	2.89	2.53	3.46	2.76
4	3.71	2.52	2.75	3.75	2.91	3.13
5	4.77	4.27	4.56	5.37	3.40	4.47
6	4.54	3.21	4.15	5.82	3.44	4.23
7	6.46	6.16	5.45	7.55	5.60	6.24

Looking at the results in Tables 1 and 2, it can be seen that, for the same target scenario, the other targets were measured to be similar. Under similar circumstances, the measurement results are expected to be obtained at constant values. In the case of the $M_{spatial}$ of Scenarios 2 and 3, we can see that there is a significantly greater difference than for Scenarios 1 and 2. Scenarios 1–3 are all cases of sedentary movement, but position change was observed as much as torso shaking in Scenario 3. Nonetheless, the sedentary movement indicator is better discriminated in Scenarios 1–3. On the other hand, the spatial movement indicator is unlikely to have a sense of value in real-time, though it statistically has a significant difference in all scenarios. Sedentary movement indicators in Scenario 7 were measured to be lower than in Scenario 6, but more than or similar to those in Scenario 4. This clearly shows that a sedentary movement indicator cannot distinguish a change in position. However, from a different point of view, it can be seen that all walking scenarios show a certain value. In other words, we can see that Scenarios 4–7 can be classified as similar situations, because they do not observe a change in position from the point of view of the sedentary movement indicator. This confirms that similar behavior is measured at similar values.

Table 3. Physical condition of the participants.

Participants	A	B	C	D	E
Gender (M/F)	M	M	M	F	M
Height (cm)	174	176	167	171	177
Weight (kg)	75	67	65	63	90

The physical conditions of the researchers participating in the measurement are shown in Table 3. Although physical conditions did not differ greatly from each other, they do not seem to have a significant effect on the results. However, it is expected that $M_{sedentary}$ will be measured as small for small children. Because their size is small, their motion is also small. Conversely, it is expected that there will be no significant difference because $M_{spatial}$ depends on position changes.

4.2. Comparison with Actigraphy

The proposed index was verified using an actigraphy sensor, which is actually used for clinical activity measurement. Similarly, we proceeded with seven scenarios, with a graph comparing the changes in real time to confirm the similarity of the data, as shown in Figure 5. Both sedentary and spatial movements can be seen to fit well with the actigraphy data. However, Figure 5b shows that, for the last 3 min, only the sedentary movement indicator is different. This is explained in Section 4.1—because the actigraphy sensor is similar to the acceleration for spatial movement index calculation, it can be seen that $M_{sedentary}$ and actigraphy are very similar.

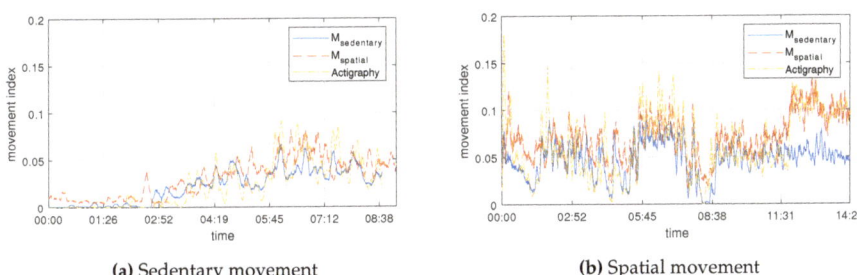

(a) Sedentary movement (b) Spatial movement

Figure 5. Graph of results when measured simultaneously with actigraphy. (a) and (b) are the results for sedentary and spatial movement, respectively. Actigraphy was scaled because the unit of the result data was not an actual physical quantity.

The limitations of the actigraphy sensor, compared with the radar sensor, can be seen in Figure 6, through an experiment of two extreme cases. In the previous 50 s, there was motion in the opposite hand and foot with the actigraphy sensor and, for the 50 s thereafter, there was movement of the hands and feet wearing the actigraphy sensor. The amount of movement measured from the radar during the entire section is largely constant, but the first 50 s has little movement measured from the actigraphy sensor. This result shows that the actigraphy sensor can only detect movement of a specific body part, but the radar sensor can detect movement of the entire body, even though it cannot distinguish each body part.

The actigraphy sensor is a contact-type sensor that should be worn on the wrists and ankles of the subject, and this can cause some pressure or stress for the subject during the experiment. In contrast, radar sensors are able to observe the patient's movements in a non-contact manner and so do not disturb the subject. It can be said that a radar sensor, which is a non-contact sensor, is effective for obtaining more detailed and reliable data in the diagnosis of movement disorders. In addition, the actigraphy sensor informs the activity amount of the subject, but it does not provide the direction of that amount of activity, the location of the actual subject, or the movement route. The position of the subject, obtained from multiple radar sensors, can provide more information than the actigraphy sensor in diagnosing a subject's specific habits, abrupt behavior, or hyperactivity. To date, there has been no index that can objectively express hyperactivity or specific movements in patients with movement disorders. However, it is possible to determine the position of the subject using the proposed algorithm and the radar sensor, and to measure the subject's degree of movement from the objectively quantified indicator.

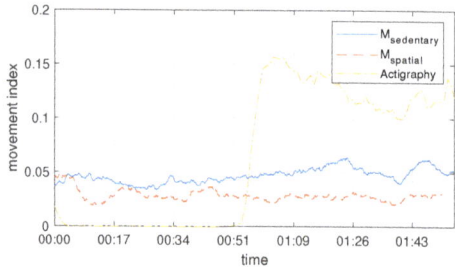

Figure 6. An extreme example of the proposed method and the use of actigraphy sensor. The actigraphy sensor is worn on a specific part of the body, so it may not reflect the whole movement (the first 50 s) or over-reflected (the last 50 s).

5. Conclusions

This paper proposed an algorithm and indicators to measure the amount of activity of people observed within certain constraints. Specifically, human activity was categorized into two types—with and without location movement—and two measurement indicators were proposed, that are specific to each activity with the signal strength and distance value data, measured by the radar sensor. Several scenarios and commercial products were used to confirm the reliability of the proposed indicators. Measurement can be performed simultaneously with the existing inspection to identify the actual activity amount, and this will be helpful in comparing the activity amount because the activity is quantified and more objective data is obtained. The IR-UWB radar sensors can measure heart rate and breathing while also recognizing gestures, making it a more practical solution in the medical field as it can be extended for additional functions in the future. Therefore, this quantitative technology for activity measurement may be useful in clinical applications as an assistive (complimentary tool) to diagnose movement disorders and to evaluate the efficacy of treatment based on direct observation by psychiatrists. It can additionally be used to overcome the limitations of conventional questionnaires by providing objective kinematic information about patients with movement disorders. Further, we can select indicators that are appropriate for our purpose because we can derive additional indicators (distance measurement, classification of standing and standing conditions, measuring activity area, among others) from our proposed algorithm.

Author Contributions: Conceptualization, D.Y. and W.H.L.; Data curation, D.Y. and W.H.L.; Formal analysis, D.Y. and W.H.L.; Investigation, D.Y. and W.H.L.; Methodology, D.Y. and W.H.L.; Resources, D.Y. and W.H.L.; Software, D.Y. and W.H.L.; Supervision, H.-K.P. and S.H.C.; Validation, J.I.K.; Writing—original draft, D.Y. and W.H.L.; Writing—review & editing, J.I.K., K.K., D.H.A., Y.-H.L., S.H.C., H.-K.P. and S.H.C.

Funding: This research was supported by Bio & Medical Technology Development Program (Next Generation Biotechnology) through the National Research Foundation of Korea (NRF) funded by the Ministry of Science, ICT, & Future Planning (2017M3A9E2064735).

Conflicts of Interest: The authors declare no conflict of interest.

References

1. FitzGerald, J.J.; Lu, Z.; Jareonsettasin, P.; Antoniades, C.A. Quantifying motor impairment in movement disorders. *Front. Neurosci.* **2018**, *12*, 202. [CrossRef]
2. American Psychiatric Association. *Diagnostic and Statistical Manual of Mental Disorders (DSM-5®)*; American Psychiatric Publishing: Washington, DC, USA, 2013.
3. Valo, S.; Tannock, R. Diagnostic instability of DSM–IV ADHD subtypes: Effects of informant source, instrumentation, and methods for combining symptom reports. *J. Clin. Child Adolescent Psychol.* **2010**, *39*, 749–760. [CrossRef]

4. Kofler, M.J.; Raiker, J.S.; Sarver, D.E.; Wells, E.L.; Soto, E.F. Is hyperactivity ubiquitous in ADHD or dependent on environmental demands? Evidence from meta-analysis. *Clin. Psychol. Rev.* **2016**, *46*, 12–24. [CrossRef] [PubMed]
5. Hult, N.; Kadesjö, J.; Kadesjö, B.; Gillberg, C.; Billstedt, E. ADHD and the QbTest: diagnostic validity of QbTest. *J. Atten. Disord.* **2015**. [CrossRef]
6. Johansson, V.; Norén Selinus, E.; Kuja-Halkola, R.; Lundström, S.; Durbeej, N.; Anckarsäter, H.; Lichtenstein, P.; Hellner, C. The Quantified Behavioral Test Failed to Differentiate ADHD in Adolescents With Neurodevelopmental Problems. *J. Atten. Disord.* **2018**. [CrossRef]
7. Hall, C.L.; Valentine, A.Z.; Walker, G.M.; Ball, H.M.; Cogger, H.; Daley, D.; Groom, M.J.; Sayal, K.; Hollis, C. Study of user experience of an objective test (QbTest) to aid ADHD assessment and medication management: a multi-methods approach. *BMC Psychiatry* **2017**, *17*, 66. [CrossRef]
8. Du, H.; Henry, P.; Ren, X.; Cheng, M.; Goldman, D.B.; Seitz, S.M.; Fox, D. Interactive 3D modeling of indoor environments with a consumer depth camera. In Proceedings of the 13th international Conference on Ubiquitous Computing, Beijing, China, 17–21 September 2011; pp. 75–84.
9. De Crescenzo, F.; Licchelli, S.; Ciabattini, M.; Menghini, D.; Armando, M.; Alfieri, P.; Mazzone, L.; Pontrelli, G.; Livadiotti, S.; Foti, F.; et al. The use of actigraphy in the monitoring of sleep and activity in ADHD: A meta-analysis. *Sleep Med. Rev.* **2016**, *26*, 9–20. [CrossRef] [PubMed]
10. Robusto, K.M.; Trost, S.G. Comparison of three generations of ActiGraph™ activity monitors in children and adolescents. *J. Sports Sci.* **2012**, *30*, 1429–1435. [CrossRef] [PubMed]
11. Kam, H.; Shin, Y.; Cho, S.; Kim, S.; Kim, K.; Park, R. Development of a decision support model for screening attention-deficit hyperactivity disorder with actigraph-based measurements of classroom activity. *Appl. Clin. Inf.* **2010**, *1*, 377–393. [CrossRef] [PubMed]
12. Schleicher, B.; Nasr, I.; Trasser, A.; Schumacher, H. IR-UWB radar demonstrator for ultra-fine movement detection and vital-sign monitoring. *IEEE Trans. Micro. Theory Tech.* **2013**, *61*, 2076–2085. [CrossRef]
13. Nguyen, V.H.; Pyun, J.Y. Location detection and tracking of moving targets by a 2D IR-UWB radar system. *Sensors* **2015**, *15*, 6740–6762. [CrossRef] [PubMed]
14. Pallesen, S.; Grønli, J.; Myhre, K.; Moen, F.; Bjorvatn, B.; Hanssen, I.; Heglum, H.S.A. A pilot study of impulse radio ultra wideband radar technology as a new tool for sleep assessment. *J. Clin. Sleep Med.* **2018**, *14*, 1249–1254. [CrossRef] [PubMed]
15. Choi, J.W.; Yim, D.H.; Cho, S.H. People counting based on an IR-UWB radar sensor. *IEEE Sens. J.* **2017**, *17*, 5717–5727. [CrossRef]
16. Lee, J.S.; Kang, S.H.; Park, E.H.; Jung, J.S.; Kim, B.N.; Son, J.W.; Park, T.W.; Kim, B.S.; Lee, Y.S.; et al. Standardization of the comprehensive attention test for the Korean children and adolescents. *J. Korean Acad. Child Adolesc. Psychiatry* **2009**, *20*, 68–75.
17. Rakibe, R.S.; Patil, B.D. Background subtraction algorithm based human motion detection. *Int. J. Sci. Res. Publ.* **2013**, *3*, 2250–3153.
18. Maali, A.; Mesloub, A.; Djeddou, M.; Baudoin, G.; Mimoun, H.; Ouldali, A. CA-CFAR threshold selection for IR-UWB TOA estimation. In Proceedings of the International Workshop on Systems, Signal Processing and their Applications, WOSSPA, Tipaza, Algeria, 9–11 May 2011; pp. 279–282.
19. Grap, M.J.; Hamilton, V.A.; McNallen, A.; Ketchum, J.M.; Best, A.M.; Arief, N.Y.I.; Wetzel, P.A. Actigraphy: analyzing patient movement. *Heart Lung J. Acute Crit. Care* **2011**, *40*, e52–e59. [CrossRef] [PubMed]
20. Yim, D.; Cho, S.H. Indoor Positioning and Body Direction Measurement System Using IR-UWB Radar. In Proceedings of the 2018 19th International Radar Symposium (IRS), Bonn, Germany, 20–22 June 2018; pp. 1–9.
21. Lee, J.M.; Choi, J.W.; Cho, S.H. Movement analysis during sleep using an IR-UWB radar sensor. In Proceedings of the 2016 IEEE International Conference on Network Infrastructure and Digital Content (IC-NIDC), Beijing, China, 23–25 September 2016; pp. 486–490.
22. Lee, W.H.; Cho, S.H.; Park, H.K.; Cho, S.H.; Lim, Y.H.; Kim, K.R. Movement Measurement of Attention-Deficit/Hyperactivity Disorder (ADHD) Patients Using IR-UWB Radar Sensor. In Proceedings of the 2018 International Conference on Network Infrastructure and Digital Content (IC-NIDC), Guiyang, China, 22–24 August 2018; pp. 214–217.

23. Merry, R.; Van de Molengraft, M.; Steinbuch, M. Velocity and acceleration estimation for optical incremental encoders. *Mechatronics* **2010**, *20*, 20–26. [CrossRef]
24. Gezici, S.; Tian, Z.; Giannakis, G.B.; Kobayashi, H.; Molisch, A.F.; Poor, H.V.; Sahinoglu, Z. Localization via ultra-wideband radios: a look at positioning aspects for future sensor networks. *IEEE Signal Process. Mag.* **2005**, *22*, 70–84. [CrossRef]

© 2019 by the authors. Licensee MDPI, Basel, Switzerland. This article is an open access article distributed under the terms and conditions of the Creative Commons Attribution (CC BY) license (http://creativecommons.org/licenses/by/4.0/).

Article

Pre-Pressure Optimization for Ultrasonic Motors Based on Multi-Sensor Fusion

Ning Chen *, Jieji Zheng and Dapeng Fan

National University of Defense Technology, Deya Road No. 109, Kaifu District, Changsha 410073, China; 1024451173@126.com (J.Z.); fdp@nudt.edu.cn (D.F.)
* Correspondence: chenning007xm@126.com

Received: 5 February 2020; Accepted: 3 April 2020; Published: 8 April 2020

Abstract: This paper investigates the pre-pressure's influence on the key performance of a traveling wave ultrasonic motor (TRUM) using simulations and experimental tests. An analytical model accompanied with power dissipation is built, and an electric cylinder is first adopted in regulating the pre-pressure rapidly, flexibly and accurately. Both results provide several new features for exploring the function of pre-pressure. It turns out that the proportion of driving zone within the contact region declines as the pre-pressure increases, while a lower power dissipation and slower temperature rise can be achieved when the driving zones and the braking zones are in balance. Moreover, the shrinking speed fluctuations with the increasing pre-pressures are verified by the periodic-varying axial pressure. Finally, stalling torque, maximum efficiency, temperature rise and speed variance are all integrated to form a novel optimization criterion, which achieves a slower temperature rise and lower stationary error between 260 and 320 N. The practical speed control errors demonstrate that the proportion of residual error declines from 2.88% to 0.75% when the pre-pressure is changed from 150 to 300 N, which serves as one of the pieces of evidence of the criterion's effectiveness.

Keywords: traveling wave ultrasonic motor; pre-pressure; contact state; power dissipation; optimization criterion

1. Introduction

Ultrasonic motors, which work based on the converse piezoelectric effect and friction drive, tend to be the choice for precise actuators in diverse areas such as aerospace robots, manufacturing facilities, and especially for biomedical devices. Ultrasonic motors show their prominent advantages in biomedical devices like cell manipulation actuators [1], ear surgical devices [2], magnetic resonance imaging systems [3] and magnetic-compatible haptic interfaces [4,5]. This is due to ultrasonic motors' features, which include excellent properties like high torque at low speed, high torque to weight ratio and no electromagnetic noise [6–8]. Pre-pressure is a non-negligible factor in all types of ultrasonic motors, for example traveling wave ultrasonic motors (TRUMs) [9], linear ultrasonic motors [10], 2-DOF ultrasonic motors [11], hybrid DOF ultrasonic motors [12] and so on. Among them, the studies of TRUMs are most representative. In TRUMs, piezoelectric ceramics are actuated by two-phase alternating voltages, the elliptical trajectories of stator particles form the traveling wave impelling the rotor revolve. Pre-pressure is applied from the rotor against the stator to assure the interface contact and friction drive.

In processor studies, the performance sensitivities to the pre-pressure are mainly analyzed from the following three aspects. The first is the frequency characteristics. The first-order resonance frequency was investigated by the distributed numerical model and finite element model by Pirrotta [13]. His work found that the first-order resonance frequency was not sensitive to the preload force until the pre-pressure reaches a considerable value. However, these results were not verified by any further experimental results. Afterwards, both the resonant frequency and anti-resonant frequency under diverse pre-pressures were tested in Oh's experiments [9]. He proved that the resonant frequency and anti-resonant frequency have a positive correlation with the preload force no matter whether the pressure is larger or smaller. Li [14] discovered that both frequencies do not increase monotonically with the pre-pressure. However, his data were collected from an impedance analyzer which only affords small-amplitude voltages (0–10 V). The second item is the contact properties. The contact state is the crucial contribution of pre-pressure for producing friction force and output torque. On this topic, Chen [15,16] developed a contact model via a semi-analytical model considering the radial slippage. The simulation results interpreted that the contact region and the driving zone become wider simultaneously when the pre-pressure increases, and the radial sliding inside the obstruction zone is also intensified. This useful yet incomplete analysis lacks any investigation on the contact state under load conditions. A contact model, which involves the distorted wave shape and standard stiffness of the contact layer, is built in [17]. This work also proves that the contact region becomes wider and the pressure at each contact point is raised with the increasing pre-pressure. Last but not least are the mechanical characteristics. Indeed, the mechanical characteristics integrate the results from the impedance characteristics and contact properties because the former determine the input power, and the latter affect the output power and torque. The most significant achievement in this aspect corresponds to Bullo [18]. He drew several three-dimension diagrams including speed, efficiency, output power both from simulations and experimental tests. However, the stator vibration amplitude was prescribed rather than adjusted by the differential functions, and the amplitude value is connected with voltage amplitude, driving frequency, applied load, and the initial pressure. His results also indicate that the incremental pre-pressure leads to a decreasing rotor speed under empty load conditions, but results in an increase of stalling torque.

Besides the above three performances concerning the preload force, this paper also proposes two new perspectives from the application viewpoint. One is the temperature rise arising from the power dissipation. As is well-known, TRUMs generate two energy conversion links, one is from electric energy to vibration energy based on the converse piezoelectric effect [19], and the other is from the vibration energy to the rotational energy via the friction between the stator and the rotor [20]. The energy losses occurring in the two processes not only reduce energy utilization but also significantly increase the internal temperature, which deteriorates TRUMs' performances. Previous authors have made several contributions [19,21] to compensate for the temperature rise to obtain a better speed control performance. However, few researchers have paid attention to the role of pre-pressure in the temperature variation. The other is the speed stability. It is known that there exist speed fluctuations due to the discontinuities in the contact distribution on stator teeth and manufacturing errors. Similarly, the role of pre-pressure in velocity stability regulation has not been explored. Besides, the sources and associated factors of fluctuation need to be better quantified. The sensitive relationships between pre-pressure and motor performances spread all over the motor, which brings difficulties in deriving a uniform optimization criterion, so a relatively effective candidate scheme for the optimization region should be assured through the joint analysis of multiple performance examples from sufficient experimental tests. Hence, accurate and digital-controllable pre-pressure adjustment devices are needed. Nonetheless, changing of the preload forces is mostly accomplished by manual feeding through a screw [14,22], which lacks guidance accuracy and regulatory flexibility. Even worse, the motor shell has been modified in some preload-adjusting apparatus [14,23], which adds the extra workloads.

In all, evaluation and optimization are the two main targets in this paper. In the evaluation process, modeling and experiment tests are synthetically implemented to evaluate the above performances under variable pre-pressure conditions. A hybrid model with the near-interface temperature rise is adopted to illustrate the frequency characteristics, contact properties, and temperature rise. A novel preload-control experimental apparatus is constructed from electric cylinder components, and a thin-film sensor is embedded inside a TRUM. In terms of the optimization process, the criterion considers the principles covering the lower temperature rise, the smaller speed stability, and the moderate mechanical properties. Finally, the optimization region of the preload force in a prototype motor is determined.

This paper is organized as follows: Section 2 focuses on the simulation model combined with power dissipation. Section 3 introduces the test bench with an electric cylinder and the thin-film temperature sensor for measuring the near-interface temperature. Section 4 analyzes the sensitivities of the main performance features in function of the preload forces from simulation and experimental tests. Section 5 proposes an optimization criterion according to the integrated results. Finally, conclusions and outlook are given in Section 6.

2. A Simulation Model with Power Dissipation

Preload force, which is imposed between motor stator and friction material, impels the motor to move. It is closely related to the performances of all motor components. Therefore a comprehensive model is required to account for the internal mechanisms in detail. In this work a TRUM60A ultrasonic motor (Chunsheng Ultrasonic Motor Co, Ltd, Nanjing, China) was investigated. Its stator operates in the B_{09} mode. In other words, the number (N) of traveling waves is equal to 9.

2.1. The Electromechanical Model

A TRUM is a typical electromechanical coupling system combining piezoelectric actuation with friction drive [24]. As shown in Figure 1a, two-phase piezoelectric ceramics are bonded symmetrically below the ring-shaped stator made of phosphor bronze material. When two-phase AC voltages are applied to the piezoelectric ceramics, the traveling wave impels the stator particles to vibrate with elliptic trajectories. Finally, the movements and accumulated energy are transferred to the rotor through a friction interaction for driving the terminate load. The friction layer of TRUM60A is polytetrafluoroethylene (PTFE), which has excellent time-varying stability [25]. What should be emphasized is that there is a gasket or disc spring between the bearing and the rotor. The component serves as the elastic element for withstanding the deformation caused by pre-pressure during assembly.

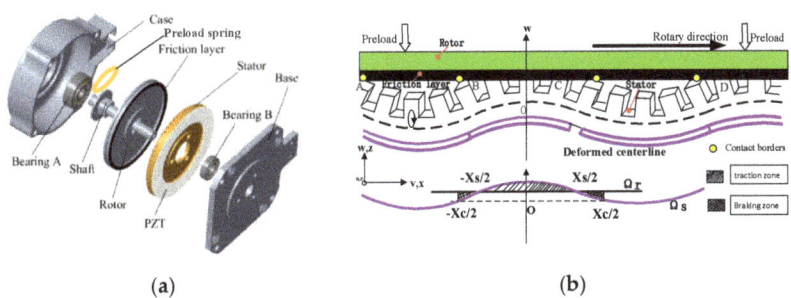

Figure 1. The mechanism of TRUM: (**a**) the motor structure [26]; (**b**) the traveling wave and contact state.

An electromechanical coupling TRUM model coalesces the models of both the driver and the motor itself. In this paper, a linear amplifying circuit is employed. After reproducing the actual circuit, the equivalent circuit model of piezoelectric ceramics is adopted. The ontology model originates from similar work by Kai [27], and the model equations and parameters are presented in Appendix A. Here, the critical parameters connected with the pre-pressure are mainly discussed and analyzed next. As shown in Figure 1b, due to the contact status and speed of the stator and the rotor, two feature points are generated on the stator teeth. One is the contact point used to distinguish whether the stator particles are embedded into the friction interface. The other is the sticking point, which is positioned at the point where the stator speed equals the rotor velocity. The stick points are applied to distinguish the driving zone and braking zone within the contact area. Usually, the contact length is defined as

Xc, and the length of the driving area is defined as Xs. Since the TRUM60A has nine traveling waves, each wave occupies 40° of circumferential space along the whole circle of the stator ring. Therefore the above two parameters yield $0 \leq Xs \leq Xc \leq 40°$. Moreover, their values can be expressed as:

$$Xc = \frac{20}{\pi}\arccos(\frac{h-z(t)}{R_{sc}\xi})$$
$$Xs = \frac{20}{\pi}\arccos(\frac{\Omega_r R_o^2}{\lambda h R_{sc}\xi f_s}) \qquad (1)$$

where ξ represents the vibration amplitude, h denotes the half-thickness of the piezoelectric laminated board, λ represents the wavelength of the traveling wave, f_s is the driving frequency and Ω_r is the angular velocity of the rotor. $R(r)$ is defined as the transverse displacement distribution function along the radial direction. The value on the middle radius at the contact surface R_0 is denoted as R_{sc}. Different from the model discussed in [27], the modal amplitude is corrected by the method proposed by Li [14], which takes the preload effect into account. When defining $k = N/R_o$, the contact parameters can be transformed into:

$$x_0 = \frac{\pi Xc}{40k} \qquad x_1 = \frac{\pi Xs}{40k} \qquad (2)$$

Thus, the overlap between the stator and the rotor defines the compression of the springs along the circumferential direction, thereby the pressure distribution function of the contact interface can be given by:

$$p(x) = K_f R_{sc}\xi[\cos(kx) - \cos(kx_0)] \qquad (3)$$

where K_f represents the equivalent stiffness of the contact area. The output friction can be calculated by the surface integral of the forces within the contact area [25]. If the friction coefficient is defined as μ, ε means the width of the contact surface. Thus, the driving friction force can be read as:

$$F_R = \mu \int_{R_o-\varepsilon/2}^{R_o+\varepsilon/2}\int_{-x_0}^{x_0} \text{sgn}(|\Omega s(x)|-|\Omega r(x)|)p(x)dxdy$$
$$= \mu\int_{R_o-\varepsilon/2}^{R_o+\varepsilon/2}\int_{-x_0}^{-x_1} p(x)dxdy + \mu\int_{R_o-\varepsilon/2}^{R_o+\varepsilon/2}\int_{-x_1}^{x_1} p(x)dxdy + \mu\int_{R_o-\varepsilon/2}^{R_o+\varepsilon/2}\int_{x_1}^{x_0} p(x)dxdy \qquad (4)$$
$$= 2\mu K_f R_{sc}\xi\varepsilon\{2[\frac{1}{k}\sin(kx_1) - x_1\cos(kx_0)] - [\frac{1}{k}\sin(kx_0) - x_0\cos(kx_0)]\}$$

Obviously, the driving torque is derived from the multiplication between the average radius and the driving force. The summarizing result is depicted as:

$$T_R = NF_R R_0 = 2\mu\varepsilon K_f R_{sc} R_o^2\xi\{2[\sin(kx_1) - kx_1\cos(kx_0)] - [\sin(kx_0) - kx_0\cos(kx_0)]\} \qquad (5)$$

Besides, the axial force, as well as the dynamic pressure on the stator surface, can be depicted as:

$$F_z = N\int_{R_o-\varepsilon/2}^{R_o+\varepsilon/2}\int_{-x_0}^{x_0} p(x)dxdy = 2NK_f R_{sc}\varepsilon\xi[\frac{1}{k}\sin(kx_0) - x_0\cos(kx_0)] \qquad (6)$$

2.2. The Power Dissipation Model

The traveling wave motor operates with several energy conversion processes from the power source to the rotor rotation. Indeed, there are three main power dissipations which are the dielectric dissipation of the piezoelectric ceramics, the vibration dissipation of the stator, and the heat dissipation of the contact interface, respectively.

At first, the dielectric dissipation of the piezoelectric ceramics yields Equation (7). Here ε_p means the dielectric constant of piezoelectric ceramic, $\tan\delta$ denotes the dielectric loss coefficient, and V_{piezo} represents the volume of the piezoelectric wafer:

$$Q_1 = 2\pi f_s \varepsilon_p \frac{U_m^2}{h_p^2} V_{piezo} \tan\delta \qquad (7)$$

Secondly, quoting from the study of Lu [28], the mechanical damping dissipation can be displayed as:

$$Q_2 = f_{n1} \int_{V_{piezo}} \delta \overline{W}_{sl} dV = 4\pi^4 \xi^2 N^4 w \left(E_s I_s \eta_s + E_p I_p \eta_p \right) / (\lambda^3) \tag{8}$$

where E_s and E_p represent the equivalent elastic modulus of the stator and the piezoelectric strip, I_s and I_p denote their inertia moment, η_s and η_p are their mechanical damping coefficient, respectively.

Last but not least is the power dissipation of the friction interface. The friction interface serves as the medium which accomplishes the transformation from the stator's vibration to the rotor's rotation. The friction drive gives rise to the radial friction losses and tangential ones between the stator and the rotor. The power losses stemming from the friction interaction can be determined by:

$$Q_3 = N \frac{\lambda}{2\pi} \int_{kx_0}^{\pi - kx_0} \mu \Delta F (v_s - v_r)(v_s - v_r) d\theta = N \xi K_f \mu [M_1 v_{sm} v_r + M_2 (v_{sm}^2)/2 + M_3 (v_r^2)] \tag{9}$$

where v_s, v_r are the linear velocity of the stator and the rotor, respectively, the v_{sm} denotes the peak stator speed of the traveling wave, and the remaining coefficients can be depicted as:

$$\begin{aligned} M_1 &= \pi - 2kx_0 + \sin(2kx_0) \\ M_2 &= -0.33 \cos(3kx_0) + 3\cos(kx_0) - M_1 \sin(kx_0) \\ M_3 &= 2\cos(3kx_0) - \pi \sin(kx_0) + 2\varphi \sin(kx_0) \end{aligned} \tag{10}$$

Among the three power losses, it is verified that the power dissipation of the friction interface is the largest. Therefore the friction interface becomes the primary heat field which has been proved by Finite Element Analysis [28], which serves as a fundament for the placement of the temperature sensor discussed in Section 3. Finally, the shell temperature function, which derives from the thermodynamics transmission processes between the air and the motor, yields:

$$T = T_{air} + \frac{Q_1 + Q_2 + Q_3}{\alpha S} \left(1 - e^{-\frac{\alpha St}{C_{usm}}} \right) \tag{11}$$

where S is the contact surface area, α represents the convective heat transfer coefficient, C_{usm} is the heat capacity of the TRUM60A device. Moreover, it is worth noting that the near-interface temperature may be larger than T owing to the weak heat transfer capacity of the friction material. The rising temperature will lead to a speed decline and motor performance deterioration.

3. Experimental Setup

Figure 2a displays the mechanical test bench which is composed of an ultrasonic motor (TRUM60A), an incremental encoder (AFS60A, 65536 lines, SICK Corp., Waldkirch, Germany), an electric cylinder (EA0400, Huitong Corp., Nanjing, China), a torque sensor (9349A, Kistler, Sindelfingen, Germany) and a DC motor (55LYX04, YGGT Corp., Beijing, China) which is capable of generating any load profile in the current closed-loop mode. The electric cylinder is first employed to regulate the preload of the TRUM. As shown in Figure 2b, the axial force generated by the cylinder is transmitted from the piston rod to the motor shell. There is a force sensor (VC20A050, Vistle Corp., Guangzhou, China) which detects the real-time pressure. The pressure value can be displayed on a monitor or transmitted through a DA channel. Besides, the piston rod, the pressure sensor, and the pressboard were all limited inside a sleeve, which can assure the guidance accuracy of the axial movement. The design not only does not modify the TRUM's structure but also assures the accurate and digital control of the pre-pressure. Moreover, an NTC film sensor (FWBM-337G104F, Fu Wen Corp., Shenzhen, China) is selected for satisfying the space restraint and response bandwidth of the temperature characteristics. The sensor measures the heating body through the polyimide whose thickness is only 15 µm, and its response time is only 0.1 s. As shown in Figure 2d, the sensor is placed against the stator's teeth gap and near the interface where the heat is most concentrated. The signal is processed through a Wheatstone bridge which has been embedded into the circuit.

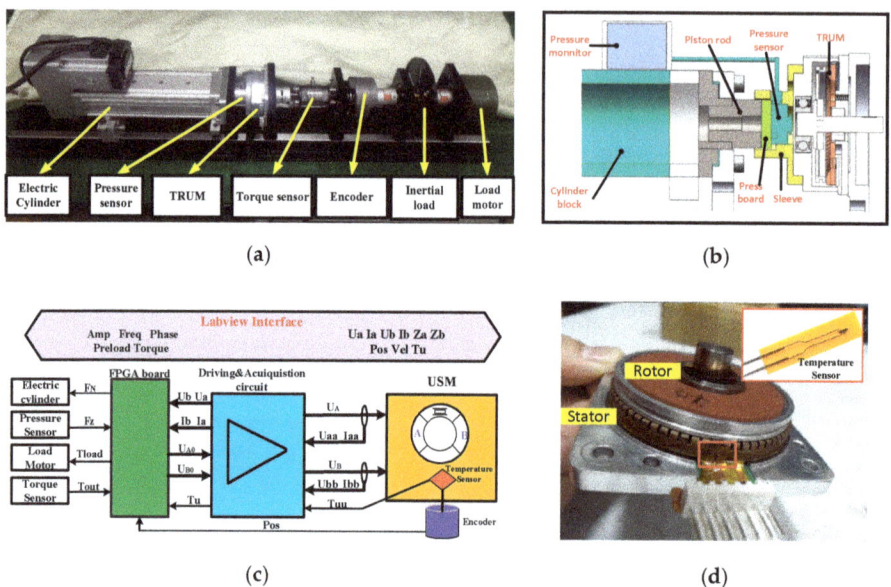

Figure 2. The integrated test system: (**a**) mechanical test bench; (**b**) structure scheme of the preload force regulating device; (**c**) framework and signal flow of the system; (**d**) the thin-temperature sensor placed near the interface.

Figure 2c introduces the framework of hardware and software. A driving & measurement circuit undertakes the function of amplifying the input signals. (U_{A0}, U_{B0}) and (U_A, U_B) are two pairs of voltages before and after amplifying, respectively. Meanwhile, the input voltages and currents are collected by voltage sensors and current sensors, respectively [29]. These channels are displayed from the signal group(U_{aa}, I_{aa}, U_{bb}, I_{bb}) to (U_a, U_b, I_a, I_b). The generation of (U_{A0}, U_{B0}) and the acquisition of (U_A, I_A, U_B, I_B) are all conducted by a Field Programmable Gate Array (FPGA) control board (PXIE 7854R, NI Corp., Los Angeles, CA, USA). The FPGA programs can be transferred from the LABVIEW software, which provides abundant interfaces for algorithm generation and data acquisition [29]. The main parameters of the LABVIEW interface are all shown on the top of Figure 2c, where the left is the input parameters, and the output ones are listed in the right. In conclusion, a test system combining the flexible control of driving parameters and the accurate collection of the key parameters is built, which builds a foundation for exploring the sensitivities of motor performances under diverse pre-pressures.

4. Simulation and Experimental Results by Varying the Preload Force

The simulation model, as well as the measurement system, are used to comprehensively analyze the pre-pressure's influences on several key performances including frequency characteristics, the stator/rotor contact states, the speed fluctuation, the temperature rise and the mechanical performance. Comparison and analysis are implemented for exploring the in-depth mechanism on how the preload force affects motion transfer and energy conversion.

4.1. The Stator/Rotor Contact

The stator/rotor contact is the premise that guarantees rotor rotation and torque output. If the driving parameters are unchanged, the contact status is mainly affected by the pre-pressure and the imposed load. Therefore, Figure 3a displays the contact length and the driving length for different cases of pre-stressing forces when the amplitude is 200 V and the frequency is 43 kHz. At first, the contact length generates a profound change when the preload force jumps away from zero value. Then the contact zone becomes more extensive with the slope 0.033°/N when the pre-pressure is increased from 100 to 570 N. Finally, the contact length increases with a higher slope until the contact area stretches over all the stator teeth. From the other curve, the length of the driving zone is equal to 40° when the stator is free from any pre-pressure. After that, the value gradually increases at a stable rate. However, it is always smaller than the contact length. This indicates that there exist driving areas and obstruction ones in the whole contact region. To be further, the proportion of the drive zone gradually decreases, which can also be verified from the proportion of the driving area in Figure 3b. These phenomena indicate that the stick-slip points gradually move toward the troughs of traveling waves, and the blocking effect is reinforced with the increasing pre-pressure, which induces more severe friction loss.

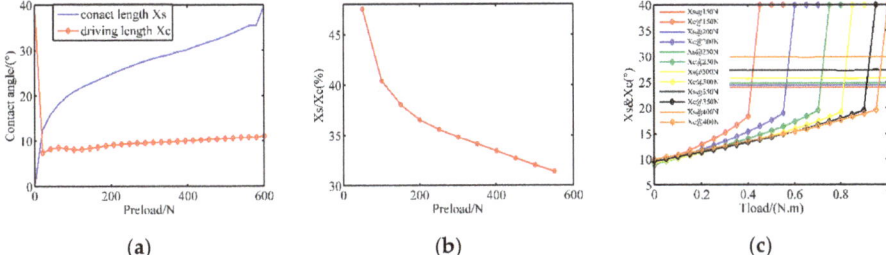

Figure 3. The contact parameters under diverse pre-pressures (simulation): (**a**) the contact angle and driving angel in no-load condition; (**b**) the proportion of driving zone in no-load condition; (**c**) the contact angle and driving angel with external load.

Once the motor operates with load, more output torque is needed to overcome the external torque, which causes the decline of rotor speed and the change of contact parameters. Figure 3c shows the contact parameters with respect to different loads from 0 to 0.9 N·m in function of pre-stressing forces. It is evident that the load has little effect on the contact length, which can also be verified by Equation (2). However, the driving length maintains a positive correlation with the load until the motor is blocked. In the blocking stage, the driving length returns to 40° again, which can be observed from the abrupt changes of the curves. In conclusion, if the preload is constant, the load torque does not affect the distribution of contact particles. And when the torque is constant, the contact area and the driving area gradually expand with the increase of the pre-pressures.

4.2. The Power Dissipation of the Motor

As discussed in Section 2, there are three main types of power dissipation inside the motor, and the temperature rise under different pre-pressures can be deduced. Figure 4 displays the respective time-variant energy loss in function of the preload forces when the amplitude and frequency are 200 V and 43 kHz, respectively. When the preload force is smaller than 460 N, the energy consumption increases as the preload force rises, which results from the gradually- widening contact area. By comparing the results of Figure 4a–c, it can be found that the dielectric loss maintains constant because the value is only related to the exciting voltage amplitude and the dielectric constant, which are both the intrinsic properties of piezoceramics. Furthermore, the stator damping loss and the friction loss follow similar positive-correlation laws until 460 N.

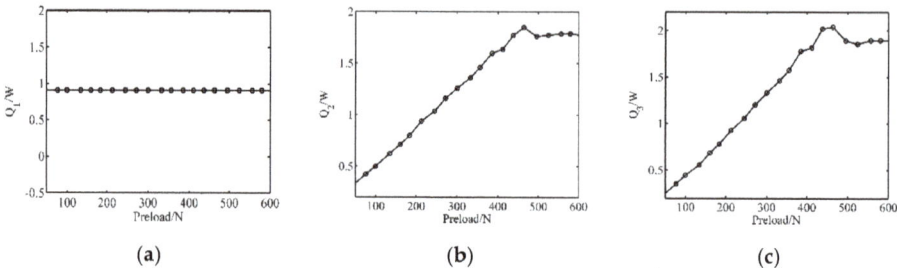

Figure 4. The respective power dissipation for the preload forces(Simulation): (**a**) Dielectric loss; (**b**) stator damping loss; (**c**) friction loss.

In order to exclude any additional factors, the starting temperature of every test should always be the same (28 °C), and the motor is turned off after the same time of operation (10 s). The results showing the motor speed and the near-interface temperature are displayed in Figure 5. The temperature rises immediately at the startup moment and drops rapidly once the driving voltages are withdrawn, which indicates that the accumulation and release of heat can be completed in a short time. Figure 5c demonstrates that the temperature rise achieves the highest value at 400 N, while the minimum temperature rise occurs when the preload is 300 N. With different temperature rise curves, the revolving speeds present different decline laws. It is evident that too fast speed decline is unfavorable for control and analysis.

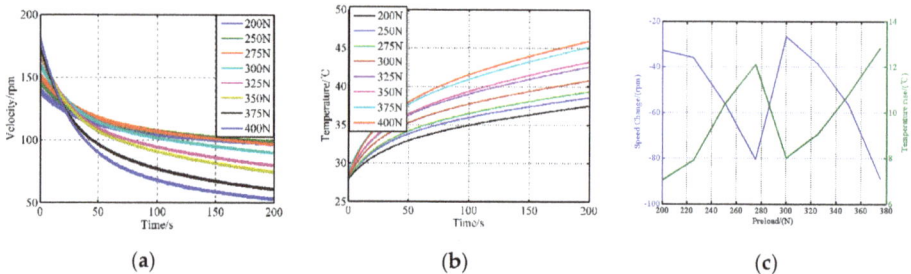

Figure 5. The startup-shutdown response under different pre-stressing forces (experimental): (**a**) the velocity response; (**b**) the temperature change; (**c**) temperature rise and speed decline.

4.3. The Speed Fluctuations

The speed fluctuations can be attributed to the errors of the shaft system during manufacturing or assembly. Whether the pre-stressing force compensates or deteriorates, the speed fluctuations should be investigated. Figure 6 displays the real-time speed with the preload force varying from 150 to 500 N in steps of 20 N. The rotor speed is recorded when the frequency is 43 kHz, and the amplitudes are 200 and 240 V, respectively. To facilitate the quantitative evaluation of the fluctuation, the standard deviation is adopted, as shown in Figure 6. The results indicate that the motor speed fluctuation becomes smaller and smaller due to the increasing embedment degree between the stator and the friction layer. It can be attributed to sufficient contact with the larger preload force. However, when the preload reaches 475 N, the reinforced radial slipping deteriorates the velocity stability. Since the simulation state ignores the machining and assembly problems, such as uneven axis alignment and surface roughness in the stator and the rotor, it is difficult to simulate the fluctuations effectively. Furthermore, the dynamic pressure applied to the rotor may be the origin of speed fluctuations. The pressure also has an impact on the contact state. Here, the fluctuations are analyzed by the measurement of the speed and axial pressure in the subsequent experiments.

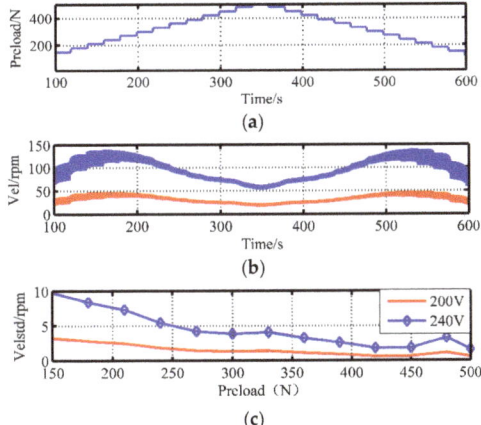

Figure 6. The speed stability under different preload forces: (**a**) staircase preloads; (**b**) output speed; (**c**) speed variance.

Based on the pressure F_z and the contact length x_0 in Equation (1) polynomial fitting is implemented to derive the mapping relationship from the axial pressure to the contact length, as shown in Equation (12). When the pre-pressure is 150 and 200 N, Figure 7 displays the real-time speed, the angular position, the corresponding pressure and the calculated contact length, respectively. We can observe that the velocity fluctuation satisfies a stretch of the 360° cycle, which demonstrates that the fluctuation mostly comes from the un-centered assembly error of the rotor shaft. The wave cycle elongates with declining speed. When the preload is 150 N, the ratio of the maximum fluctuant zone (1.01°) inside the contact zone (24.15°) is 2.53%. If the pre-tightening force is 200 N, the proportion becomes 3.11% as the fluctuation range and the contact angle are respectively 0.9° and 28.94°. In conclusion, the speed fluctuation comes from both the pressure change and the axial assembly error. The change of contact parameters is also accompanied by a pressure change:

$$X_c = 2.7 \times 10^{-16} F_z^7 - 6 \times 10^{-13} F_z^6 + 5.4 \times 10^{-10} F_z^5 - 2.5 \times 10^{-7} F_z^4 \\ + 6.4 \times 10^{-5} F_z^3 - 0.0088 F_z^2 + 0.65 F_z + 0.72 \tag{12}$$

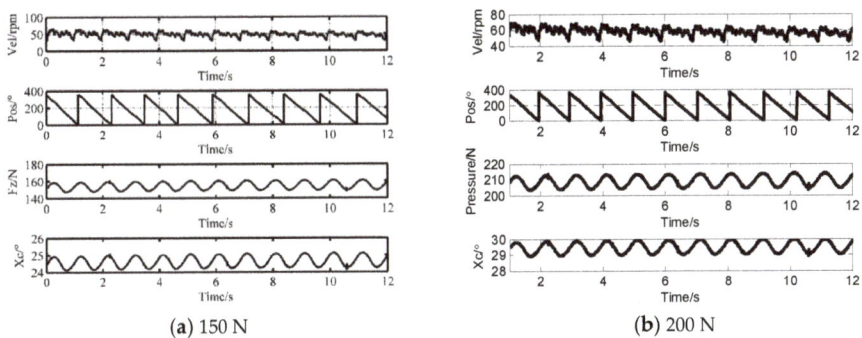

Figure 7. Time-domain analysis of dynamic pressure fluctuations and calculated contact angle.

4.4. The Mechanical Characteristics

With the aid of the electric cylinder and current sensors, a three-dimensional representation of the speed and the efficiency in function of the torque is shown in Figure 8 within the preload ranging from 150 to 400 N. The curves are obtained when the amplitude is 200 V and the frequency is 42 kHz. The dark blue areas are the stalling-torque areas, while the dark red ones are the maximum value zones of the respective curves. There exists a limit of the pre-stressing forces from which the motor performances fall. Figure 8b indicates the peak efficiency moves towards the larger-torque direction with the change of the preload forces. Deriving from the pictures, the blocking torque and maximum efficiency in function of the pre-pressures are displayed in Figure 8c. The blocking torque achieves the peak value in the moderate pre-pressure (260 N), so does the mechanical efficiency. The pre-pressure at the maximum efficiency point is 320 N. This is because the motor is close to the rotor-locked state when the pre-pressure is substantial, therefore the energy utilization is lower.

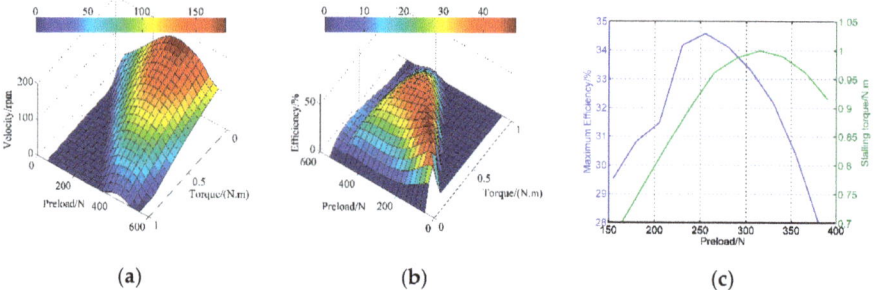

Figure 8. Simulation results of mechanical characteristics for different cases of pre-stressing forces: (**a**) revolving speed versus torque; (**b**) efficiency versus torque; (**c**) the calculated maximum efficiency and stalling torque.

5. Discussion and Verification of the Optimal Preload Force

The above simulation and experimental results reveal how the performances are sensitive to the pre-stressing force. In this section, these key performances are reviewed, and an optimal criterion is proposed.

5.1. Optimization Criterion

Some conclusions can be obtained from the above test results:

(1) The velocity increases first and then decreases as the pre-pressure increases, and a low pre-pressure cannot provide sufficient friction force while a higher one causes more tangential friction zones;
(2) When the pre-stressing force is lower, the velocity stability deteriorates because of the weakened constraints applied on the stator and the stability improves as the preload force gradually increases;
(3) With the increase of the pre-pressure, the points both from the resonant frequency and the anti-resonant frequency gradually shift to the right due to the increasing stiffness;
(4) The blocking torque achieves the peak value in the moderate pre-pressure, like the mechanical efficiency, however, the apexes are different from each other.

When limiting the pre-pressure range between 200 and 400 N, the key performances are summarized and arranged in Figure 9. Aiming at the lower open-loop speed stability, the preload force is supposed to be large enough for minimizing the dynamic fluctuations across the stator/rotor contact interface. Nonetheless, the moderate pressure is chosen based on the minimum temperature rise, which can decrease the undesirable changes in the tracking performances in the servo-control scheme. Based on the above analysis, the selected region is the yellow rectangle drawn in Figure 9.

The region can not only meet the requirements of low-speed stability and small temperature rise but also the blocking torque and mechanical efficiency are in an ideal range. This method is different from others in that it gives priority to speed smoothness and temperature rise as reference points for pre-pressure assessments.

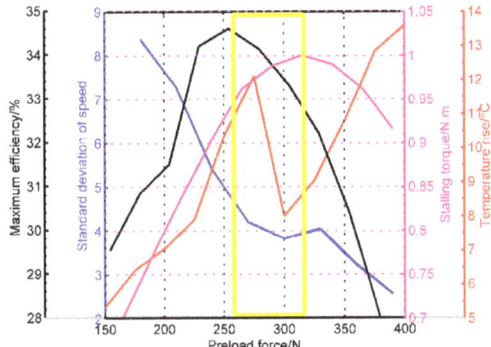

Figure 9. The diagram which interprets the optimization rule of the preload force.

5.2. Speed Control Performances

The validity of the optimization criterion is proved by the speed control performance, which not only stems from speed fluctuations but also reflects the effect of temperature rise. The speed closed-loop control with the target 72°/s is implemented under different preloads from 200 to 400 N. Considering the control target is located in the low-speed section, we select the voltage amplitude as the control parameter, the PID control method is adopted for every control process. Figure 10 displays the respective controlling results. It can be seen that residual control error decreases with the increasing preload forces when the preload force is less than 300 N and the error begins to rise once the pre-pressure exceeds 300 N. These results verify the effectiveness of pre-pressure optimization from one aspect, while other aspects will be further investigated.

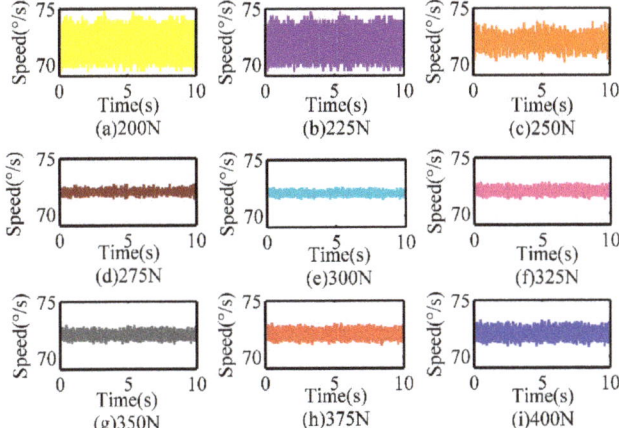

Figure 10. Comparison of the speed control performance under different pre-pressures.

6. Conclusions

Pre-pressure is one of the critical parameters that restrict the performance of an ultrasonic motor. The target of this paper is to comprehensively analyze the sensitivities of motor performances on the pre-pressure and to put forward a targeted optimization method. A simulation model with power dissipation and an integrated experimental facility with the preload adjustment device is adopted to analyze the laws from multiple perspectives. The preload adjustment is mainly designed by an electric cylinder and a pressure sensor, which is first employed in the preload change of the TRUM. Besides the stator/rotor contact state, the inspected properties cover rotor speed stability, near-interface temperature rise and mechanical properties. From these indicators, the speed stability, the temperature rise, the stalling torque, and the maximum efficiency are derived from the test results and drawn in a picture together. The operation makes these four indicators in function of the pre-pressure easy to distinguish in the candidate region according to different application targets. The optimization criterion in this paper contains three principles, which are the lower temperature rise, the smaller speed stability, and the moderate mechanical properties, respectively. Finally, the optimization region of the preload force is determined as [260, 320], and the smaller speed stability error in the optimization area verifies the criterion's validity.

Besides, the contact state under different preloads is interpreted in simulation and experimental tests. This paper first proposes a polynomial formula for transferring the pressure into the contact angle, which helps us measure the interface contact properties indirectly. Therefore the speed fluctuations can be attributed to the contact variation. What should be mentioned that slight fluctuations of the compression preload force will occur during a TRUM's physical life, especially in the application combining high speed with high load. The traditional preload component is a disc spring or thin-copper sheet, which cannot assure the correct imposing pressure in the total life-cycle. Luckily, the zero-stiffness structure similar to that of a PCB motor [30] can be used to ensure the pre-pressure, which will be the future optimization direction for all types of ultrasonic motors.

Author Contributions: N.C. contributed to this work by building the integrated measurement system and the analysis from simulation and experiment; J.Z. contributed to this work with the performance tests of the motor, D.F. contributed to the guidance of the research and the revision of the manuscript. All authors have read and agree to the published version of the manuscript.

Funding: The authors acknowledge the financial supports from the National Basic Research Program of China (973 Program, Grant No.2015CB057503).

Conflicts of Interest: The authors declare no conflict of interest.

Appendix A

The simulation model displayed as a state-space function:

$$\dot{x} = Ax + Bu \tag{A1}$$

In this function:

$$A = \begin{bmatrix} O_{2\times 2} & I_{2\times 2} & O_{2\times 2} & O_{2\times 2} \\ \begin{matrix} -k_o/m_o & 0 \\ 0 & -k_o/m_o \end{matrix} & \begin{matrix} -d_o/m_o & 0 \\ 0 & -d_o/m_o \end{matrix} & O_{2\times 2} & O_{2\times 2} \\ O_{2\times 2} & O_{2\times 2} & \begin{matrix} 0 & 1 \\ 0 & -d_r/J_r \end{matrix} & O_{2\times 2} \\ O_{2\times 2} & O_{2\times 2} & O_{2\times 2} & \begin{matrix} 0 & 0 \\ 0 & -d_z/m_r \end{matrix} \end{bmatrix} \tag{A2}$$

$$B^T = \begin{bmatrix} O_{2\times 2} & \begin{array}{cc} 1/m_o & 0 \\ 0 & 1/m_o \end{array} & O_{2\times 2} & O_{2\times 2} \\ O_{2\times 2} & O_{2\times 2} & \begin{array}{cc} 0 & 0 \\ 1/J_r & 0 \end{array} & \begin{array}{cc} 0 & 0 \\ 0 & 1/m_r \end{array} \end{bmatrix} \quad (A3)$$

$$u = \begin{bmatrix} k_c U_A + \varepsilon k_c U_B + F dn_1 + F dt_1 \\ \varepsilon k_c U_A + k_c U_B + F dn_2 + F dt_2 \\ T_R - T_{load} \\ F_Z - F_N \end{bmatrix} \quad (A4)$$

The critical simulation parameters are listed in the following table.

Table A1. The TRUM60A simulation parameters.

	Description	Value (Units)
N	Number of wavelengths	9
Ro	The average radius of the stator ring	0.02625 (m)
λ	Traveling wavelength	0.0187 (m)
mo	Modal mass of stator elastic body	0.005 (kg)
ko	Modal stiffness of stator elastic body	4.56×10^9 (kg.m^2)
do	Modal damping of stator elastic body	0.05 (N·s/m)
dz	Damping in the axial direction of the rotor	1.5×10^4 (N·s/m)
dr	Damping in the tangential direction of the rotor	5×10^{-4} (N·m·s)
kc	Force factor of piezoelectric ceramics	0.4147 (N/V)
ε	Unbalance coefficient between two-phase voltage	0.02
mr	Rotor mass	0.03 (kg)
Jr	Rotor inertia	7.2×10^{-6} (kg/m^2)
μ	Friction coefficient	0.3
Cp	Capacitance	5.41 (nF)
Rm	Dynamic resistor	149.82 (Ω)
Lm	Dynamic inductance	0.102 (H)
Cm	Dynamic capacitance	16.63 (pF)
Rd	Resumption resistor	31.15 (KΩ)
σ	Speed drop coefficient with the decreasing torque	9.9484 (rad/(N·m·s)
h	The thickness of the friction layer	0.5 (mm)
S	Motor surface area	0.02366 (m^2)
α	Coefficient of heat transfer	10

References

1. Qiu, Y.Q.; Wang, H.; Demore, C.E.M.; Hughes, D.A.; Glynne-Jones, P.; Gebhardt, S.; Bolhovitins, A.; Poltarjonoks, B.; Weijer, K.; Schönecker, A.; et al. Acoustic Devices for Particle and Cell Manipulation and Sensing. *Sensors* **2014**, *14*, 14806–14838. [CrossRef] [PubMed]
2. Liang, W.; Ma, J.; Tan, K.K. Contact force control on soft membrane for an ear surgical device. *IEEE Trans. Ind. Electron.* **2018**, *65*, 9593–9603. [CrossRef]
3. Shokrollahi, P.; James, M.D.; Goldenberg, A.A. Measuring the Temperature Increase of an Ultrasonic Motor in a 3-Tesla Magnetic Resonance Imaging System. *Actuators* **2017**, *62*, 20. [CrossRef]
4. Flueckiger, M.; Bullo, M.; Chapuis, D.; Gassert, R.; Perriard, Y. FMRI compatible haptic interface actuated with traveling wave ultrasonic motor. In Proceedings of the IEEE Industry Applications Conference Fortieth IAS Annual Meeting, Kowloon, Hong Kong, China, 2–6 October 2015.
5. Chapuis, D.; Gassert, R.; Burdet, E.; Bleuler, H. Hybrid ultrasonic motor and electrorheological clutch system for MR-compatible haptic rendering. In Proceedings of the IEEE/RSJ International Conference on Intelligent Robots & Systems, Beijing, China, 9–15 October 2006.
6. Liu, Y.X.; Xu, D.M.; Yu, Z.Y.; Yan, J.P.; Yang, X.H.; Chen, Y.S. A Novel Rotary Piezoelectric Motor Using First Bending Hybrid Transducers. *Appl. Sci.* **2015**, *5*, 472–484. [CrossRef]
7. Li, H.Y.; Wang, L.; Cheng, T.H. A High-Thrust Screw-Type Piezoelectric Ultrasonic Motor with Three-Wavelength Exciting Mode. *Appl. Sci.* **2016**, *12*, 442. [CrossRef]

8. Shao, S.J.; Shi, S.J.; Chen, W.S.; Liu, J.K. Research on a Linear Piezoelectric Actuator Using T-Shape Transducer to Realize High Mechanical Output. *Appl. Sci.* **2016**, *6*, 103. [CrossRef]
9. Oh, J.H.; Yuk, H.S.; Lim, J.N.; Lim, K.J.; Park, D.H.; Kim, H.H. An analysis of the resonance characteristics of a traveling wave types ultrasonic motor by applying a normal force and input voltage. *Ferroelectrics* **2009**, *378*, 135–143. [CrossRef]
10. Wang, L.; Liu, J.K.; Lin, Y.X.; Tian, X.Q.; Yan, J.P. A novel single-mode linear piezoelectric ultrasonic motor based on asymmetric structure. *Ultrasonics* **2018**, *89*, 137–142. [CrossRef] [PubMed]
11. Wang, J.; Hu, X.X.; Wang, B.; Guo, J.F. A novel two-degree-of-freedom spherical ultrasonic motor using three traveling-wave type annular stators. *J. Cent. South Univ.* **2015**, *22*, 1298–1306. [CrossRef]
12. Zhang, X.; Zhang, G.; Nakamura, K.; Ueha, S. A robot finger joint driven by the hybrid multi-DOF piezoelectric ultrasonic motor. *Sens. Actuators A Phys.* **2011**, *169*, 206–210. [CrossRef]
13. Pirrotta, S.; Sinatra, R.; Meschini, A. Evaluation of the effect of preload force on resonance frequencies for a traveling wave ultrasonic motor. *IEEE Trans. Ultrason. Ferroelectr. Freq. Control* **2006**, *53*, 746–753. [PubMed]
14. Li, J.; Liu, S.; Qu, J.; Cui, Y.; Liu, Y. A contact model of traveling-wave ultrasonic motors considering preload and load torque effects. *Int. J. Appl. Electromagnet Mech.* **2018**, *56*, 151–164. [CrossRef]
15. Chen, C.; Zhao, C.S. A novel semi-analytical model of the stator of TRUM based on dynamic substructure method. In Proceedings of the USE 2005, The 26th Symposium on ULTRASONIC ELECTRONICS, Tokyo, Japan, 16–18 November 2005; pp. 17–25.
16. Zhao, C.S.; Chen, C.; Zeng, J.S. A New Modeling of Traveling Wave Type Rotary Ultrasonic Motor Based on Three-Dimension Contact Mechanism. In Proceedings of the 2005 IEEE International Ultrasonic Symposium, Rotterdam, The Netherlands, 15–16 October 2005.
17. Zhang, D.; Wang, S.; Xiu, J. Piezoelectric parametric effects on wave vibration and contact mechanics of traveling wave ultrasonic motor. *Ultrasonics* **2017**, *81*, 118–126. [CrossRef] [PubMed]
18. Bullo, M.; Perriard, Y. Influences to the mechanical performances of the traveling wave ultrasonic motor by varying the prestressing force between stator and rotor. In Proceedings of the IEEE Symposium on Ultrasonics, Honolulu, HI, USA, 5–8 October 2003.
19. Tavallaei, M.; Atashzar, S.F.; Drangova, M. Robust Motion Control of Ultrasonic Motors under Temperature Disturbance. *IEEE Trans. Ind. Electron.* **2016**, *63*, 2360–2368. [CrossRef]
20. Moal, P.L.; Cusin, P. Optimization of traveling wave ultrasonic motors using a three-dimensional analysis of the contact mechanism at the stator–rotor interface. *Eur. J. Mech. A. Solids* **1999**, *18*, 1061–1084. [CrossRef]
21. Yano, Y.; Iida, K.; Sakabe, T.; Yabugami, K.; Nakata, Y. Approach to speed control using temperature characteristics of ultrasonic motor. In Proceedings of the 2004 47th Midwest Symposium On Circuits And Systems, MWSCAS '04, Hiroshima, Japan, 25–28 July 2004.
22. Mashimo, T. Miniature preload mechanisms for a micro ultrasonic motor. *Sens. Actuators A Phys.* **2017**, *257*, 106–112. [CrossRef]
23. Zhao, C. *Ultrasonic Motors: Technologies and Applications*; Springer Science & Business Media: Berlin/Heidelberg, Germany, 2011.
24. Abdullah, M.; Takeshi, M. Efficiency optimization of rotary ultrasonic motors using extremum seeking control with current feedback. *Sens. Actuators A Phys.* **2019**, *289*, 26–33.
25. El, G.N. Hybrid Modelling of a Traveling Wave Piezoelectric Motor. Ph.D. Thesis, Aalborg University, Aalborg, Denmark, 2000.
26. Ren, W.H.; Lin, Y.; Ma, C.C.; Li, X.; Zhang, J. Output performance simulation and contact analysis of traveling wave rotary ultrasonic motor based on ADINA. *Comput. Struct.* **2019**, *216*, 15–25. [CrossRef]
27. Kai, J. Dynamic Modeling of Traveling Wave Motor and the Optimal Control of Vibration Mode Vector. Ph.D. Thesis, Hebei University of Technology, Tianjin, China, 2016. (In Chinese).
28. Lu, X.; Hu, J.; Zhao, C. Analyses of the temperature field of traveling-wave rotary ultrasonic motors. *IEEE Trans. Ultrason. Ferroelectr. Freq. Control* **2011**, *58*, 2708–2719. [CrossRef] [PubMed]

29. Chen, N.; Qi, C.; Zheng, J.J.; Fan, D.; Fan, S. Impedance Characteristics Test of Ultrasonic Motor Based on Labview and FPGA. In Proceedings of the CSAA/IET International Conference on Aircraft Utility Systems (AUS 2018), Guiyang, China, 19–22 June 2018.
30. PCB Motor Inc. PCB Motor Starter Kit User Guide. Available online: http://ww1.microchip.com/downloads/en/devicedoc/mcskuserguide75015a.pdf (accessed on 4 April 2020).

© 2020 by the authors. Licensee MDPI, Basel, Switzerland. This article is an open access article distributed under the terms and conditions of the Creative Commons Attribution (CC BY) license (http://creativecommons.org/licenses/by/4.0/).

Article

On-Line Monitoring of Pipe Wall Thinning by a High Temperature Ultrasonic Waveguide System at the Flow Accelerated Corrosion Proof Facility

Se-beom Oh [1,2,*], Yong-moo Cheong [1], Dong-jin Kim [1] and Kyung-mo Kim [1]

1. Nuclear Materials Research Division, Korea Atomic Energy Research Institute, 989-111, Daedeok-daero, Yuseong-gu, Daejeon 34057, Korea; ymcheong@kaeri.re.kr (Y.-m.C.); djink@kaeri.re.kr (D.-j.K.); kmkim@kaeri.re.kr (K.-m.K.)
2. Department of Materials Science and Engineering, Dankook University, 119, Dandae-ro, Dongnam-gu, Cheonan 31116, Korea
* Correspondence: sbfull1269@gmail.com

Received: 10 March 2019; Accepted: 8 April 2019; Published: 12 April 2019

Abstract: Pipe wall thinning and leakage due to flow accelerated corrosion (FAC) are important safety concerns for nuclear power plants. A shear horizontal ultrasonic pitch/catch technique was developed for the accurate monitoring of the pipe wall-thickness. A solid couplant should be used to ensure high quality ultrasonic signals for a long operation time at an elevated temperature. We developed a high temperature ultrasonic thickness monitoring method using a pair of shear horizontal transducers and waveguide strips. A computer program for on-line monitoring of the pipe thickness at high temperature was also developed. Both a conventional buffer rod pulse-echo type and a developed shear horizontal ultrasonic waveguide type for a high temperature thickness monitoring system were successfully installed to test a section of the FAC proof test facility. The overall measurement error was estimated as ±15 μm during a cycle ranging from room temperature to 150 °C. The developed waveguide system was stable for about 3300 h and sensitive to changes in the internal flow velocity. This system can be used for high temperature thickness monitoring in all industries as well as nuclear power plants.

Keywords: high temperature pipe; pipe wall thinning; flow accelerated corrosion

1. Introduction

During the operation of nuclear power plants, the thickness of the piping decreases over time which is known as pipe wall thinning. If the reduced thickness is concentrated on one side, the piping could be damaged by the pressure in the pipe, and an internal solution may cause a leak. Pipe wall thinning is mainly caused by FAC (flow accelerated corrosion). FAC occurs mostly in carbon steel pipes, in which the pipe thickness gradually decreases as the Fe ions on the surface of the carbon steel pipe are released. Because pipe wall thinning due to FAC is very slow (a few tens of μm per one year), it is necessary to monitor the piping walls for delamination, cracks and leaks as well as the piping thickness with very high accuracy. This is one of the important issues in the structural stability of a system, which requires continuous monitoring [1–5]. Presently, an ultrasonic method is used, which is one of the nondestructive inspection techniques for measuring the piping wall thickness. The ultrasonic technique is widely used to assess the safety of nuclear piping and to measure the piping wall thickness. Manual ultrasonic methods are generally used to measure the pipe thickness. However, the manual ultrasonic technique has several disadvantages in nuclear power plants. First, the inspection areas of a nuclear power plant are at high temperatures and highly radioactive. Second, the power plant must be shut down and the insulator removed before the thickness of a pipe can be measured because the transducer must be in direct contact with the pipe surface. When the measurement is completed, it is necessary to install the insulation again, so the shutdown time will be longer. Therefore, if the shutdown of the power plant time is prolonged, it will lead to a loss in power production. This process is inconvenient because it is repeatedly measured during a period of time to assess its structural health. Third, the manual ultrasonic thickness measurement method has a low reliability. The accuracy decreases depending on the operator's skill or condition, the measuring instrument, temperature, the ultrasonic coupling, and the difference in data reading conditions. Therefore, it is necessary to have a stable thickness measurement method for a high temperature and highly radioactive environment without having to stop the operation of the nuclear power plant.

Measuring the thickness of high-temperature piping using ultrasonic waves has several problems that have not been solved. In the case of a nuclear power plant, the temperature of the fluid flowing inside the pipe increases to about 200 °C. At this time, the piping temperature is up to about 150 °C, and the difference in thermal expansion between each of the piezoelectric and coupling materials may cause errors in the thickness measurements. Conventional piezoelectric materials depolarize if they rise above the Curie temperature; thus, current ultrasonic thickness measurement techniques cannot be used at high temperatures above 200 °C [6–8].

To solve this problem, a method for installing a buffer rod system and a waveguide method are currently being studied. The buffer-rod system has the advantage of using the existing longitudinal wave transducer; however, it has a disadvantage that it cannot perfectly protect the piezoelectric element from the high temperature piping because the buffer-block is short, and the distance from the specimen is not long. The ultrasonic waveguide system can be used to protect the piezoelectric element from high temperature piping by using a long thin plate to keep the piezoelectric element away from the test specimen [9–14]. In this method, the ultrasonic dispersion characteristics in the guide should be considered because the ultrasonic wave propagates through the waveguide [15]. The shear horizontal vibration mode can be used because there is no dispersion characteristic when the wave propagates in the plate. Based on these techniques, we developed a pipe wall thinning monitoring system using a shear horizontal ultrasonic transducer and waveguide strip. Clamping devices were designed and installed with a solid coupling material for safe acoustic contact between the waveguide strip and the pipe surface. The shear horizontal waveguide and clamping device provided an excellent S/N ratio and high measurement accuracy for long time exposure at high temperature conditions.

2. Issue of High-Temperature Ultrasonic Thickness Measurements

The ultrasonic thickness measurement principle is usually performed by measuring the flight time between continuous echoes in the time domain. The thickness of the specimen can be determined by calculating the material flight time of the waves and the known ultrasonic velocity values. Assuming that there is minimal ultrasonic dispersion, a sharper ultrasonic signal will increase the resolution of the measurement, and in general is the most accurate way to perform the temporal measurement by measuring the peak to peak time or the perform pulse-echo overlap [16]. Pipe wall thinning of carbon steel pipes in nuclear power plants occurs at several tens of micrometers per year, thus the measurement errors should be minimized. Several factors have been discussed as ways to overcome these errors [17]. First, environmental factors cause errors due to the geometric factors of the piping such as surface roughness, specimen curvature and contact pressure between the coupling material and the transducer as well as ultrasonic velocity errors in the specimen due to changes in the temperature. Thus, the device should be calibrated, and the surface condition should be maintained to minimize measurement errors. Second, there are errors due to the transducer performance and signal processing, such as measurement conditions between the transducer and the specimen, the performance of the analog-to-digital converters and delays caused by digital signal processing, respectively [18].

Typical piezoelectric ceramic elements are exposed to temperatures higher than the Curie temperature resulting in depolarization of the element, which then loses its piezoelectric properties making it difficult to accurately measure the signal. The signal quality of the piezoelectric vibrator degrades, and the error in determining the peak position of the signal may increase as the temperature varies. It is necessary to improve the acoustic contact condition between the transducer and the specimen by minimizing the deterioration of the probe at high temperatures. Third, there is a problem with the couplant. For a high temperature pipe, the couplant used at room temperature will evaporate, resulting in an error on the contact surface. Stable ultrasonic sound effect was maintained by using a special high temperature couplant that does not evaporate even in an environment of 150 °C. A dry couplant (gold plate) was used between the waveguide strip and the surface of pipe to minimize the error due to thermal expansion. The gold plate can minimize errors due to thermal expansion between the two objects by constantly maintaining its shape of the coefficient at high temperatures.

3. Ultrasonic On-Line Monitoring System for Measuring Wall Thinning of High Temperature Pipe

The conventional buffer-rod type system is widely used as a technique for measuring the thickness of high temperature test specimens. It can be used by inserting a buffer block between the ultrasonic transducer and the specimen shown in Figure 1a. The material of the buffer block should be acoustically stable, should not deform at high temperatures, and should protect the transducer. Glycerin or machine oil used in conventional ultrasonic couplings does not work properly at high temperatures. Therefore, a special solid material coupling, such as a thin gold plate, was used to maintain good acoustic contact between the buffer rod and the specimen. The advantage is that it minimizes the difference in thermal expansion between the buffer rod materials and the specimen, which can be maintained for long periods of the test. Figure 1b shows a buffer-rod type high temperature ultrasonic transducer assembled in a test pipe for thickness monitoring.

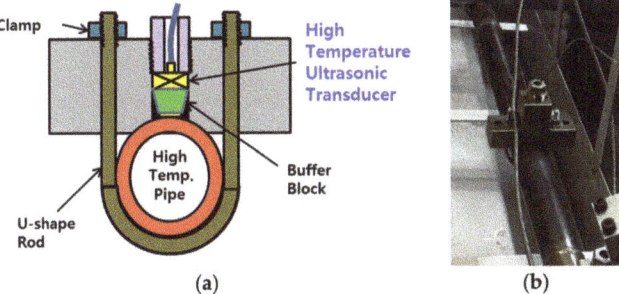

Figure 1. (a) An assembly drawing diagram of a high temperature pipe using a buffer-rod type measurement system; (b) an installed buffer-rod type system for thickness monitoring on a pipe.

Another approach to measure pipe thickness at high temperatures is to use an ultrasonic waveguide strip. This improved method was attempted using a waveguide strip to reduce the acoustic parameters between the ultrasonic transducer and the specimen. A pair of shear horizontal transducers and a long waveguide strip were designed and manufactured. The shear horizontal vibration mode was chosen to ensure that there was proper ultrasonic wave transmission at the thin strips. The shear horizontal mode had sharp and clear ultrasonic signals within a certain frequency range because there was low dispersion in the plate. This vibration mode is advantageous for obtaining sensitive and accurate experimental data at high temperatures [19]. The shear horizontal wave transducer was attached to the edge of the waveguide strip shown in Figure 2. When the transducer and the waveguide strip contacted each other exactly at a perpendicular level, the shear horizontal mode was stably transmitted to the waveguide. On the opposite side of the waveguide strip, a clamping device was designed and fabricated to precisely hold the specimen, and two waveguide strips were installed in parallel to divide the transmitter and receiver. A thin solid plate (gold plate) was used as a couplant between the waveguide strip and the test specimen similar to the buffer-rod system. The transducer used a waveguide strip that was far from the specimen, which was maintained at about 35 °C when the temperature of the pipe was 150 °C. This meant that the developed system completely freed the transducer from the constraints of high temperatures. The waveguide pitch/catch method using two waveguide strips can increase the S/N ratio compared to the pulse/echo technique. Two strips are used to set up the waveguide system device. A transducer was connected to each strip, one on the transmitting transducer and the other on the receiving transducer. The two strips were spaced 1 mm apart to filter the noise received at the surface of the pipe to be measured. If the strip is placed at 1 mm, it measured only the wanted signal because of the time difference between the signal transmitted to the pipe surface, and the signal reflected from the opposite side of the pipe. Because it received a signal only from the piping specimen, noise from undesired reflections at the end of the strip were prevented. This method had no main bang signal, and the signal reflected from the end of the waveguide strip was very small. Additionally, the multiple reflected signals on the back wall of the pipe, which was to be inspected, had a high S/N ratio.

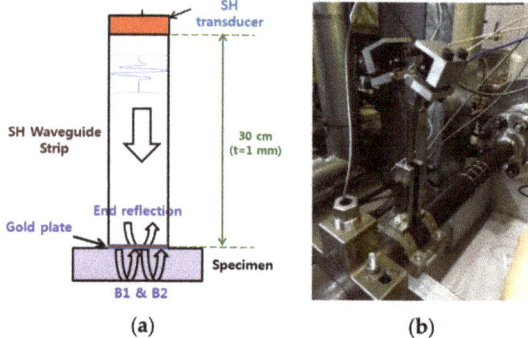

Figure 2. (a) Conceptual diagram of a pair of waveguide strips for high temperature thickness monitoring; (b) installed waveguide system for thickness monitoring on a pipe.

4. High Temperature Pipe Thickness Measuring Program

The thickness measurement program was designed as a moving gate in real time to accurately measure the reflected flight time. The first gate was set to the signal from the end of the transmitting waveguide strip shown in Figure 3. The second gate was set to the first back wall echo signal, and the third gate was set to the second back wall echo signal. The second and third gates were set as the moving gates to follow the first gate setting. The moving gate moves along with the movement of the rf signal according to the noise or the temperature change, making it possible to measure the peak to peak signal stably. The peak of the first and second back wall echoes were automatically determined by the flight time and denoted as t_1 and t_2 shown in Figure 3. The ultrasonic wave velocity was constant and the path length of the actual reciprocating wave can be seen by the strip and pipe wall thickness. Thus, the time of the received rf signal can be calculated. The shear horizontal wave velocity of the carbon steel is about 3300 m/s, and the flight time reflected by the waveguide strip 300 mm in length is calculated to be 170 μs. The flight time between the first back wall and the second back wall of a 5.54 mm thick pipe is estimated to be about 3.2 μs. All ultrasonic rf (radio frequency) waveforms are in the time domain and displayed on the PC screen. Moreover, this system inspects the signal quality and is designed to display an alarm indicator on the screen when receiving unwanted signals. Between t_1 and t_2, the flight time was automatically calculated on average, hundreds of times to obtain accurate thickness data. Because ultrasonic velocity is a function of temperature, variation in the ultrasonic velocity at high temperatures can be a main problem in terms of measurement data errors. Therefore, to measure the thickness in real time at high temperature, pre-calibration is required to reflect the relationship between the ultrasonic velocity and the temperature.

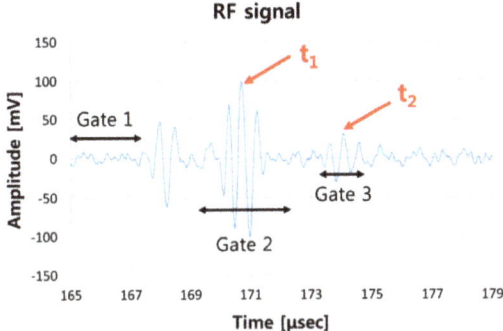

Figure 3. Typical ultrasonic rf signals by a developed waveguide with the pitch/catch method at 150 °C.

The ultrasonic velocity can be determined as a function of the temperature by measuring the variation in the velocity with a temperature change in the material to be measured. Figure 4 shows the measurement data of the shear mode wave velocity in a carbon steel pipe based on a variation in the temperature. The velocity of the ultrasonic waves was measured by heating the same materials pipe as the pipe used in the FAC proof facility to the furnace from 0 to 250 °C. The pipe thickness did not change when measuring the sound velocity change by calculating the time of the received signal according to the temperature change [20]. This function was entered into the thickness measurement program to reduce the error due to the temperature variation.

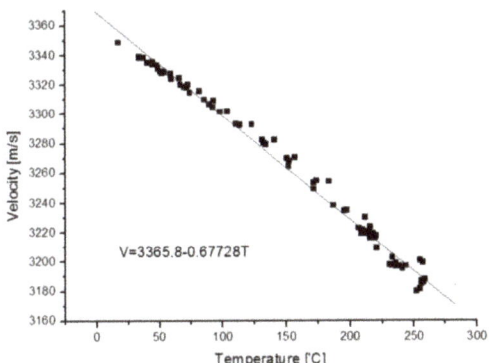

Figure 4. Calibration of the shear horizontal wave velocity with varying temperatures for a carbon steel pipe SA 106.

5. Verification Experiment in the FAC Proof Facility and Results

The FAC demonstration test facility was manufactured to operate in the same environment as a nuclear power plant. This facility is designed to operate for 1200 h at high temperatures at more than 150 °C, per one cycle and with the adjustable pH, DO, and flow rate for the fluid flowing inside the piping. To measure pipe wall thinning, we prepared a test section with the insulation removed from part of the facility. This section is made of carbon steel (SA 106), and the pipe has an outer diameter of 60.4 mm, a wall thickness of 5.54 mm and a length of 750 mm. The chemical composition of the materials is shown in Table 1. The pipe thickness was measured by the buffer-rod type system and an ultrasonic waveguide high temperature thickness monitoring system. The two systems were installed on the surface of the carbon steel pipe to compare the signals from each other, as shown in Figure 5.

Prior to the experiment, the thickness measurement error was confirmed according to the temperature variation for the stability and accuracy of the system. Figure 6 shows the actual measured waveguide system data. A provisional thickness error range was determined by increasing the temperature of the same material pipe to 200 °C and measuring the change in thickness during the cooling process. The range of the thickness measurement error was ±15 µm while the pipe was heated to 0~150 °C and then cooled. These devices were able to acquire ultrasonic signals which were reliable for a long time at high temperatures, and the flight times of the signals through calculations were converted to the thickness of the pipe. Minimizing the measurement errors was possible with normalization of the signal amplitude, the automatic setting of the ultrasonic flight time and moving the gate control by applying a temperature compensation factor. The pipe thickness data were directly compared with the measured data at room temperature during the rest period to determine the reliability of the measured data.

Table 1. The chemical composition of the material (SA 106 Gr. B).

Material	Cr	Mo	Cu	Mn	Ni	Si	C	P	S
SA106 Gr. B	0.02	0.01	0.04	0.37	0.02	0.22	0.19	0.008	0.006

Figure 5. Test section in the flow accelerated corrosion (FAC) facility (**left**) and another type system installed at a test section (**right**).

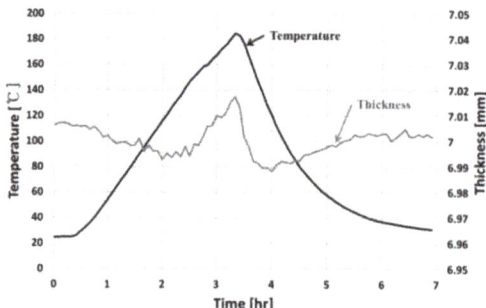

Figure 6. The measurement error by temperature variation in the developed system.

Figure 7 shows the change in the slope of the wall thickness reduction with the flow rate in the waveguide system. This developed system measured the tendency of pipe wall thinning by changing the flow rates from 7 to 10 to 12 m/s every 1100 h. The blue line is the data measured in real time once every hour. The red line slope indicated the FAC rate in each section. As the flow rate increased, the FAC rate became faster and the FAC rate decreased as the flow rate decreased. As the flow rate increased from 7 m/s to 12 m/s, the wall thickness reduction rate also increased. This result meant that is possible to predict the change in the flow rate inside the pipe by analyzing the rate of pipe wall thinning.

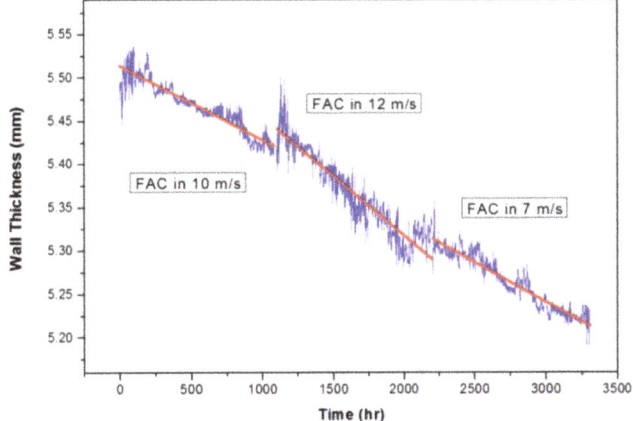

Figure 7. Pipe wall-thickness monitoring of carbon steel piping in the flow accelerated corrosion proof test facility: Different wall-thinning ratios observed depending on the flow velocities.

Figure 8 shows the pipe wall thinning measurement data from approximately 3300 h of operation using the buffer-rod and developed waveguide systems. The temperature of the fluent flowing inside the pipe was maintained at 150 °C, and both devices measured the wall thickness reduction of about 260 µm. The blue line in Figure 8 is the data of thickness thinning of buffer-rod type commercial equipment. From the point at which the flow rate increased from 10 m/s to 12 m/s, it can be seen that the thickness thinning rate tendency was less than that of the developed equipment. Furthermore, although the flow velocity slowed to 7 m/s at a large velocity of 12 m/s, but the buffer-rod type system was measured to show that the thickness reduction rate was increasing. On the other hand, the red line is measuring data by the developed waveguide system. The thickness reduction rate was measured to suit the change in the flow rate. It is possible to compare more clearly converted (mm/year) for the thickness reduction for each period. As the flow rate inside the piping increased, the rate of the pipe wall thinning increased. The developed waveguide system showed more accurate reduction trends as the flow rate changed. The developed waveguide system operated stably at a temperature cycle of 150 °C for a long time, and the measurement error was about ±20 µm.

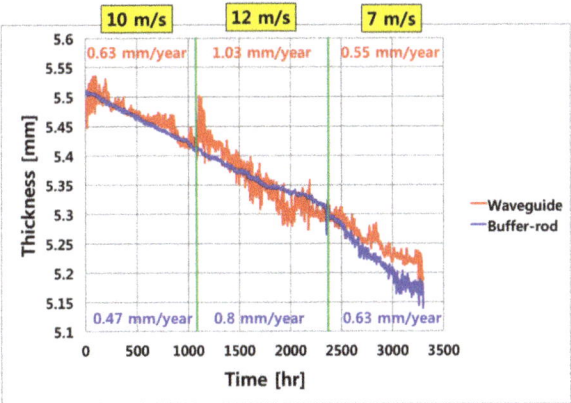

Figure 8. The pipe thickness measurement data using the buffer-rod and waveguide type systems.

Figure 9 shows the result of the manual UT measurement at room temperature. A total of six points were set in the same direction as the position where the two systems were installed, and measurements were made at room temperature. Reliable comparative data measures exactly at the same point, but there was an error generated by removing and reinstalling real-time measuring equipment. Measuring the six zones from A to F next to the point where two pieces of equipment were installed (see Figure 10). A total of six points were measured 10 times and averaged in order to increase the accuracy. The equipment used for the measurement was 38DL PLUS (OLYMPUS). The tendency of the thickness reduction measured with the manual room temperature UT was very similar to that for the high temperature on-line UT system. The reliability of the developed waveguide system was verified by comparing the conventional buffer-rod system and the manual room temperature UT.

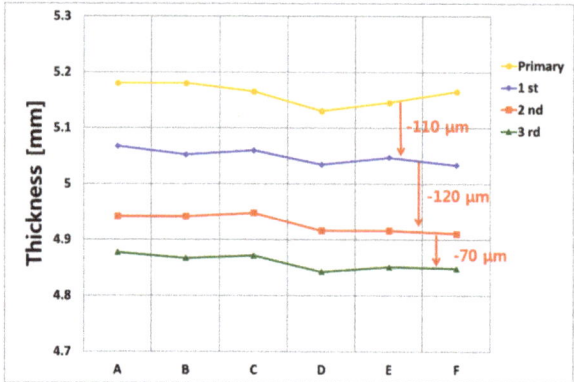

Figure 9. The thickness measurement result of the manual ultrasonic testing (UT) at room temperature.

Figure 10. A grid for measuring ultrasonic testing at room temperature in the same direction next to two installed systems.

6. Conclusions

A shear horizontal ultrasonic pitch/catch waveguide system was developed for the accurate online monitoring of the pipe wall thickness in the FAC certification test facility. A clamping device was designed and installed for the gold-plate contact between the end of the waveguide strip and the pipe surface. A computer program was developed for online monitoring of the pipe thickness at high temperatures. The system minimized measurement errors by controlling the moving gate with temperature deviation, normalizing the signal amplitude, automatically determining the ultrasonic flight time and including a temperature compensation function. The buffer-rod and ultrasonic waveguide high temperature ultrasonic thickness monitoring systems were successfully installed in the test section of the FAC test facility. These systems were confirmed as a stable operation with an error of ±20 μm in temperature cycles up to 150 °C for 3300 h, and performed better than other similar measurement systems. In addition, the developed waveguide system was able to predict the velocity of the fluid flowing inside the pipe by analyzing the thickness reduction rate.

Finally, it was confirmed that the thickness reduction measurement was very accurate in comparison with the room temperature UT results. This result demonstrates that a waveguide system is sensitive to the flow of internal fluids and can measure thickness better than that of commercial systems. This system can be applied to high temperature thickness monitoring in all industries as well as nuclear power plants.

Author Contributions: S.-b.O. conducted the signal measurement, data analysis and wrote the manuscript; Y.-m.C. provided analysis cooperation and reviewed the manuscript; D.-j.K. supervised this study; K.-m.K supervised this study.

Funding: This work was supported by a Korean Atomic Energy Research Institute (KAERI) Grant funded by the Korean government (Ministry of Science, ICT, and Future Planning).

Acknowledgments: This work was supported by a Korean Atomic Energy Research Institute (KAERI) Grant funded by the Korean government (Ministry of Science, ICT, and Future Planning) and the National Research Foundation of Korea (NRF) Grant funded by the Korean government (2017M2A8A4015158).

Conflicts of Interest: The authors declare no conflict of interest.

References

1. Cheong, Y.M.; Kim, K.M.; Kim, D.J. High-Temperature Ultrasonic Thickness Monitoring for a Pipe Thinning in FAC Proof Test Facility. *Nuclear Eng. Technol.* **2017**, *49*, 1463–1471. [CrossRef]
2. Dooley, R.B.; Chexal, V.K. Flow-accelerated corrosion of pressure vessels in fossil plants. *Int. J. Press. Vessels Pip.* **2000**, *77*, 85–90. [CrossRef]
3. Choi, Y.H.; Kang, S.C. Evaluation of Piping Integrity in Thinned Main Feedwater Pipes. *Nucl. Eng. Technol.* **2000**, *32*, 67–76.
4. Kim, D.; Cho, Y.; Lee, J. Assessment of Wall-Thinning in Carbon Steel Pipe by Using Laser-generated Guided Wave. *Nucl. Eng. Technol.* **2010**, *42*, 546–551. [CrossRef]
5. Chung, H.S. A Review of CANDU Feeder Wall Thinning. *Nucl. Eng. Technol.* **2010**, *42*, 568–575. [CrossRef]
6. Baba, A.; Searfass, C.T.; Tittermann, B.R. High Temperature Ultrasonic Transducer up to 1000 °C using Lithium Niobate Single Crystal. *Appl. Phys. Lett.* **2010**, *97*, 232901.
7. Wu, K.T.; Kobayashi, M.; Jen, C.K. Integrated High-temperature Piezoelectric Plate Acoustic Wave Transducers Using Mode conversion. *IEEE Trans. Ultrason. Ferroelectr. Freq. Control.* **2009**, *56*, 1218–1224. [PubMed]
8. Megriche, A.; Lebrun, L.; Troccaz, M. Materials of $Bi_4Ti_3O_{12}$ Type for High Temperature Acoustic Piezo-sensors. *Sens. Actuators A* **1999**, *78*, 88–91. [CrossRef]
9. Carandente, R.; Lovstad, A.; Cawley, P. The Influence of Sharp Edges in Corrosion Profiles on the Reflection of Guided Waves. *NDT&E Int.* **2012**, *52*, 57–68.
10. Zhou, W.J.; Ichchou, M.N.; Mencik, J.M. Analysis of Wave Propagation in Cylindrical Pipes with Local Inhomogeneities. *J. Sound Vib.* **2009**, *319*, 335–354. [CrossRef]
11. Nurmalia, N.; Nakamura, H.; Ogi, M.; Hirao, K. Nakahata, Mode Conversion Behavior of SH Guided Wave in a Tapered Plate. *NDT&E Int.* **2012**, *45*, 156–161.
12. Honarvar, F.; Salehi, F.; Safavi, V.; Mokhtari, A.; Sinclair, A.N. Ultrasonic Monitoring of Erosion/Corrosion Thinning Rates in Industrial Piping systems. *Ultrasonics* **2013**, *53*, 1251–1258. [CrossRef] [PubMed]
13. Kazys, R.; Voleisis, V.; Voleisiene, B. High Temperature Ultrasonic transducer: Review. *Ultrasound* **2008**, *63*, 7–17.
14. Francis, L.A.; Friedt, J.M.; Bertrand, P. Influence of Electromagnetic Interferences on the Mass Sensitivity of Love Mode Surface Acoustic Wave Sensors. *Sens. Actuators A* **2005**, *123*, 360–369. [CrossRef]
15. Cheong, Y.M.; Kim, H.P.; Lee, D.H. A Shear Horizontal Waveguide Technique for Monitoring of High Temperature Pipe Thinning. In Proceedings of the 2014 Spring Meeting of the KNS, Jeju, Korea, 28–30 May 2014.
16. Dixon, S.; Edwards, C.; Palmer, S.B. High Accuracy Non-contact Ultrasonic Thickness Gauging of Aluminum Sheet using electromagnetic Acoustic Transducers. *Ultrasonics* **2001**, *39*, 445–453. [CrossRef]

17. Maio, L.; Ricci, F.; Memmolo, V.; Monaco, E.; Boffa, N.D. Application of laser Doppler vibrometry for ultrasonic velocity assessment in a composite panel with defect. *Compos. Struct.* **2018**, *184*, 1030–1039. [CrossRef]
18. Cheong, Y.M.; Kim, H.N.; Kim, H.P. An Ultrasonic Waveguide Technique for On-line Monitoring of the High Temperature Pipe Thinning. In Proceedings of the Transactions of the Korean Nuclear Society Spring Meeting, Jeju, Korea, 29–30 May 2014.
19. Celga, F.B.; Cawley, P.; Allin, J.; Davies, J. High-temperature (>500 °C) Wall Thickness Monitoring Using Dry-coupled Ultrasonic Waveguide transducers. *IEEE Trans. Ultrason. Ferroelectr. Freq. Control.* **2011**, *58*, 156–167.
20. Cheong, Y.M.; Kim, J.H.; Hong, J.H.; Jeong, H.K. Measurement of Dynamic Elastic Constants of RPV Steel Weld due to Localized Microstructural Variation. *J. Korean Soc. Non-Destruct. Test.* **2001**, *20*, 559–564.

© 2019 by the authors. Licensee MDPI, Basel, Switzerland. This article is an open access article distributed under the terms and conditions of the Creative Commons Attribution (CC BY) license (http://creativecommons.org/licenses/by/4.0/).

Article

A Low-Cost Continuous Turbidity Monitor

David Gillett [1] and Alan Marchiori [2],*

1. Department of Chemical Engineering, Bucknell University, Lewisburg, PA 17837, USA
2. Department of Computer Science, Bucknell University, Lewisburg, PA 17837, USA
* Correspondence: amm042@bucknell.edu; Tel.: +1-570-577-1751

Received: 31 May 2019; Accepted: 8 July 2019; Published: 10 July 2019

Abstract: Turbidity describes the cloudiness, or clarity, of a liquid. It is a principal indicator of water quality, sensitive to any suspended solids present. Prior work has identified the lack of low-cost turbidity monitoring as a significant hurdle to overcome to improve water quality in many domains, especially in the developing world. Low-cost hand-held benchtop meters have been proposed. This work adapts and verifies the technology for continuous monitoring. Lab tests show the low-cost continuous monitor can achieve 1 nephelometric turbidity unit (NTU) accuracy in the range 0–100 NTU and costs approximately 64 USD in components to construct. This level of accuracy yields useful and actionable data about water quality and may be sufficient in certain applications where cost is a primary constraint. A 38-day continuous monitoring trial, including a step change in turbidity, showed promising results with a median error of 0.45 and 1.40 NTU for two different monitors. However, some noise was present in the readings resulting in a standard deviation of 1.90 and 6.55 NTU, respectively. The cause was primarily attributed to ambient light and bubbles in the piping. By controlling these noise sources, we believe the low-cost continuous turbidity monitor could be a useful tool in multiple domains.

Keywords: turbidity; low-cost; continuous water quality monitor; water

---

1. Introduction

The United Nations states that high-quality drinking water is at the core of sustainable development and is critical for socioeconomic development, healthy ecosystems, and for human survival itself. It is vital for reducing the global burden of disease and improving the health, welfare, and productivity of human populations. It is central to the production and preservation of a host of benefits and services for people. Water is also at the heart of adaptation to climate change, serving as the crucial link between the climate system, human society, and the environment [1].

Water quality monitoring is the process by which critical characteristics of water (physical, chemical, biological) are measured. Turbidity is one of the most universal metrics of water quality. It is a measure of the cloudiness (the inverse of clarity) of water. In watersheds, the presence of high turbidity can be indicative of both organic and inorganic materials. In the case of organic materials, high turbidity can indicate problems such as increased algae growth caused by fertilizer run-off. In the case of inorganic materials, high turbidity can indicate problems such as high suspended sediment caused by erosion during a rainstorm or water churn caused by high winds. Turbidity is a non-specific measure and therefore alone cannot identify the root cause of water cloudiness. However, under certain conditions, it can be used to estimate certain quantitative parameters such as stream loading, total suspended solids, and soil loss. There is a variety of published research on the effect of turbidity on different organisms and the implications on human drinking water [2–4].

Therefore, turbidity is a useful measure for many water resource management applications. This monitoring can help inform decisions regarding the allocation of funds and what future actions would be the best for a watershed. Presently, the sensors that are used are expensive, typically costing thousands of dollars. This causes most of the sensors to be owned by companies that communities hire to take samples a small number of times a year. This is far from the best approach as rapid changes in turbidity are indicative of problems. The best time to tackle these problems is right when they occur. With current sampling frequency, these rapid changes are unlikely to be measured and extremely unlikely to have proper actions taken to deal with the root cause of these changes. The key to efficient and proactive water resource management is continuous and accurate monitoring. However, the cost and complexity of deploying such monitoring systems presently limit their use. It is critical that the cost of individual sensors be decreased to make widespread implementations of these monitoring systems feasible. Also, it is critical that the accuracy of these sensors be high enough to provide useful water quality data. Automated continuous sensing would allow the labor cost of water monitoring to decrease substantially as after the initial setup, except for minor ongoing maintenance, the sensors run continuously without human intervention. An automated sensor platform could also be used by people with little, if any, formal training in water monitoring.

Open-source technologies have been identified as the most promising solution to this challenge [5]. As a result, some groups have begun developing their own low-cost monitoring solutions [6–9]. However, these prior works for turbidity monitoring focus on hand-held meters and leave continuous monitoring for future work. Lambrou et al. [10] built a complete continuous monitoring system using off-the-shelf sensors without addressing cost or complexity concerns. Lorena Parra et al. built a low-cost water quality monitoring system for fish farming that contained a simple turbidity sensor [11]. Kofi Sarpong Adu-Manu et al. [12] broadly review methods for water quality monitoring focusing on technologically advanced methods employing wireless sensor networks. In this paper, we present the development of a low-cost continuous turbidity sensor. Our goal is a sensor that could be used in both watershed and drinking water continuous monitoring applications.

2. Related Work

Standard laboratory methods to measure turbidity are well understood and the most commonly used standard is maintained as method 180.1 by the U.S. EPA [13]. This method specifies a tungsten lamp illuminating a sample from not more than 10 cm away with a photo-electric detector oriented 90° from the source. This method is specified from 0–40 nephelometric turbidity units (NTU) with instrument sensitivity of at least 0.02 NTU in water under 1.0 NTU. The NTU units themselves are defined by the response of the nephelometric sensor to known standards. There is no mathematical definition of NTU.

There are at least four other standards for measuring turbidity using nephelometry (ISO 7027, GLI Method 2, Hatch Method 101033, and Standard Methods 2130B) [14]. These variants specify different light sources and detector arrangements. However, none of these standard methods lend themselves to low-cost continuous water quality monitoring. In this work, we follow the general approach of using a light source with a detector located at 90° built using only commonly available electronic components, 3D printable structures, and open-source software with the goal of determining if such a low-cost sensor could be suitable for continuous water quality monitoring applications.

To our knowledge, Christopher Kelly and his team proposed the first low-cost turbidity sensor [8]. This project represents the first publicly available peer-reviewed characterization of an affordable nephelometric turbidimeter. The team set out to create a battery-powered, high accuracy turbidity meter for drinking water monitoring in low-resource communities. This goal required a few design constraints that they set out to meet: run on a single set of batteries for weeks to months of regular use, a high measurement accuracy, and the ability to differentiate small changes in turbidity especially over the range of 0–10 NTU, the sensor must have all of its parts documented and be able to be made by non-experts who want to create their own version of the sensor.

The developed system is a cuvette-based turbidity meter using a single near infrared light emitting diode and a TSL230R light-to-frequency sensor set at 90° apart in a single beam design. This is where there are a single LED emitter and a single receiver perpendicular to the light beam from the LED.

The receiver converts light intensity to a signal that can be read by a microcontroller. The theory behind this design is that the clearer the solution, the more light that makes it straight through the solution. The more turbid the solution, the more light that is reflected perpendicular to the light beam. The meter does not store the data but rather displays it on a LED display for manual recording. Using turbidity standards created using cutting oil and water, the team tested a known turbidity meter next to the created turbidity sensor and measured the readings from both. This data was used to create four calibration curves (each for a different range) that are used to convert the light-to-frequency sensor output from the created turbidity meter to the turbidity reported by the commercial sensor.

The study showed the created turbidity meter had an accuracy within 3% of the commercial sensor or 0.3 NTU whichever is larger over the range of 0.02 NTU to 1100 NTU. They reported that in 8 trials results were within 0.01 NTU for the four turbidity standards under 0.5 NTU. These results support the notion that a low-cost turbidity meter is a possibility; however, more tests to evaluate and verify these results are needed. The proposed next steps as of when the paper was written were to account for thermal fluctuation effects on the turbidity of a solution, minimizing the light leakage into the sensor housing through the external casing, investigating the use of GSM data transmission, and investigating an inline immersible version of the turbidity meter.

Closely related is the optical sensor for sea quality monitoring by Filippo Attivissimo et al. [9]. This sensor is designed to take in situ continuous measurements of chlorophyll fluorescence and turbidity. The sensor has a blue LED, red LED, and photodiode evenly spaced surrounding the water sample. It achieves high sensitivity by using a numeric lock-in amplifier. Preliminary turbidity measurements were made using a solution of milled flour in seawater. Although promising, the results were limited, and the precision of the turbidity measurement was not presented.

Optical fibers can also by employed in the measurement of turbidity and to detect specific organic molecules [15]. Ahmad Fairuz Bin Omar and Mohd Zubir Bin MatJafri present a good overview of the optical properties important for turbidity measurement and a design of an optical fiber turbidimeter [16]. While their design provides laboratory support for the device, the authors state that continued development is needed before it is appropriate for field measurements.

Kevin Murpty et al. also developed a low-cost autonomous optical sensor for water quality monitoring [17]. The device is similar to commercially available turbidity sondes. It contains five color LEDs and a photodiode in a cylindrical sensor body to enable spectral analysis of water quality by submerging the device. Laboratory tests and a field deployment verified the operation of the device. The measurements had high correlation with a commercial turbidity sonde. The system component cost was approximately €650 which is significantly higher than other low-cost systems proposed by others.

3. Appliance Sensors

As a first step to the development of a low-cost continuous turbidity sensor, we evaluated existing commercial low-cost appliance turbidity sensors. These sensors are used in dishwasher and clothes washing machines typically to determine when the contents of the appliance are clean. It was hoped that they would be able to sufficiently determine differences in water clarity to provide useful data for water management applications. Three different turbidity sensors from Amphenol were tested (TST-10, TSD-10, and TSW-10) pictured in Figure 1. All models contain an LED emitter and a phototransistor oriented directly across (180°) from the LED. The output is proportional to the amount of light traveling through the sample and arriving at the phototransistor instead of to the measurement of the scattered light provided by a nephelometric meter. The primary difference between the various models is the mechanical enclosure. The TST-10 is a flow-through design while the TSD-10 is designed to be inserted into the water flow. Either of these could be adapted for continuous monitoring applications.

Figure 1. Amphenol TST-10 (**left**) and TSD-10 (**right**). TSW-10 is similar to the TSD-10 (not pictured) (images from Amphenol).

Each sensor was tested using the reference circuit specified in the datasheet [18–20] shown in Figure 2 and recording the voltage output of the sensor using an Arduino Mega's internal analog to digital converter. The more light that is transmitted through the sample to the receiver the higher the output voltage. This higher voltage means the solution is clearer which is equivalent to saying that it has lower turbidity.

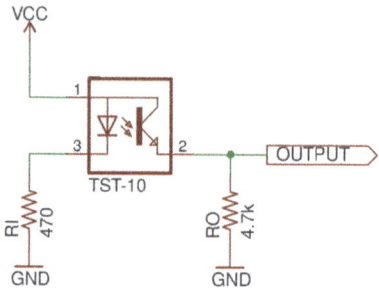

Figure 2. Amphenol (TST-10) appliance turbidity test circuit where VCC = 5 V. Other Amphenol models use the same circuit.

To test the hardware variation between sensors, we created test solutions by adding a small amount of cutting oil to water and tested four appliance sensors of the same model in the same solution. Ideally, the sensors should output the same voltage in the same solution. We performed a simple linear conversion from voltage to approximate NTU using the output curve specified in the data sheet for each sensor. Table 1 shows the observed variation between the sensors in this experiment. The result shows the actual variation is less than the worst-case value calculated from the curve in the data sheet. The TST-10 performed best with 50 NTU difference; however, for most water management applications this variation is far too large to be useful.

Table 1. Variation between the appliance sensors of the same model.

Sensor	Specified Variation (NTU)	Observed Variation (NTU)
TSD-10	305	162
TST-10	325	50
TSW-10	748	348

To improve accuracy, we can individually calibrate each sensor. According to the TST-10 datasheet, the useful range of the sensor is 0–4000 NTU with a voltage differential of 2.7 V. We used tap water (NTU $\approx$ 0) and recorded the sensor's maximum voltage. The minimum voltage is specified at 4000 NTU with output voltage 2.7 V less.

To estimate the sensor's precision, we can use a first-order linear approximation of the output over the full 4000 NTU range of the sensor. Therefore, the maximum resolution of the sensor using the

Arduino's 10-bit analog to digital converter is 7.25 NTU per analog-to-digital converter (ADC) count. As the last bit of ADC output is typically noisy, we expect the best possible result using this approach to be ±7.25 NTU with slightly better results under 1000 NTU and slightly worse results over 1000 NTU due to the non-linear output of the sensor. For most water management applications, ±1 NTU is useful, therefore, we conclude that directly connecting the sensors to the ADC cannot provide the needed resolution for water management applications even without noise or other sources of error.

4. Validation of the Low-Cost Nephelometric Sensor

From our previous experiments with the appliance sensors and the Arduino's ADC, we conclude a nephelometric sensor with higher resolution ADC is necessary to achieve the precision necessary for water management applications. To explore this design space, we first constructed a sample-based sensor similar to the one developed by Kelley et al. [8].

This design overcomes the ADC precision by using a TAOS TSL235R light-to-frequency converter, shown in Figure 3, to measure light intensity rather than providing an analog output. Internally the device has a photodiode sensitive to light in the range 320 nm–1050 nm. The diode current is converted to a square wave with 50% duty cycle where the output frequency is proportional to the light intensity. The range of frequencies that the converter outputs are from 0–800 kHz. Using the Arduino's onboard Timer/Counter and Paul Stoffregen's FreqCount library [21], we can measure the average frequency over a short interval (e.g., 1 s) with very high accuracy and precision. This approach to measuring light intensity results in far greater resolution than what is possible using the Arduino's ADC. As a result, the sensor has a much larger dynamic range yielding higher resolution readings that are no longer strongly limited by the ADC resolution.

Figure 3. TAOS TSL235R light-to-frequency converter.

To evaluate the sensor, we constructed a simple test tube-based design that was 3D printed shown in Figure 4. The test tube holder allowed the 100 mA IR LED and TAOS TSL234R to be mounted securely in both 90° and 180° configurations. The IR LED was driven by an Arduino GPIO pin through a series 1 kΩ resistor. The frequency count was read using FreqCount on an Arduino Mega 2560.

Figure 5 shows the results from several validation tests of the light-to-frequency sensor. These tests begin without a test tube inserted (Air) and three different empty test tubes. Then we test two solutions, distilled water (≈0 NTU) and a 126 NTU calibration solution. The figure shows a box plot of the measured output frequency measurements for each test on the X-axis. Each frequency measurement was generated by averaging 10 samples on the Arduino Mega and was repeated to produce 50–100 data points. From these results, we see little variation between measurements, we can clearly identify empty test tubes with frequency around 52–54 kHz, and an empty test chamber (i.e., no test tube inserted) with frequency around 47 kHz. The median of the 126 NTU calibration solution and distilled water yielded 1329.2 Hz difference. A two-point linear calibration from these values suggests sensing resolution better than 0.1 NTU per Hz may be possible (i.e., 126 NTU/1329.2 Hz = 0.095 NTU/Hz).

Although further tests would need to be performed with more NTU standards to classify the accuracy and ensure a linear response.

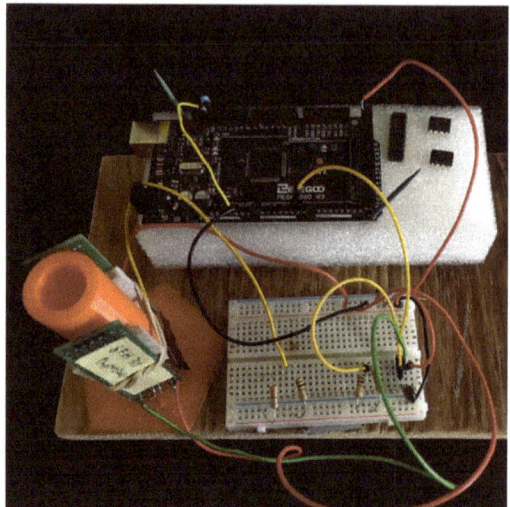

Figure 4. Circular sample holder and test circuit.

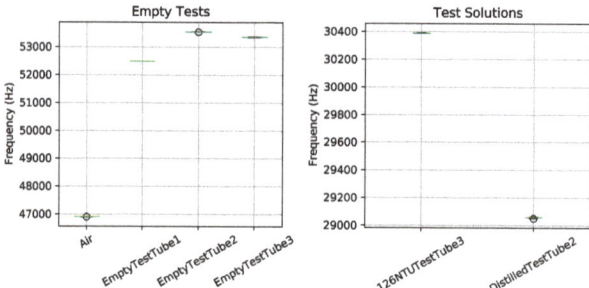

Figure 5. Light-to-frequency initial results using standard test tubes.

However, these results are promising but not as good as those reported by Kelly et al. [8]. We suspect some of the error is due to the large reflections and optical impurities in the test tube. Because of the circular shape of the test tube, it is nearly impossible to keep the IR LED exactly perpendicular to its surface. As a result, we decided to switch to plastic cuvettes as they are similar but lower cost than the quartz cuvettes used by Kelly et al. [8]. Cuvettes have straight sides and are typically used in spectrophotometry where optical clarity is important.

The housing was redesigned to have a square shape with internal walls to block any light from getting to the receiver unless it first went through the sample as shown in Figure 6a. We tested the sample holder with distilled water and a calibration solution. The test solutions were measured with a calibrated Hach 2100P turbidity meter before the experiment and measured 0.39 and 86 NTU, respectively. Figure 6b shows the observed frequency output from the light-to-frequency converter. The median difference between the samples was 581.45 Hz. Assuming a linear response, this would yield a resolution of 0.15 NTU per Hz (85.61 NTU/581.45 Hz). However, there is some overlap in the measured results between the samples and the standard deviation of the samples was large at 142.2 and 254 Hz. Moreover, this resolution is slightly worse than the test-tube-based design.

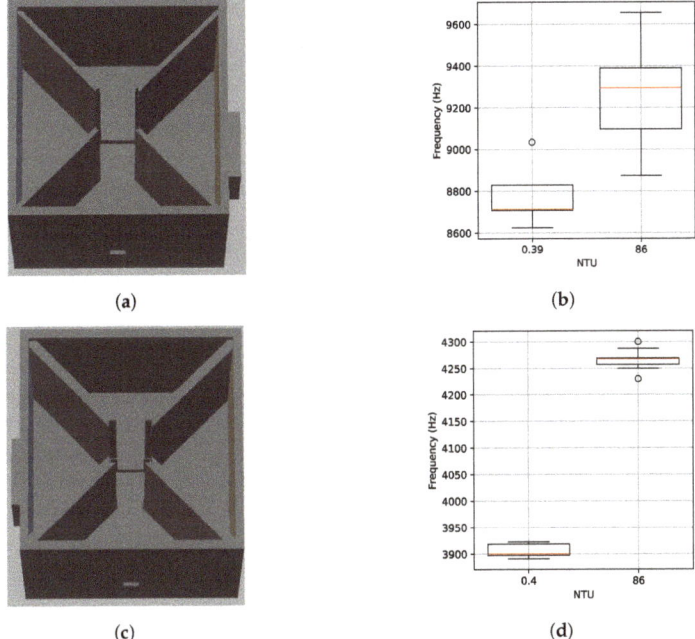

Figure 6. Square cuvette sample holder designs with a light-to-frequency converter at 90° from the IR LED as well as the results from validation tests. (**a**) Initial square design. (**b**) Frequency output for initial square sample holder. (**c**) Revised square sample holder with thicker walls and tighter fit to cuvette. (**d**) Frequency output for revised square sample holder.

After investigation, we found that the cuvettes could rotate slightly in the sample holder and that external ambient light was causing variation in the output frequency. To rectify these problems, we revised the design to have a tighter fit to the cuvette to eliminate rotation and increased the wall thickness to reduce the effect of external light. The revised sample holder is shown in Figure 6c. We repeated the experiment with distilled water and our calibration solution. The results in Figure 6d show a significant reduction in frequency at both readings and significantly reduced variation. This result is consistent with the reduction of external light and more constant cuvette position. Although the average frequency difference was reduced to 367 Hz (0.23 NTU per Hz), the noise was greatly reduced with standard deviation of 12.0 and 19.0 Hz yielding statistically different readings in all cases.

With these results, we conclude that a sample-based low-cost nephelometric turbidity sensor using a light-to-frequency converter can provide the needed resolution needed for many water quality monitoring applications. Additionally, our revised cuvette-based sample holder successfully reduced variation in the readings. With further study and improvement, such as increasing LED brightness, we believe the performance could be improved further. This general design will be used to inform the future development of a low-cost continuous turbidity sensor.

5. Low-Cost Continuous Turbidity Sensing

From our previous experiments, we have validated that a low-cost nephelometric turbidity sensor can meet the requirements (i.e., better than 1 NTU resolution) needed for providing useful data for water quality monitoring applications. To provide continuous turbidity data, we will adapt the basic sensor design for flow-through applications. Many applications, such as drinking water and agriculture use commonly available pipes to transport water, such as PVC. In the U.S., schedule 40 and

80 are common specifications of PVC pipe which are available in a variety of colors and importantly for this application, clear.

Our approach to the continuous low-cost turbidity monitor is to attach an LED and a light sensor on the outside of a clear PVC pipe segment oriented 90° apart in the nephelometric configuration. In the previous tests, the separation between the LED and sensor was proportional to the width of the cuvette, which is 10 mm. In the piped configuration, this distance will be proportional to the pipe size, which could be several inches. Because the LED will be illuminating a much larger volume of water through a much thicker wall, we surmise it is useful to increase the brightness. High-powered IR LEDs (several watts) are not readily available and specialty IR LEDs are expensive. However, high-powered white LEDs are common. As a result, we replaced the IR LED with a commonly available Cree XLamp white LED (4000 K). To properly drive the LED, we use a commonly used constant current LED driver (Diodes Incorporated AL8805) configured to deliver up to 500 mA of current to the LED via a PWM control signal. This allows us to also replace the IR light-to-frequency converter with a low-cost ambient light sensor (TSL4531). These sensors are commonly used to control display brightness and provide a digital i2c output of light intensity as an integer value that is calibrated to lux. To support wireless data collection, we connect the LED driver and light sensor to an ESP32 Wi-Fi-enabled microcontroller. A diagram of the complete low-cost continuous turbidity sensing system is shown in Figure 7.

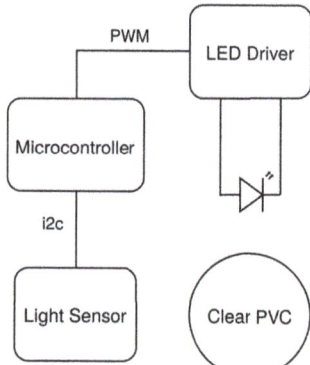

Figure 7. Low-cost continuous turbidity monitoring system diagram.

To provide consistent contact with the PVC pipe, we designed a 3D-printable mounting ring to mechanically fix the LED and sensor to the pipe. Different pipe diameters can be accommodated by adjusting the dimensions of the mounting ring. Figure 8 shows a rendering of (a) our initial design and (b) revised mounting ring. With the initial design, the LED and ambient light detector were mounted to a small PCB and glued to the mounting ring. Because the PCB used through-hole connections, solder joins on the bottom of the PCB caused an uneven fit with the ring. This mechanical ring was also narrow (1 inch) and allowed ambient light to reach the light sensor. As a result, the design was revised to include PCB standoffs, recessed areas to accommodate solder joints, screw holes were added to the ring, and the height of the ring increased to block more ambient light. The mounting ring was sized to tightly fit over a section of 2-inch schedule 40 clear PVC pipe and printed in black ABS on an Ultimaker 2+ 3D printer. Black was selected to minimize reflected light in the device. Although we did not characterize this effect, we tested other colors and found black to have the lowest light level with the LED on. This suggests that reflections are minimized as desired.

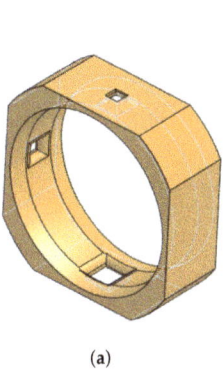 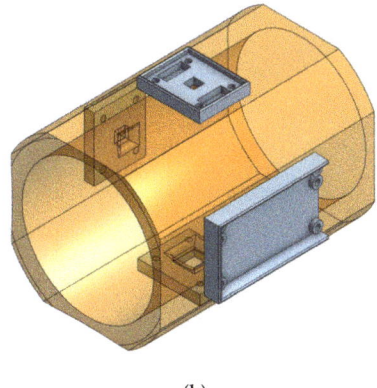

(a) (b)

Figure 8. Nephelometric sensor mounts for clear pipes. (**a**) Initial design. (**b**) Revised design including PCB mounts and reduced ambient light.

The approximate unit component cost of the completed sensor is 64 USD. The component costs are shown in Table 2. This makes the low-cost continuous turbidity monitor more expensive than the affordable open-source turbidimeter ($25–$35) by Kelley et al. [8]. However, much of the cost difference is due to the WiFi microcontroller and printed circuit boards which add significant value by providing wireless communication and improved reliability over a breadboard circuit. This also compares favorably to commercial turbidity probes for water quality from manufacturers such as YSI, In situ, and Eureka where prices range from $1000 to $5000, depending on the specific model. Even compared to the low-cost autonomous optical sensor by Murphy et al. at €650 [17], we see a significant reduction in cost.

Table 2. Unit component cost of the prototype sensor.

Component	Approximate Cost (USD)
ESP32 microcontroller	20
TSL4531 light sensor	1
XLamp MX-6 LED (2)	2
Printed circuit boards	10
PVC pipe 2 in × 12 in	16
Black ABS filament (83 g)	5
Miscellaneous components	10
Total	64

5.1. Laboratory Calibration

Four low-cost continuous turbidity monitors were constructed and tested over the range of 0–100 NTU to explore the variation that exists in the different devices made from the same components. The devices are labeled with the last two digits of their ESP32 WiFi MAC address. For calibration, the devices were oriented vertically over a short section of clear PVC pipe with silicone caulk securing the mounting ring to the pipe and a Qwik Cap sealing the bottom as shown in Figure 9. Test solutions were added to fill the PVC pipe and a cover was placed over the top to block ambient light.

Figure 9. Sensor 18 configured for laboratory calibration.

The test solutions were created by diluting a 4000 NTU formazin standard with deionized water ($\approx$0.20 NTU) to produce solutions with values of (0.20, 5, 20, 40, and 100 NTU) [22]. The test solutions were made, and the devices were filled with deionized water and allowed to collect data for 12 h. This allowed the components and test solutions to reach a constant temperature and any air bubbles to dissipate. The devices were rinsed thoroughly with deionized water between different samples to clean any residual sample out of the pipe. Each of the samples was tested in each device for at least 10 min where the light intensity at 90-degrees, 180-degrees and the dark reading (90-degree sensor reading without the LED on) was measured every 6 s during the sampling interval. If more than 100 samples were collected for a given test solution, only the middle 100 samples were used in the analysis. Manual turbidity readings were also made of the test sample every 2 min using a Hach 2100P turbidity meter. This was done as a single reading from the meter can vary. We do not believe that the standards degraded during the testing.

After the laboratory sampling was complete, the data from each device was fit to a model of the form:

$$NTU = c_1 \times d_0 + c_2 \times d_{90} + c_3 \times d_{180} + \epsilon \tag{1}$$

where d_0 is the light intensity at 90 degrees from the LED with the LED off in lux, d_{90} is the light intensity with the LED on at 90 degrees from the LED in lux, d_{180} is the light intensity with the LED on at 180 degrees from the LED in lux, and ϵ is the y-intercept. These values were computed using ordinary least squares linear regression comparing the predicted NTU to the average manual NTU reading of the sample. Models were generated for each device individually in addition to a combined model using data from all the devices. To explore the impact of each sensor (90 degrees and 180 degrees from the LED), models were generated with each sensor individually as well as both of the sensors. Table 3 shows all computed model parameters, the R^2 measure, and variance (σ^2) of the residual.

Table 3. Individual and combined model parameters, R^2, and residual variance (σ^2) for using the 90- and 180-degree sensors, only the 90 degree, and only the 180 degree sensor respectively. The units of c_1, c_2 and c_3 are NTU/lux. The unit of ϵ and σ^2 are NTU and NTU^2 respectively.

Device	Sensor (s)	c_1	c_2	c_3	ϵ	R^2	σ^2
Device 18	d_{90}, d_{180}	−0.2956	0.1608	−0.0046	41.2997	1.0000	0.0093
Device 18	d_{90}	−0.2970	0.2088	0.0000	−16.0266	0.9999	0.0316
Device 18	d_{180}	−0.3372	0.0000	−0.0200	232.7049	0.9989	0.2582
Device 8C	d_{90}, d_{180}	0.1112	0.2313	0.0018	−38.2779	0.9998	0.2446
Device 8C	d_{90}	0.1055	0.2136	0.0000	−15.4716	0.9998	0.2570
Device 8C	d_{180}	0.3477	0.0000	−0.0212	259.3874	0.9984	2.3556
Device 94	d_{90}, d_{180}	−0.0995	0.3770	0.0117	−174.0959	0.9996	0.5432
Device 94	d_{90}	−1.1277	0.2398	0.0000	−17.6853	0.9993	1.0413
Device 94	d_{180}	−2.5629	0.0000	−0.0204	254.4228	0.9972	4.2926
Device B8	d_{90}, d_{180}	−0.5757	0.3274	0.0059	−100.8465	0.9999	0.1515
Device B8	d_{90}	−0.9409	0.2537	0.0000	−21.2899	0.9997	0.4269
Device B8	d_{180}	−1.9002	0.0000	−0.0201	251.4739	0.9957	5.5707
Combined	d_{90}, d_{180}	1.6055	0.2252	−0.0003	−13.6549	0.9934	8.0515
Combined	d_{90}	1.6541	0.2286	0.0000	−17.8185	0.9934	8.0627
Combined	d_{180}	1.2328	0.0000	−0.0202	246.3589	0.9558	53.9348

The value of c_2 is expected to be positive as more light reflected at 90 degrees would be more indicative of a turbid solution and all the models follow this. The value of c_3 is expected to be negative as the more turbid the solution, the less light that would pass straight through it to reach the d_{180} sensor. Device 8C, 94, and B8's models including both sensors (d_{90}, d_{180}) do not follow this expectation and have a reduced magnitude, suggesting the d_{180} sensor provides inconsistent data in this NTU range.

From these results, we see that in general, the computed model fits the data well for the d_{90} and d_{180} as well as the d_{90} only device-specific models. The variance using both sensors was slightly smaller, suggesting that the d_{180} sensor does provide some information when used with the d_{90} sensor. The d_{180} only models have significantly larger variance when compared to the other models. The combined model also has variance that is more than an order of magnitude larger, even with both sensors, indicating there is variation between devices. This variation could be caused by the mechanical assembly construction, different brightness of the LED resulting from manufacturing differences in the LED itself or LED driver circuit, and manufacturing differences that impact the clarity of the PVC pipe. The idea of device-specific calibration is not unique to low-cost sensing. It is widely used in many manufacturing processes to compensate for errors caused by real-world manufacturing constraints without having to determine each source of variance. As a result, we omit the combined model from further analysis.

To visually explore the results, Figure 10 shows several plots for the predicted NTU vs. measured NTU using the low-cost turbidity monitor. Each row presents results from a specific device while each column shows increasing NTU ranges. Missing plots result from not testing every sample on every device. The results with only the 180-degree sensor are omitted for clarity as this case performed significantly worse than the others. If the model were to produce the exact same value as the measured readings, the points would fall on the dotted diagonal line. The further the points fall from the diagonal line the larger the error of the prediction. For example, the B8 device model can predict the NTU readings for all the samples within about ±1.0 NTU. The B8 model overpredicts the NTU for the 38 to 40 NTU sample and underpredicts the NTU for the 20 to 22 NTU sample. The diagonal line bisects the plotted points for the 0 to 1 NTU sample and the 95 to 100 NTU standard, suggesting that the model, on average, predicts these cases well.

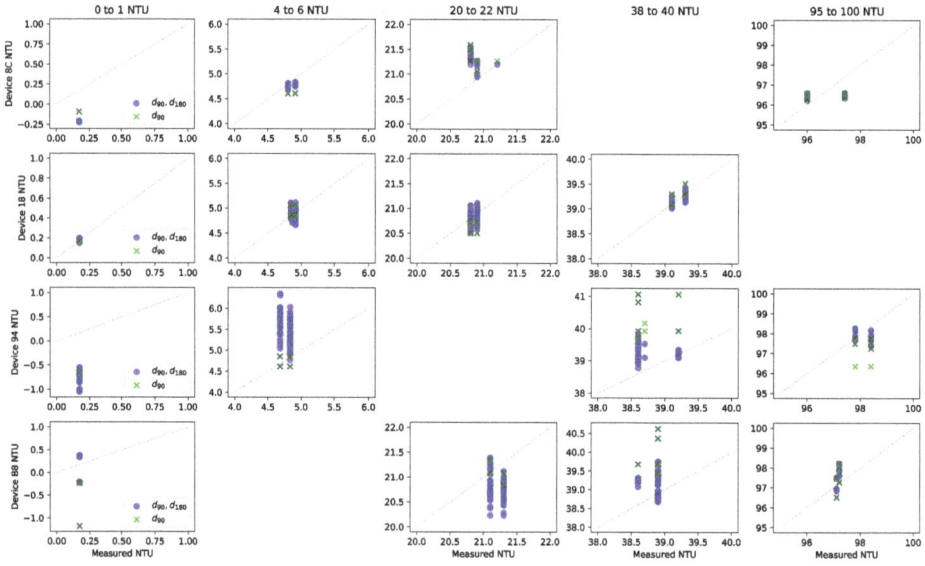

Figure 10. Predicted NTU vs. measured NTU for the individual models on the 5 tested NTU ranges.

To better understand these results, Figure 11 shows the cumulative distribution function (CDF) of the absolute value of the residual using the computed models. The residual is the difference between the estimated value and measured value computed individually for both sensor configurations. The range is limited to $(0, 1)$ for clarity. Residuals from all device-specific models are combined in the final "All Devices" plot. The d_{180} model is not shown because it performed significantly worse than the others. We see the devices perform similarly with most of the residuals less than 1 NTU. Device 18 does the best and 94 does the worst while 8C and B8 are in the middle. For all devices, with both sensors, the median residual was -0.0032 NTU with a standard deviation of 0.4870 NTU.

Figure 12 shows the CDF of the combined absolute value of the residuals from all device-specific models at each tested NTU range. This confirms that the device can be used over this whole range with consistent performance. However, this uniformity is mostly a result of performing linear regression with an equal number of data points (100) in each range. By oversampling any particular NTU range, the generated model would produce a better fit in that range at the expense of performance in the other ranges. This could be useful in certain applications to better detect small variations from an expected turbidity value.

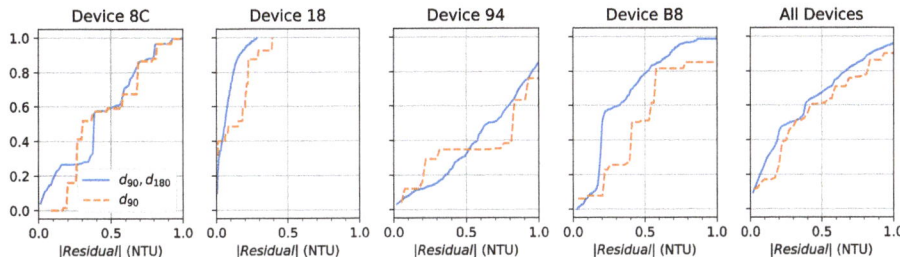

Figure 11. Cumulative distribution function of the absolute value of the residual by device.

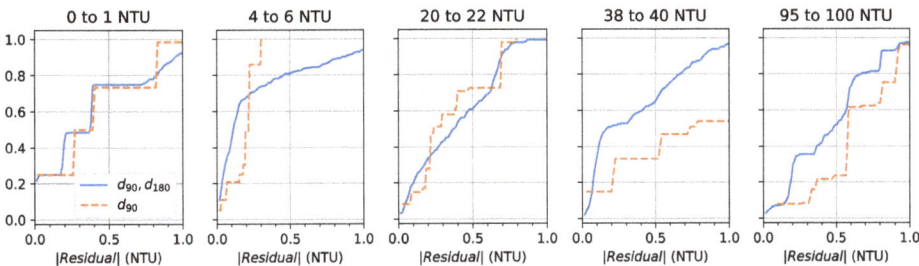

Figure 12. Cumulative distribution function of the absolute value of the residual by NTU range.

These results show that with device-specific calibration the device will achieve accuracy closer to 1.0 NTU than our goal of 0.10 NTU over the range of 0–100 NTU. However, since the median residual was near zero, averaging multiple samples could reduce noise to approach this goal. This dataset did not have enough samples to fully investigate this question, so we will explore this in the next section.

5.2. Pumped Tank Test

Having calibrated and explored the performance of the low-cost continuous turbidity monitor in a laboratory setting, we now move to a simulated real-world test. For this test, we used a 1000-gallon water tank and a 1000 GPH pool pump to circulate the water. To explore if the device should be on the pump inlet or outlet, we installed a device on both. Device B8 was installed on the pump inlet and device 94 was installed on the pump outlet. Figure 13 shows a diagram of the pumped tank test. The sensors were installed using three-foot pipe segments between the tank and the pump.

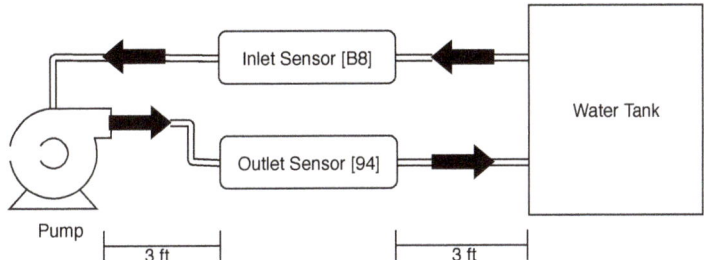

Figure 13. Diagram of pumped tank test.

The tank was filled with fresh drinking-quality water and manual turbidity measurements were made daily with the Hatch 2100P turbidimeter. These measurements were linearly interpolated between samples to produce a continuous turbidity value in the tank for analysis. The low-cost continuous turbidity monitor readings were made once every 6 s. Timestamps for each sample were recorded by the device and the clock was synchronized with a public NTP server at the start of the experiment. The timestamp and raw sensor values were then transmitted over a WiFi network to a database for storage. For analysis, the raw sensor values were linearly interpolated to a constant 1 Hz rate and a 20-min moving average of 1200 samples at 1 Hz containing about 200 raw samples was computed over 5-min periods, resulting in 288 samples per day. We chose these values to reduce the amount of data as we expect turbidity to change relatively slowly and simultaneously reduce sensor noise by averaging multiple readings. Experimentally we found that averaging over 1-min periods (10 raw samples) was sufficient to eliminate most of the sensor noise but we elected to use longer periods in our analysis to produce the desired sample rate.

The filtered sensor readings were then used in the device-specific laboratory models (Table 3) to estimate the NTU reading in the tank. A small offset was present at installation, so we adjusted each device's ϵ parameter after making the first manual reading to remove this error. Shortly after the installation, device 94 failed and the LED and light sensor was replaced with the components from device 18. Data is reported as device 94; however, the model generated by device 18 is used to predict NTU. Figure 14 shows results for 38 days of measurements. During two intervals between days 5 and 7, the data collection failed, and no samples were recorded. For analysis, the missing data were linearly interpolated between the available samples.

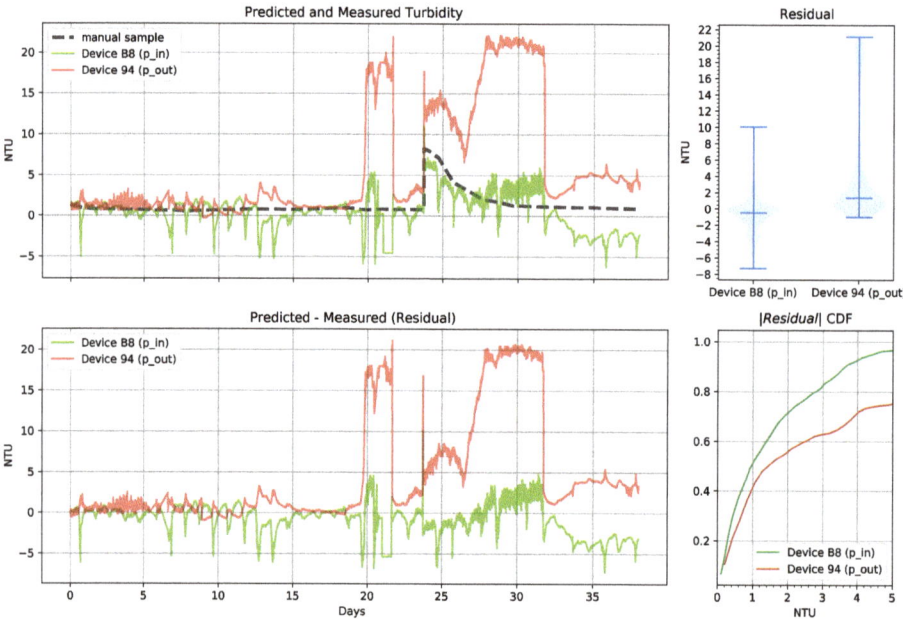

Figure 14. Pumped tank low-cost continuous turbidity monitor predictions from the pump inlet (p_in) and outlet (p_out) and manual measurements from a hand-held turbidimeter showing all collected data (days 0 through 37).

Initially, through day 5, the sensor on the outlet had significant noise. On day 6 we discovered that bubbles were present in the pipes near the outlet sensor and we purged the air from the pipes. On day 9 we discovered that a small hole was allowing air into the pipes. We sealed the hole and both sensors showed significantly reduced noise after this. In sealing the hole, we repositioned the outlet sensor, which caused an offset in the readings. At day 20 there was another air leak that caused a significant error and was sealed by day 22.

To investigate the response of the sensor to changing turbidity, we continued our measurements and added one quarter cup of Coffee Mate® powdered coffee creamer to the 1000-gallon tank on day 23 at the pump inlet. This quickly increased the tank turbidity to about 8 NTU. The inlet sensor closely tracked this change demonstrating the impulse response of the sensor. On day 27 both sensors NTU reading begin increasing and we discovered the patch to the pipe had failed. We let this continue until day 32 when it was patched again.

The median residual and standard deviation for the inlet and outlet low-cost continuous turbidity monitors over the entire test were −0.4507, 1.9063 NTU and 1.3997, 6.5511 NTU, respectively. The cumulative distribution function of the absolute value of the residual shows that overall only slightly more than 50% of the predictions are within 1 NTU as expected from the laboratory tests even with the better performing inlet monitor. However, the large errors resulted from air in the pipes. When no air was present on days 15 through 17, the median residual and standard deviation was −0.1881, 0.2643 NTU and 0.2266, 0.1138 NTU respectively and all of the predictions were within 1 NTU with 80% of them being within 0.5 NTU.

Comparing the inlet and outlet monitors in the presence of bubbles on days 20 and 27, we see the inlet is much less sensitive to the presence of bubbles. We speculate that by going through the pump, the relatively large bubbles passing through the inlet monitor were broken up into many more small bubbles before passing through the outlet monitor. This resulted in a correspondingly larger estimated turbidity on the outlet monitor.

Both sensors showed patterns of daily periodic errors. These patterns are more apparent when examining the relatively stable period between days 15 through 17, shown in Figure 15. On this plot we added the measured solar radiation (W/m^2) from a nearby weather station for reference. The inlet sensor has a more negative residual during the day while the outlet sensor has a more positive residual and the Pearson correlation coefficients between the residual and the solar radiation was −0.58 and 0.17 respectively. To explain the difference in sign, we reexamine the model parameters from Table 3 and recall that the parameter c_3 has the opposite sign between these two devices. Furthermore, the absolute value of the parameter is larger for the inlet device. This parameter corresponds to the 180 degree sensor. As a result, we conclude that this phenomenon is almost certainly caused by ambient light. While both sensors and devices were affected, the 180 degree sensor on device B8 was the most sensitive to ambient light. The position of the devices in the test also contributed to these differences. Because our laboratory experiments were taken in relatively dark conditions, the model did not properly account for the influence of ambient light even with the dark term in the model.

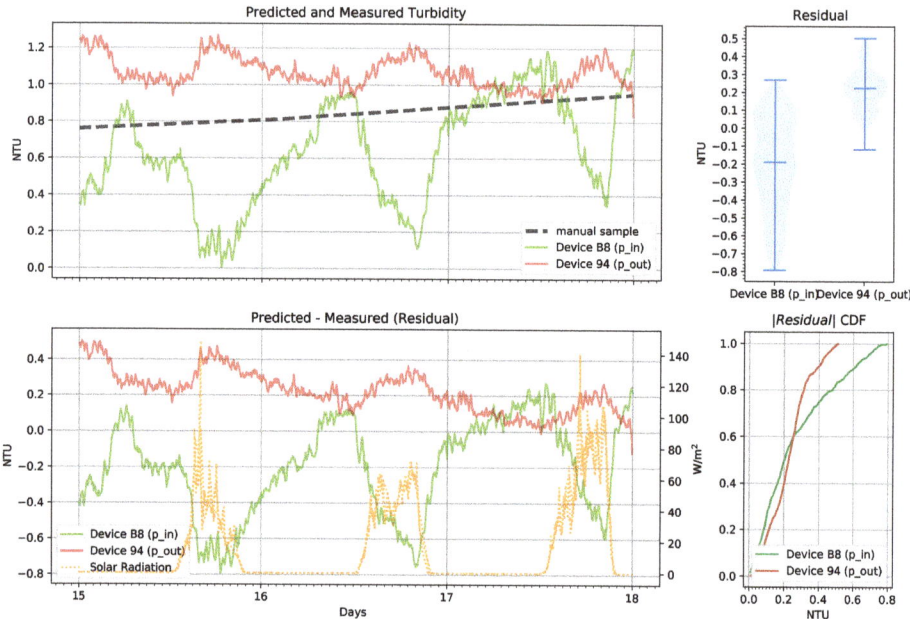

Figure 15. Pumped tank low-cost continuous turbidity monitor predictions from the pump inlet (p_in) and outlet (p_out) and manual measurements from a hand-held turbidimeter showing days 15 through 17.

5.3. Discussion

In this section, we describe the creation of a low-cost continuous nephelometric turbidity monitor built using commonly available components. The turbidity monitor is designed to fit over a short section of clear PVC pipe. This approach reduces the mechanical complexity of the system since no sensing components are ever in direct contact with water.

Laboratory experiments demonstrated the model was able to reliably estimate turbidity within 1 NTU with some noise present in the readings. Since the median residual was near 0, averaging multiple readings could approach our goal of 0.1 NTU resolution under well-controlled conditions.

The pumped tank test demonstrated that the sensor can continuously predict turbidity installed either on the inlet and outlet of a pump. However, the inlet sensor had better impulse response to a turbidity change. The inlet sensor showed more interference from ambient light, but we attribute this to sensor positioning and not an artifact of the pump position. The outlet of the pump was more affected by bubbles, this was attributed to the fact the pump formed some bubbles during operation. These bubbles likely dissipated when they reached the tank resulting in the inlet sensor not seeing the same level of bubbles. Overall, neither sensor achieved the same accuracy as in the lab experiment, even with averaging many sensor readings. However, on days 15–17 when no air was present, the performance of both sensors was equal to the lab experiments. By eliminating ambient light and bubbles we believe this level of performance can be maintained over longer periods. Furthermore, even at the current level of performance, many applications could benefit from low-cost continuous turbidity monitoring by detecting larger changes in turbidity (e.g., >1 NTU). Results from the last days of the experiment showed a significant offset was present suggesting that periodic calibration may be required. We plan to explore the long-term stability of the low-cost continuous turbidity monitor in future work.

6. Conclusions and Future Work

In this paper, we explored the development of a low-cost continuous turbidity monitor. We began by evaluating readily available appliance turbidity sensors. While inexpensive, in our tests they do not have the required accuracy for water quality monitoring applications. They were also prone to a large amount of noise and are difficult to precisely calibrate. Examining prior work on low-cost turbidity sensors, we verified that accurate low-cost sample-based turbidity sensors can be constructed. In our tests, the main source of error was the imprecision of the sample holder (Cuvette or Test Tube) in the sensor apparatus. Using this design as a starting point, we adapted the sensor for use in piped-water applications. Lab tests verified that with individual calibration, accuracy better than 1 NTU is possible over the range 0–100 NTU. A 38-day long experiment was performed with the low-cost continuous turbidity monitor in a piped-water application. The monitor showed more error than in the lab experiments, yielding $\approx$5 NTU accuracy and good response to changes in turbidity. The primary source of error was attributed to bubbles in the liquid and ambient light. This may be sufficient for some continuous monitoring applications. For other applications where higher accuracy is needed, we believe that by reducing ambient light on the sensor and eliminating all air from in the pipes will yield accuracy better than 1 NTU. Like all other turbidity sensors, periodic calibration is necessary to maintain the accuracy of the low-cost continuous turbidity monitor.

As we found that device-specific calibration significantly improves performance, a simpler way to calibrate the sensor is recommended as lab-made turbidity standards are not commonly available by citizen scientists. There are other processed liquids that have consistent turbidity such as apple juice and tea which could be used for calibration. A validated procedure to calibrate the sensor with these liquids could be developed. We also plan longer trials to verify the long-term behavior of the low-cost continuous turbidity monitor. One long-term concern is if and when to remove and clean the clear PVC section. Since PVC can develop a static charge, contaminants may be attracted to the clear pipe segment. It is not clear if the pumped liquid is sufficient to remove these contaminants. We plan to redesign the mounting ring to simplify removal for inspection and cleaning.

Author Contributions: Conceptualization, Methodology, Software, Formal Analysis, Investigation, Data Curation, Writing—Original Draft Preparation, Writing—Review & Editing, Visualization, A.M. and D.G.; Validation, D.G.; Resources, Supervision, Project Administration, Funding Acquisition, A.M.

Funding: This research received no external funding.

Conflicts of Interest: The authors declare no conflict of interest.

References

1. United Nations Educational, Scientific and Cultural Organization (UNESCO). Water in the post-2015 development agenda and sustainable development goals. In *International Hydrological Programme (IHP)*; UNESCO: Paris, France, 2014.
2. Duan, W.; Takara, K.; He, B.; Luo, P.; Nover, D.; Yamashiki, Y. Spatial and temporal trends in estimates of nutrient and suspended sediment loads in the Ishikari River, Japan, 1985 to 2010. *Sci. Total Environ.* **2013**, *461–462*, 499–508. [CrossRef] [PubMed]
3. Duan, W.; He, B.; Nover, D.; Yang, G.; Chen, W.; Meng, H.; Zou, S.; Liu, C. Water Quality Assessment and Pollution Source Identification of the Eastern Poyang Lake Basin Using Multivariate Statistical Methods. *Sustainability* **2016**, *8*, 133. [CrossRef]
4. Duan, W.; He, B.; Chen, Y.; Zou, S.; Wang, Y.; Nover, D.; Chen, W.; Yang, G. Identification of long-term trends and seasonality in high-frequency water quality data from the Yangtze River basin, China. *PLoS ONE* **2018**, *13*, 1–18. [CrossRef] [PubMed]
5. Burke, D.G.; Allenby, J. Low Cost Water Quality Monitoring Needs Assessment. Available online: https://chesapeakeconservancy.org/what-we-do/innovate/water-quality-monitoring/low-cost-water-quality-monitoring/ (accessed on 1 February 2019).

6. Metzger, M.; Konrad, A.; Blendinger, F.; Modler, A.; Meixner, A.; Bucher, V.; Brecht, M. Low-Cost GRIN-Lens-Based Nephelometric Turbidity Sensing in the Range of 0.1–1000 NTU. *Sensors* **2018**, *18*, 1115. [CrossRef] [PubMed]
7. Hicks, S.; Aufdenkampe, A.; Montgomery, D. Creative Uses of Custom Electronics for Environmental Monitoring at the Christina River Basin CZO. In Proceedings of the American Geophysical Union Annual Fall Meeting, San Francisco, CA, USA, 3–7 December 2012.
8. Kelley, C.; Krolick, A.; Brunner, L.; Burklund, A.; Kahn, D.; Ball, W.; Weber-Shirk, M. An Affordable Open-Source Turbidimeter. *Sensors* **2014**, *14*, 7142–7155. [CrossRef]
9. Attivissimo, F.; Carducci, C.G.C.; Lanzolla, A.M.L.; Massaro, A.; Vadrucci, M.R. A Portable Optical Sensor for Sea Quality Monitoring. *IEEE Sens. J.* **2015**, *15*, 146–153. [CrossRef]
10. Lambrou, T.P.; Panayiotou, C.G.; Anastasiou, C.C. A low-cost system for real time monitoring and assessment of potable water quality at consumer sites. In Proceedings of the SENSORS, Taipei, Taiwan, 28–31 October 2012; pp. 1–4. [CrossRef]
11. Parra, L.; Sendra, S.; García, L.; Lloret, J. Design and Deployment of Low-Cost Sensors for Monitoring the Water Quality and Fish Behavior in Aquaculture Tanks during the Feeding Process. *Sensors* **2018**, *18*, 750. [CrossRef] [PubMed]
12. Adu-Manu, K.S.; Tapparello, C.; Heinzelman, W.; Katsriku, F.A.; Abdulai, J.D. Water Quality Monitoring Using Wireless Sensor Networks: Current Trends and Future Research Directions. *ACM Trans. Sens. Netw.* **2017**, *13*, 4:1–4:41. [CrossRef]
13. O'Dell, J.W. *Method 180.1 Determination of Turbidity by Nephelometry*; Environmental Monitoring Systems Laboratory Office of Research and Development U.S. Environmental Protection Agency: Cincinnati, OH, USA, 1993.
14. Fondriest Environmental, Inc. Fundamentals of Environmental Measurements: Measuring Turbidity, TSS, and Water Clarity. 2014. Available online: https://www.fondriest.com/environmental-measurements/measurements/measuring-water-quality/turbidity-sensors-meters-and-methods/ (accessed on 1 May 2018).
15. Mizaikoff, B. Infrared optical sensors for water quality monitoring. *Water Sci. Technol.* **2003**, *47*, 35–42. [CrossRef] [PubMed]
16. Bin Omar, A.; Bin MatJafri, M. Turbidimeter Design and Analysis: A Review on Optical Fiber Sensors for the Measurement of Water Turbidity. *Sensors* **2009**, *9*, 8311–8335. [CrossRef] [PubMed]
17. Murphy, K.; Heery, B.; Sullivan, T.; Zhang, D.; Paludetti, L.; Lau, K.T.; Diamond, D.; Costa, E.; O'Connor, N.; Regan, F. A low-cost autonomous optical sensor for water quality monitoring. *Talanta* **2015**, *132*, 520–527. [CrossRef]
18. Amphenol. *TST-10 Turbidity Sensor Datasheet*; Amphenol Thermometrics, Inc.: St. Marys, PA, USA, 2014.
19. Amphenol. *TSD-10 Turbidity Sensor Datasheet*; Amphenol Thermometrics, Inc.: St. Marys, PA, USA, 2014.
20. Amphenol. *TSW-10 Turbidity Sensor Datasheet*; Amphenol Thermometrics, Inc.: St. Marys, PA, USA, 2014.
21. Stoffregen, P. FreqCount Library. Available online: https://github.com/PaulStoffregen/FreqCount (accessed on 1 May 2018).
22. Sadar, M. *Turbidity Standards*; Technical Information Series—Booklet No. 12; Hach Company: Loveland, CO, USA, 2003.

© 2019 by the authors. Licensee MDPI, Basel, Switzerland. This article is an open access article distributed under the terms and conditions of the Creative Commons Attribution (CC BY) license (http://creativecommons.org/licenses/by/4.0/).

Article

Laboratory Calibration and Field Validation of Soil Water Content and Salinity Measurements Using the 5TE Sensor

Nessrine Zemni [1,2,*], Fethi Bouksila [1], Magnus Persson [3], Fairouz Slama [2], Ronny Berndtsson [3,4] and Rachida Bouhlila [2]

1. National Institute for Research in Rural Engineering, Water, and Forestry, Box 10, Ariana 2080, Tunisia; bouksila.fethi@iresa.agrinet.tn
2. Laboratory of Modelling in Hydraulics and Environment, National Engineering School of Tunis, University of Tunis El Manar (ENIT), Box 37, Le Belvédère Tunis 1002, Tunisia; fairouz.slama@enit.utm.tn (F.S.); rachida.bouhlila@enit.utm.tn (R.B.)
3. Department of Water Resources Engineering, Lund University, Box 118, SE-221 00 Lund, Sweden; magnus.persson@tvrl.lth.se (M.P.); ronny.berndtsson@tvrl.lth.se (R.B.)
4. Centre for Middle Eastern Studies, Lund University, Box 201, SE-221 00 Lund, Sweden
* Correspondence: nessrine.zemni@enit.utm.tn; Tel.: +216-28-083-156

Received: 8 October 2019; Accepted: 27 November 2019; Published: 29 November 2019

Abstract: Capacitance sensors are widely used in agriculture for irrigation and soil management purposes. However, their use under saline conditions is a major challenge, especially for sensors operating with low frequency. Their dielectric readings are often biased by high soil electrical conductivity. New calculation approaches for soil water content (θ) and pore water electrical conductivity (ECp), in which apparent soil electrical conductivity (ECa) is included, have been suggested in recent research. However, these methods have neither been tested with low-cost capacitance probes such as the 5TE (70 MHz, Decagon Devices, Pullman, WA, USA) nor for field conditions. Thus, it is important to determine the performance of these approaches and to test the application range using the 5TE sensor for irrigated soils. For this purpose, sandy soil was collected from the Jemna oasis in southern Tunisia and four 5TE sensors were installed in the field at four soil depths. Measurements of apparent dielectric permittivity (Ka), ECa, and soil temperature were taken under different electrical conductivity of soil moisture solutions. Results show that, under field conditions, 5TE accuracy for θ estimation increased when considering the ECa effect. Field calibrated models gave better θ estimation (root mean square error (RMSE) = 0.03 m^3 m^{-3}) as compared to laboratory experiments (RMSE = 0.06 m^3 m^{-3}). For ECp prediction, two corrections of the Hilhorst model were investigated. The first approach, which considers the ECa effect on K′ reading, failed to improve the Hilhorst model for ECp > 3 dS m^{-1} for both laboratory and field conditions. However, the second approach, which considers the effect of ECa on the soil parameter K_0, increased the performance of the Hilhorst model and gave accurate measurements of ECp using the 5TE sensor for irrigated soil.

Keywords: soil salinity; soil water content; FDR sensor; soil pore water electrical conductivity; sensor calibration and validation; real time monitoring

1. Introduction

In arid and semiarid countries, such as Tunisia, irrigation is necessary for improved agricultural production. Water resources with good quality are limited, resulting in the use of low-quality irrigation water. This can induce soil salinization, leading to crop yield reduction, decreasing the agricultural productivity, and causing general income loss [1,2]. Thus, accurate monitoring of soil salinity in time and space is of great importance for precision irrigation scheduling to save water and avoid soil degradation. Over the last decades, soil dielectric sensors have been developed to measure apparent electrical conductivity (ECa) from which real soil salinity, the soil pore electrical conductivity (ECp), can be estimated [3]. Time domain reflectometry (TDR) has been established as the most accurate dielectric technique to estimate both volumetric water content (θ) and ECp in soils providing automatic, simultaneous, and continuous readings [4]. The efficiency of the TDR method has led to development of other techniques based on similar principles, such as capacitance methods. Some examples are the WET (Delta-T Devices Ltd., Cambridge, UK) and the 5TE (Decagon Devices Inc., Pullman, WA, USA) sensors, both based on frequency domain reflectometry (FDR). Compared to TDR, FDR sensors use a fixed frequency wave instead of a broad-band signal that makes them cheaper and smaller [5]. Dielectric methods are based on determination of apparent soil electrical conductivity (ECa) and soil apparent dielectric permittivity (Ka) [6]. Many models for the relationships between Ka and θ [4,7], ECa-θ, and ECa-ECp-Ka have been proposed in recent research [3,8–10]. However, dielectric properties are affected by physical and chemical soil properties. For example, high ECa affects the wave propagation, leading to errors in the estimation of Ka [11,12]. Thus, it is important to improve θ and ECp prediction models.

Hilhorst [8] presented a theoretical model describing a linear relationship between ECa and Ka to predict ECp. This linear model can be used in a wide range of soil types without soil-specific calibration. Persson [13] evaluated the Hilhorst model using TDR in three sandy soils and confirmed the accuracy of the linear model with significant dependency on soil type. Many researchers [14–17] have tested the Hilhorst model using the WET sensor and showed that it can be improved with soil specific calibration. Using the WET sensor, improved correction of the Hilhorst model was proposed by Bouksila et al. [18], using loamy sand soil with about 65% gypsum. They found that the accuracy of ECp prediction is very poor when using standard soil parameters (K_0). Thus, they proposed a correction by introducing a third-order polynomial fitted to the K_0–ECa relationship instead of using the default K_0. Kargas et al. [6] introduced a linear permittivity corrected model, proposed by Robinson et al. [5], in the Hilhorst relationship. They found that the correction depends on soil characteristics and that it is valid for ECa close to 2 dS m^{-1}. These approaches consider the ECa effect on the prediction of ECp. However, research has not been performed using simultaneous controlled laboratory and field-scale experiments where effects of heterogeneity, root density, insect burrowing, etc., affect the observations [19]. Ideally, sensor calibration should be performed in structured soils due to its importance for pore size distribution and associated matrix potential [20]. Research has shown that calibration in repacked soil columns differs from calibration in disturbed soil used in laboratory experiments [21]. In addition, intrinsic soil factors such as soil temperature, presence of gravel, and microorganisms affect the soil structure and porosity contributing to the variability in ECa and Ka measurements under field conditions as compared to measurements in the laboratory [19].

Nowadays, farmers are embracing precision agriculture using sensors with high accuracy and low cost to increase yields and maintain the sustainability of irrigated land. The 5TE dielectric soil sensor, which also uses the Hilhorst model for ECp estimation, was introduced in 2007 and it is much cheaper than the WET sensor [22]. Several recent studies have investigated the 5TE probe in agricultural applications [2,23,24]. The 5TE sensor has electrodes at the end of the probe that are influenced by soil density making them sensitive to any variation in soil structure and θ content [25]. Despite this fact, most studies on the 5TE sensor performance [16,26,27] have been carried out under laboratory conditions. Thus, almost no research has been done in the field for testing its performance for ECp estimation, neither with the most used linear Hilhorst model nor with the more recent ECp approach proposed in literature. Another important practical aspect is to determine the application range of these sensors for irrigated soils under saline conditions. For example, it is important to determine at what ECa threshold the dielectric losses are no longer negligible and need to be corrected for. Furthermore, there is a lack of understanding of how laboratory calibration can be translated into field conditions.

Thus, the sensors must be calibrated and validated under both conditions in order to assess the errors associated with translating one to the other [28].

In view of the above, the objective of the present study was to assess the performance of the 5TE sensor to estimate soil water content and soil pore electrical conductivity for a representative sandy soil used for cultivation of date palms. Both standard models and a novel approach using corrected models to compensate for high electrical conductivity were used. Results from both field and laboratory experiments were compared. The location of the field experiments was the Jemna oasis, southern Tunisia.

2. Materials and Methods

Soil parameter acronyms, data source, sensor specification and models used in the present work were presented in Appendix A.

2.1. Theoretical Considerations

Any porous medium, such as soils, can be characterized by its permittivity, which is a complex quantity (K) composed of a real part (K') describing energy storage, and an imaginary part (K") describing energy loss:

$$K = K' - j K'' \quad \text{with } j = \sqrt{-1} \tag{1}$$

For soils with low salinity, it is often assumed that the polarization and conductivity effects can be neglected [4]. Under such conditions, the effect of K" is eliminated and K' becomes equal to K, represented by Ka as the apparent dielectric constant [4]. Under saline conditions, the imaginary part of the dielectric permittivity increases with ECa, leading to error in the permittivity measurement. This problem becomes important for frequencies lower than 200 MHz [6]. According to Campbell [29], for a frequency range of 1–50 MHz, conductivity is the most important mechanism related to energy loss. However, using the hydra impedance probe, Kelleners and Verma [30] found that, in general, the total energy loss is related to relaxation loss except for fine sandy soil, where it is equal to zero at 50 MHz.

2.1.1. Permittivity-Corrected Linear Model

Many researchers [5,17,31,32] have studied how well low-frequency capacitance sensors measure Ka and to what degree it is affected by K". In general, it has been shown that the most important factor to consider is the conductivity effect on Ka, whereas the effect of relaxation losses appears to be small [4,6]. Thus, it is possible to correct the Ka reading by introducing a term for the ECa effect. Based on the work of Whalley [32], Robinson et al. [5] proposed a permittivity-corrected linear model where the theoretical permittivity can be considered equivalent to the refractive index of measurements by the TDR. Robinson et al. [5] conducted experiments using TDR and capacitance dielectric sensor in sandy soils with high ECa levels (up to 2.5 dS m^{-1}) and they proposed a linear model that includes the ECa effect on the Ka prediction according to:

$$\sqrt{K'} = \sqrt{Ka} - 0.628\, ECa \tag{2}$$

From this equation, we notice that the increase of ECa (dS m^{-1}) leads to an increase in Ka. Using Equation (2), a corrected permittivity K' can be determined eliminating the ECa effect [6].

2.1.2. Water Content Model

The dielectric constant is about 80 for water (at 20 °C), 2 to 5 for dry soil, and 1 for air. Therefore, Ka is highly dependent on θ. Various equations for the Ka vs. θ relationship have been published. The most used θ-model is a third-order polynomial [4]. However, Ledieu et al. [7] showed that there is a simpler linear relationship for the θ prediction with only two empirical parameters, of the form:

$$\theta = a\sqrt{Ka} + b \quad (3)$$

where a and b are fitting parameters.

Figure 1 shows a schematic of calibration and validation possibilities for θ estimations that were used in the present study. The calibration consisted of fitting of parameters in different models (Figure 1). Optimal values for a and b, vs. a' and b' were determined by linear regression in the relationship $\sqrt{Ka}$-θ_m denoted as the CAL-Ka model (Figure 1, Step-A.1) and $\sqrt{K'}$-θ_m denoted as CAL-Kar model (Figure 1, Step-A.2), respectively. The θ_m was measured in experiments for different salinity levels. The standard Ledieu et al. [7] model (Figure 1) was used for comparison purposes as it is the simplest known model for mineral soil. The different steps (A.1 and A.2) were first completed using laboratory experiments (laboratory calibration) and then using field data (field calibration). The laboratory and field calibrated models were then compared with each other (Figure 1, Step-A.3). Finally, we used field data (step B.1, B.2, and B.3) to validate the laboratory experiments (laboratory model validation).

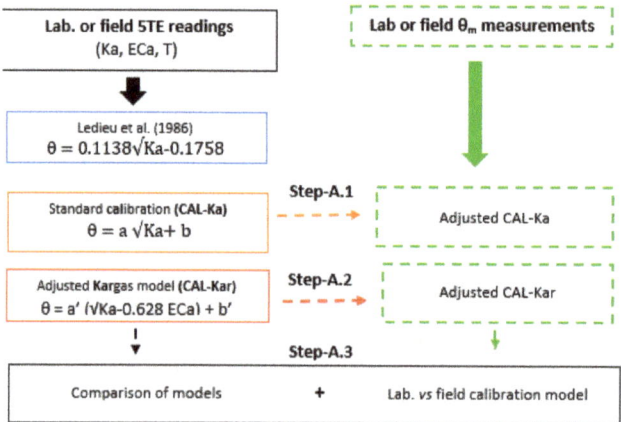

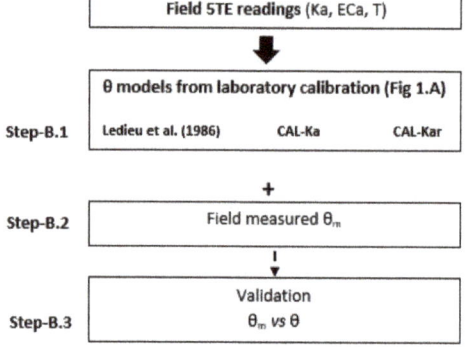

Figure 1. Schematic of θ calibration and validation possibilities investigated in the present study.

2.1.3. Pore Water Electrical Conductivity Model

Different studies [33,34] have shown that ECa depends on both θ and ECp. Malicki et al. [35] and Malicki and Walczak [9] found that for Ka > 6 and when ECp is constant, the relationship between Ka and ECa is linear. An empirical ECp–ECa–Ka model has, thus, been proposed. Based on their results, Hilhorst [8] presented the following equation applicable when $\theta \geq 0.10$ m^3 m^{-3}:

$$\text{ECp} = \left(\frac{K_w}{(K_a - K_0)}\right) \times \text{ECa} \tag{4}$$

where K_w is the dielectric constant of the pore water (equal to 80.3) and K_0 is a soil parameter equal to Ka when ECa = 0 (see [8], for details). According to Hilhorst [8], the K_0 parameter depends on soil texture but is independent of ECa. He found the range of K_0 to be between 1.9 and 7.6. For best results, this should be determined experimentally for each soil type. For most soils, a value of 4.1 has been recommended. One should notice, that in the Hilhorst model (Equation (4)), the Ka, K_w, and K_0 represent the real part of the dielectric constant only. From the linear relationship ECp = f (ECa), the slope that is inversely proportional to ECp and intercept K_0 can be determined.

In the present study, the Hilhorst model (Figure 2, Step-C.1) was tested using varying K_0 soil parameters (4.1, 6 and 3.3). The K_0 = 4.1 is the default value recommended by Hilhorst, K_0 = 6 is the recommended value in the 5TE manual [36] while K_0 = 3.3 is the value measured with distilled water according to the WET sensor manual [37].

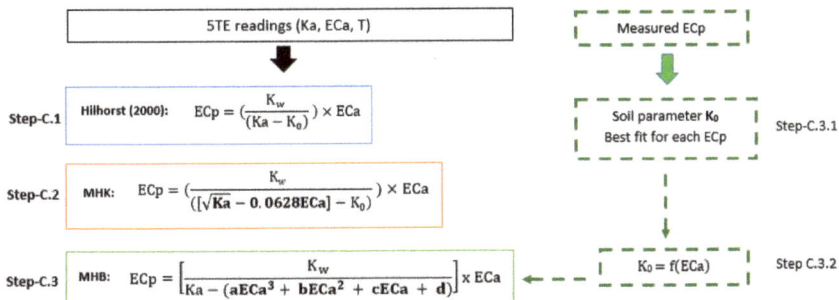

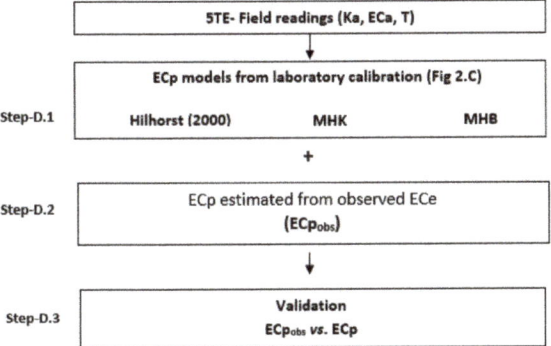

Figure 2. Schematic of electrical conductivity (ECp) calibration and validation used in the present paper.

Inspired by Bouksila et al. [18] and Kargas et al. [6], a modification of the Hilhorst model was investigated. Accordingly, a permittivity-corrected linear equation (Equation (2)) can be introduced in the Hilhorst model (Figure 2, Step-C.2) and ECp is predicted with two different K_0 values (K_0 = 4.1 and K_0 = 3.3). Beside this, the soil fit parameter K_0 is calculated for each salinity level by minimizing the mean square error (MSE) of the estimated ECp in the Hilhorst model (Step-C.3.1). The best fit K_0 parameters are then plotted against ECa for the seven different ECp and a third-order polynomial function is determined (Step-C.3.2), and introduced in the Hilhorst model (Step-C.3). Finally, we used field data (step D.1, D.2, and D.3) to validate the laboratory experiments (laboratory model validation).

The temperature is an important factor influencing the electrical conductivity measurements; indeed, all ECa reading were adjusted in the present work using Equation (5). Besides, during experiments the temperature effect on K_w parameter was considered using the recommended temperature correction equation in the 5TE manual [36].

$$ECa_{25} = ECa \left[1 - ((T - 25) \times 0.02)\right] \quad (5)$$

Measured Ka, ECa, and T in laboratory and field experiments are converted to ECp using the Hilhorst [8] model (Step-C.1), Kargas et al. [6] approach (Step-C.2), and Bouksila et al. [18] approach (Step-C.3), denoted as H, MHK, and MHB, respectively.

The different approaches in Figures 1 and 2 have not been tested before using the 5TE sensor. The approaches CAL-Kar, MHK and MHB have previously only been tested once under controlled laboratory condition using the WET sensor. The novelty of the present work is to validate these approaches under field condition using the low cost capacitance sensor 5TE. In addition, the MHB approach developed by Bouksila et al. [18], used an experimentally determined K_0 = f (ECa) relationship. Our new approach instead uses a K_0 derived from best-fit parameter for each ECp level, which make the application of MHB approach much easier since there is no need for the K_0 laboratory experiment.

Model performance for θ and ECp, was evaluated using both the root mean square error (RMSE) and coefficient of determination (R^2). In addition, mean relative error (MRE) and coefficient of variation (CV) were used for ECp and θ, respectively.

2.2. Study Area

The field study was conducted in the Jemna oasis (33°36'15."N, 9°00'39."E), belonging to the Agricultural Extension and Training Agency (AVFA) located in the Kebeli Governorate, southern Tunisia. The oasis is equipped with a micro-irrigation system. The main crop is adult date-palm trees. The climate is arid with an annual rainfall of less than 100 mm, which is insufficient to sustain agriculture. The annual potential evapotranspiration is about 2000 mm [38]. Groundwater, situated at 17 m soil depth, with an electrical conductivity (ECiw) of about 3.5 dS m^{-1}, is used for irrigation. The pH of groundwater is 7.8 and the geochemical facies is sodium chloride. Soil samples were collected from the top soil at 0–0.5 m depth. The soil was leached with distilled water in order to remove soluble salts and oven dried (105 °C) for 24 h. Then, the soil was passed through a 2 mm sieve. Soil particle size distribution was determined using the sedimentation method (pipette and hydrometer) and the electrical conductivity of saturated soil paste extract (ECe) was measured according to the United States Department of Agriculture (USDA) [39]. A summary of soil properties is presented in Table 1.

Table 1. Particle size percentage, pH and electrical conductivity of saturated soil paste extract (ECe) of investigated soil samples.

Depth (m)	Clay (%)	Fine Silt (%)	Coarse Silt (%)	Fine Sand (%)	Coarse Sand (%)	pH	ECe (dS m^{-1})
0–0.5	5	3	4	22	65	8.5	1.8

2.3. Laboratory Experiments

Seven NaCl solutions with different electrical conductivity (0.02, 0.2, 0.5, 3.6, 5.3, 7.2, and 8.2 dS m^{-1}) were prepared for the infiltration experiments. The soil was initially mixed with a small amount (about 0.05 m^3 m^{-3}) of the same water as used in the infiltration experiments to prevent water repellency. The soil was repacked into a plexiglas soil columns, 0.12 m in diameter and 0.15 m long (Soil Measurement System, Tucson, Arizona), to the average dry bulk density encountered in the field (about 1450 kg m^{-3}).

The 5TE sensor was used for observations [23]. It is a multifunctional sensor measuring Ka, ECa, and T (for more details, see Appendix A). The measuring frequency is 70 MHz and it is a three-rod type sensor with 0.052 m long prongs and 0.01 m spacing between adjacent prongs [23,40]. The 5TE probe was inserted vertically in the center of the column. Upward infiltration experiments were carried out by stepwise pumping a known volume of a NaCl solution (45 mL) with a precise syringe pump from the bottom of the column. Twenty minutes after each injection, three measurements of Ka, ECa, and temperature were taken and averaged. This procedure was repeated until saturation (0.40 m^3 m^{-3}) was reached. Four hours after reaching saturation, measurements were again taken and pore water was extracted from the bottom of the column with a manual vacuum pump. Electrical conductivity of extracted pore water ECp$_m$ was measured with a conductivity meter. In total, seven upward infiltration experiments were conducted, one for every NaCl solution.

2.4. Field Measurements

Four 5TE sensors were installed between date-palm trees at four soil depths (0.10, 0.15, 0.30, and 0.45 m). The 5TE probes were connected to a Decagon Em50 data logger. The DataTrac3 software version 3.15 [23] was used to download collected data from the Em50. Volumetric soil water content and pore electrical conductivity were estimated using standard parameters of the Ledieu et al. [7] and Hilhorst [8] models, respectively. In addition, soil samples were taken by hand auger at the same depth of sensor installation on 24 April and 3 October 2018. Gravimetric water content θ_m and electrical conductivity of saturated soil paste extract (ECe) were measured in laboratory according to USDA standards. The soil dry bulk density (Bd) was measured in the field using the cylinder method at five soil depths (0.1 m depth intervals to 0.5 m). During April 2018, the average soil Bd was equal to 1.43 g cm^{-3} and varied from 1.3 to 1.6 g cm^{-3}.

3. Results

3.1. Soil Water Content

Figure 3 presents the relationship between Ka and observed θ_m with different salinity levels (ECp, dS m^{-1}) measured during the upward infiltration experiments. For largest ECp, ECa did not exceed 2.5 dS m^{-1}. It is seen that ECa considerably affects the Ka readings, especially for high ECp. This can lead to significant errors for both Ka and ECa, indicating that 5TE probe readings need to be corrected when used in saline soils. The overestimation of Ka as ECa increases has been described by several authors (e.g., [19,27]).

In Figure 4, Ka and K' (corrected with Equation (2)) for two ECp levels (3 and 9.8 dS m^{-1}) are plotted against measured θ_m. K' values are very close to Ka when ECp ≤ 3 dS m^{-1}, especially at low θ ($\theta \leq 0.15$ m^3 m^{-3} and ECa ≤ 0.43 dS m^{-1}). However, for ECp = 9.8 dS m^{-1}, the difference between Ka and K' is more pronounced, especially for $\theta \geq 0.15$ m^3 m^{-3} and ECa ≥ 0.75 dS m^{-1}.

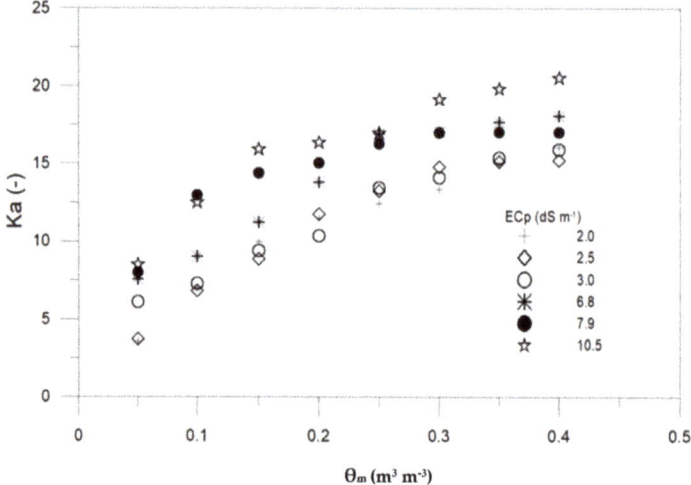

Figure 3. Apparent dielectric permittivity (Ka) vs. measured volumetric water content (θ_m) for various pore electrical conductivity (ECp) levels (dS m^{-1}).

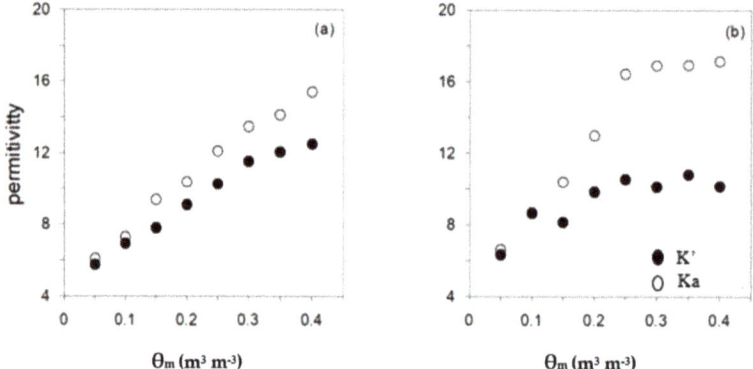

Figure 4. Relationship Ka-θ_m (open circles) and K'-θ_m (filled circles) using the 5TE sensor for ECp = 3 dS m^{-1} (**a**) and ECp = 9.8 dS m^{-1} (**b**).

The calibrated parameters using laboratory data for CAL-Ka and CAL-Kar approaches are presented in Table 2. For all models tested under laboratory conditions, RMSE increased with ECp. Soil water content from CAL-Kar approach matched well measured θ_m for ECp ≤ 3 dS m^{-1} (ECa < 0.7 dS m^{-1}) and gave the best θ estimation compared to the Ledieu et al. [13] model and the soil-specific calibration CAL-Ka. However, for ECp ≥ 6.8 dS m^{-1}, the CAL-Ka approach gave lower RMSE compared to the CAL-Kar model. For high ECp (≥ 6.8 dS m^{-1}), the performance of the CAL-Kar model deteriorated.

Table 2. Root mean square error (RMSE, m³ m³), determination coefficient (R²) and coefficient of variation (CV,%) of estimated soil water content using Ledieu et al. [7], standard calibration (CAL-Ka) and permittivity corrected model (CAL-Kar) for different water pore electrical conductivity (ECp).

		Laboratory Calibration		
ECp (dS m^{-1})		Ledieu et al. (1986)	CAL-Ka	CAL-Kar
	Fit	Equation (4)	$\theta = 0.16\sqrt{Ka^1} - 0.30$	$\theta = 0.18\sqrt{K'^2} - 0.33$
ECp ≤ 3	RMSE	0.06	0.05	0.04
	R²	0.93	0.95	0.95
ECp = 6.8	RMSE	0.08	0.06	0.10
	R²	0.73	0.87	0.50
6.8 < ECp ≤ 10.5	RMSE	0.09	0.07	0.13
	R²	0.77	0.85	0.39
Mean RMSE		0.08	0.06	0.09
Mean R²		0.8	0.9	0.6
CV (%)		26.5	20	19.8
		Field calibration		
	Fit		$\theta = 0.15\sqrt{Ka} - 0.26$	$\theta = 0.20\sqrt{K'} - 0.37$
ECa³ ≤ 0.7 and 1.7 ≤ ECe⁴ ≤ 4.1	RMSE (m³ m^{-3})	-	0.04	0.03
	R²	-	0.94	0.97
	CV (%)	-	23	24
		Field validation		
ECa ≤ 0.7 and 1.7 ≤ ECe ≤ 4.1	RMSE (m³ m^{-3})	0.1	0.060	0.060
	R²	0.80	0.88	0.97
	CV (%)	27	21	24

[1] Apparent soil permittivity, [2] Corrected apparent soil permittivity, [3] Soil apparent electrical conductivity, [4] Electrical conductivity of saturated soil paste extract.

3.2. Field Validation of Soil Water Content Models

During field experiments, Ka measured by the four 5TE probes varied from 6.5 to 11, ECa from 0.17 to 0.75 dS m^{-1}, and measured soil moisture (θ_m) from 0.10 to 0.24 m³ m^{-3}. According to R² of field validation results (Table 2), the best model to predict θ under field conditions is CAL-Kar followed by CAL-Ka. However, RMSE analysis indicates that there is no significant difference between observed and estimated θ using both approaches, implying that both predicted θ accurately for ECa ≤ 0.7 dS m^{-1}.

From Figure 5, a slight underestimation of the different models is observed and this is more pronounced for the Ledieu et al. [7] model. The underestimation can be related to adsorbed water, resulting in a lower amount of mobile water in the soil, thus reducing the Ka readings (detection) by the 5TE sensor and eventually resulting in underestimation of Ka [41,42]. The difference between observed and predicted θ may also be attributed to variability in soil structure, bulk density, presence of stones, roots, and other inert material in the core samples. The difference may also be linked to the spatial variability of θ between sampled and monitored soils. Similar findings have been reported for mineral soils using the 5TE sensor [41], for Luvisol using the 5TM capacitance sensor [42], and using the ECH2O sensor in sandy soil [43]. The success of CAL-Ka and CAL-Kar models to calculate θ at field conditions is closely linked to the low range of ECa data measured by the 5TE sensor, below 0.7 dS m^{-1}, during the period of investigation.

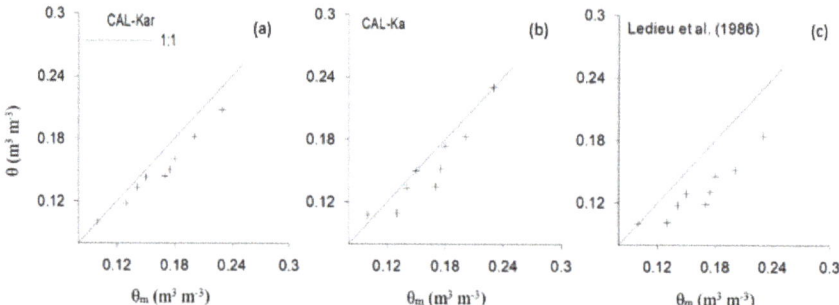

Figure 5. Estimated soil water content (θ) vs. measured (θ_m) using CAL-Kar approach (**a**), CAL-Ka approach (**b**) and Ledieu et al. [13] model (**c**) under field conditions, solid line gives the 1:1 relationship.

For the same range of soil salinity, RMSE was higher for the field as compared to laboratory data. For laboratory experiments, soil was crushed, washed, and passed through a 2 mm sieve. This means that its structure was changed as well as the pore size distribution, and some of the organic matter may have been removed. This allows more mobile water compared to field conditions [44]. As well, for field conditions, observed Bd profiles are not uniform and may vary with time. In contrast to the controlled laboratory experiments (e.g., constant Bd), the field Bd spatial and temporal variation will induce an additional error when laboratory models are used to estimate θ.

We used the field data to calibrate the CAL-Ka and CAL-Kar models, the calibrated parameters for the models are presented in Table 2 (Field calibration). The RMSE decreased from 0.06 to 0.04 m³m^{-3} and from 0.06 to 0.03 m³ m^{-3} for CAL-Ka and CAL-Kar, respectively. Thus, the CAL-Kar approach gave better field predictions of θ. Similarly, Kinzli et al. [45] reported that field calibration was most successful for sandy soils. According to this finding, we may support the earlier conclusion that the permittivity corrected (CAL-Kar) model is recommended under field conditions if ECa is below 0.75 dS m^{-1}. However, the Ledieu et al. [7] model cannot be used safely under field conditions in the case when soil specific calibration is not available.

3.3. Soil Pore Electrical Conductivity (ECp)

3.3.1. ECp Laboratory Calibration

Table 3 presents the RMSE for the different models. All models showed good performance in the 0–3 dS m^{-1} range, except Hilhorst with ($K_0 = 6$) and MHK with $K_0 = 4.1$. Moreover, RMSE results (Table 3), showed an increase of the range of default H model validity until ECp = 6.8 dS m^{-1}. This finding can be linked to the higher operating frequency of 5TE (70 MHz) compared to the capacitance sensor used by Hilhorst (30 Mhz). Hilhorst reported that the model assumption ceases to be accurate at higher salinity as ECp significantly deviates from that of free water.

Table 3. Root mean square error (RMSE, dS m^{-1}) of estimated pore electrical conductivity (ECp) using Hilhorst (K_0 = 4.1, 3.3, and 6), modified Hilhorst according to Kargas et al. [6] (MHK) (K_0= 4.1 and 3.3), and modified Hilhorst according to Bouksila et al. [18] (MHB) models.

ECp (dS m^{-1})	Hilhorst (2000)			MHK		MHB
Soil Parameter-K_0	$K_0 = 4.1$	$K_0 = 3.3$ [1]	$K_0 = 6$	$K_0 = 4.1$	$K_0 = 3.3$ [1]	Best Fit K_0 = f (ECa²)
ECp ≤ 3	0.29	0.14	0.83	0.88	0.34	0.044
ECp = 6.8	0.57	0.21	1.7	6.3	3.8	0.050
6.8 < ECp ≤ 10.5	1.48	0.99	3.06	-	-	0.054

[1] K_0 soil parameter determined experimentally according to the method in the Wet sensor manual using distilled water. [2] Soil apparent electrical conductivity.

From the results presented in Table 3, the ECp limit for accurate measurements seems to be 6.8 dS m^{-1}. Similar results were reported by Scudiero et al. [40], using the 5TE sensor and ECp limit <10 dS m^{-1} with RMSE equal to 0.68 dS m^{-1}. Using the H model with K_0 value recommended in the Decagons manual (K_0 = 6) showed a larger RMSE for all salinity levels compared the default parameter (K_0 = 4.1). The H model with K_0 = 3.3 (determined experimentally according to the WET manual) gave better results for the three salinity ranges. Persson [13] stated that the H model using a fitted soil parameter gives ECp values statistically similar to other model results (e.g., [3,10,46]).

Focusing on the modified Hilhorst model using the MHK approach with K_0 = 4.1, one can observe that the RMSE is at maximum, especially for ECp ≥ 6.8 dS m^{-1}. Kargas el al. [6] validated this approach using a lower salinity level (ECp ≤ 6 dS m^{-1}). According to our results (Figure 7), an overestimation of the H model, especially at ECp ≥ 3 dS m^{-1}, is observed. Similarly, Visconti et al. [19] showed an overestimation of ECp in the range of 0–10 dS m^{-1} and Scudiero et al. [40] showed an overestimation of ECp in the range 3–10 dS m^{-1}, both working with the 5TE sensor and the H model. In the present study, the H model overestimated ECp, thus using the MHK approach will not improve results.

The observed overestimation by the H model might be due to K_0, which was assumed to be equal to 4.1. In addition, one should note that the H model does not consider solid particle surface conductivity, which could contribute to the ECp error [17]. From Table 3, decreasing K_0 from 4.1 to 3.3 for both the H and MHK model leads to a significant decrease of RMSE, two times lower than the default. The H model seems to be more dependent on the soil parameter K_0 than on Ka and ECa.

K_0 estimated from the best fit approach for the different salinity levels is plotted against ECa in Figure 6. The K_0 range varied between 1.29 and 3.2 with a mean of 3.0, which is similar to the K_0 determined experimentally using distilled water (K_0 = 3.3).

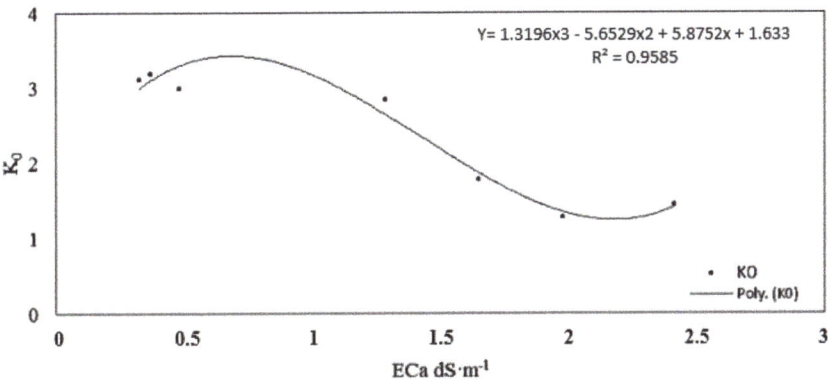

Figure 6. Best fit soil parameter (K_0) vs. bulk soil electrical conductivity (ECa).

At saturation, ECa was equal to 0.32 dS m^{-1} and 2.4 dS m^{-1} and Ka was equal to 15 and 19 for the lowest (2 dS m^{-1}) and the highest (10.5 dS m^{-1}) observed ECp, respectively. According to Figure 6, K_0 decreases with increasing salinity. Similar to [18], our results showed that K_0 is not constant, but depends on ECa and that a third-order polynomial fitted the K_0–ECa relationship rather well (R^2 ≥ 0.95). K_0 = f (ECa) in Figure 6, was used in the H model to predict ECp. Compared to the H model, for the individual ECp levels, using the MHB model, RMSE decreased significantly.

Figure 7 shows observed and predicted ECp using the H model with three different K_0 and the MHK and MHB approaches, respectively. All model performances, are approximately the same for ECp ≤ 3 dS m^{-1}, except when using K_0 = 6 and K_0 = 4.1 for H and MHK models, respectively.

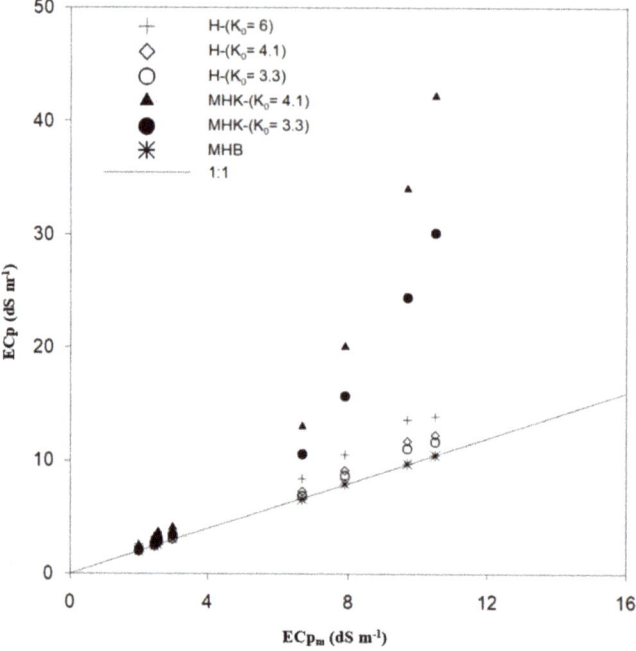

Figure 7. Estimated pore electrical conductivity (ECp) vs. measured for different model tested for laboratory conditions.

Based on the laboratory results, the MHB approach improved the H model and gave accurate estimation of ECp with $R^2 = 0.99$ for all salinity levels. Thus, for high soil salinity (6.8 dS m^{-1} ≤ ECp ≤ 10.5 dS m^{-1}), the MHB approach is recommended for achieving optimal accuracy of ECp measurements. For lower ECp (≤3 dS m^{-1}), the standard H model is sufficient. For high ECp, the MHK approach failed to reproduce the observed ECp correctly and the approach is not recommended based on the results of our study. Further studies for different soil types are needed so that this combined approach in predicting ECp can be validated.

3.3.2. Field Validation of ECp Models

Unfortunately, we do not have field observed ECp to validate and statistically compare the different models. Instead, we determined a linear relationship (ECp = f (ECe)) for different calculated ECp, using the H, MHK, and MHB models and 5TE measurements, with observed field ECe. Several researchers have studied relationships between ECe and ECp, e.g., [3], showing that the relationship is strongly linear. The relationship (ECp = f (ECe)) with the highest $R^2 = 0.9$ was chosen to predict the field ECp values (ECp$_{obs}$). During the investigation period, ECe was determined from soil samples, according to the USDA standard (collected at the same depth as the location of the 5TE sensors), ranging between 1.7 and 4.1 dS m^{-1}. The relatively low soil salinity is due to a rainfall observed in the field one day before soil sampling.

The observed ECp$_{obs}$ obtained from the best fit relationship is plotted against the estimated ECp for the different models in Figure 8. The H model with $K_0 = 6.6$ was not included in the figure since it gave out of range values. The ECp estimation with MHB approach appears uniformly scattered about the 1:1 line. On the other hand, the H model with $K_0 = 3.3$ shows a cloud of points near the 1:1 line.

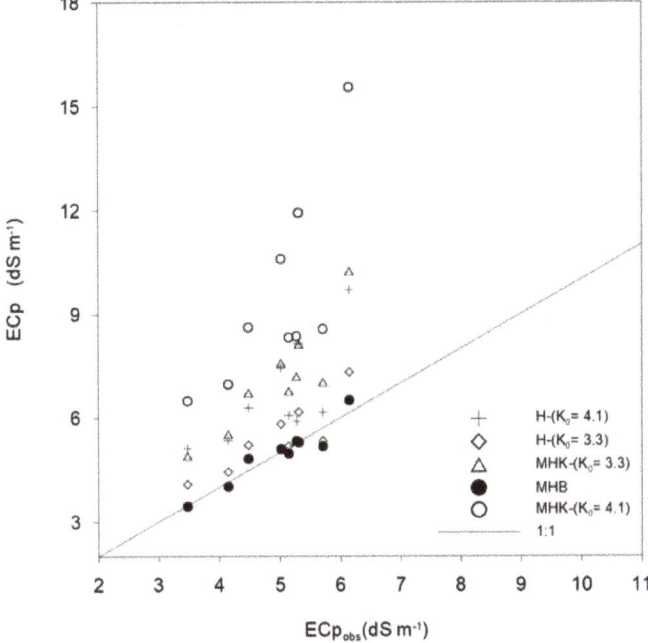

Figure 8. Estimated ECp vs. observed under field conditions.

Compared to laboratory results, for the same ECa range (ECa ≤ 0.7 dS m^{-1}) (Table 4), observed errors are higher for the field validation. The RMSE increased for all models. Errors are mainly related to a number of factors absent in the laboratory but present under field conditions. Due to this reason, a methodological approach composed by laboratory calibration and field validation is optimal.

Table 4. Root mean square error (RMSE, dS m^{-1}) and determination coefficient (R^2) of Hilhorst (K_0 = 4.1, 3.3, and 6), modified Hilhorst according to Kargas et al. [6] (MHK) (K_0 = 4.1 and 3.3) and modified Hilhorst according to Bouksila et al. [18] (MHB) models field validation.

	Hilhorst (2000)			MHK		MHB
ECa2 ≤ 0.7 and 1.7 ≤ ECe3 ≤ 4.1	K_0 = 4.1	K_0 = 3.3[1]	K_0 = 6	K_0 = 4.1	K_0 = 3.3[1]	Best Fit K_0 = f (ECa)
RMSE (dS m−1)	0.82	0.70	10	1.8	1.34	0.30
R^2	0.53	0.73	0.26	0.56	0.77	0.90

[1.] K_0 soil parameter determined experimentally according to the method in the Wet sensor manual using distilled water. [3] Soil apparent electrical conductivity, [4] Electrical conductivity of saturated soil paste extract.

The MHB approach presents a significant improvement of the H model, especially at high ECp (Table 4). The H and MHK model fit is acceptable for field and laboratory conditions only for ECp ≤ 3dS m^{-1} while the MHB approach is acceptable for field conditions and it can be safely used for sandy soil and ECp ≤ 7 dS m^{-1}.

Since variation and uncertainties in the field are higher, it is recommended to validate the calibrated models with field data. According to our results, the H model with K_0 = 6 is not recommended either with laboratory nor field data. However, the reduction of K_0 to 3.3 increased the performance of the model and it can be safely used for ECp < 3 dS m^{-1}. For ECp > 3 dS m^{-1}, the MHK approach did not improve the H model with RMSE more than 1 dS m^{-1} and it is not recommended. Thus, for achieving optimal accuracy of ECp measurements, the MHB approach is recommended for ECp ≤ 7dS m^{-1}.

4. Conclusions

In this study, the 5TE sensor performance for volumetric soil water content (θ) and soil pore electrical conductivity (ECp) estimation was investigated under laboratory and field conditions. First, two procedures for θ estimation based on a linear relationship of $\sqrt{Ka}$-θ_m (CAL-Ka approach) and $\sqrt{K'}$-θ_m (CAL-Kar approach) were investigated. Using the CAL-Kar approach, the effect of soil apparent electrical conductivity (ECa) on the real part of the complex dielectric permittivity (K') was considered. In addition, the Ledieu et al. [7] relationship was used for comparison purposes. A site-specific validation of CAL-Ka and CAL-Kar models using 5TE field subset data and θ from soil samples at different depth was performed. Secondly, 5TE performance for soil salinity assessment was investigated using the H linear model according to correction proposed by Kargas et al. [6] (MHK model), and Bouksila et al. [17] (MHB model). The default value of soil parameter $K_0 = 4.1$ and $K_0 = 6$ recommended in the 5TE manual was used for comparison.

For soil water content, calibration considering the ECa effect on K' increased the performance of the 5TE sensor under field conditions for ECa ≤ 0.75 dS m^{-1} ($R^2 = 0.97$, RMSE = 0.06 m^3 m^{-3}). However, the error in predicting θ was highest (0.10 m^3 m^{-3}) when the Ledieu et al. [7] model was used. Indeed, this model cannot be safely used under field conditions. Thus, we conclude that field calibration of the 5TE sensor is recommended for accurate soil water content estimation. Soil pore electrical conductivity calibration results, show that the 5TE sensor limit using the default H model is equal to 6.8 dS m^{-1} with RMSE = 0.57 dS m^{-1} and MRE = 9%. The 5TE sensor manual value ($K_0 = 6$) is not recommended. However, $K_0 = 3.3$ increases model performance over the investigated salinity range. The MHK approach, introducing the permittivity correction in the H model, failed to reproduce the observed ECp correctly and it is not recommended. In the next step, considering the effect of ECa on the K_0 soil parameter in the H model (MHB approach), it was found that the standard model improves and gives accurate estimation of ECp with R^2 equal to 0.99 for all salinity levels. Under field conditions, the MHB approach gives the best results for sandy soils.

It is a challenge to perform real-time monitoring of irrigated land under high-saline conditions to provide sustainable agriculture and farmer income increase. Using θ and ECp observations, it was shown that a methodological approach composed of a laboratory calibration and field validation is necessary. Further studies, for different soil types, are needed to validate this combined approach in predicting ECp.

Author Contributions: N.Z. was the main author executing the experiments, data curation, formal analysis and writing. F.B. assisted in the execution of experiments. F.B. and M.P. contributed in data curation, formal analysis and writing original draft. Investigation was carried by N.Z., F.B., and F.S. R.B. (Ronny Berndtsson), F.S. and R.B. (Rachida Bouhlila) provided advice and assisted in reviewing and editing the final document. Funding acquisition and resources were made by F.B., M.P. and R.B. (Ronny Berndtsson). F.B., M.P. and R.B. (Rachida Bouhlila) supervised this work. All authors provided assistance in reviewing and editing the manuscript. All authors contributed to the conceptualization, methodology and validation of the work.

Funding: This research was funded by the Tunisian Institution of Agricultural Research and Higher Education (IRESA) through the SALTFREE project (ARIMNET2/0005/2015, grant agreement N° 618127) and the European Union Horizon 2020 program, under Faster project, grant agreement N° [810812].

Acknowledgments: The authors acknowledge support received from Soil department (DGACTA, Tunisia).

Conflicts of Interest: The funders had no role in the design of the study; in the collection, analyses, or interpretation of data, in the writing the manuscript, or in the decision to publish the results.

Appendix A

Table A1. Soil parameter acronyms, data source, sensor specification and models used in the present work.

Soil Parameter	Acronym	Data Source	Sensor/Method
Soil dry bulk density	Bd	Measured	Cylinder method- United States Department of Agriculture (USDA)
Soil pH	pH	Measured	pH-meter
Apparent soil permittivity	Ka	Measured	5TE-probe
Soil parameter	K_0	Estimated	5TE-probe
Dielectric constant of pore water	K_w	Estimated	5TE-probe
Corrected apparent soil permittivity	K'	Estimated	5TE-probe
Soil temperature	T	Measured	5TE-probe
Electrical conductivity of saturated soil paste extract	ECe	Measured	EC-meter/USDA method
Soil apparent electrical conductivity	ECa	Measured	5TE-probe
Irrigation water electrical conductivity	ECiw	Measured	EC-meter
Measured soil water content	$θ_m$	Measured	Gravimetric method-USDA
Estimated volumetric water content	θ	Estimated	θ –Models (see Figure 1)
Laboratory measured pore water electrical conductivity	ECp_m	Measured	EC-meter
Field observed pore water electrical conductivity	ECp_{obs}	Measured	$ECp_{obs} = a\ ECe + b$ (see Figure 2)
Pore water electrical conductivity	ECp	Estimated	ECp-Models (see Figure 2)
5TE sensor specification		Specifics	
Type			
Sensor type		FDR (Frequency Domain Reflectometry)	
Power supply		+3.6 to +15 V	
Frequency		70 MHz	
Size		Length 10.9 cm (4.3 in) Width 3.4 cm (1.3 in) Height 1.0 cm (0.4 in)	
Measurement volume		300 cm^3	
Direct output data		Ka, ECa, and T	
Indirect output data		θ and ECp	
Range (Ka, ECa)		1–80, 0–7 dS m^{-1}	
Resolution (Ka, ECa)		0.1, 0.01 dS m^{-1}	
Accuracy (Ka, ECa)		±3%, ±10%	
Models			
CAL-Ka (see Figure 1)		Calibration of soil water content model without permittivity correction	
CAL-Kar (see Figure 1)		Calibration of soil water content model with permittivity correction according to Kargas et al. (2017)	
H (see Figure 2)		Standard Hilhorst (2000) model for ECp prediction	
MHK (see Figure 2)		Modified Hilhorst model according to Kargas et al. (2017) for ECp prediction	
MHB (see Figure 2)		Modified Hilhorst model according to Bouksila et al. (2008) for ECp prediction	
Model performance statistic tool			
RMSE		Root Mean Square Error	
R^2		Coefficient of determination	
MRE		Mean Relative Error	
CV		Coefficient of Variation	

References

1. Selim, T.; Bouksila, F.; Berndtsson, R.; Persson, M. Soil Water and Salinity Distribution under Different Treatments of Drip Irrigation. *Soil Sci. Soc. Am. J.* **2013**, *77*, 1144–1156. [CrossRef]
2. Slama, F.; Zemni, N.; Bouksila, F.; De Mascellis, R.; Bouhlila, R. Modelling the Impact on Root Water Uptake and Solute Return Flow of Different Drip Irrigation Regimes with Brackish Water. *Water* **2019**, *11*, 425. [CrossRef]
3. Rhoades, J.D.; Manteghi, N.A.; Shouse, P.J.; Alves, W.J. Soil Electrical Conductivity and Soil Salinity: New Formulations and Calibrations. *Soil Sci. Soc. Am. J.* **1989**, *53*, 433–439. [CrossRef]
4. Topp, G.C.; Davis, J.L.; Annan, A.P. Electromagnetic determination of soil water content: Measurements in coaxial transmission lines. *Water Resour. Res.* **1980**, *16*, 574–582. [CrossRef]
5. Robinson, D.A.; Gardner, C.M.K.; Cooper, J.D. Measurement of relative permittivity in sandy soils using TDR, capacitance and theta probes: Comparison, including the effects of bulk soil electrical conductivity. *J. Hydrol.* **1999**, *223*, 198–211. [CrossRef]
6. Kargas, G.; Persson, M.; Kanelis, G.; Markopoulou, I.; Kerkides, P. Prediction of Soil Solution Electrical Conductivity by the Permittivity Corrected Linear Model Using a Dielectric Sensor. *J. Irrig. Drain. Eng.* **2017**, *143*, 04017030. [CrossRef]
7. Ledieu, J.; Ridder, P.D.; Clerck, P.D.; Dautrebande, S. A method of measuring soil moisture by time-domain reflectometry. *J. Hydrol.* **1986**, *88*, 319–328. [CrossRef]
8. Hilhorst, M.A. A Pore Water Conductivity Sensor. *Soil Sci. Soc. Am. J.* **2000**, *64*, 1922–1925. [CrossRef]
9. Malicki, M.A.; Walczak, R.T. Evaluating soil salinity status from bulk electrical conductivity and permittivity. *Eur. J. Soil Sci.* **1999**, *50*, 505–514. [CrossRef]
10. Mualem, Y.; Friedman, S.P. Theoretical Prediction of Electrical Conductivity in Saturated and Unsaturated Soil. *Water Resour. Res.* **1991**, *27*, 2771–2777. [CrossRef]
11. Dalton, F.N. Development of Time-Domain Reflectometry for Measuring Soil Water Content and Bulk Soil Electrical Conductivity. In *Advances in Measurement of Soil Physical Properties: Bringing Theory into Practice*; Topp, G.C., Reynolds, W.D., Green, R.E., Eds.; Soil Science Society of America: Madison, WI, USA, 1992; pp. 143–167. [CrossRef]
12. Nadler, A.; Gamliel, A.; Peretz, I. Practical Aspects of Salinity Effect on TDR-Measured Water Content A Field Study Contribution from the Agricultural Research Organization, Volcani Center, Bet Dagan, 50-250, Israel; No 611/98 1998 series. *Soil Sci. Soc. Am. J.* **1999**, *63*, 1070–1076. [CrossRef]
13. Persson, M. Evaluating the linear dielectric constant-electrical conductivity model using time-domain reflectometry. *Hydrol. Sci. J.* **2000**, *47*, 269–277. [CrossRef]
14. Hamed, Y.; Samy, G.; Persson, M. Evaluation of the WET sensor compared to time domain reflectometry. *Hydrol. Sci. J.* **2006**, *51*, 671–681. [CrossRef]
15. Inoue, M.; Ould Ahmed, B.A.; Saito, T.; Irshad, M.; Uzoma, K.C. Comparison of three dielectric moisture sensors for measurement of water in saline sandy soil. *Soil Use Manag.* **2008**, *24*, 156–162. [CrossRef]
16. Kargas, G.; Soulis, K.X. Performance Analysis and Calibration of a New Low-Cost Capacitance Soil Moisture Sensor. *J. Irrig. Drain. Eng.* **2012**, *138*, 632–641. [CrossRef]
17. Regalado, C.M.; Ritter, A.; Rodríguez-González, R.M. Performance of the Commercial WET Capacitance Sensor as Compared with Time Domain Reflectometry in Volcanic Soils. *Vadose Zone J.* **2007**, *6*, 244–254. [CrossRef]
18. Bouksila, F.; Persson, M.; Berndtsson, R.; Bahri, A. Soil water content and salinity determination using different dielectric methods in saline gypsiferous soil. *Hydrol. Sci. J.* **2008**, *53*, 253–265. [CrossRef]
19. Visconti, F.; de Paz, J.M.; MartÃnez, D.; Molina, M.J. Laboratory and field assessment of the capacitance sensors Decagon 10HS and 5TE for estimating the water content of irrigated soils. *Agric. Water Manag.* **2014**, *132*, 111–119. [CrossRef]
20. Nimmo, J.R. Porosity and Pore Size Distribution. *Encycl. Soils Environ.* **2004**, *3*, 295–303.
21. Iwata, Y.; Miyamoto, T.; Kameyama, K.; Nishiya, M. Effect of sensor installation on the accurate measurement of soil water content. *Eur. J. Soil Sci.* **2017**, *68*, 817–828. [CrossRef]
22. Pardossi, A.; Incrocci, L.; Incrocci, G.; Malorgio, F.; Battista, P.; Bacci, L.; Rapi, B.; Marzialetti, P.; Hemming, J.; Balendonck, J. Root Zone Sensors for Irrigation Management in Intensive Agriculture. *Sensors* **2009**, *9*, 2809–2835. [CrossRef] [PubMed]

23. Baram, S.; Couvreurd, V.; Harter, T.; Read, M.; Brown, P.H.; Kandelous, M.; Smart, D.R.; Hopmans, J.W. Estimating Nitrate Leaching to Groundwater from Orchards: Comparing Crop Nitrogen Excess, Deep Vadose Zone Data-Driven Estimates, and HYDRUS Modeling. *Vadose Zone J.* **2016**, *15*. [CrossRef]
24. Gamage, D.N.V.; Biswas, A.; Strachan, I.B. Field Water Balance Closure with Actively Heated Fiber-Optics and Point-Based Soil Water Sensors. *Water* **2019**, *11*, 135. [CrossRef]
25. Evett, S.R.; Tolk, J.A.; Howell, T.A. Soil Profile Water Content Determination. *Vadose Zone J.* **2006**, *5*, 894–907. [CrossRef]
26. Schwartz, R.C.; Casanova, J.J.; Pelletier, M.G.; Evett, S.R.; Baumhardt, R.L. Soil Permittivity Response to Bulk Electrical Conductivity for Selected Soil Water Sensors. *Vadose Zone J.* **2013**, *12*. [CrossRef]
27. Varble, J.L.; Chavez, J.L. Performance evaluation and calibration of soil water content and potential sensors for agricultural soils in eastern Colorado. *Agric. Water Manag.* **2011**, *101*, 93–106. [CrossRef]
28. Jae-Kwon, S.; Won-Tae, S.; Jae-Young, C. Laboratory and Field Assessment of the Decagon 5TE and GS3 Sensors for Estimating Soil Water Content in Saline-Alkali Reclaimed Soils. *Commun. Soil Sci. Plant Anal.* **2017**, *48*, 2268–2279.
29. Campbell, J.E. Dielectric Properties and Influence of Conductivity in Soils at One to Fifty Megahertz. *Soil Sci. Soc. Am. J.* **1990**, *54*, 332–341. [CrossRef]
30. Kelleners, T.J.; Verma, A.K. Measured and Modeled Dielectric Properties of Soils at 50 Megahertz. *Soil Sci. Soc. Am. J.* **2010**, *74*, 744–752. [CrossRef]
31. Jones, S.B.; Blonquist, J.M.J.; Robinson, D.A.; Philip Rasmussen, V.; Or, D. Standardizing Characterization of Electromagnetic Water Content Sensors: Part 1. Methodology. *Vadose Zone J.* **2005**, *4*, 1048–1058. [CrossRef]
32. Whalley, W.R. Considerations on the use of time-domain reflectometry (TDR) for measuring soil water content. *J. Soil Sci.* **1993**, *44*, 1–9. [CrossRef]
33. Persson, M. Soil Solution Electrical Conductivity Measurements under Transient Conditions Using Time Domain Reflectometry. *Soil Sci. Soc. Am. J.* **1997**, *61*, 997–1003. [CrossRef]
34. Rhoades, J.D.; van Schilfgaarde, J. An Electrical Conductivity Probe for Determining Soil Salinity. *Soil Sci. Soc. Am. J.* **1976**, *40*, 647–651. [CrossRef]
35. Malicki, M.W.R.; Walczak, R.; Koch, S.; Fluhler, H. Determining soil salinity from simultaneous readings of its electrical conductivity and permittivity using TDR. In Proceedings of the Time Domain Reflectometry in Environmental, Infrastructure, and Mining Applications United States Department of Interior Bureau of Mines, the Time Domain Reflectometry in Environmental, Infrastructure, and Mining Applications United States Department of Interior Bureau of Mines, 7–9 September 1994; Northwestern University: Evanston, IL, USA; pp. 328–336.
36. 5TE. *Sensor Manual. 5TE- Water Content, Electrical Conductivity (EC) and Temperature sensor*; Decagon Devices Inc.: Pullman, WA, USA, 2016; Available online: https://www.decagon.com/ (accessed on 15 March 2016).
37. WET. *Sensor Manual-UTM-1.6*; Delta-T Devices Ltd.: Cambridge, UK, 2019; Available online: https://www.delta-t.co.uk/ (accessed on 10 July 2019).
38. Zammouri, M.; Siegfried, T.; El-Fahem, T.; Kriaca, S.; Kinzelbach, W. Salinization of groundwater in the Nefzawa oases region, Tunisia: Results of a regional-scale hydrogeologic approach. *Hydrogeol. J.* **2007**, *15*, 1357–1375. [CrossRef]
39. USDA. Diagnostic and improvement of saline and alkali soil. In *Agriculture Handbook No. 60*; US Department of Agriculture: Washington, DC, USA, 1954. Available online: https://www.ars.usda.gov (accessed on 20 September 2019).
40. Scudiero, E.; Berti, A.; Teatini, P.; Morari, F. Simultaneous Monitoring of Soil Water Content and Salinity with a Low-Cost Capacitance-Resistance Probe. *Sensors* **2012**, *12*, 17588–17607. [CrossRef]
41. Bircher, S.; Demontoux, F.O.; Razafindratsima, S.; Zakharova, E.; Drusch, M.; Wigneron, J.-P.; Kerr, Y. L-Band relative permittivity of organic soil surface layers: A new dataset of resonant cavity measurements and model evaluation. *Remote Sens.* **2016**, *8*, 1–17. [CrossRef]
42. Parvin, N.; Degra, A. Soil-specific calibration of capacitance sensors considering clay content and bulk density. *Soil Res.* **2016**, *54*, 111–119. [CrossRef]
43. Cardenas-Lailhacar, B.; Dukes, M.D. Precision of soil moisture sensor irrigation controllers under field conditions. *Agric. Water Manag.* **2010**, *97*, 666–672. [CrossRef]
44. Kassaye, K.; Boulange, J.; Saito, H.; Watanabe, H. Calibration of capacitance sensor for Andosol under field and laboratory conditions in the temperate monsoon climate. *Soil Tillage Res.* **2019**, *189*, 52–63. [CrossRef]

45. Kinzli, K.-D.; Manana, N.; Oad, R. Comparison of Laboratory and Field Calibration of a Soil-Moisture Capacitance Probe for Various Soils. *J. Irrig. Drain. Eng.* **2012**, *138*, 310–321. [CrossRef]
46. Heimovaara, T.J.; Focke, A.G.; Bouten, W.; Verstraten, J.M. Assessing Temporal Variations in Soil Water Composition with Time Domain Reflectometry. *Soil Sci. Soc. Am. J.* **1995**, *59*, 689–698. [CrossRef]

© 2019 by the authors. Licensee MDPI, Basel, Switzerland. This article is an open access article distributed under the terms and conditions of the Creative Commons Attribution (CC BY) license (http://creativecommons.org/licenses/by/4.0/).

Article

Quantitative Analysis of Elements in Fertilizer Using Laser-Induced Breakdown Spectroscopy Coupled with Support Vector Regression Model

Wen Sha [1], Jiangtao Li [1], Wubing Xiao [1], Pengpeng Ling [1] and Cuiping Lu [2,*]

[1] Key Laboratory of Intelligent Computing and Signal Processing of Ministry of Education, School of Electric Engineering and Automation, Anhui University, Hefei 230061, China
[2] Laboratory of Intelligent Decision, Institute of Intelligent Machines, Chinese Academy of Sciences, Hefei 230031, China
* Correspondence: cplu@iim.ac.cn; Tel.: +86-551-6559-5025

Received: 21 June 2019; Accepted: 22 July 2019; Published: 25 July 2019

Abstract: The rapid detection of the elements nitrogen (N), phosphorus (P), and potassium (K) is beneficial to the control of the compound fertilizer production process, and it is of great significance in the fertilizer industry. The aim of this work was to compare the detection ability of laser-induced breakdown spectroscopy (LIBS) coupled with support vector regression (SVR) and obtain an accurate and reliable method for the rapid detection of all three elements. A total of 58 fertilizer samples were provided by Anhui Huilong Group. The collection of samples was divided into a calibration set (43 samples) and a prediction set (15 samples) by the Kennard–Stone (KS) method. Four different parameter optimization methods were used to construct the SVR calibration models by element concentration and the intensity of characteristic line variables, namely the traditional grid search method (GSM), genetic algorithm (GA), particle swarm optimization (PSO), and least squares (LS). The training time, determination coefficient, and the root-mean-square error for all parameter optimization methods were analyzed. The results indicated that the LIBS technique coupled with the least squares–support vector regression (LS-SVR) method could be a reliable and accurate method in the quantitative determination of N, P, and K elements in complex matrix like compound fertilizers.

Keywords: fertilizer; support vector regression; laser-induced breakdown spectroscopy; grid method; genetic algorithm; particle swarm optimization; least squares

1. Introduction

The foundation of precise fertilization is accurately obtaining the content of elements in compound fertilizers to maximize their benefits. At present, the main sensing methods used by fertilizer manufacturers are national standard methods [1], inductively coupled plasma–atomic emission spectroscopy (ICP-AES) [2], flame atomic absorption spectrometry (FAAS) [3], atomic absorption spectroscopy (AAS) [4], near infrared reflectance spectroscopy (NIRS) [5], etc. These sensing methods need field samples and pretreatment before laboratory analysis, which are time-consuming, labor-intensive, and expensive requirements. Meanwhile, due to the contents of N, P, and K in compound fertilizer being typically high, fertilizer must be diluted several times before measuring, which leads to increase systematic errors. Further, these sensing methods cannot provide real-time information during the production process. At present, the pass rate of compound fertilizer products is only 97% [6], and non-qualifying products need to be returned to the factory for reprocessing, causing huge economic losses. Therefore, a technology that can quickly and accurately detect the elemental content in compound fertilizers is urgently needed.

Laser-induced breakdown spectroscopy (LIBS) is an ideal laser high-temperature ablation spectroscopy technique because it provides fast (in the order of milliseconds), insitu (smaller sample size), non-destructive, safe, environmentally friendly (no secondary pollution), multi-element analysis, and direct analysis of any state of matter. It has been successfully used in applications relating to water pollution [7], coal combustion [8], agriculture [9], space exploration [10], etc. In recent years, some studies have used the LIBS technique to detect the components of fertilizer. Andrade et al. turned liquid fertilizer into a solid state and detected its content using the LIBS technique. The levels of Cu, K, Mg, Mn, Zn, As, Cd, Cr and Pb in the fertilizer were analyzed, and the detection error range was ~0.02%–0.06%, which demonstrated that this method provides accurate measurement of liquid fertilizer [11]. Nicolodelli et al. used single-pulse and dual-pulse LIBS technology to measure phosphate rock and organic phosphate fertilizer. The samples were identified by principal component analysis (PCA) and partial least squares regression (PLSR), and the recognition result was able to reach a 95% confidence level, which showed that LIBS technology can be used to rapidly classify phosphate fertilizers in situ [12]. Yao et al. analyzed phosphorus and potassium elements in compound fertilizers. A quantitative analysis model was established by using the partial least squares (PLS) method in Unscrambler software. The obtained results were superior to those obtained using traditional methods, but the accuracy of detection still needed improvement [13]. Marangoni et al. used LIBS technology to analyze phosphorus in 26 different organic and inorganic fertilizers. Baseline correction and peak intensity normalization were used to pre-process the spectrum. The correlation between the measured value of LIBS and the true value was improved, but the absolute error of the two verification samples was close to 5% [14]. Andradeet al. optimized the LIBS system parameters and directly analyzed the levels of Cd, Cr, Pb, B, Cu, Mn, Na, Zn, Ca, and Mg elements in solid compound fertilizer. The quantitative analysis results were compared to those obtained using ICP-AES, and the correlation was good. The detection limit of the above elements was determined to be ~2 ppm–1%. These results demonstrated the ability of LIBS to be used for rapid analysis of fertilizer [15]. Liao et al. used LIBS to analyze the phosphorus content in compound fertilizer. The correlation coefficient increased from 0.83 to 0.98 when considering the influence of the oxygen characteristic line, and the relative error was only ~0.38%–1.70%. However, the number of samples was too small, so further modeling should be completed if the method is to be applied to actual field detection [16].

The above studies all showed that LIBS technology can be used to detect the types and contents of elements in chemical fertilizers. However, the accuracy and reliability of the quantitative analysisis still a shortcoming of LIBS, which greatly limits its practical application. Thus, there are many problems to be solved before it can be used in actual applications in rapid sensing.

Support vector regression (SVR) is a machine learning method based on statistical learning theory. It uses interval maximization to carry out model training by mapping difficult problems in the original space to a higher-dimensional space and seeking interval maximization to calculate the optimal linear hyper plane. The number of compound fertilizer samples collected in this experimental method was small, which is suitable for the statistical analysis of small samples [17]. In small samples, nonlinear pattern recognition has certain advantages. Zhang et al. employed a LIBS technique coupled with SVR and PLS methods to perform quantitative and classification analysis of 20 slag samples. The results showed that the SVR model could eliminate the influence of nonlinear factors due to self-absorption in the plasma and provide a better predictive result. It has been confirmed that the LIBS technique coupled with the SVR method is a promising approach to achieving the online analysis and process control of slag [18]. Shi et al. compared PLSR and SVR methods for quantitative analysis of the concentrations of five main elements (Si, Ca, Mg, Fe, and Al) in sedimentary rock samples. The parameter optimization method used was genetic algorithm (GA). The results demonstrated that the SVR model performed better, with more satisfactory accuracy and precision under the optimized conditions [19]. He et al. employed single-pulse and double-pulse LIBS to analyze nutrient elements in soil. Good performance was obtained using the PLSR and LS-SVR calibration model, with R^2 greater than 0.95 in both the calibration and prediction sets for all nutrient elements. The results indicated that LIBS combined with PLSR and LS-SVR could be a good method for detecting nutrient elements in soil [20]. Liu et al. also used the PLSR and LS-SVR quantitative analysis method to detect the Cd content in soil. The results showed that the LS-SVR model under an Ar atmosphere obtained the best performance. The root-mean-square error for calibration (RMSEC) and the root-mean square error for prediction (RMSEP) were only 0.026 and 0.034, respectively, which demonstrated the ability of LIBS for the accurate quantitative detection of Cd in soil [21]. To the best of our knowledge, the simultaneous quantitative detection of N, P, and K elements based on LIBS coupled with SVR models by different parameter optimization methods has not been investigated.

The purpose of this paper was to explore the detection ability of LIBS for N, P, and K elements in fertilizer, and to find a fast and accurate quantitative analysis method. A selection of 58 fertilizer samples were provided by Huilong Chemical Fertilizer Plant, Anhui, China. The contents of all three elements in compound fertilizer were determined by ICP-AES. Four different parameter optimization methods were employed to establish the SVR model for quantitative analysis of all elements. The accuracy levels of the four SVR models were compared based on the performance of each model.

2. Materials and Methods

2.1. Sample Preparation

In this study, 58 compound fertilizer samples were collected and placed into sealed plastic bags to avoid contamination. As the compound fertilizer products had been pelletized, they had to be broken into powder before spectral scanning. All fertilizer powders were then sieved through a 60 mesh screen. A total 2 g of powder from each of the 58 fertilizer samples was weighed and pressed into tablets 30 mm in diameter and 2 mm in thickness, using 5 MPa force for 60 s (769YP-40C, KQ, Tianjin, China). The reference concentrations of N, P, and K in these samples were analyzed by ICP-AES. The statistics of the N, P, and K concentrations in the compound fertilizer samples are listed in Table 1.

Table 1. Statistics of the effective constituents of compound fertilizer samples.

Properties	Total Nitrogen (TN/%)	P_2O_5 (%)	K_2O (%)
Minimum value	13.60	14.50	14.40
Maximum value	15.60	16.70	16.40
Mean value	14.42	15.79	15.39
Standard deviation values	2.86	2.97	3.29

2.2. Experimental Setup

The self-built LIBS system used in this experiment is shown in Figure 1. A Q-switched Nd: YAG pulsed laser (ICE450, 1064 nm, 6 ns pulse duration, Big Sky Laser Technologies, Morgan Hill, CA, USA; note that the company has changed its name to Quantel Laser) was used to generate the plasma on the compound fertilizer pellet. The pulse laser energy was 100 mJ, and it was focused with a 2.54 cm diameter, 4.5 cm focal length convex lens onto the fertilizer sample. The spot diameter size of the pulsed laser was approximately 0.5 mm, and the peak power density on the compound fertilizer sample was able to reach 2.2 GW/cm^2. The light emitted from the plasma was collected via a quartz lens with 3.5 cm focal length and transmitted via an optical fiber with a diameter of 200 μm to a spectrometer (Avantes-ULS2048-USB2, Avantes, Apeldoorn, The Netherlands). The spectrometer used has four channels containing separate gratings and a charge-coupled device array, and all spectra were taken simultaneously in the wavelength ranges of 190–510 and 690–890 nm. The resolution of the spectrometer was approximately 0.1 nm. The spectrometer was triggered by the laser Q-switch output, and it had a digital delay generator which can control the gate delay. In this experiment, the delay time and the integration time were set for spectra acquisition at 1.28 μs and 1.05 ms (spectrometer minimum integration time), respectively [22]. Sample tablets were placed on the X-Y rotary stage, the speed of which can be adjusted by stepper motor, and the laser beam was adjusted to focus 3 mm below the sample surface to acquire LIBS spectra [23]. Argon was then passed through the cylinder, draining the air and forming an Ar atmosphere on the surface of the compound fertilizer sample, as shown in

Figure 1. Thus, the sample was immersed in an Ar atmosphere. In order to eliminate shot-to-shot fluctuation, each sample was measured eight times, and each spectrum was collected with an average of 20 laser shots.

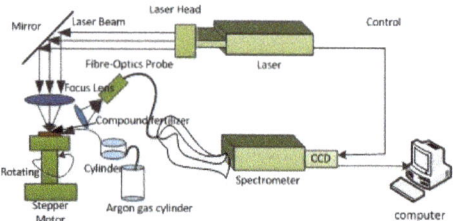

Figure 1. Schematic diagram of the laser-induced breakdown spectroscopy (LIBS) system for fertilizer samples.

2.3. SVR Algorithm Model Establishment

Under local thermal equilibrium (LTE) conditions and ignoring self-absorption effects, the measured characteristic atomic spectrum of ionic line intensity in LIBS spectroscopy can be expressed as

$$I_{k,i} = FC_s \frac{A_{k,i} g_k}{U_s(T)} \exp(-E_k/k_B T) \tag{1}$$

where the subscripts k and l indicate the upper and lower energy levels of the transition line, respectively; F is an instrumental constant for fixed experimental conditions; C_s is the atomic or ionic number density of the specific element; g, A, and $U(T)$, are the statistical weight, transition probability and partition function at temperature T, respectively; k_B and E represent the Boltzmann constant and excitation energy, respectively; and I is the spectrally integrated line intensity. When the plasma is in the local thermal equilibrium state, the plasma temperature can be approximated as a constant. Equation (1) can be simplified as,

$$C_S = A I_{k,i} \tag{2}$$

The concentration of the element to be tested in the sample can be calculated according to Equation (2). However, due to the influence of the matrix effect, parameter A is difficult to determine experimentally. Moreover, when the concentration of the element increases, as self-absorption effect occurs, and the relationship between C_s and $I_{k,i}$ can be expressed as,

$$C_S = K_b (I_{k,i})^b \tag{3}$$

where K_b is the proportionality factor and b is the absorption coefficient. We simplified $I_{k,i}$, to I, and substituted it as a variable to support vector machine regression objective function,

$$C_S = \Sigma_{i \in v} \partial_i k_{libs}(I_i, I) + b \tag{4}$$

where v is the set of support vectors; ∂_i is the Lagrange multiplier; $k_{libs}(I_i, I)$ is the kernel function; and b is the constant.

A hybrid kernel function for LIBS is obtained by considering the relationship between the element concentration and spectrum line intensity (Equations (2) and (3)):

$$k_{libs}(I_i, I) = cII_i + (1-c)\exp(\frac{-\|I - I_i\|}{2g^2}) \tag{5}$$

The mixed kernel function consists of two parts; the former is a linear kernel function cII_i, and the latter is a radial basis kernel function. The support vector machine kernel function is often used to solve nonlinear mapping problems in data. A large number of experiments and data have shown that the radial basis kernel function has high fitting and prediction accuracy, so it is usually selected as a kernel function for research. The adjustment of the parameters in the SVR largely determines the regression effect. When the radial basis kernel function is selected as the kernel function, the penalty coefficient c and the kernel parameter g are mainly optimized in the quantitative model of the compound fertilizer element analysis [24].

2.4. Parameter Optimization Methods

In related research, SVR optimization methods have mainly been summarized as non-heuristic and heuristic. The traditional grid search method (GSM) and experimental method are non-heuristic, but the experimental method uses several experiments to compare parameters and determine the optimal ones. It is time-consuming and not easy to find the optimal parameters, so this research did not use the experimental method.

In the GSM method, X.L. Liu et al. proposed that the parameters c and g should be divided into an equal grid within a certain spatial range [25]. Every grid node then represents a set of parameters. In the optimization process, the optimal parameters were found by gradual approximation of all the nodes in the grid, and the c and g parameters with the smallest regression mean square error were taken as the optimal parameters. A large number of experimental studies have shown that the parameters c and g have interval sensitivity. If the parameter optimization interval can first be roughly determined, then an accurate search can be performed to reduce unnecessary calculations. First, a large step size should be used to perform a rough search in a large range and to select a set of c, g values for the minimum regression mean square error. If multiple sets of c and g values correspond to the minimum regression mean square error in the parameter selection process, then the group of c and g values with the smallest parameter c should be selected as the best parameters. Too high a value of parameter c would lead to an over-learning state, that is, a state where the RMSEC is small, but that of the prediction set is large, and the generalization ability of the SVR is reduced. After finding the local optimal parameters, we selected a cell in the vicinity of this group parameters and used a small step size to perform the second, finer search to find the final optimal parameters. Chen P W et al. proposed a global probability search algorithm based on biological mechanisms such as natural selection and genetic variation [26]. As with other heuristic search methods, the evolutionary mechanisms of organisms are simulated during evolutionary computation, starting from a set of solutions and evaluating their performance. Hybridization and mutational gene manipulations are then performed to generate a group of next-generation solutions with better performance metrics, until the final search for the global optimal solution. The particle swarm optimization (PSO) algorithm is proposed by J. Kennedy and R.C. Eberhart et al. and based on the study of the predation behavior of birds [27]. The solution of each problem is regarded as a bird in the search space, denoted as particles. In each iteration, the particle will track two "extreme values" to update itself; one is the optimal solution found by the particle itself, and the other is the current optimal solution found by the entire population. This extreme value is the global extreme.

In 1999, Suyken et al. added the squared error term to the standard SVR objective function and proposed the LS-SVR method [28]. In this method, the observed value is the sample value, and the theoretical value is the assumed fitting function. The fitting function model is then obtained when the objective function is the smallest. We set the objective function as

$$h_\theta(x_1, x_2, \ldots x_n) = \theta_0 + \theta_1 x_1 + \ldots + \theta_{n-1} x_{n-1} \tag{6}$$

and the loss function as,

$$J(\theta) = \frac{1}{2}(X\theta - Y)^T(X\theta - Y) \tag{7}$$

After derivation and sorting out the parameters as,

$$\theta = (X^T X)^{-1} X^T Y \tag{8}$$

It can be seen that the LS algorithm is simple and efficient. The constraint avoids the quadratic programming in the objective function and solves the problems of robustness, sparseness, and large-scale operation, which greatly shortens the optimization time. However, there are also some limitations: (1) when the inverse matrix of $X^T X$ does not exist, the LS algorithm is no longer applicable, and data processing needs to remove redundant features; (2) the fitting function must be a linear function, which must be converted before use; and (3) when the sample feature number N is large, it takes a lot of time to calculate the inverse matrix, and it may not be able to be calculated. In this study, the number of compound fertilizer samples was small, and the characteristic dimension of SVR was declining. Thus, the LS algorithm could be used for parameter optimization.

3. Results

3.1. Spectral Analysis

Due to the existence of a large number of matrix element emission lines in the compound fertilizer, many characteristic lines interfered with each other. When selecting the characteristic line of an element, an unsaturated line with a high signal-to-noise ratio should be selected. According to the National Institute of Standards and Technology (NIST) database, the characteristic lines of elemental phosphorus are 213.5 nm, 214.9 nm, 215.4 nm, 253.4 nm, 253.6 nm, 255.3 nm, and 255.5 nm; the characteristic lines of elemental K are 404.7 nm, 766.5nm, and 769.9 nm; and the characteristic lines of elemental N are 742.4 nm, 744.2 nm, 746.8 nm, 856.7nm, 859.4 nm, 862.9 nm, 870.3 nm, 871.2 nm, and 871.8 nm. Figure 2 shows the spectrum of a compound fertilizer sample (Sample No.1) in the ranges of 210–405 nm and 740–890 nm. Although the two characteristic lines of P at 253.4 nm and 253.6 nm were strong, they were easily interfered with by the characteristic line of iron (Fe). In addition, the two lines at 255.3 nm and 255.5 nm were too close to distinguish. The characteristic lines at the wavelengths of 213.5 nm, 214.9 nm, and 215.4 nm were not interfered by other elements. It can be seen from Figure 2 that the characteristic lines of N were observed easily without any interference by other spectral lines, and the intensity at 746.8 nm was the strongest. The characteristic lines of elemental K in the compound fertilizer were not rich. Only three characteristic lines at 404.4 nm, 766.5 nm, and 769.9 nm were observed. The line at 404.4 nm was possibly interfered with by the characteristic line of Fe at 404.8 nm. However, the lines at 766.5 nm and 769.9 nm were too strong to be due to self-absorption. Elemental oxygen is also one of the main ingredients of compound fertilizer; the characteristic lines of O are 777.2 nm, 844.6 nm, and 882.0 nm.

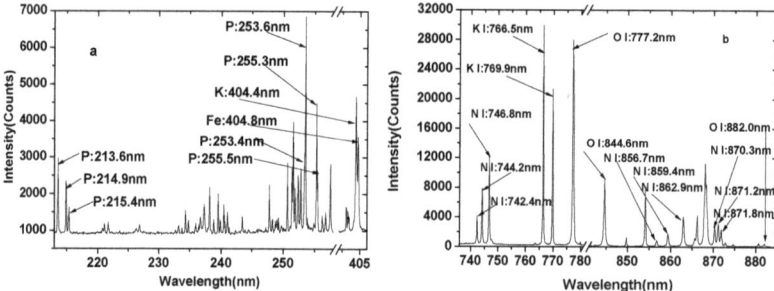

Figure 2. LIBS spectra of a compound fertilizer sample in the ranges of (**a**) 210–405 nm and (**b**) 740–890 nm.

3.2. Univariate Analysis

Before modeling, the 58 fertilizer samples were divided into a calibration set (43 samples) and a prediction set (15 samples) using the Kennard–Stone (KS) method. The univariate calibration models were constructed using the line intensities (the height of Lorentz fits) of N at 746.8 nm, P at 213.6 nm, and K at 404.4 nm versus the corresponding contents. Figure 3a–c shows the calibration curves of N, P, and K elements, respectively. Figure 3a indicates the linear trend between the N line intensity and content, with a coefficient of correlation of 0.809. For P, the coefficient of correlation is 0.909, while it is 0.857 for K. For all three elements, the correlation coefficients cannot meet practical measurement needs and should be improved for further quantitative analysis.

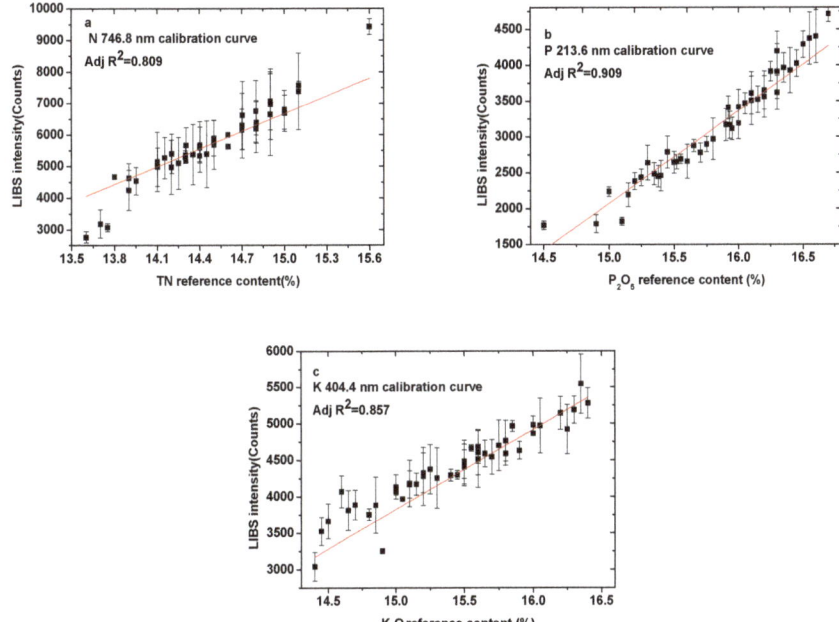

Figure 3. Calibration curves of the elemental spectral lines: (**a**) N: 746.8 nm, (**b**) P: 213.6 nm, (**c**) K: 404.4 nm.

3.3. SVR Analysis Models of Compound Fertilizer

SVR is a multivariate analytical technique which can make full use of spectral information and improve the accuracy of quantitative analysis by reducing the matrix effect. Calibration sets were used to construct the SVR model using the contents of N, P, and K elements and the LIBS spectral signal. These correlations were used to predict the contents of the prediction set. For each element, a proper spectral range was selected for modeling analysis in order to avoid over-fitting of the model. The reduced spectral ranges of 740–890 nm for N, 210–260 nm and 770–885 nm for P, and 400–410 nm and 770–885 nm for K were used to obtain the calibration model. MATLAB software was used for SVR model construction. The statistical parameters that determine the capacity of the regression model are the training time, the determination coefficients of the calibration set (R^2_C) and prediction set (R^2_P), and the RMSEC and RMSEP, which are given in this paper.

3.3.1. Particle Swarm Optimization

Figure 4a–f shows the calibration and prediction results of SVR models using the PSO algorithm for N, P, and K elements, respectively. All of the parameters for both the calibration and prediction sets are presented in Table 2. The training times for N, P, and K were 2.98 s, 3.31 s, and 4.32 s, respectively. The determination coefficients for the calibration sets (R^2_C) were 0.930 for N, 0.980 for P, and 0.979 for K. Those for the prediction sets (R^2_P) were 0.923, 0.964, and 0.952. Meanwhile, for N, P, and K, respectively, the values of the RMSEC were 0.0996, 0.0701, and 0.0894, and those of the RMSEP were 0.0952, 0.0677, and 0.0921. The PSO-SVR optimization data showed that there was little difference between the three optimization times. The correlation coefficient between the N element calibration set and the prediction set was small and the error was large. The determination coefficients R^2_C and R^2_P for N indicated that the correlation needed to be improved. Meanwhile, the RMSEC and RMSEP for elements N and K were large.

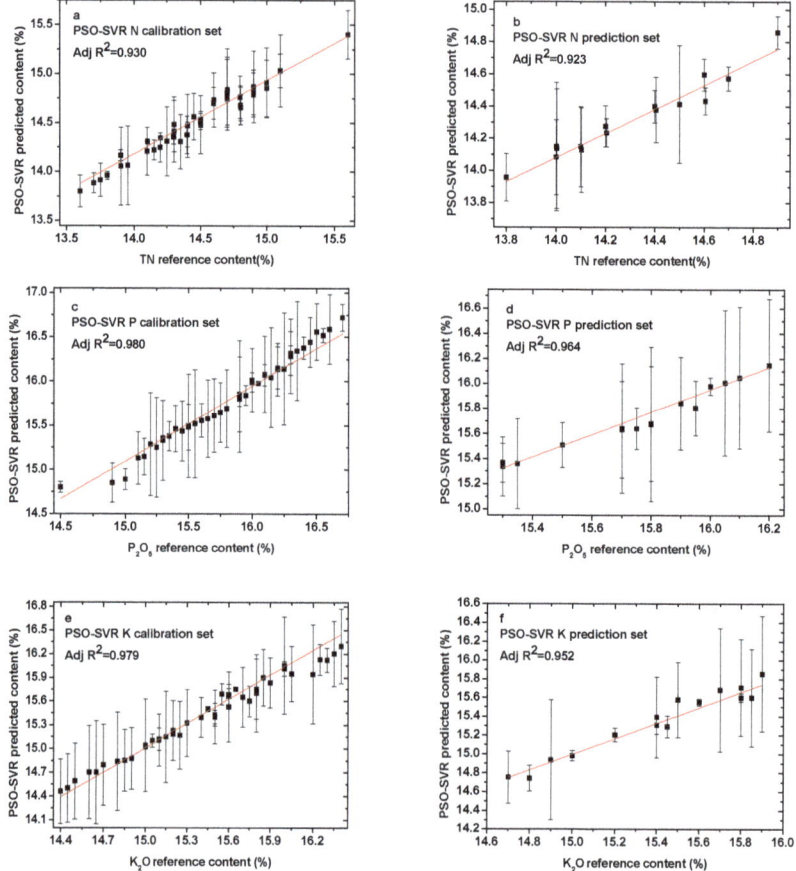

Figure 4. Comparison between PSO-SVR predicted content and reference content present in the (**a**) N calibration set; (**b**) N prediction set; (**c**) P calibration set; (**d**) P prediction set; (**e**) K calibration set; and (**f**) K prediction set.

Table 2. The results of the particle swarm optimization–support vector regression (PSO-SVR) model for elements N, P, K.

Element	t/s	R^2_C	RMSEC	R^2_P	RMSEP
N	2.98	0.930	0.0996	0.923	0.0952
P	3.31	0.980	0.0701	0.964	0.0677
K	4.32	0.979	0.0894	0.952	0.0921

3.3.2. Genetic Algorithm

The optimal calibration and prediction results of N, P, and K elements based on parameters obtained using the genetic algorithm are shown in Figure 5a–f. Table 3 presents the detailed parameter results. The training times increased dramatically for all three elements. They were 5.67 s, 5.09 s, and 12.37 s, for N, P, and K, respectively. The R^2_C of N increased from 0.930 to 0.948, but the R^2_P

changed from 0.923 to 0.936. The values of RMSEC and RMSEP for N were reduced to 0.0688 and 0.0694, which was better than the PSO algorithm. However, for elements P and K elements, the parameter optimization results of PSO and GA were basically the same, with only the R^2_P value of P increased, from 0.964 to 0.985. This indicates that GA and PSO had many similarities, but GA was less inefficient due to random variation.

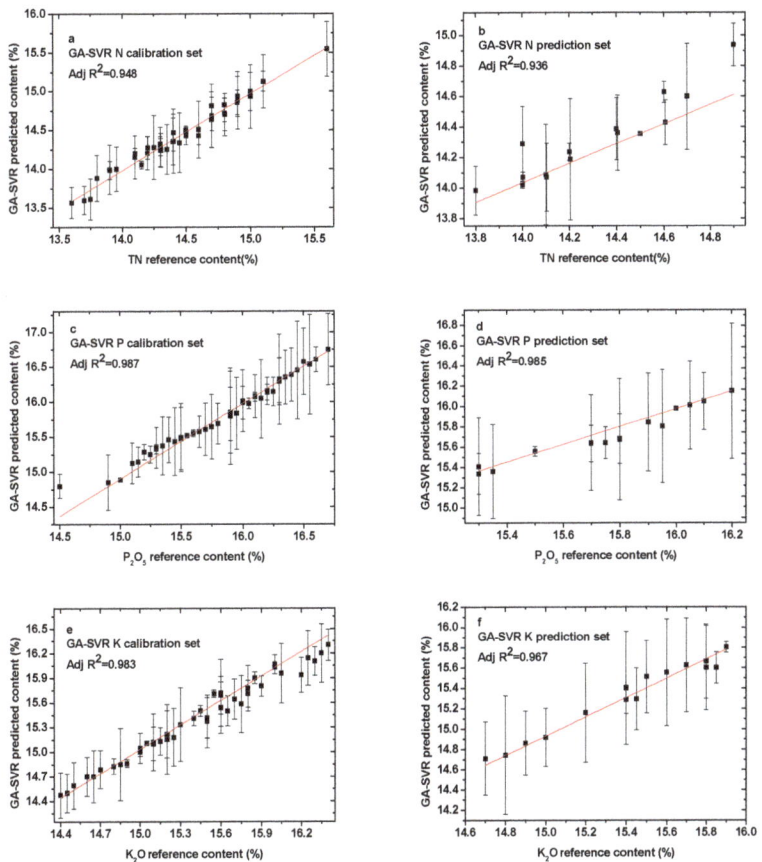

Figure 5. Comparison between GA-SVR predicted content and reference content present in the (**a**) N calibration set; (**b**) N prediction set; (**c**) P calibration set; (**d**) P prediction set; (**e**) K calibration set; and (**f**) K prediction set.

Table 3. The results of the genetic algorithm–support vector regression (GA-SVR) model for N, P, and K.

Element	t/s	R^2_C	RMSEC	R^2_P	RMSEP
N	5.67	0.948	0.0688	0.936	0.0694
P	5.09	0.987	0.0692	0.985	0.0680
K	12.37	0.983	0.0775	0.967	0.1007

3.3.3. Grid Search Method

A quantitative analysis SVR model of elements N, P, and K in the compound fertilizer was established based on parameter optimization by GSM. Figure 6a–f shows the calibration and prediction results of GSM parameter optimization for the elements N, P, and K. All of the results are presented in

Table 4. For N, the training times, RMSEC, and RMSEP were almost identical to the results of the GA algorithm, but the R^2_C and R^2_P increased greatly. Among the three parameter optimization methods, for the P element, the results obtained by the GSM method were the best, and the training time was only 1.76 s. However, the best results for the K element were obtained by using the GA algorithm.

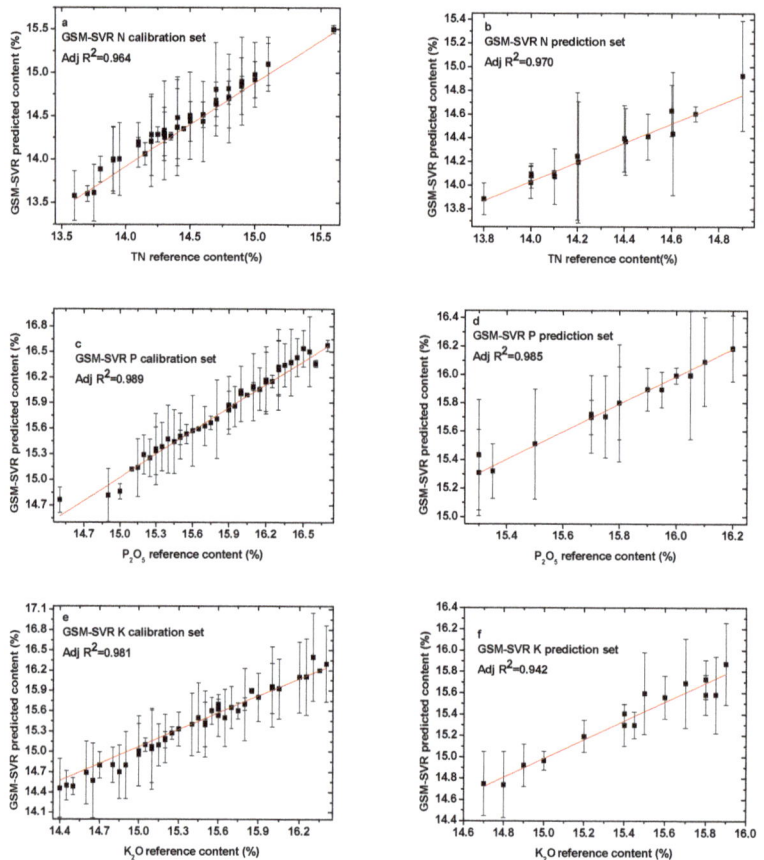

Figure 6. Comparison between GSM-SVR predicted content and reference content present in the (**a**) N calibration set; (**b**) N prediction set; (**c**) P calibration set; (**d**) P prediction set; (**e**) K calibration set; and (**f**) K prediction set.

Table 4. The results of grid search method–support vector regression (GSM-SVR) model for elements N, P, and K.

Element	t/s	R^2_C	RMSEC	R^2_P	RMSEP
N	4.89	0.964	0.0685	0.970	0.0712
P	1.76	0.989	0.0632	0.985	0.0576
K	4.21	0.981	0.0942	0.942	0.0969

It can be seen that the quantitative analysis of the elements N, P, and K in compound fertilizer from the above three parameter optimization results was better than that by traditional methods [23], but it should be further improved.

3.3.4. Least Squares

The calibration and prediction results of the LS parameter optimization models for all analyzed elements are provided in Figure 7a–f. All of the specific parameter optimization results are stated in Table 5. It can be observed in Figure 6 that the calibration and prediction data points fitted well, indicating that the LS-SVR model had reliable prediction power for the quantitative analysis of compound fertilizer. In addition, the R^2 values of the calibration and prediction sets for all elements were obviously improved. All the values of R^2 were greater than 0.99. Meanwhile, for all elements, the values of RMSEC and RMSEP were significantly reduced. The values of RMSEC were reduced to 0.0240, 0.0258, and 0.0248 for elements N, P, and K, respectively, and the values of RMSEP were only 0.0218 for N, 0.0261 for P, and 0.0248 for K. The training time was significantly reduced; all the values of t were smaller than 0.3 s, which is more suitable for the rapid quantitative analysis of elements in compound fertilizer. Thus, it was demonstrated that the LS-SVR model can be developed to predict the content of unknown samples.

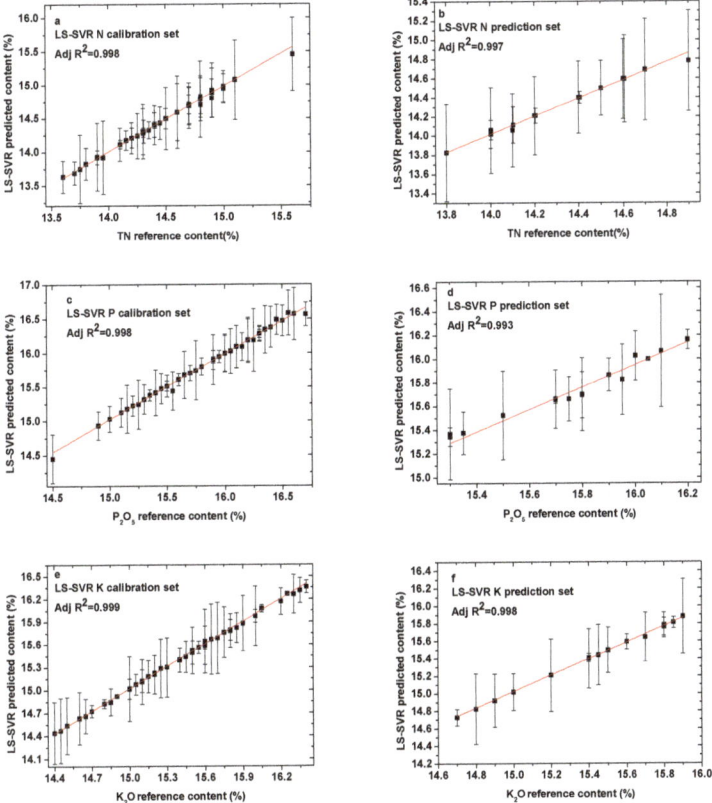

Figure 7. Comparison between the LS-SVR predicted content and reference content present in the (**a**) N calibration set; (**b**) N prediction set; (**c**) P calibration set; (**d**) P prediction set; (**e**) K calibration set; and (**f**) K prediction set.

Table 5. The results of the least squares–support vector regression (LS-SVR) model for elements N, P, and K.

Element	t/s	R^2_C	RMSEC	R^2_P	RMSEP
N	0.23	0.998	0.0240	0.997	0.0218
P	0.02	0.998	0.0258	0.993	0.0261
K	0.02	0.999	0.0239	0.998	0.0248

4. Conclusions

In summary, we demonstrated that LIBS coupling with the SVR method can provide a robust and accurate technology for the analysis of compound fertilizers. Four parameter optimization SVR models—the PSO model, the GA model, the GSM model, and the LS model were employed to quantitatively analyze elements N, P, and K in fertilizer. In general, the complex element composition of fertilizer causes difficulty for traditional calibration methods. For the conventional PSO, GA, and GSM parameter optimization methods, the determination coefficients for all three elements were greater than 0.92, and the root-mean-square errors were less than 0.101. However, the best parameter optimization model was the GSM method for N, GSM for P, and GA for K. A parameter optimization method suitable for quantitative analysis of all three elements was still needed. A LS-SVR model was then used to establish a quantitative analysis model for the three elements. From the results of the LS-SVR model, calibration and prediction models were obtained for the three elements with determination coefficients close to 1. For elements N, P, and K, respectively, the values of RMSEC were 0.0240, 0.0258, and 0.0239, and those of RMSEP were 0.0218, 0.0261, and 0.0248. After considering the evaluation indicators of the model comprehensively, the LS-SVR model is regarded as the most suitable for quantitative analysis of the three elements, with robust and satisfactory modeling performance. This model could therefore provide a basis for real-time analysis of N, P, and K elements in compound fertilizers. Furthermore, methods for improving the accuracy of the LIBS technique in the rapid detection of compound fertilizer on production lines will be the focus of future work.

Author Contributions: J.L., C.L. and W.S. conceived and designed the experiment; P.L. performed the experiments; J.L. and W.X. contributed to data analysis; C.L. and W.S. analyzed the data and wrote the whole paper; all authors reviewed the manuscript.

Funding: This research was funded by National Natural Science Foundation of China (61505001) and Science Technology Service for Regional Key Projects of Chinese Academy of Sciences (CAS) (KFJ-STS-QYZD-106).

Acknowledgments: The fertilizer samples tested in this research were provided by the compound fertilizer production line of Bengbu Huilong Chemical Fertilizer Plant, Anhui, China.

Conflicts of Interest: The authors declare no conflict of interest.

References

1. Cui, S.L. Effective phosphorus detection in compound fertilizers by weight method. *Mod. Agric. Sci. Technol.* **2016**, *5*, 231–232.
2. Wang, M.R.; Yuan, Y.M.; Tao, N.L. Determination of available phosphate content in calcium magnesium phosphate by ICP-AES. *Mod. Agric. Sci. Technol.* **2012**, *7*, 20–21.
3. Yuan, C.; Lv, G.L. Determination of chromium content in compound fertilizer by flame atomic absorption spectrometry. *Chem. Fertil. Ind.* **2018**, *45*, 17–18.
4. Xiao, Z.M.; Liang, H.J. Improvement of rapid determination method of available magnesium content in compound fertilizer. *Chem. Fertil. Ind.* **2015**, *42*, 12–15.
5. Song, L.; Zhang, H.; Ni, X.-Y.; Wu, L.; Liu, B.-M.; Yu, L.-X.; Wang, Q.; Wu, Y.-J. Quantitative analysis of contents in compound fertilizer and application research using near infrared reflectance spectroscopy. *Guang Pu Xue Yu Guang Pu Fen Xi* **2014**, *34*, 73–77. [PubMed]
6. Cai, Y.Q.; Wang, J.; Wang, Z.Q.; Meng, F.M.; Liu, J.X. A production method of granular urea formaldehyde slow-release compound fertilizer. *Phosphate Compd. Fertil.* **2017**, *32*, 18–19.
7. Zhang, D.C.; Hu, Z.Q.; Su, Y.B.; Hai, B.; Zhu, X.L.; Zhu, J.F.; Ma, X. Simple method for liquid analysis by laser-induced breakdown spectroscopy (LIBS). *Opt. Express* **2018**, *26*, 18794–18802. [CrossRef]

8. Yan, C.; Qi, J.; Liang, J.; Zhang, T.; Li, H. Determination of coal properties using laser-induced breakdown spectroscopy combined with kernel extreme learning machine and variable selection. *J. Anal. At. Spectrom.* **2018**, *33*, 2089–2097. [CrossRef]
9. Jull, H.; Künnemeyer, R.; Schaare, P. Nutrient quantification in fresh and dried mixtures of ryegrass and clover leaves using laser-induced breakdown spectroscopy. *Precis. Agric.* **2018**, *19*, 823–839. [CrossRef]
10. Cho, Y.; Horiuchi, M.; Shibasaki, K.; Kameda, S.; Sugita, S. Quantitative Potassium Measurements with Laser-Induced Breakdown Spectroscopy Using Low-Energy Lasers: Application to In Situ K-Ar Geochronology for Planetary Exploration. *Appl. Spectrosc.* **2017**, *71*, 1969–1981. [CrossRef]
11. Andrade, D.F.; Sperança, M.A.; Pereira-Filho, E.R. Different sample preparation methods for the analysis of suspension fertilizers combining LIBS and liquid-to-solid matrix conversion: Determination of essential and toxic elements. *Anal. Methods* **2017**, *9*, 5156–5164. [CrossRef]
12. Nicolodelli, G.; Senesi, G.S.; Perazzoli, I.L.D.O.; Marangoni, B.S.; Benites, V.D.M.; Milori, D.M.B.P. Double pulse laser induced breakdown spectroscopy: A potential tool for the analysis of contaminants and macro/micronutrients in organic mineral fertilizers. *Sci. Total. Environ.* **2016**, *565*, 1116–1123. [CrossRef] [PubMed]
13. Yao, S.; Lu, J.; Li, J.; Chen, K.; Li, J.; Dong, M. Multi-elemental analysis of fertilizer using laser-induced breakdown spectroscopy coupled with partial least squares regression. *J. Anal. At. Spectrom.* **2010**, *25*, 1733. [CrossRef]
14. Marangoni, B.S.; Silva, K.S.G.; Nicolodelli, G.; Senesi, G.S.; Cabral, J.S.; Villas-Boas, P.R.; Nogueira, A.R.A.; Teixeira, P.C.; Benites, V.M.; Milori, D.M.B.P. Phosphorus quantification in fertilizers using laser induced breakdown spectroscopy (LIBS): A methodology of analysis to correct physical matrix effects. *Anal. Methods* **2016**, *8*, 78–82. [CrossRef]
15. Andrade, D.F.; Pereira-Filho, E.R. Direct Determination of Contaminants and Major and Minor Nutrients in Solid Fertilizers Using Laser-Induced Breakdown Spectroscopy (LIBS). *J. Agric. Food Chem.* **2016**, *64*, 7890–7898. [CrossRef] [PubMed]
16. Liao, S.Y.; Wu, X.L.; Li, G.H.; Wei, M.; Zhang, M. Multi-element nonlinear quantitative analysis of phosphorus in compound fertilizer by laser induced breakdown spectroscopy. *Spectrosc. Spectr. Anal.* **2018**, *38*, 271–275.
17. Chen, L. PSO-SVM Learning Algorithm and Its Application in Spatial Data Analysis. Master's Thesis, Xi'an Polytechnic University, Xi'an, China, 2012.
18. Zhang, T.; Wu, S.; Dong, J.; Wei, J.; Wang, K.; Tang, H.; Yang, X.; Li, H. Quantitative and classification analysis of slag samples by laser induced breakdown spectroscopy (LIBS) coupled with support vector machine (SVM) and partial least square (PLS) methods. *J. Anal. At. Spectrom.* **2015**, *30*, 368–374. [CrossRef]
19. Niu, G.; Lin, Q.; Shi, Q.; Xu, T.; Li, F.; Duan, Y. Quantitative analysis of sedimentary rocks using laser-induced breakdown spectroscopy: Comparison of support vector regression and partial least squares regression chemometric methods. *J. Anal. At. Spectrom.* **2015**, *30*, 2384–2393.
20. He, Y.; Liu, X.; Lv, Y.; Liu, F.; Peng, J.; Shen, T.; Zhao, Y.; Tang, Y.; Luo, S. Quantitative Analysis of Nutrient Elements in Soil Using Single and Double-Pulse Laser-Induced Breakdown Spectroscopy. *Sensors* **2018**, *18*, 1526. [CrossRef]
21. Liu, X.; Liu, F.; Huang, W.; Peng, J.; Shen, T.; He, Y. Quantitative Determination of Cd in Soil Using Laser-Induced Breakdown Spectroscopy in Air and Ar Conditions. *Molecules* **2018**, *23*, 2492. [CrossRef]
22. Sha, W.; Niu, P.; Zhen, C.; Lu, C.; Jiang, Y. Analysis of Phosphorus in Fertilizer Using Laser-Induced Breakdown Spectroscopy. *J. Appl. Spectrosc.* **2018**, *85*, 653–658. [CrossRef]
23. Zhang, B.; Ling, P.; Sha, W.; Jiang, Y.; Cui, Z. Univariate and Multivariate Analysis of Phosphorus Element in Fertilizers Using Laser-Induced Breakdown Spectroscopy. *Sensors* **2019**, *19*, 1727. [CrossRef] [PubMed]
24. Qu, J.; Chen, H.Y.; Liu, W.Zh.; Li, Zh.B.; Zhang, B.; Ying, Y.H. Application of support vector machine based on improved grid search in quantitative analysis of gas. *Chin. J. Sens. Actuators* **2015**, *28*, 774–778.
25. Liu, X.L.; Jia, D.X.; Li, H. Research on kernel parameter optimization of support vector machine in speaker recognition. *Sci. Technol. Eng.* **2010**, *10*, 1669–1673.
26. Chen, P.W.; Wang, J.Y.; Lee, H. Model selection of SVMs using GA approach. In Proceedings of the 2004 IEEE International Joint Conference on Neural Networks, Budapest, Hungary, 25–29 July 2004.
27. Lu, J.H. Research on Particle Swarm Optimization Algorithm for Solving Automatic Cotton Matching Problem. Ph.D. Thesis, Zhejiang University, Hangzhou, China, 2011.

28. Castro-Garcia, R.; Agudelo, O.M.; Tiels, K.; Suykens, J.A.K. Hammerstein system identification using LS-SVM and steady state time response. In Proceedings of the 2016 European Control Conference (ECC), Ålborg, Denmark, 29 June–1 July 2016; pp. 1063–1068.

© 2019 by the authors. Licensee MDPI, Basel, Switzerland. This article is an open access article distributed under the terms and conditions of the Creative Commons Attribution (CC BY) license (http://creativecommons.org/licenses/by/4.0/).

Article

Univariate and Multivariate Analysis of Phosphorus Element in Fertilizers Using Laser-Induced Breakdown Spectroscopy

Baohua Zhang [1,*], Pengpeng Ling [2], Wen Sha [2], Yongcheng Jiang [2] and Zhifeng Cui [3]

1. School of Electronics and Information Engineering, Anhui University, Hefei 230061, China
2. School of Electric Engineering and Automation, Anhui University, Hefei 230601, China; ppling@yeah.net (P.L.); ahu001@163.com (W.S.); ycjiang@126.com (Y.J.)
3. Institute of Atomic and Molecular Physics, Anhui Normal University, Wuhu 241000, China; zfcui@mail.ahnu.edu.cn
* Correspondence: qinji1983@126.com; Tel.: +86-551-6386-1237

Received: 5 March 2019; Accepted: 9 April 2019; Published: 11 April 2019

Abstract: Rapid detection of phosphorus (P) element is beneficial to the control of compound fertilizer production process and is of great significance in the fertilizer industry. The aim of this work was to compare the univariate and multivariate analysis of phosphorus element in compound fertilizers and obtain a reliable and accurate method for rapid detection of phosphorus element. A total of 47 fertilizer samples were collected from the production line; 36 samples were used as a calibration set, and 11 samples were used as a prediction set. The univariate calibration curve was constructed by the intensity of characteristic line and the concentration of P. The linear correlation coefficient was 0.854 as the existence of the matrix effect. In order to eliminate the matrix effect, the internal standardization as the appropriate methodology was used to increase the accuracy. Using silicon (Si) element as an internal element, a linear correlation coefficient of 0.932 was obtained. Furthermore, the chemometrics model of partial least-squares regression (PLSR) was used to analysis the concentration of P in fertilizer. The correlation coefficient was 0.977 and 0.976 for the calibration set and prediction set, respectively. The results indicated that the LIBS technique coupled with PLSR could be a reliable and accurate method in the quantitative determination of P element in complex matrices like compound fertilizers.

Keywords: fertilizer; phosphorus element; laser-induced breakdown spectroscopy; chemometrics

---

1. Introduction

The use of compound fertilizers in agriculture to improve soil quality is very common. China's compound fertilizer use ranks first in the world according to statistics. Fertilization in some areas is extremely unreasonable, causing serious environmental pollution [1]. Phosphorus(P) element is a major nutrient element for crops and is very important in agriculture. Quality control is very important for compound fertilizer manufacturers, and can help guarantee the quality of products. At present, the real-time sensing rapid detection of compound fertilizer production has been mainly manual sampling, sample preparation and laboratory testing. The traditional sensing method of P in compound fertilizers is the phosphomolybdate quinoline gravimetric method [2], which is mature and has high accuracy. However, this sensing detection technique commonly requires the dissolution of the solid sample, which involves the use of high temperatures and strong oxidants. With the development of sensing analysis technology, optical detection methods are increasingly used for the detection of compound fertilizer components, such as flame atomic absorption spectrometry (FAAS), and inductively coupled plasma-mass spectrometry (ICP-MS) [3,4], but, because of a typically high concentration in compound fertilizer, the sensing measurement of P needs to be diluted several times. The aforementioned sensing detection methods are time-consuming, labor intensive, and expensive. As a consequence, the use of the traditional detection techniques increases the systematic errors besides producing large volumes of chemical residues [5].

Laser-induced breakdown spectroscopy (LIBS) is an emerging analytical technique in the current spectroscopic field. The LIBS technique has many advantages, such as in situ detection, real-time, remote sensing capability, multi-elemental analysis, minimal sample preparation, and direct analysis of any state of matter [6]. It has been successfully used in chemical and biological testing [7], water pollution [8], coal combustion [9], agriculture [10], artifacts and jewelry identification [11], space exploration [12], etc. Some works have used the LIBS technique for the analysis of the component of compound fertilizer. Farooq et al. determined P, Mg, and Mn in the fertilizer using LIBS [13]. Quantitative LIBS analysis of phosphorus in 26 different organic and inorganic fertilizers has been reported by Bruno S. Marangoni et al., however, the absolute error of the measurement for the two verification samples is close to 5% [14]. The elements of Cu, K, Mg, Mn, Zn, As, Cd, Cr and Pb in liquid fertilizers were analyzed with LIBS technique by Daniel Fernandes Andrade et al. [15]. S.C. Yao et al. detected phosphorus and potassium elements in the compound fertilizer using LIBS, and the PLS quantitative analysis model was established by using Unscrambler software [16]. Daniel Fandrade et al. have reported an application of LIBS for quantification of the metal elements in solid compound fertilizers [17]. However, quantitative aspects have generally been considered a shortcoming of LIBS, which greatly limits its application. Thus, there are still many problems to be solved prior to routine practical applications.

The univariate calibration considers the emission intensities of excited element and its concentration. However, the fertilizer was a complex sample, which contains many elements of Fe, Si, Mg, Al, and O. All of these elements may produce matrix effects, and also, due to the fluctuations observed in LIBS technique associated with the instruments and sample non-uniformity, many strategies are used for the calibration methods, such as different spectral preprocessing and multivariate calibration models. Internal standardization is a common method used to minimize fluctuations in LIBS technique, which consists of normalizing the analytical signal by an internal signal. Usually, the internal element concentration must be nearly constant [18]. However, the internal element concentration may slightly change from sample to sample, thus, the accuracy of quantitative analysis results still needs to be improved. In chemometrics, partial least-squares regression (PLSR) is one of the multivariate analytical techniques. It is very crucial to reduce the matrix effect when dealing with complex sample [19].

In this work, 47 fertilizer samples provided by the compound fertilizer production line of Hefei Hongsifang Chemical Fertilizer Plant, Anhui, China were used as the testing samples. The concentration of phosphorus in compound fertilizerswas determined by inductively coupled plasma (ICP). The univariate calibration was established by the LIBS intensity and the concentration of P. Then, the internal standardization method and PLSR were used to quantitatively analyze the phosphorus concentration. The main goal of this research is to prove that LIBS technique can be used for on-line rapid detection of phosphorus element in compound fertilizer.

2. Materials and Methods

2.1. Sample Preparation

In this study, 47 compound fertilizer samples were collected and placed into sealed plastic bags so as to avoid contamination by manufacturer. Since these samples had been pelletized, these samples were smashed by using a grinder and sieved through a 60-mesh screen. 2 g powders from each of 47 fertilizer samples was weighed. All fertilizer powders were pressed into tablets with 25 mm diameter and 5 mm in thickness, using 5 MPa pressure for 1 min (769YP-40C, KQ, Tianjin, China).

The actual concentration of P in these samples were analyzed by inductively coupled plasma (ICP). The statistics of the P concentrations in compound fertilizer samples was listed in Table 1.

Table 1. Statistics of the Effective Constituents of Compound Fertilizer Samples.

Properties	P_2O_5 (%)
Minimum value	46.27
Maximum value	49.22
Mean value	47.731
Standard deviation values	0.617

2.2. Experimental Setup

Figure 1 gave out the spectral acquisition system used in this experiment. The laser pulses were generated by using a Q-switched Nd: YAG laser (ICE450, 1064 nm, 6 ns pulse duration, Big Sky Laser Technologies, Morgan Hill, CA, USA; Note that the company has changed its name to Quantel Laser). The laser pulse energy was 100 mJ and focused onto the sample through a lens of focal length 5 cm. The spot size of beam was approximately 0.5 mm and the peak power density on the surface of the compound fertilizer sample reached 2.2 GW/cm^2. When the plasma generated from the fertilizer sample, a quartz lens with 3.5 cm focal length, which was connected to a four-channel spectrometer (Avantes-ULS2048-USB2, Avantes, Apeldoorn, The Netherlands) via a 200-μm diameter optical fiber, was used to collect the spectra from the plasma. The spectrograph signal was integrated with a charge-coupled device detector. This spectrometer can simultaneously take all spectra in the wavelength ranges of 190–510 and 690–890 nm, and the resolution of the spectrometer was approximately 0.1 nm. The laser Q-switch output was used to trigger the spectrometer, and the spectrometer has a digital delay generator, which can control the gate delay. Here, the Q-switched delay time selected for spectra acquisition was 1.28 μs and the integration time was 1.05ms (spectrometer minimum integration time) [20]. A rotary platform on which the fertilizer sample was placed was rotated uniformly to avoid continuous ablation of the same spot. During the experiment, each sample was measured eight times, and each spectrum collected was an average of 20 laser spot, and during each measurement, the fertilizer sample rotated once by adjusting the speed of the stepper motor.

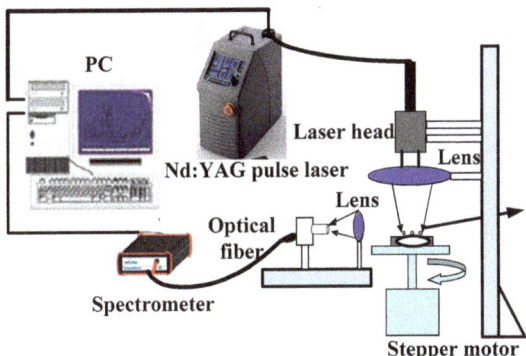

Figure 1. Schematic of LIBS experimental system.

2.3. Chemometrics Methods

In LIBS technique, the calibration curve method, also called univariate analysis, is a traditional quantitative analysis method. The element characteristic line intensity is proportional to the concentration in the sample, when there is no self-absorption [21]. Therefore, the calibration curve can be established by the element concentration and the intensity of LIBS signal. Then, the concentration of an unknown sample can be calculated according to the calibration curve. However, due to the matrix effect, this method is not suitable for quantitative analysis of complex sample. Internal standardization is usually used for quantitative analysis in LIBS technique, which can improve the accuracy of LIBS technique and reduce the fluctuations observed in LIBS measurements, but, there are some principles for selecting the internal standard element [22]. The concentration of the internal standard element should be approximately constant, and the wavelength of the internal standard element should be close to the analytical element. In addition, the excitation potential should be similar for the internal standard element and the analytical element. The coefficient of determination and the root mean squared error, for the calibration set and validation set, were adopted to evaluate the performance of the internal standardization model.

PLSR is widely used for quantitative analysis of LIBS spectra in recent years [23,24]. This method performs quantitative spectral analysis by selecting latent variables [25,26]. Therefore, it is very important to select the latent variables, which directly determined the predictive performance of the calibration model. The PLSR model was established by the LIBS signal intensity and the concentration of P for fertilizer samples. In order to avoid the overfitting of the PLSR model, and also to obtain a reliable and robust PLSR model, full cross-validation was applied. The number of latent variables was determined when the mean squared error was minimum. Furthermore, the statistic parameters for evaluating the performance of PLSR model include the determination coefficient for calibration (R_C^2) and prediction (R_P^2), the root mean square error for calibration (RMSEC) and prediction (RMSEP), and residual predictive deviation (RPD) [27]. All data processing procedures were compiled with MATLAB.

3. Results

3.1. Spectral Analysis

The LIBS spectrum of the compound fertilizer pellet (No.1 sample) in the ranges of 210–220 and 250–260 nm is shown in Figure 2, which includes the emission lines of silicon (Si) and P. Compound fertilizer production enterprises generally use phosphate ore as raw material. Thus, Si is one of the main ingredients of compound fertilizers, and the characteristic lines of Si are 212.4 nm, 221.1 nm, and 221.7 nm according to the National Institute of Standards and Technology (NIST) database. It can be seen from Figure 2 that the compound fertilizer sample contains abundant characteristic lines of P element. The feature spectral lines of P element detected by LIBS were 213.6 nm, 214.9 nm, 215.4 nm, 253.4 nm, 253.6 nm, 255.3 nm, and 255.5 nm. The characteristic lines of 253.4 nm and 253.6 nm are interfered by the characteristic line of Fe element in the compound fertilizer. The adjacent peaks of the two characteristic lines of 255.3 nm and 255.5 nm can be clearly distinguished, and Lorentz double peak fitting is needed when fitting the line intensity. The characteristic lines of 213.6 nm, 214.9 nm and 215.4 nm are not disrupted by other elements.

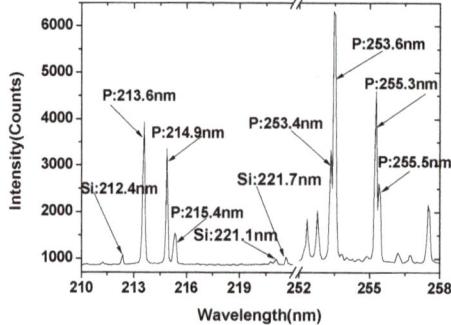

Figure 2. LIBS spectrum of compound fertilizer sample in the ranges of 210–222 and 252–258 nm (n = 20).

In order to obtain a stable signal, the focus of laser beam was adjusted at the position of the fertilizer sample. When the laser focus located on the surface of the fertilizer sample, the distance was recorded as d = 0 mm. Then, the laser focusing system was adjusted, each time moving the laser focus to the surface of the fertilizer sample 1 mm, up to 8mm. The P: 213.6 nm was selected as the analytical line, the relationship of the line intensity and the signal-to-background ratio (SBR) with the laser focus position was shown in Figure 3. The maximum value of the line intensity and the SBR of P were all located at 3 mm below the surface of the fertilizer sample. When the focus of the laser beam gradually moved downward from the surface of the fertilizer sample, the laser pulse energy was more absorbed by the sample, so that the ablation amount was gradually increased, more atoms and ions were in an excited state. But, as the distance between the laser pulse focus and the surface of the fertilizer sample further increased, the radiant power of the laser pulse on the surface of the fertilizer sample gradually decreased, so that the ablation amount of the composite fertilizer sample decreased. However, when the focus of the laser pulse moved down to a certain distance, it was basically difficult to break down the fertilizer sample, so the line intensity and the SBR of the phosphorus element tended to be stable.

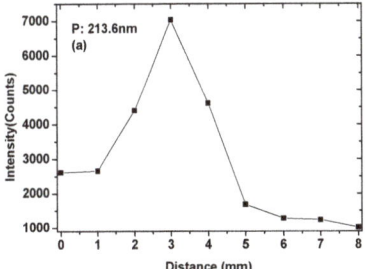 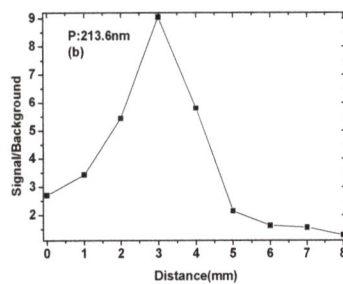

Figure 3. The line intensity of P 213.6 nm (a) and signal to background ratio (b) as a function of detection distance.

3.2. Univariate Analysis

Before modeling, 47 fertilizer samples were split into a calibration set (36 samples) and a prediction set (11 samples) based on the K-S method. The univariate calibration model was constructed by the line intensity (the height of Lorentz fits) of P versus the corresponding concentration [28]. The content of P is in the range of 46.27–49.22%. Figure 4a–c show the calibration curve of three characteristic lines of P. The spectral line (P: 213.6 nm) that obtained the best modeling result was applied for subsequent analysis. Figure 4a indicates the linear trend between line intensity and concentration with a coefficient of correlationof 0.854. For the prediction set, 11 fertilizer samples were used to estimate the prediction accuracy of the LIBS technology. The predicted content of samples can be obtained by taking the line intensity into the calibration fitting curve. The relation between the reference content and LIBS-predicted content for P is shown in Figure 4d, with an R^2 value of 0.923. Although the relative error of the prediction set is not very large, but the correlation coefficient should be improved for further quantitative analysis.

3.3. Multivariate Analysis

According the principles for selecting the internal standard element, the characteristic line of Si 212.4 nm was selected as the internal standard line. The internal standard curve was constructed by calculating the lineintensity ratio of the analytical element and the internal standard element. The ratio of P line intensity (213.6 nm) to that of Si (212.4 nm) versus the concentration of P in fertilizers was used for calibration. Figure 5a shows the internal standard curve for P element. The value of R^2 was obviously improved to 0.932 from 0.854. Similar to the univariate calibration method, the sample content of prediction set were calculated. Figure 5b shows the relation between the reference concentration and LIBS predictedconcentration for P, with R^2 changing from 0.923 to 0.946. The range of relative error was 0.04–0.65%, which was improved with the univariate calibration method. These results indicate that the internal standard method can partly eliminate the instability of the LIBS signal, but the detection sensitivity and the prediction accuracy still need to be improved.

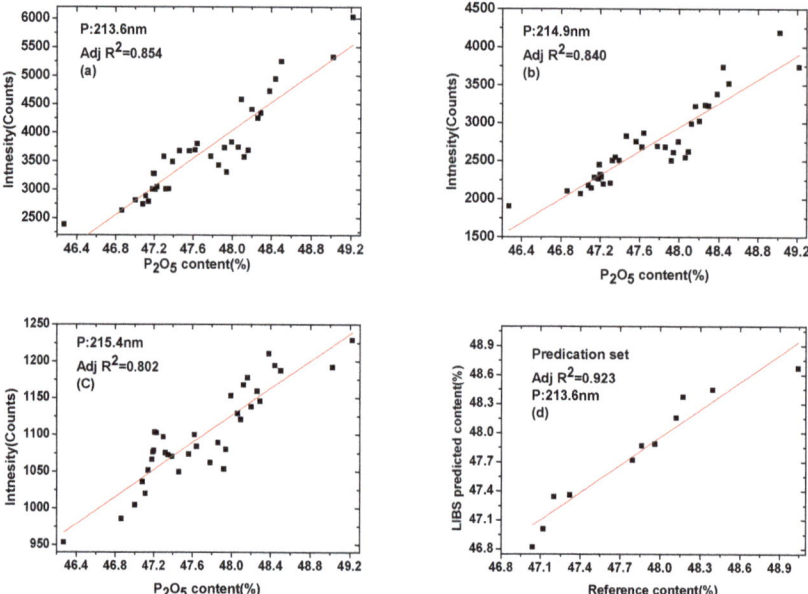

Figure 4. Calibration curves of P spectral line: (**a**) P: 213.6 nm, (**b**) P: 214.9 nm, (**c**) P: 215.4 nm, and (**d**) the relation of LIBS predicted value and reference value for the prediction set.

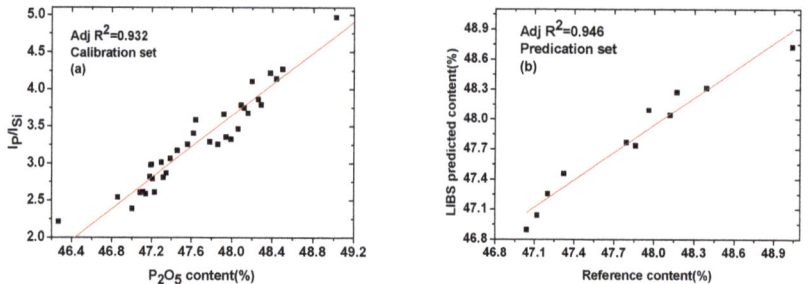

Figure 5. Internal standard method: (**a**) calibration curve using Si as an internal standardelement and (**b**) comparison of P content predicted by LIBS and the referencevalue (ICP).

Because of the complex fertilizer matrix, the analysis focusing only on one line intensity of an element might result in the loss of valid information, which cannot meet the requirements of the quantitative analysis of LIBS. PLSR is one of the multivariate analytical techniques, which can make full use of the spectral information, reduce the matrix effect and improve the accuracy of quantitative analysis. Sample sets were the same as those of the above univariate model and internal standard model. Calibration set was used to construct a model correlating the LIBS signal and the concentration of P; this correlation can later be used to predict concentrations of prediction set. In order to improve the processing speed and avoid overfitting of the model, a proper spectral range was selected for modeling analysis. A reduced spectral range from the full spectrum, which included most of the strong lines of P, was taken into account for PLSR model. The reduced wavelength ranges from 210 to 260 nm for P was used to obtain calibration model. The MATLAB software was used for PLSR. Seven principal components are used to construct the PLSR model. Figure 6a–b shows the calibration and prediction results of PLSR model for P, respectively.

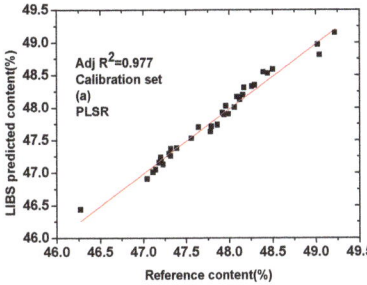 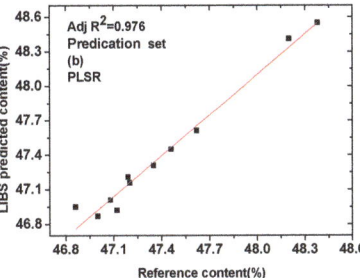

Figure 6. Comparison between LIBS predicted value (PLSR model) and reference value presented in (**a**) thirty-six calibration samples and (**b**) eleven prediction samples.

It can be seen from Figure 6 that most of the calibration and prediction data points were distributed around the fitting curve, which indicating that the PLSR model performed well in predicting of P content. The major statistic parameters that determine capacity of the regression model are R_C^2, R_P^2, RMSEC, RMSEP and RPD. All of the parameters for both calibration and prediction sets are presented in Table 2. The determination coefficient for calibration set (R_C^2) was changed from 0.932 to 0.977, while for prediction (R_P^2) set was improved to 0.976 from 0.946. The slope for both the calibration and prediction sets was close to one, which showed a strong correlation between predicted and reference values. Meanwhile, the value of RMSEC and RMSEP were 0.117 and 0.113, respectively, which was better than the results reported by S.C. Yao et al. [16]. In [16], the RMSEC was 0.234, and in this case, RPD value exceeded 5, which suggesting that the established prediction models can be employed for robust quantitative analysis. In real agricultural applications, the measured value of P content in compound fertilizer is 1.5% plus or minus the standard value, and the absolute difference between different laboratory measurements is not more than 0.5% (ISO 5315: 1984, MOD). In this paper, the difference between the predicted value of LIBS and the reference value isin the range of 0.02%–0.32%, which can fit the requirement. The total relative error for calibration and prediction set was 5.83% and 2.08%, respectively, which was better than the results reported by Bruno S. Marangoni et al. [14]. In [14], the average error of 15% found in cross-validation of LIBS quantification appeared feasible for P quantification in fertilizers. It is demonstrated that PLSR model can be developed to predict the concentrations of unknown samples. Thus, the measurement accuracy of this result can meet the measurement requirements [29].

Table 2. R^2 Regression Coefficients, RMSEC, RMSEP, RPD for Calibration and Prediction Curve.

Parameters	R_C^2	RMSEC	R_P^2	RMSEP	RPD
Values	0.977	0.117	0.976	0.113	5.31

4. Discussion

In this paper, the LIBS technique was used for univariate and multivariate analyses of P element in compound fertilizers. Forty-seven samples were provided by Hongsifang production. The calibration curve was established based on the three selected emission lines and the concentration of P. The results showed that the characteristic line of 213.6 nm was most suitable for establishing calibration curves. Due to the occurrence of matrix effects, the prediction accuracy of the method could not be achieved by applying the univariate calibration method, which only using I_P as the variable. The internal standardization method based on Si was naturally present in the samples, which showed the proper correction of the P signal. The internal standard method was found to be better than the calibration method because the correlation coefficient for the calibration set was changed from 0.854 to 0.932. Moreover, the range of relative errors are 0.04–0.65%. Thus, the internal standard method can improve the accuracy of the measurements in some extent. Then, PLSR was used as a multivariate analytical technique for analysis of compound fertilizers in pellet form. From the results of PLSR regression, calibration and prediction models were obtained for P element with very good correlation coefficients. The values of RMSEC and RMSEP were 0.117 and 0.113, with RPD value of 5.31. All of these results demonstrated that the PLSR regression method can improve the accuracy of LIBS measurement, and the results in this study can provide the basis of real-time analysis of P in compound fertilizer.

Author Contributions: B.Z., Z.C. and W.S. conceived and designed the experiment; P.L. performed the experiments; Y.J. and W.S. contributed to data analysis; B.Z. and W.S. analyzed the data and wrote the whole paper; all authors reviewed the manuscript.

Funding: This research was funded by National Natural Science Foundation of China (61505001).

Acknowledgments: The fertilizer samples tested in this research was provided by the compound fertilizer production line of Hefei Hongsifang Chemical Fertilizer Plant, Anhui, China.

Conflicts of Interest: The authors declare no conflict of interest.

References

1. Niu, X.Y. Research on Variable-Rate Fertilizer Application Technology and Its Applying System in Precision Agriculture. Ph.D. Thesis, Hebei Agriculture University, Baoding, Hebei, China, 2005.
2. Cui, S.L. Effective Phosphorus Detection in Compound Fertilizers by Weight Method. *Mod. Agric. Technol.* **2016**, *5*, 231–232.
3. Yang, Q.Q.; Zhou, C.H. The Test of Potassium Oxide on Mixed and Compound Fertilizer by Flame Atomic Absorption Spectrometry. *Contemp. Chem. Ind.* **2007**, *36*, 209–211.
4. Wang, M.-R.; Yuan, Y.-M.; Tao, N.-L. Determination of Available Phosphate Content in Calcium Magnesium Phosphate by ICP-AES. *Mod. Agric. Sci. Technol.* **2012**, *7*, 20–21.
5. Anastas, P.T. Green chemistry and the role of analytical methodology development. *Crit. Rev. Anal. Chem.* **1999**, *29*, 167–175. [CrossRef]
6. Peng, J.; He, Y.; Ye, L.; Shen, T.; Liu, F.; Kong, W.; Liu, X.; Zhao, Y. Moisture influence reducing method for heavy metals detection in plant materials using laser-induced breakdown spectroscopy: A case study for chromium content detection in rice leaves. *Anal. Chem.* **2017**, *89*, 7593–7600. [CrossRef] [PubMed]
7. Gottfried, J.L. Discrimination of biological and chemical threat simulants in residue mixtures on multiple substrates. *Anal. Bioanal. Chem.* **2011**, *400*, 3289–3301. [CrossRef]
8. Zhang, D.C.; Hu, Z.Q.; Su, Y.B.; Hai, B.; Zhu, X.L.; Ma, X. Simplemethod for liquid analysis by laser-induced breakdown spectroscopy (LIBS). *Opt. Express* **2018**, *26*, 18794–18802. [CrossRef]
9. Yan, C.H.; Qi, J.; Liang, J.; Zhang, T.L.; Li, H. Determination of coal properties using laser-induced breakdown spectroscopy combined with kernel extreme learning machine and variable selection. *J. Anal. At. Spectrom.* **2018**, *33*, 2089–2097. [CrossRef]
10. Jull, H.; Kunnemeyer, R.; Schaare, P. Nutrient quantification in fresh and dried mixtures of ryegrass and clover leaves using laser-induced breakdown spectroscopy. *Precis. Agric.* **2018**, *19*, 823–839. [CrossRef]
11. Anglos, D.; Detalle, V. *Cultural Heritage Applications of LIBS in Laser-Induced Breakdown Spectroscopy*; Springer: Berlin/Heidelberg, Germany, 2014; pp. 531–554.
12. Cho, Y.; Horiuchi, M.; Shibasaki, K.; Kameda, S.; Sugita, S. Quantitative Potassium Measurements with Laser-Induced Breakdown Spectroscopy Using Low-Energy Lasers: Application to In Situ K-Ar Geochronology for Planetary Exploration. *Appl. Spectrosc.* **2017**, *71*, 1969–1981. [CrossRef]
13. Farooq, W.A.; Al-Mutairi, F.N.; Khater, A.E.M.; Al-Dwayyan, A.S.; AlSalhi, M.S.; Atif, M. Elemental analysis of fertilizer using laser induced breakdown spectroscopy. *Opt. Spectrosc.* **2012**, *112*, 874–880. [CrossRef]
14. Marangoni, B.S.; Silva, K.S.G.; Nicolodelli, G.; Senesi, G.S.; Cabral, J.S.; Villas-Boas, P.R.; Silva, C.S.; Teixeira, P.C.; Nogueira, A.R.A.; Benites, V.M.; et al. Phosphorus quantification in fertilizers using laser

induced breakdown spectroscopy (LIBS): A methodology of analysis to correct physical matrix effects. *Anal. Methods* **2016**, *8*, 78–82. [CrossRef]

15. Andrade, D.F.; Pereira-Filho, E.R. Direct determination of contaminants and major and minor nutrients in solid fertilizers using laser-induced breakdown spectroscopy (LIBS). *J. Agric. Food Chem.* **2016**, *64*, 7890–7898. [CrossRef]

16. Yao, S.C.; Lu, J.D.; Li, J.Y.; Chen, K.; Li, J.; Dong, M.R. Multi-elemental analysis of fertilizer using laser-induced breakdown spectroscopy coupled with partial least squares regression. *J. Anal. At. Spectrom.* **2010**, *25*, 1733–1738. [CrossRef]

17. Andrade, D.F.; Sperança, M.A.; Pereira-Filho, E.R. Different sample preparation methods for the analysis of suspension fertilizers combining LIBS and liquid-to-solid matrix conversion: Determination of essential and toxic elements. *Anal. Methods* **2017**, *9*, 5156–5164. [CrossRef]

18. Bechlin, M.A.; Ferreira, E.C.; Neto, J.A.G.; Ramos, J.C.; Borges, D.L.G. Contributions on the use of bismuth as internal standard for lead determinations using ICPbased techniques. *J. Braz. Chem. Soc.* **2015**, *26*, 1879–1886.

19. Clegg, S.M.; Sklute, E.; Dyar, M.D.; Barefield, J.E.; Wiens, R.C. Multivariate analysis of remote laser-induced breakdown spectroscopy spectra using partial least squares, principal component analysis, and related techniques. *Spectrochim. Acta B* **2009**, *64*, 79–88. [CrossRef]

20. Sha, W.; Niu, P.; Zhen, C.; Lu, C.; Jiang, Y. Analysis of phosphorus in fertilizer using laser-induced breakdown spectroscopy. *J. Appl. Spectrosc.* **2018**, *84*, 653–658. [CrossRef]

21. Miziolek, A.W.; Palleschi, V.; Schechter, I. *Laser-Induced Breakdown Spectroscopy (LIBS): Fundamentals and Applications*; Cambridge University Press: New York, NY, USA, 2006.

22. Cui, Z.; Zhang, X.; Yao, G.; Wang, X.; Xu, X.; Zheng, X.; Feng, E.; Ji, X. Quantitative analysis of the trace element in Cu-Pb alloy by the LIBS. *Acta Phys. Sin.* **2006**, *55*, 4506–4513.

23. Nie, P.; Dong, T.; He, Y.; Xiao, S. Research on the effects of drying temperature on nitrogen detection of different soil types by near infrared sensors. *Sensors* **2018**, *18*, 391. [CrossRef]

24. He, Y.; Xiao, S.; Nie, P.; Dong, T.; Qu, F.; Lin, L. Research on the optimum water content of detecting soil nitrogen using near infrared sensor. *Sensors* **2017**, *17*, 2045. [CrossRef]

25. Martín, M.E.; Hernández, O.M.; Jiménez, A.I.; Arias, J.J.; Jiménez, F. Partial least-squares method in analysis by differential pulse polarography. Simultaneous determination of amiloride and hydrochlorothiazide in pharmaceutical preparations. *Anal. Chim. Acta* **1999**, *381*, 247–256. [CrossRef]

26. Sjöström, M.; Wold, S.; Lindberg, W.; Persson, J.Å.; Martens, H. A multivariate calibration problem in analytical chemistry solved by partial least-squares models in latent variables. *Anal. Chim. Acta* **1983**, *150*, 61–70. [CrossRef]

27. Rossel, R.A.V.; McGlynn, R.N.; McBratney, A.B. Determining the composition of mineral-organic mixes using UV-vis–NIR diffuse reflectance spectroscopy. *Geoderma* **2006**, *137*, 70–82. [CrossRef]

28. Amamou, H.; Bois, A.; Ferhat, B.; Redon, R.; Rossetto, B.; Ripert, M. Correction of the self-absorption for reversed spectral lines: Application to two resonance lines of neutral aluminium. *J. Quant. Spectrosc. Radiat. Transf.* **2003**, *77*, 365–372. [CrossRef]

29. Sirven, J.B.; Bousquet, B.; Canioni, L.; Sarger, L. Laser-induced Breakdown Spectroscopy of CompositeSamples: Comparison of Advanced Chemometrics Methods. *Anal. Chem.* **2006**, *78*, 1462–1469. [CrossRef]

© 2019 by the authors. Licensee MDPI, Basel, Switzerland. This article is an open access article distributed under the terms and conditions of the Creative Commons Attribution (CC BY) license (http://creativecommons.org/licenses/by/4.0/).

Article

The Efficiency of Color Space Channels to Quantify Color and Color Intensity Change in Liquids, pH Strips, and Lateral Flow Assays with Smartphones

Joost Laurus Dinant Nelis [1,*], **Laszlo Bura** [2], **Yunfeng Zhao** [1,3], **Konstantin M. Burkin** [4], **Karen Rafferty** [3], **Christopher T. Elliott** [1] **and Katrina Campbell** [1,*]

[1] Institute for Global Food Security, School of Biological Sciences, Queen's University of Belfast, 19 Chlorine Gardens, Belfast BT9 5DL, UK; Y.Zhao@qub.ac.uk (Y.Z.); chris.elliott@qub.ac.uk (C.T.E.)
[2] Department of Food and Drug, University of Parma, Parco Area delle Scienze 27/A, 43124 Parma, Italy; laszlo.bura@studenti.unipr.it
[3] School of Electronics, Electrical Engineering and Computer Science, Queen's University Belfast, 125 Stranmillis Road, Belfast BT9 5AH, UK; K.Rafferty@ee.qub.ac.uk
[4] Faculty of Chemistry, Lomonosov Moscow State University, 1-3 Leninskiye Gory, GSP-1, Moscow 119991, Russia; burkin-kost@yandex.ru
* Correspondence: J.Nelis@qub.ac.uk (J.L.D.N.); katrina.campbell@qub.ac.uk (K.C.)

Received: 25 October 2019; Accepted: 19 November 2019; Published: 21 November 2019

Abstract: Bottom-up, end-user based feed, and food analysis through smartphone quantification of lateral flow assays (LFA) has the potential to cause a paradigm shift in testing capabilities. However, most developed devices do not test the presence of and implications of inter-phone variation. Much discussion remains regarding optimum color space for smartphone colorimetric analyses and, an in-depth comparison of color space performance is missing. Moreover, a light-shielding box is often used to avoid variations caused by background illumination while the use of such a bulky add-on may be avoidable through image background correction. Here, quantification performance of individual channels of RGB, HSV, and LAB color space and ΔRGB was determined for color and color intensity variation using pH strips, filter paper with dropped nanoparticles, and colored solutions. LAB and HSV color space channels never outperformed the best RGB channels in any test. Background correction avoided measurement variation if no direct sunlight was used and functioned more efficiently outside a light-shielding box (prediction errors < 5%/35% for color/color intensity change). The system was validated using various phones for quantification of major allergens (i.e., gluten in buffer, bovine milk in goat milk and goat cheese), and, pH in soil extracts with commercial pH strips and LFA. Inter-phone variation was significant for LFA quantification but low using pH strips (prediction errors < 10% for all six phones compared). Thus, assays based on color change hold the strongest promise for end-user adapted smartphone diagnostics.

Keywords: smartphone colorimetrics; lateral flow assay quantification; color space; image correction; food contaminant screening; allergens; background correction; point of site analyses

1. Introduction

1.1. General Introduction

Detecting, quantifying, and mitigating against contamination in the food supply chain is paramount to global food security. High-end laboratory equipment such as mass spectrometry is often used for this purpose [1]. Unfortunately, such equipment is often unavailable in the developing world [2]. Moreover, contamination and fraud often go undetected due to a lack of surveillance. A recent report from the European Rapid Alert System for Food and Feed (RASFF) showed that alert notifications in 2017 increased 26% compared to 2016 and reported on various outbreaks (with pathogens and mycotoxins being most prominent) [3]. The main increase in alerts were follow up alerts representing additional testing on products that might have been placed on the market in another country [3]. This increase in follow up alerts might indicate that the system suffers from limited traceability in the industrialized global market, as has been reported previously [4,5]. Use of on-site detection methods (performed by primary producers, supermarkets, or even consumers) can complement the current systems and provide a means for developing countries to enhance food security. Indeed, a plethora of smartphone-based devices has been developed to complement laboratory-based analyses in several sectors including the food sector. Some examples are the fluorescent detection of antibodies against recombinant bovine growth hormone in milk [6,7], fluorescent detection of *Escherichia coli* in yoghurt and egg [8], colorimetric detection of marine toxins (okadaic acid and saxitoxin) in shellfish [9], aflatoxin B1 in maize [10], hazelnut allergen in cookies [11], and peanut allergen in cookies [12]. Such systems have great potential to influence the future market of food quality analyses particularly if analyses are rapid and straightforward allowing uptake at the consumer level. To this end, especially paper-based colorimetric assays such as the lateral flow assays have great potential since test results can be quantified rapidly with a smartphone [13]. A good example is the system reported by Ross et al., which enabled smartphone-based hazelnut allergen quantification within 2 min after applying the extract to the assay [11]. Other systems that may become rapid after optimization are liquid-based assays such as ELISA for which several examples exist of commercial diagnostic tests, such as various assays for mycotoxin analyses, that can be performed in under 10 min [14]. This being said, the food sector is relatively behind in implementation of smartphone-based technology for rapid/real time detection of contaminants when compared to other sectors such as medicine and environmental contaminant detection and it may be interesting to piggyback on such systems for food contaminant analyses [15].

1.2. Hyphenating Lateral Flow and Other Colorimetric Assays with Smartphones

A LFA is a paper-based platform consisting of a sample pad, a conjugate release pad, a membrane with a test and control line and an absorbent pad, all attached to a backing card. If a liquid sample containing a target is loaded onto the sample pad, it will run over the conjugate release pad and membrane to the absorbent pad by capillary force. In this process, the target can bind/form a complex with a labelled immunoreagent (often gold-nanoparticle, carbon black, or latex bead conjugated antibody) present in the conjugate release pad of the LFA. When this complex arrives at the test line other immunoreagents immobilized there (often antibodies in a sandwich assay set-up) can catch the complex which causes a colored line to appear. A control line is equally formed by a similar immunoreaction of other immunoreagents present in the conjugation pad which can complex specifically with other immunoreagents immobilized on the control line to ensure the assay functioned properly [16]. Thus, LFAs are rapid colorimetric tests with color intensity variation in relation to the target concentration and results can generally be read within 5–10 min [14]. Simply photographing such a test and quantifying the color by image analyses on a user friendly app with a smartphone without compromising the functionality of the phone is attractive [13]. Moreover, this combination has merit since smartphones are ubiquitous, allow for user independent quantification of the LFA and enable real-time and place stamped reporting of the results using the smartphone's wireless connectivity and build-in GPS. Moreover, many LFA have already been commercialized for on-site, non-expert use for food contaminant screening [14]. However, only one commercial smartphone-based LFA reader (RIDA Smart app; R-Biopharm) was identified for food contaminant analyses from the > 300 commercial assays included in the mentioned database [14]. This assay quantifies various LFAs testing for mycotoxins but only works for those specific assays and on a few android-based smartphones. Evidently, consumer friendly rapid on-site analyses would greatly benefit from an app that is compatible with a large variety of phone-models and allows rapid quantification of a larger variety of (commercially available) colorimetric assays with phone A by interpolating on a calibration curve made in the laboratory with phone B. To this end, a universal approach for color quantification is needed. Color quantification with a smartphone using red green blue (RGB), hue saturation value (HSV), or lightness and chromatic axes A/B (LAB) color spaces have been reported for this purpose and were reviewed recently in the context of smartphone-based biosensors [17], smartphone-based food diagnostics [18], and quantitative LFA [19]. Roda et al. [17] outlines LAB and HSV color spaces as superior to RGB space for measuring small changes in color. However, which of the individual channels should be used or why HSV or LAB color spaces are better than RGB are not mentioned. Moreover, superior performance of the L channel of LAB is equally reported for the smartphone-based detection of hazelnut allergen with LFAs using carbon black based labeling [11] although no comparison with other channels was presented. In another work [20] paper-based detection of the mycotoxin producing black mold (*Stachybotrys chartarum*) is detected by using the R channel. However, no comparison with the B or G channel or other color spaces was reported. In these previous works [18,19] no recommendations on color space use were given although it has been reported that the R channel instead of the combined weighed RGB values lead to background reduction [21]. In another study, the performance of a ΔRGB or ΔLAB system was compared and ΔRGB outperformed ΔLAB for the quantification of plasmonic-ELISA assays [22] and found to perform well in a colorimetric paper-based assay [23]. To obtain Δ values a resultant of R, G, B or L, A, B vectors relative to control values was calculated [22]. However, no comparison between the single R, G, B or L, A, B channels with ΔRGB was performed in either study. Thus, there currently is no consensus regarding which color space/channel should be used for the smartphone image analyses needed for optimal functioning of smartphone hyphenated colorimetric assays such as ELISA, LFA, and other paper-based systems for food contaminant analyses. However, color space/channel choice clearly effects the performance of smartphone-based assays since several of the studies mentioned above reported substantial differences in assay performance in function of the color space/channel used. As a result, conversions to color spaces other than RGB are perhaps executed unnecessarily or suboptimal channels/color spaces may be chosen as a starting point for smartphone-based image analyses based on incomplete recommendations found in the literature. Other issues with current systems are (i) background illumination variation and (ii) inter-phone channel values variation. A light-shielding box is often used to tackle variation in background illumination [9,17–19], however, this detracts from the opportuneness of smartphone-based analyses if add-on items are required. Inter-phone channel value variation is especially important since it is undesirable to develop any system to be used by consumers if bespoke calibration is required for each phone. Unfortunately, most systems described previously were only tested with a single phone [9,11,17–20,24]. Overall, in the scientific literature there appears to be large variations in opinions regarding the use of color spaces and channel combinations. Moreover, no extensive comparison of the individual channels of RGB, LAB, HSV color space, and ΔRGB values with various phones has been identified.

1.3. Workflow Reported in This Study

In the present study, prediction accuracy of individual channels of RGB, HSV and LAB color space, and ΔRGB was determined for the quantification of color variation, using pH strips, and color intensity variation, using filter paper with dropped nanoparticles often used for LFA (i.e., gold, latex or carbon black nanoparticles) and nanoparticle and oxidized tetramethylbenzidine (TMB) solutions in ELISA wells. Background correction was shown to avoid measurement variation in the absence of direct sunlight more efficiently than a light-shielding box. Inter-phone (n = 6) variation was limited for color change quantification, permitting the quantification of color with phone A using a calibration curve constructed with phones B–F. The optimized system was validated using various phones with model food application exemplars in the quantification of gluten in buffer, bovine milk in goat milk and goat cheese, and pH determination in soil extracts with commercial pH strips and LFA.

2. Materials and Methods

2.1. Materials

Dibasic and monobasic sodium phosphate, sodium carbonate/bicarbonate, HCl (37%), 3,3′,5,5′-tetramethylbenzidine (TMB), NaOH, $HAuCl_4$, sodium citrate, $AgNO_3$, L-ascorbic acid, blue latex beads (LNPs), horseradish peroxidase (HRP), gluten from wheat (crude, ≥ 75% protein), and grade 1 Whatman filter paper were purchased from Sigma-Aldrich (Irvine, UK). Carbon black (N220) (CB) was obtained from Cabot Corporation (Ravenna, Italy), ZEU Proteon Gluten Express ZE/PR/GL25 and ZEU IC-BOVINO lateral flow assays for gluten and cow milk protein detection and ZEU Proteon ELISA were obtained from Zeulab (Zaragoza, Spain). Cow milk and goat cheese were purchased in the local market. Pure goat milk was home produced. DUS alkaline pH strips (5.00–8.50) were purchased from DFI (Gimhae, Korea). Soil was collected in Queen's University garden (Belfast, UK). UV-Vis measurements were performed using a Tecan Safire IIplate reader. Smartphone measurements were performed with a Huawei P8 Lite (12 megapixels (MP)), iPhone 7 (12 MP), Samsung Galaxy Tab E (5 MP), Xiaomi mi5 (16 MP), HTC One M7 (4 MP) and a Samsung Galaxy J7 (13 MP), and a Smartscope Xscience (Ravensburger, The Netherlands) smartphone loop with 3D printed lenses for magnification where mentioned.

2.2. Nanoparticle Synthesis

Glassware was cleaned with piranha solution and aqua regia to remove all residues. The Turkevich method [25] was used for gold nanoparticle (GNP) synthesis. Briefly, 500 µL of 100 mM $HAuCl_4$ together with 194.5 mL MQ was brought to boil, while stirring, in a round bottom flask equipped with a condenser. At boiling point, 5 ml sodium citrate solution (1% (w/v)), was added and the mixture was left boiling for 30 min then cooled down gradually. Stock GNP concentration (2.8 nM) was estimated using a protocol detailed in [26] and size estimations were reported elsewhere [27].

2.3. pH System (Color Change)

Citrate-phosphate buffers (0.1 M; pH 5.0, 5.5), phosphate buffer (0.1 M; pH 6.0–7.5), and carbonate/bicarbonate buffers (0.1 M; pH 8.0, 8.5) were used for calibration (pH 5.0, 6.0, 6.5, 7.0, 7.5, 8.0, 8.5) and prediction (pH 5.5, 6.25, 6.75, 7.25, 7.75, 8.25) curves. For buffered soil extracts, 5 g of soil was added to 25 mL of buffer and vortexed. The pH was adjusted to match the buffer pH described above. Samples were allowed to settle 30 min and directly used for pH measurements. Images taken with the Huawei of pH strips dipped in various buffers are shown in Figure 1.

Figure 1. Exemplary images taken with the Huawei of pH strips that were used to build calibration curves.

2.4. Nanoparticle Suspensions and Filter Paper Preparation

A 40X concentration was obtained from a stock GNP solution by centrifuging (13,000 RCF; 30 min; 20 °C). Concentrates were used to prepare a 2/3 dilution series from 37.5 up to 2.195 nM to construct calibration curves and a 2/3 dilution series from 31.25 up to 2.74 nM for predictions. For CB, a 10 mg/mL dispersion was sonicated 30 min then diluted to a 2 mg/mL dispersion and sonicated for another 10 min. Serial dilutions (2X) were made until 0.0078125 mg/mL. For predictions, a 1.5 mg/ml CB solution was made and diluted in 2X steps until 0.0117 mg/mL. LNP stock concentration (2.5%) was 2X diluted until

0.00977% to construct calibration curves. For the prediction set a 1.875% solution was diluted in 2X steps until 0.00732422%. Images taken with the Huawei of the nanomaterials on filter paper are shown in Figure 2.

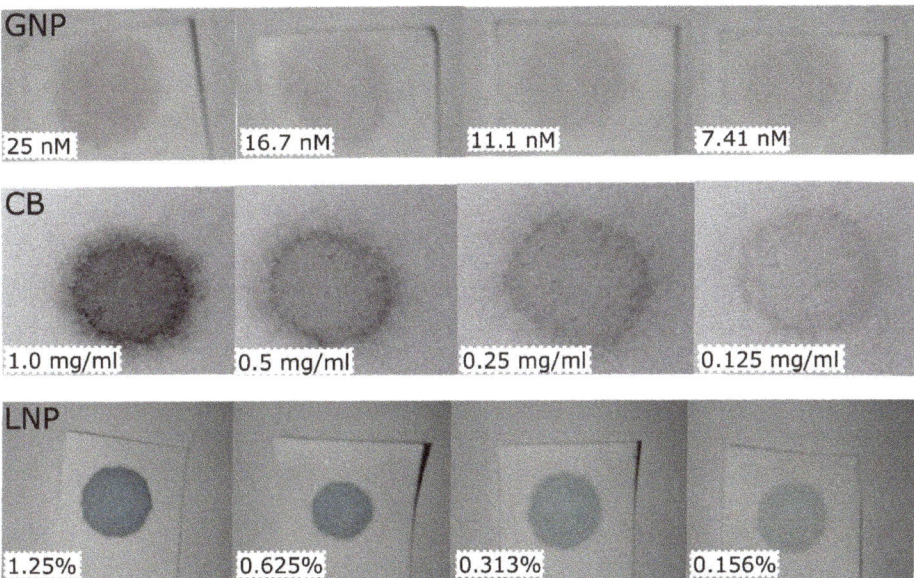

Figure 2. Exemplary images taken with the Huawei of the filter papers with nanomaterials on them at various concentrations. Top row are gold nanoparticles (GNP). Middle row are carbon black nanoparticles (CB). Bottom row are latex nanoparticles (LNP). Particle concentration of the solutions used is indicated in the left bottom corner of each image.

2.5. Liquid Assays Preparation

Colloid GNPs (200 µL) at varying concentration (8 step, 2X serial dilution from 84 nM for calibration and 7 step 2X dilution from 42 nM for prediction) or 150 µL of TMB with HRP at varying concentration (8 step 2X serial dilution from 60 pM for calibration and 6 step 2X dilution from 40 pM for prediction) were pipetted into transparent 96-well plates. TMB enzymatic reaction was stopped after 30 min with 4N H_2SO_4 (50 µL/well). Absorbance was read at 450 nm for plates with HRP and 513 nm (plasmon peak) for plates with colloid GNPs. Smartphone pictures were taken thereafter. An exemplary image taken with the Xiaomi of the colloid GNPs is shown in Figure 3.

2.6. Sample Preparation and Picture Capturing

From GNP, CB, and LNP dilutions 5 µL was dropped on filter paper (n = 3), dried, and photographed. pH strips were immersed in buffer, dried, and photographed after 40 s. LFA test strips were photographed 10 min after exposure to extracts. For liquid assays, a phone with white screen was placed under the 96-well plate to provide counter illumination and avoid reflections in the images. All pictures were taken from 5 cm distance with the flashlight on (Figure 4). For background illumination experiments the following light changes were tried: Dark background in a closed windowless laboratory, normal room light (TL light), indirect sunlight in a windowsill, direct sunlight in a windowsill. A black cardboard box (11 × 11 × 5 cm) was made for the Huawei (Figure 4). The box had a hole precisely in the center for the camera and flash. Prediction images taken under varying background illumination were interpolated on calibration curves constructed at room light conditions to test robustness of the background correction applied. All images taken in the box for prediction were interpolated on a calibration curve constructed with images taken in the box at identical light conditions (room light).

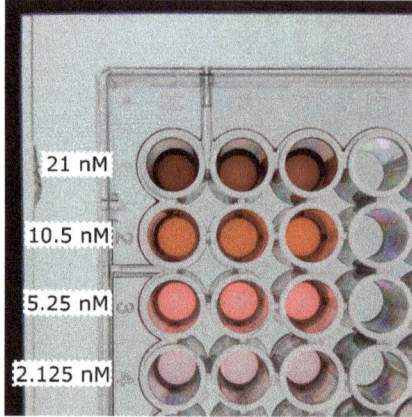

Figure 3. Image taken with the Xiaomi of colloid gold nanoparticles in a 96-well plate. Concentrations vary per row (as indicated). Each row (1–4) contains three replicas (H–F) used to construct the calibration curves. Backlight is provided using the white screen of the Huawei placed under the 96- well plate.

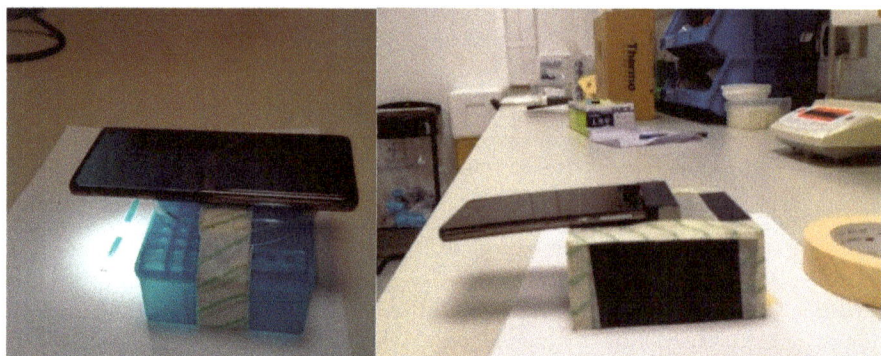

Figure 4. Left; an image of the Samsung taking an image of a LFA (the ZEU Proteon Gluten Express) under room-light conditions with the flashlight on at 5 cm distance. Right; an image of the light-shielding box used. The dimensions of the box are 11 × 11 × 5 cm to maintain the standard 5 cm as a distance for the photo capturing. The box had a hole in the center precisely for the camera and flash. The phone displayed on the box is the Huawei.

2.7. Commercial Assays

LFA for gluten, bovine milk, and cheese detection were used according to the instructions of the manufacturer. Briefly, (for estimation of gluten with the commercial LFA assay; ZEU Proteon Gluten Express ZE/PR/GL25), 1 g of gluten was extracted with 10 mL of the given extraction solution, vortexed and centrifuged for 10 min at 3500 RCF. The supernatant was diluted with analysis buffer to obtain 10, 5, 2.5, 1.25, 0.75, 0.5, 0.025, 0.01, and 0.005 ppm of gluten for calibration curve construction and 2, 0.5, 0.3, 0.15, 0.02, and 0.0075 ppm for predictions. The test strip was immersed in 250 µL of these dilutions for 10 min and photographed. For the estimation of cow milk/cheese in goat milk/cheese the ZEU-IC-BOVINO LFA assay was used. Cow milk was spiked into pure, home produced goat milk at 5%, 2.5%, 1.25%, 0.625%, 0.3125%, 0.15625%, 0.078125%, 0.04%, and 0.02% for calibration curve construction and at 3.75%, 1.875%, 0.46875%, 0.234375%, and 0.117188% for predictions. LFAs were immersed in two drops of these extracts diluted with three drops dilution solution. Purity of the goat cheese used was tested using the RC-BOVINO ELISA kit specific for the IgG of cow milk following the manufacturer's instructions. Next extracts of cow cheese and pure goat cheese were prepared by adding 5 g of homogenized cheese to 10 ml H_2O and vortexing. Extracts were centrifuged (10 min; 3000 RCF) and goat cheese supernatant was spiked with cow cheese supernatant (final concentrations 8%, 4%, 2%, 1%, 0.75%, 0.5%, and 0.1% of cow cheese) and used for calibration curve generation. Final concentrations of 6%, 3%, 1.5%, and 0.375% cow cheese were used for predictions. LFAs were immersed in one drop of extract diluted with two drops of the given dilution solution and photographed after 10 min. For all commercial assays duplicates were used. Figure 5 shows an image of a set of LFAs used to build a calibration curve for gluten quantification.

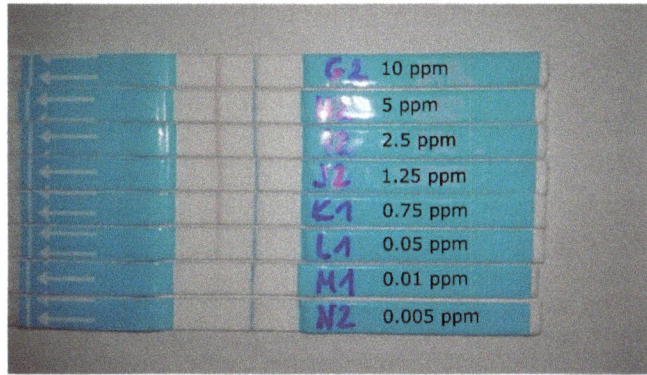

Figure 5. A set of LFAs used to build a calibration curve for gluten quantification with various smartphones. The concentrations of gluten used are indicated.

2.8. Scoring System

A scoring system was used to rate the performance of the channels which allowed to generate calibration curves with an $R^2 > 0.80$ (R, G, B, V, L, H, and V). B of LAB (called C of LAC from hereon to avoid confusion with the B channel of RGB) was not considered since it performed only for pH and even there was outperformed by all other channels. R^2 values on the regression functions fitted to the predicted concentrations were ranked from highest to lowest. Lowest being 1 point, highest 7 points. The same scoring system was used for the slopes of the regression fits. The channel with a slope closest to 1 got 7 points the next in line 6, etc. If no regression line could be fitted through the predictions the score was 0 points both for prediction and slope. Slope and R^2 scores were summed up to get a final score for each channel.

2.9. Software and Data Treatment

Standard RGB values were detected with a free app from Google Play (RGB Android) (n = 3 per image) and converted to HSV and Cielab with the open source library of colormine.org (last accessed 11 August 2019). S values were rescaled to have a 1–100 range (instead of 0–1) to match RGB scale for visualization. For A and B values 128 points were added to allow a scale from 0–256 and avoid negative numbers complicating the background correction applied.

ΔRGB was calculated as specified in [22]. The applied image background correction was as follows:

Corrected channel values = (Raw signal value)/(Raw background value) × Raw signal value   (1)

For paper-based assays white paper was used as the background except for the LFA where control lines were used instead. For liquid assays, wells filled with water adjacent to the tested wells were used as background. All channel values mentioned in the results are corrected channel values. Data analyses was performed with GraphPad Prism 6 Software. For post-hoc analyses, a one-way ANOVA Tukey's multiple comparison test was used. For two-way ANOVA Sidak's multiple comparison test was used. p-values were corrected for multiplicity. For calibration curves, a four-parameter dose-response curve was used unless mentioned otherwise. Mean average error percentile (MAE) was calculated on the totality of predictions over the concentration ranges mentioned above. LOD, IC_{50}, and linear range were obtained by interpolating 90%, 50%, and 20%–80% signal values from fitted normalized curves.

3. Results and Discussion

3.1. Comparing Channel Performance on a Huawei P8

Various concentrations (n = 3) of GNP, CB, and LNP nanoparticles were dropped on filter paper and photographed. R, G, B, values of all pictures were extracted and used to calculate L, A, B, H, S, V and ΔRGB values. The values were fitted to either a four-parameter dose-response curve or a two-phase decay function, which ever resulted in the best fit in terms of R^2 value (Table 1). The fitted calibration curves (with an $R^2 > 0.8$) are depicted in Figure 6, left column. Of the LAB color space (called LAC to avoid confusion of B channels with RGB); only L generated adequate calibration curves in all color systems. A curve in all color systems had an $R^2 < 0.8$ and were not used for predictions. C allowed generating a curve for pH determination only and symmetry was observed which made it only possible to predict pH below pH 7. Thus, C was not used for predictions. However, combining A and C might enable color change quantification, as shown previously [28]. From the HSV color space S never reached R^2 values above 0.8 and was not used. H was effective in generating calibration curves for pH (as previously shown [29]) and CB, although errors were large in the latter color system with a mean average error (MAE) on all predictions (n = 18) of 90% ± 113%. The V channel worked well, producing calibration curves with $R^2 > 0.95$ in all color systems except for GNP where R^2 was 0.88. Prediction images of various pH, GNP, CB, and LNP concentrations were taken and interpolated using the fitted calibration curves of Figure 6. Predicted values are plotted in a scatter plot (Figure 6; right column). The R^2 and slope values of the regressions are shown Table 2. For pH predictions, most channels shown in Figure 6 worked for predictions between pH 8.25 and 6.5. At the lowest pH tested (5.5) only R and ΔRGB allowed good predictions. For variation in nanoparticle concentrations low concentrations caused higher variance in all channels. H performed particularly badly in CB predictions. Larger variation in prediction was observed for the color intensity variations as for color change (MAE for the best channels were typically around 25%–35% ± 20%–40% and 1%–2.5% ± 1%–1.5%, respectively). See Section 3.2 for more detail and MAEs for various phones. This might be explained partly by the greater variance in the hand-made replicas for the latter system and partly by the potentially greater effects of background illumination variation on these tests (which was investigated in Section 3.3). To tease apart differences a scoring system (see Section 2.8) was adapted to rank the performance of the channels for each color system individually as well as for all color systems tested (Tables 2 and 3). B, R, and ΔRGB scored the highest overall score, followed by V, L, G, and finally H. H underperformed since that channel was only effective in pH predictions. Moreover, in three out of four color systems either R, B, or ΔRGB had a top score (albeit shared with L in some cases, Table 2). For LNP predictions, the V channel scored highest. However, the R^2 values and slopes of V, L, G, and B were close to each other (Table 2) and differences may only reflect random variation. In any case, this comparison shows that conversion to HSV and LAC color space did not increase performance compared to RGB in all color systems tested.

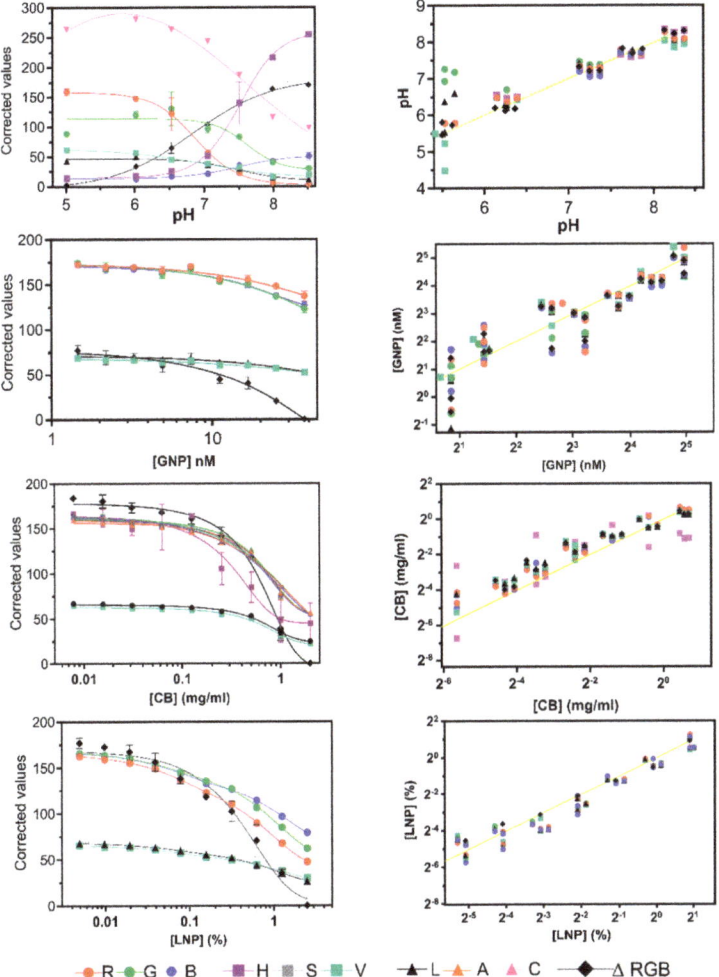

Figure 6. The left column shows calibration curves (with an $R^2 > 0.8$) obtained from background corrected channel values from the RGB, HSV, and LAB (Called LAC to avoid confusion of the B channels of RGB and LAB) color spaces for various pH values, GNP, CB, and LNP concentrations. The right column shows scatter plots of predictions obtained with those calibration curves. The yellow line represents a perfect correlation (slope = 1). Color and symbol codes are indicated.

Table 1. Function types used to fit data points of the papers with varying GNP, CB, LNP concentration, and pH strips measured in various buffers. R^2 values for each fit are mentioned. The fits with $R^2 > 0.80$ were used as calibration curves in Figure 6. C stands for the B channel of LAB which is called LAC throughout the manuscript to avoid confusion of the B channel of LAB with the B channel of RGB. Sigmoidal stands for four-parameter dose-response.

Channel	GNP R^2	Function	CB R^2	Function	LNP R^2	Function	pH R^2	Function
R	0.8871	Sigmoidal	0.9856	Sigmoidal	0.9847	Two phase decay	0.9795	Sigmoidal
G	0.9365	Sigmoidal	0.9859	Sigmoidal	0.9830	Two phase decay	0.8350	Sigmoidal
B	0.9383	Sigmoidal	0.9877		0.9818	Two phase decay	0.9531	Sigmoidal
ΔRGB	0.9374	Sigmoidal	0.9870	Sigmoidal	0.9867	Two phase decay	0.9958	Sigmoidal
H	-	-	0.8862	Sigmoidal	0.6042	Two phase decay	0.9844	Sigmoidal
S	0.6895	Sigmoidal	-	-	0.7707	Sigmoidal	0.1878	Sigmoidal
V	0.8794	Sigmoidal	0.9836	Sigmoidal	0.9803	Two phase decay	0.9600	Sigmoidal
L	0.9330	Sigmoidal	0.9861	Sigmoidal	0.9811	Two phase decay	0.9610	Sigmoidal
A	-	-	-	-	-	-	0.7979	Sigmoidal
C	-	-	-	-	-	-	0.9941	Sigmoidal

Table 2. Slopes and R^2 scores of log–log regression functions fitted to predictions of GNP, CB, LNP concentrations, and pH values shown in the scatter plots of Figure 6. A scoring system was used to attribute scores to R^2 and slope values for each channel (see Section 2.8). Scores (Table 3) determined ranks for each color system, as well as total scores per channel. Δ is ΔRGB. Ch is channel.

Rank	GNP Ch	R^2	Slope	CB Ch	R^2	Slope	Latex Ch	R^2	Slope	pH Ch	R^2	Slope	Total Scores Ch	Score
1	Δ	0.9173	1.008	B	0.979	0.801	V	0.951	1,012	Δ	0.987	0.974	B	37
2	L	0.9008	0.996	R	0.977	0.814	L	0.949	1,049	R	0.996	0.860	R	36
3	G	0.9164	0.972	G	0.973	0.795	G	0.947	1.028	H	0.977	0.874	Δ	35
4	B	0.8856	0.996	L	0.974	0.658	B	0.944	1.051	V	0.927	1.043	V	34
5	R	0.895	1.168	Δ	0.971	0.677	R	0.927	1.074	B	0.973	1.239	L	34
6	V	0.8678	1.058	V	0.972	0.676	Δ	0.762	1.178	L	0.898	0.603	G	32
7	H	-	-	H	0.462	0.354	H	-	-	G	0.712	0.307	H	12

Table 3. Scores of the individual channels in each color system, as well as overall scores for all color systems. Shared scores are in bold.

GNP Channel	Score	CB Channel	Score	LNP Channel	Score	pH Channel	Score	Total Channel	Score
ΔRGB	**12**	B	**13**	V	14	ΔRGB	13	B	37
L	**12**	R	**13**	L	**11**	R	**11**	R	36
G	10	G	9	G	**11**	H	10	ΔRGB	35
B	9	L	7	B	8	V	9	V	**34**
R	6	ΔRGB	6	R	6	B	7	L	**34**
V	5	V	6	ΔRGB	4	L	4	G	32
H	0	H	2	H	0	G	2	H	12

3.2. Comparing Channel Performance between Phones

The experiments detailed in Section 3.1 were repeated on a tablet and iPhone to test if channel performance behaved similarly over various digital camera devices. Figure 7 shows a comparison of the total mean average percentile error (MAE) for the predictions (pH n = 15; GNP, LNP, and CB n = 18) over various concentrations. MAEs for pH calculations were low and stayed below 2.5% for R and ΔRGB for all phone models. For color intensity, low-end MAEs were typically 20%–30% ± 20%–30% for GNP and LNP predictions. For CB higher MAEs were observed although best functioning channels (B and ΔRGB) showed MAEs below 50%. Two-way ANOVAs (Table 4) show that for pH the variance in MAE caused by channels was significant ($p = 0.0002$), as well as interaction between channels and phone models ($p < 0.006$) although phone models alone did not cause significant variance in MAE ($p = 0.24$). For (LNP), (GNP), and (CB) predictions the phone model caused significant differences ($p < 0.0001$) as well as channel and interaction ($p < 0.0001$) in all but the GNP dataset, where only the phone model effects on MAEs was significant ($p < 0.0001$). Thus, choosing any specific channel as universally ideal for smartphone colorimetric analyses or even for a color specific system seems challenging, especially for color intensity change. Nonetheless, Sidak's post-hoc simple effects within rows of multiple comparisons showed that most of the different effects between phones was limited to one or another channel of the RGB color system, or the H channel (for CB predictions). For ΔRGB no significant effects were observed in this test and thus shows that an error in one of the RGB channels can be compensated in this model, which makes it an interesting option. L and V equally show little variation between phones. However, variance on the MAE as well as absolute errors in the L and V values, although not significant, was larger when compared to the best functioning channel of RGB or ΔRGB in each color system (for L in pH predictions the highest MAE is 3.8% ± 8.4% versus 1.8% ± 1.7% for ΔRGB and for CB predictions the MAE was approximately 40%–50% ± 30% for B and ΔRGB and 50%–80% ± 30%–70% for L and V). Overall, H channel performed poorly in the color intensity experiments and did not outperform R, G, B, V, or L channels for color change predictions. Thus, H was no longer used in the following experiments.

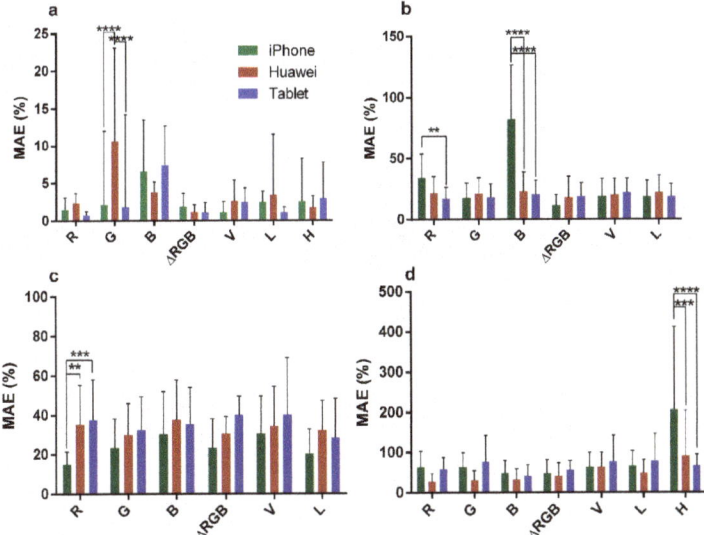

Figure 7. Mean average percentile errors (MAEs) for pH (**a**), (LNP) (**b**), (GNP) (**c**), and (CB) (**d**) predictions using an iPhone (dark green), Huawei (dark red), or Tablet (blue). Significant post-hoc Sidak multi comparisons of two-way ANOVAs are indicated. Stars indicate *p*-values with *p*-value correction for multiplicity. ** = $p < 0.01$, *** = $p < 0.001$, **** = $p < 0.0001$.

Table 4. Two-way ANOVA on phone model and channel for GNP, CB, LNP, and pH predictions.

	GNP		CB		LNP		pH Original	
Source	Var	p-Value	Var	p-Value	Var	p-Value	Var	p-Value
Interaction	2.8	0.45	8.7	0.0001	25.7	<0.0001	8.1	0.0058
Channel	2.6	0.10	7.1	<0.0001	15.0	<0.0001	7.8	0.0002
Phone	7.9	<0.0001	5.7	<0.0001	5.2	<0.0001	0.8	0.24

3.3. Comparing Box/No-Box Effects on Predictions in Various Background Settings

The effectiveness of the internal background correction was tested, with a Huawei, by comparing MAEs for pH and [GNP] prediction under varying background illumination with predictions when a light shielding box was used. Channels R, G, B, ΔRGB, V, and L were used since H has proven only functional for pH measurements and even for that application, performance was suboptimal. Little variation in MAEs was observed for pH estimation using the channels R and ΔRGB, even at illumination in direct sunlight (Figure 8). Moreover, two-way ANOVA was only significant for channels and not for interaction or background illumination (Table 5). Interestingly, for R and ΔRGB predictions variation was bigger using the box compared to all background illumination conditions. This may be explained by extensive scatter and unequal light distribution within the box compared to when no box was used. If the background varies between two images but stays equal throughout the individual images then internal background correction should largely correct for it. However, the error can be introduced if background illumination within a picture is patchy. This may also explain the observation by Masawat et al. that a larger box produces less error on predictions as a small one [24]. For [GNP] predictions, no difference was observed between box and no box over the conditions dark, room light, and indirect sunlight for all channels. However, in direct sunlight the background correction applied ceases to function and two-way ANOVA and post-hoc multiple comparisons were highly significant ($p < 0.0001$) for background illumination and direct sunlight, respectively (Table 5). At all other background illumination conditions, the MAE and variation on the MAE were similar in all conditions, including using the box. Thus, the simple background illumination correction applied eliminates the necessity to use a box for all color systems tested if measuring in direct sunlight is avoided.

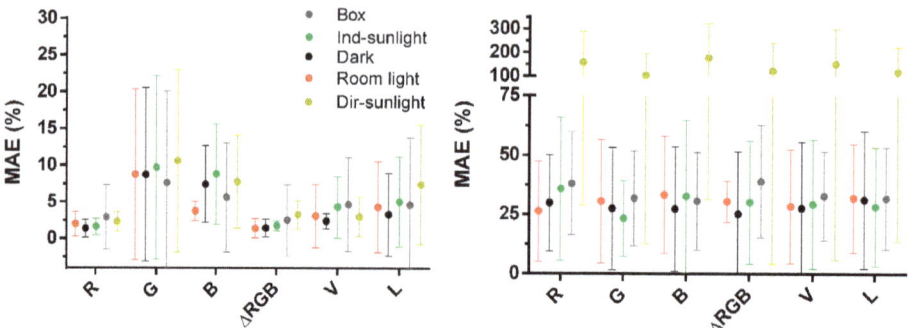

Figure 8. Left, MAEs for pH predictions (n = 15), right, MAEs for [GNP] predictions (n = 18), at various illumination conditions for R, G, B, ΔRGB, V, and L channels without use of a box and in a box at room illumination conditions. Images under dark (black balls) background were taken in a dark room; room background (red balls) in a windowless laboratory illuminated with a tube light bulb; indirect sunlight (green balls) in a windowsill at a cloudy day; direct sunlight (dark yellow balls) in a windowsill in full sunlight.

Table 5. Two-way ANOVA on background illumination and channel for GNP, and pH predictions.

	pH		GNP	
Source	Var	p-Value	Var	p-Value
Interaction	1.652	0.9893	2.177	0.6248
Channels	13.24	<0.0001	0.6425	0.4002
Light conditions	0.9718	0.3046	39.7	<0.0001

3.4. Channel Performance for Colored Liquids

Channel performance to quantify color intensity in liquid solutions (as typically done for ELISA) was investigated using various concentrations of colloid GNPs (mimicking plasmonic ELISA) and various amounts of HRP to oxidize TMB (mimicking standard HRP-based ELISA). Color change was measured using a benchtop spectrometer and a smartphone (Xiaomi) using R, G, B, ΔRGB, V, and L channels. All four-parameter dose response curves showed good fits ($R^2 > 0.8$) and were used to predict GNP and HRP concentrations. Predictions and linear regressions are shown in Figure 9a,b. Slopes and R^2 values of the calibration curves and R^2 and slope values of the linear regression functions are shown in Table 6. For [HRP] prediction, the spectrometer, B, G, and ΔRGB channels were acceptable. For [GNP] prediction the spectrometer and G, B, ΔRGB, and L channels were acceptable (Figure 9a; Table 6). For comparison, the calibration curves of these channels and the spectrometer were normalized (Figure 9c,d). LOD, linear range, and IC_{50} for both assays are shown in Table 6. Interestingly, G, B, and ΔRGB channels for colloid [GNP] had a ~3X lower LOD as the absorption curve of the spectrometer (~1.5 versus ~7.5 nM, respectively). For the [HRP] curves, B and ΔRGB channels gave lower LODs as the spectrometer (~2.3 versus ~3.4 pM, respectively). Thus, the smartphone-based system was slightly more sensitive than the spectrometer. However, the linear range was reduced compared to the spectrometer. MAEs were calculated for G, B, ΔRGB, and spectrometer [HRP] predictions. For G and B only three concentrations (n = 3 × 3) were used since the linear range was reduced. For ΔRGB and spectrometer four concentrations (n = 4 × 3) were used. One-way ANOVA showed that MAEs for [HRP] prediction using G and B channels did not significantly differ from the MAEs for [HRP] prediction when the spectrometer was used (although with reduced range). ΔRGB had slightly higher MAEs ($p < 0.05$ Tukey post-hoc; Figure 9c inset). This is probably due to the error introduced into ΔRGB from the R channel which varied significantly and could not be used to build a calibration curve. For [GNP] predictions ΔRGB and L channels had significantly higher MAEs as G and B and the spectrometer ($p < 0.01$; Tukey post-hoc; Figure 9d inset). Again, the R channel could not be used to build a calibration curve.

3.5. Channel Performance Comparison Using Commercial LFA for Milk Allergen Detection

Commercial LFA for the quantification of bovine milk in goat milk and cow cheese in goat cheese (Figure 10a,b) were used to further challenge the smartphone-based quantification using the Huawei. For the LFA used to quantify bovine milk in goat milk (Figure 10a) G, B, L, and V channels showed promising calibration curves ($R^2 > 0.9$). However, the curve determined with the V channel values was flat (min–max difference about 25 corrected value units). Equally, the L channel was quite flat and had slightly lower R^2 values (0.91) as B and G (0.93 and 0.94, respectively). ΔRGB did not allow construction of a calibration curve, which was most likely caused by variations observed in the R channel values. Thus, only B, G, and L curves were used for predictions. Linear regressions on predictions were good for each channel ($R^2 > 0.95$). However, the L channel did not predict at 0.23% cow milk (Figure 11). Goat cheese extract was spiked with cow cheese and used to test the applicability of the assay for the identification of cow protein in goat cheese. Here, only the B channel gave a satisfactory calibration curve ($R^2 = 0.951$; Figure 10b). Predictions showed excellent linear regression ($R^2 > 0.98$) (Figure 10b, inset).

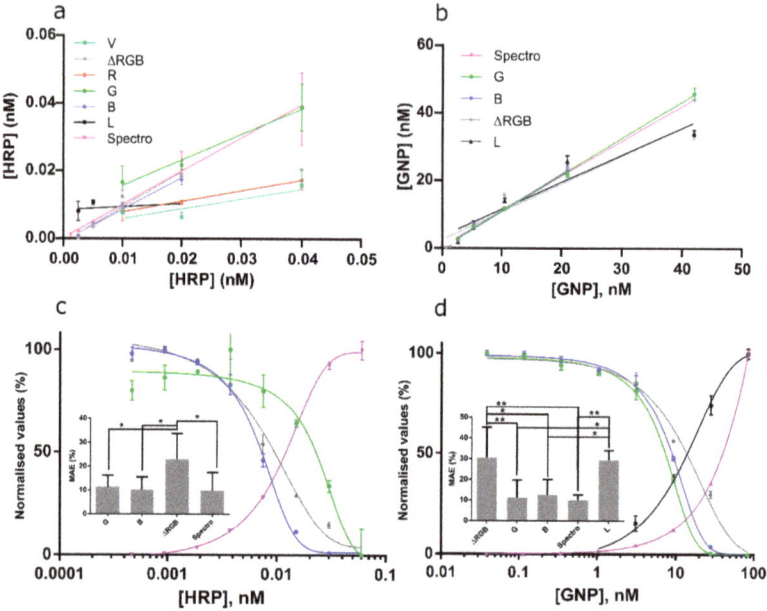

Figure 9. (a) Linear regression functions for oxidation of tetramethylbenzidine (TMB) with horseradish peroxidase (HRP) in ELISA wells for the smartphone analyses using various channels and a benchtop ELISA plate spectrometer. (b) Linear regression functions for colloid GNP at various concentrations in ELISA wells for the smartphone analyses using various channels and a benchtop ELISA plate spectrometer. Color codes and channels used are indicated. (c,d) normalized four-parameter dose-response curve fits for HRP oxidation of TMB (a) and colloid GNP (b) in ELISA wells. Green and blue circles stand for G and B channels, grey diamonds ΔRGB, magenta triangles absorption values measured by a spectrometer. (c,d inset) MAEs calculated from the predictions shown in (a,b). Stars indicate p-values from post-hoc analyses with p-value correction for multiplicity. * = $p < 0.05$, ** = $p < 0.01$.

Table 6. Analytical parameters of prediction and calibration curves for GNP and TMB solution ELISA. R^2 and slopes given for prediction curves are from linear regression lines. All calibration curves were prepared using four-parameter dose-response functions. LOD, linear range, and IC_{50} values were interpolated from the normalized calibration curves at 90%, 80%–20%, and 50% for R, G, B, and ΔRGB channels and 10%, 20%–80%, and 50% for spectrometer values, respectively.

		Prediction Curves		Calibration Curves (nM)			
Solution Type	Channel	R^2	Slope	R^2	LOD	Linear Range	IC_{50}
GNP	G	0.9966	1.086	0.9969	1.5	3.1-12.8	7.4
GNP	B	0.9890	1.089	0.9973	1.8	3.7–16.2	9.0
GNP	ΔRGB	0.9094	0.824	0.9890	1.6	4.1–32.7	13.8
GNP	L	0.913	0.8009	0.8947	-	-	-
GNP	spectro	0.9989	1.039	0.9996	7.5	15.3–65.7	39.7
HRP	B	0.9742	0.932	0.9957	0.0025	0.0038–0.012	0.0073
HRP	R	0.6087	0.601	0.829	NA	0.010–0.038	0.027
HRP	G	0.8365	0.506	0.9377	NA	0.0083–0.036	0.023
HRP	ΔRGB	0.9099	0.734	0.9804	0.0023	0.0039–0.020	0.010
HRP	L	0.07789	0.08	0.9307	-	-	-
	V	0.532	0.29	0.8992	-	-	-
HRP	spectro	0.9752	1.153	0.9982	0.0034	0.0057–0.021	0.012

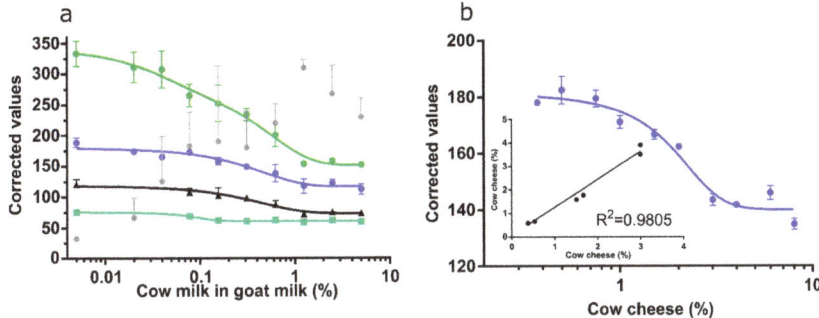

Figure 10. (**a**) G (green balls), B (blue balls), L (black triangles), and V (turquoise squares) channel values were fitted to a calibration curve for LFA quantification of cow milk spiked into pure goat milk. ΔRGB values (grey diamonds) did not allow to fit to a curve. (**b**) B channel values fitted to a calibration curve and, (inset), linear regression on predictions for cow milk in goat milk at three concentrations (n = 6).

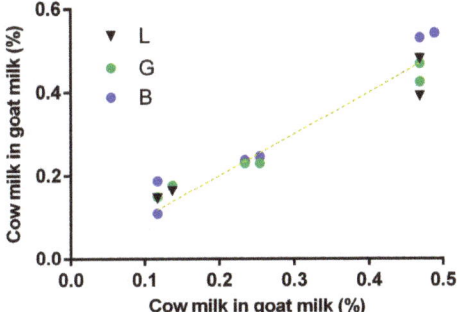

Figure 11. Predictions of cow milk (x-axis) in goat milk using L, G, and B channels. Each replica (n = 2) is indicated. Dashed line represents a curve at 45 degrees.

3.6. Channel Performance Comparison Using Commercial LFA for Gluten and pH Strips for Soil pH Prediction

Next, the ability to use various phones to quantify gluten with a commercial LFA, as well as pH strip quantification in buffered soil was tested (Figure 12). B values performed again optimally for the LFA quantification and were used to construct calibration curves for the quantification of gluten in a buffer using four different phones (Xiaomi, Huawei, iPhone, and Tablet) (Figure 12a). All phones enabled the construction of calibration curves (R^2 = 0.93; 0.88; 0.94; 0.86 for Xiaomi, Huawei, iPhone, and Tablet, respectively). Unfortunately, the curves did not overlap, except for the Tablet and Huawei, and cannot be directly used by end-users without calibration of individual phones. Magnification of LFA strips was attempted to correct inter-phone variation. LFA were photographed under the lens of a low-cost instrument (the smartscope; Ravensburger). Here, calibration curves with G channel values showed the highest R^2 values (0.87, 0.95, 0.93, and 0.92 for Xiaomi, Huawei, iPhone, and Tablet, respectively) (Figure 12b). Although the distance between the curves was narrower as in the previous set-up, the overlap of all curves was insufficient for the use of various phones to predict values on the same calibration curve. Thus, for color intensity variation using LFA, it seems that two- point calibration or camera calibration is necessary. Some suggested methods exist to obtain such calibration and improve sensitivity [30–33]. One interesting concept is to use a color reference chart in order to stabilize color variations caused by built-in automatic image correction operations [30]. Another option would be adjusting the white balance of the phones to a standardized value. This may be done by using a reference grey card or by locking the exposure and gain while selecting a preset white balance in a control window that is illuminated with an external phone-independent constant light source [32]. Setting the white balance, gain, and exposure can be done in various manual camera applications (e.g., Open Camera, ProShot) and using such a system in combination with a greycard or external constant light source (flashlights of various phones can be different in light output) may decrease inter-phone variation. Moreover, higher sensitivity may be reached by adjusting the exposure time of the assay in such manual camera applications as was previously shown for other smartphone-based LFA quantification for the detection of bacterial fruit blotch [33]. However, the four-parameter dose-response curves did overlap for all four phones for pH determination of buffered soil samples using ΔRGB (Figure 12c) and R (Figure 12d) using the Xiaomi, Huawei, iPhone, and Tablet ($R^2 > 0.99$) showing that the proposed method worked without additional adjustments for this assay. Overlap was slightly less when using ΔRGB values, thus, the R channel was used from hereon for this application. Another two phone models (HTC and Samsung) were included to check the universality of the system. Four-parameter dose-response curves showed again excellent fits ($R^2 > 0.99$) and overlap with the other curves (Figure 12d). Finally, MAEs of predictions (n = 15) were calculated for each phone using its own calibration curve or a calibration curve constructed with all phone models except the phone model used to take the images for the predictions (Figure 12d; inset). Two-way ANOVA analyses of this data was significant for the type of curve ($p < 0.001$). MAEs increased significantly when predictions were performed on phone A with calibration curves prepared using phones B–F. However, for all phones except the tablet this increase in MAE was below 3% which explains why post-hoc simple effects within phone models and multiple comparisons was only significant for the tablet ($p < 0.001$). Thus, direct quantification of color change by end-users with the pH strips without the need of phone calibration seems possible if slightly increased MAEs are acceptable.

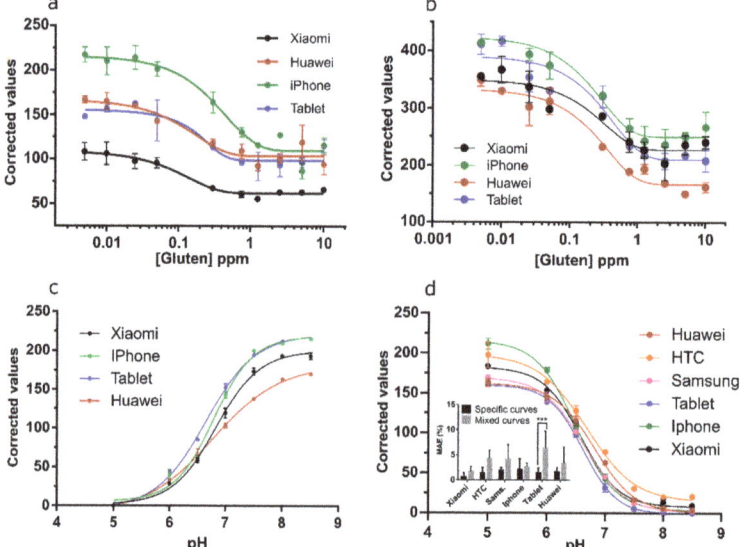

Figure 12. (a) B channel values fitted to four calibration curves for LFA quantification of gluten spiked into buffer with four phones (models indicated). (b) Calibration curves for LFA quantification using the smartscope and corrected G values. (c) Calibration curves for pH in buffered soil extract using corrected ΔRGB values. (d) R channel values fitted to calibration curves for pH estimation of buffered soil extracts with six phones (models indicated). Inset, the results of post-hoc analyses on the MAEs for pH predictions (n = 15) using the phone's specific calibration curve or a mixed curve of all phone models except the one used for prediction. Stars indicate p-values with p-value correction for multiplicity. *** = $p < 0.001$. All calibration curves were made using a four-parameter dose-response function.

4. Conclusions

The ability of all channels of the RGB, HSV, and LAB (named LAC here) color spaces to quantify color intensity and color change variations was tested in paper based and liquid assays. Channel performance was compared under varying background illumination using various phones. A channel of LAC proved not suitable for color quantification and C channel performance was suboptimal. L channel showed good performance in most systems tested (as did V) but both never outperformed the best RGB channel for a specific test. S performed suboptimal in all systems tested. H channel performed satisfactorily for color change but not for color intensity quantification. Moreover, H channel never outperformed the best RGB channel for a specific test. Overall, R functioned best for color change while B and G worked best for color intensity variation. ΔRGB values did show some robustness towards errors in an individual channel of RGB thus providing more universal applicability. However, ΔRGB stops to perform if an error in an individual channel of the RGB color space is too high. Thus, all RGB channels and ΔRGB should be initially plotted for assay development to determine the optimum channel. However, conversions to LAC or HSV color space are unnecessary. These results are specific for colorimetric assays and the implications might not hold for fluorescence or chemiluminescence-based assays where using luminance values seems a more logical choice. This being said, for such assays the light measured will equally pass the Bayer filter of the phone and is converted to RGB values. Mathematically converting these RGB values to LAB values may lead to similar results as reported here. However, a detailed characterization of the performance of RGB, LAB, and HSV color spaces/channels of such luminance-based assays with various phones and fluorophores is needed to test this assumption. Using a light-shielding box to prevent the error caused by background variation was less efficient than internal background correction if images were not taken in direct sunlight. Thus, use of a box is superfluous for colorimetric analysis. For color intensity variation it was shown that images taken with phone A could not be used on a calibration curve taken with phone B. Magnification did improve this situation but did not completely resolve the problem. Thus, camera calibration, white balance adjustment, and exposure time adjustment should be considered for LFA quantification. For color change quantification (pH determination of buffered soil extracts) calibration curves of the six phone models tested overlapped significantly. Thus, color change quantification by end-users without a light-shielding box or phone specific calibration seems feasible. Future research will focus on reducing prediction error and inter-phone variation for color change and intensity-based assays by combining channels from RGB, HSV, and CieLAB color spaces with machine learning algorithms.

Author Contributions: J.L.D.N. was the main author executing the experiments, data curation, and writing. L.B. assisted in the execution of experiments and data curation. Y.Z. and K.M.B. assisted in executing the experiments. C.T.E. and K.R. provided advice and assisted in reviewing and editing the final document. K.C. contributed to the conceptualization, supervised this work, and provided assistance in reviewing and editing the manuscript.

Funding: This project has received funding from the European Union's Horizon 2020 research and innovation program under the Marie Sklodowska-Curie grant agreement No 720325.

Acknowledgments: We further want to thank Master Sparrow Nelis for providing his smartscope for image magnification and Ilyana Nelis for providing home produced pure goat milk.

Conflicts of Interest: The authors declare no conflict of interest. The funders had no role in the design of the study; in the collection, analyses, or interpretation of data; in the writing of the manuscript, or in the decision to publish the results.

References

1. Malik, A.K.; Blasco, C.; Picó, Y. Liquid chromatography-mass spectrometry in food safety. *J. Chromatogr. A* **2010**, *1217*, 4018–4040. [CrossRef] [PubMed]
2. Matumba, L.; Van Poucke, C.; Njumbe Ediage, E.; De Saeger, S. Keeping mycotoxins away from the food: Does the existence of regulations have any impact in Africa? *Crit. Rev. Food Sci. Nutr.* **2017**, *57*, 1584–1592. [CrossRef] [PubMed]

3. The Rapid Alert System for Food and Feed (2017 Annual Report). Available online: https://ec.europa.eu/food/sites/food/files/safety/docs/rasff_annual_report_2017.pdf (accessed on 20 November 2019).
4. Sivadasan, S.; Efstathiou, J.; Calinescu, A.; Huatuco, L.H. Advances on measuring the operational complexity of supplier-customer systems. *Eur. J. Oper. Res.* **2006**, *171*, 208–226. [CrossRef]
5. Knowles, T.; Moody, R.; McEachern, M. European food scares and their impact on EU food policy. *Br. Food J.* **2007**, *109*, 43–67. [CrossRef]
6. Ludwig, S.K.; Tokarski, C.; Lang, S.N.; van Ginkel, L.A.; Zhu, H. Calling Biomarkers in Milk Using a Protein Microarray on Your Smartphone. *PLoS One* **2015**, *10*, e0134360. [CrossRef] [PubMed]
7. Ludwig, S.K.J.; Zhu, H.; Phillips, S.; Shiledar, A.; Feng, S.; Tseng, D.; van Ginkel, L.A.; Nielen, M.W.F.; Ozcan, A. Cellphone-based detection platform for rbST biomarker analysis in milk extracts using a microsphere fluorescence immunoassay. *Anal. Bioanal. Chem.* **2014**, *406*, 6857–6866. [CrossRef]
8. Zeinhom, M.M.A.; Wang, Y.; Song, Y.; Zhu, M.-J.; Lin, Y.; Du, D. A portable smart-phone device for rapid and sensitive detection of *E. coli* O157:H7 in Yoghurt and Egg. *Biosens. Bioelectron.* **2018**, *99*, 479–485. [CrossRef]
9. Fang, J.; Qiu, X.; Wan, Z.; Zou, Q.; Su, K.; Hu, N.; Wang, P. A sensing smartphone and its portable accessory for on-site rapid biochemical detection of marine toxins. *Anal. Methods* **2016**, *8*, 6895–6902. [CrossRef]
10. Lee, S.; Kim, G.; Moon, J. Performance improvement of the one-dot lateral flow immunoassay for aflatoxin b1 by using a smartphone-based reading system. *Sensors* **2013**, *13*, 5109–5116. [CrossRef]
11. Ross, G.M.S.; Bremer, M.G.E.G.; Wichers, J.H.; Van Amerongen, A.; Nielen, M.W.F. Rapid antibody selection using surface plasmon resonance for high-speed and sensitive hazelnut lateral flow prototypes. *Biosensors* **2018**, *8*, 130. [CrossRef]
12. Coskun, A.F.; Wong, J.; Khodadadi, D.; Nagi, R.; Tey, A.; Ozcan, A. A personalized food allergen testing platform on a cellphone. *Lab Chip* **2013**, *13*, 636–640. [CrossRef] [PubMed]
13. Tsagkaris, A.S.; Nelis, J.L.D.; Ross, G.M.S.; Jafari, S.; Guercetti, J.; Kopper, K.; Zhao, Y.; Rafferty, K.; Salvador, J.P.; Migliorelli, D.; et al. Critical assessment of recent trends related to screening and confirmatory analytical methods for selected food contaminants and allergens. *TrAC Trends Anal. Chem.* **2019**, *121*, 115688. [CrossRef]
14. Nelis, J.L.D.; Tsagkaris, A.S.; Zhao, Y.; Lou-Franco, J.; Nolan, P.; Zhou, H.; Cao, C.; Rafferty, K.; Hajslova, J.; Elliott, C.T.; et al. The end user sensor tree: An end-user friendly sensor database. *Biosens. Bioelectron.* **2019**, *130*, 245–253. [CrossRef] [PubMed]
15. Nelis, J.; Elliott, C.; Campbell, K. "The Smartphone's Guide to the Galaxy": In Situ Analysis in Space. *Biosensors* **2018**, *8*, 96. [CrossRef]
16. Koczula, K.M.; Gallotta, A. Lateral flow assays. *Essays Biochem.* **2016**, *60*, 111–120.
17. Roda, A.; Michelini, E.; Zangheri, M.; Di Fusco, M.; Calabria, D.; Simoni, P. Smartphone-based biosensors: A critical review and perspectives. *TrAC Trends Anal. Chem.* **2016**, *79*, 317–325. [CrossRef]
18. Rateni, G.; Dario, P.; Cavallo, F. Smartphone-Based Food Diagnostic Technologies: A Review. *Sensors* **2017**, *17*, 1453. [CrossRef]
19. Urusov, A.E.; Zherdev, A.V.; Dzantiev, B.B. Towards Lateral Flow Quantitative Assays: Detection Approaches. *Biosensors* **2019**, *9*, 89. [CrossRef]
20. Suaifan, G.A.R.Y.; Zourob, M. Portable paper-based colorimetric nanoprobe for the detection of Stachybotrys chartarum using peptide labeled magnetic nanoparticles. *Microchim. Acta* **2019**, *186*, 230. [CrossRef]
21. Zhang, X.; Zhi, X.; Zhang, C.; Wang, K.; Cui, D.; Li, D.; Wang, C. A CCD-based reader combined quantum dots-labeled lateral flow strips for ultrasensitive quantitative detection of anti-HBs antibody. *J. Biomed. Nanotechnol.* **2012**, *8*, 372–379. [CrossRef]
22. Murdock, R.C.; Shen, L.; Griffin, D.K.; Kelley-Loughnane, N.; Papautsky, I.; Hagen, J.A. Optimization of a paper-based ELISA for a human performance biomarker. *Anal. Chem.* **2013**, *85*, 11634–11642. [CrossRef] [PubMed]
23. Kong, T.; You, J.B.; Zhang, B.; Nguyen, B.; Tarlan, F.; Jarvi, K.; Sinton, D. Accessory-free quantitative smartphone imaging of colorimetric paper-based assays. *Lab Chip* **2019**, *19*, 1991–1999. [CrossRef] [PubMed]
24. Masawat, P.; Harfield, A.; Namwong, A. An iPhone-based digital image colorimeter for detecting tetracycline in milk. *Food Chem.* **2015**, *184*, 23–29. [CrossRef] [PubMed]
25. Turkevich, J. Colloidal gold. Part I. *Gold Bull.* **1985**, *18*, 125–131. [CrossRef]
26. Haiss, W.; Thanh, N.T.K.; Aveyard, J.; Fernig, D.G. Determination of Size and Concentration of Gold Nanoparticles from UV. *Anal. Chem.* **2007**, *79*, 4215–4221. [CrossRef]

27. Mcvey, C.; Logan, N.; Thanh, N.T.K.; Elliott, C.; Cao, C. Unusual switchable peroxidase-mimicking nanozyme for the deter- mination of proteolytic biomarker. *Nano Res.* **2019**, *12*, 1–8. [CrossRef]
28. Shen, L.; Hagen, J.A.; Papautsky, I. Point-of-care colorimetric detection with a smartphone. *Lab Chip* **2012**, *12*, 4240–4243. [CrossRef]
29. Cantrell, K.; Erenas, M.M.; de Orbe-Payá, I.; Capitán-Vallvey, L.F. Use of the Hue Parameter of the Hue, Saturation, Value Color Space as a Quantitative Analytical Parameter for Bitonal Optical Sensors. *Anal. Chem.* **2010**, *82*, 531–542. [CrossRef]
30. Kim, S.D.; Koo, Y.; Yun, Y. A smartphone-based automatic measurement method for colorimetric pH detection using a color adaptation algorithm. *Sensors* **2017**, *17*, 1604. [CrossRef]
31. Zhao, Y.; Choi, S.Y.; Nelis, J.L.D.; Zhou, H.; Cao, C.; Campbell, K.; Elliott, C.; Rafferty, K. Smartphone Modulated Colorimetric Reader with Color Subtraction. In Proceedings of the IEEE Sensors 2019 Conference, Montreal, QC, Canada, 27–30 October 2019.
32. Skandarajah, A.; Reber, C.D.; Switz, N.A.; Fletcher, D.A. Quantitative imaging with a mobile phone microscope. *PLoS ONE* **2014**, *9*, e96906. [CrossRef]
33. Saisin, L.; Amarit, R.; Somboonkaew, A.; Gajanandana, O.; Himananto, O.; Sutapun, B. Significant Sensitivity Improvement for Camera-Based Lateral Flow Immunoassay Readers. *Sensors* **2018**, *18*, 4026. [CrossRef] [PubMed]

© 2019 by the authors. Licensee MDPI, Basel, Switzerland. This article is an open access article distributed under the terms and conditions of the Creative Commons Attribution (CC BY) license (http://creativecommons.org/licenses/by/4.0/).

Article

An Innovative Ultrasonic Apparatus and Technology for Diagnosis of Freeze-Drying Process

Chin-Chi Cheng [1,*], Yen-Hsiang Tseng [2] and Shih-Chang Huang [1]

1. Department of Energy and Refrigerating Air-Conditioning Engineering,
   National Taipei University of Technology, Taipei 10608, Taiwan; albertschuang@gmail.com
2. Tai Yiaeh Enterprise Co., Ltd., New Taipei City 23942, Taiwan; yhtntut@gmail.com
* Correspondence: newmanch@ntut.edu.tw; Tel.: +886-2-27712171 (ext. 3527)

Received: 17 April 2019; Accepted: 8 May 2019; Published: 11 May 2019

Abstract: The freeze-drying process removes water from a product through freezing, sublimation and desorption procedures. However, the extreme conditions of the freeze-drying environment, such as the limited space, vacuum and freezing temperatures of as much as −50 °C, may block the ability to use certain diagnostic sensors. In this paper, an ultrasonic transducer (UT) is integrated onto the bottom of a specially designed frozen bottle for the purpose of observing the freeze-drying process of water at varying amounts. The temperatures and visual observations made with a camera are then compared with the corresponding ultrasonic signatures. Among all of the diagnostic tools and technologies available, only ultrasonic and visual records are able to analyze the entire progression of the freeze-drying process of water. Compared with typical experiment settings, the indication of drying point for water by the amplitude variations of ultrasonic L^3 echo could reduce the process period and energy consumption. This study demonstrates how an innovative frozen bottle, an integrated ultrasonic sensor and diagnostic methods used to measure and optimize the freeze-drying process of water can save energy.

Keywords: freeze-drying process diagnosis; ultrasonic transducer (UT); freezing/drying point; drying period

1. Introduction

Dehydration extends food's usage period longer than that of fresh food by preserving it in a stable and safe condition [1,2]. The conventional methods of drying include solar drying, air drying, spray drying [3], microwave drying [4], infrared drying, fluidized bed drying [5], spouted bed drying, vacuum drying and freeze-drying [6]. Drying methods can be separated into natural and artificial categories. Artificial methods are more advantageous than the natural methods [7] because they can remove large amounts of moisture efficiently by being able to control the different parameters involved such as the temperature, drying air flux and time of drying and so forth. [8]. For long-term storage of food, drug and biopharmaceutical products, most manufacturers utilize the freeze drying process, due to its advantages including better stability, easy handling and storage, as well as better overall product quality [9,10].

The freeze-drying process is comprised of the freezing, primary drying and then secondary drying stages [11] to remove the water from a product. The water contained in food is cooled down and becomes ice during the freezing stage. This stage governs the sublimation and desorption rates and the quality of the lyophilized product [12]. Then, in the primary drying stage, the air in the vacuum chamber is exhausted and the chamber pressure is reduced below the vapor pressure of ice. Meanwhile, the shelf temperature increases gradually to sublimate the ice. The residual water inside the food will be desorbed thoroughly in the secondary drying stage [13].

In the freezing stage, the accuracy of the freezing point data is related to the water activity, frozen water [14], freezing and the thawing of frozen food. The freezing point is also important in order to estimate the freezing time, the end point of freezing and the fraction of unfrozen water in food. Then, in the primary drying stage, the chamber pressure and food temperature [15] decides the sublimation rate of the ice, primary drying period, dried pore structure and product quality. A lower product temperature and the corresponding lower vapor pressure of the ice can result in longer primary drying times and higher manufacturing costs [16]. However, if the product temperature increases above the "critical formulation temperature," this may lead to losing the pore structure, to shrinkage or to fully collapsing. In the secondary drying stage, the product temperature controls the rate of desorption and the obtainable moisture level. Further reduction of the chamber pressure is not necessary in this stage [17]. Secondary drying times are usually designed to achieve a reduction of moisture content within the dried product to less than 1% [18]. Due to the importance of these parameters and its wider applications to the freeze drying process, the estimated freezing/drying points and the ability to model these properties is crucial in food processing (freezing and drying) and food stability during storage [10,19].

Being able to perform real-time process diagnosis will be beneficial to understanding the complex interplay between the different elements during the freeze-drying process to better enhance quality control procedures. However, the extreme conditions of the freeze-drying environment, such as the limited space, vacuum and freezing temperatures as low as −50 °C, may block the ability to use certain diagnostic sensors [6,20]. The typically used diagnostic tools for real-time measurement of the freeze-drying process are temperature and pressure sensors [20,21]. In order to provide temperature information, a temperature sensor is put in direct contact with a frozen sample [21]. However, inconveniences arise because it is often not easy to remove the sensor after the freeze-drying process. Additionally, using this diagnostic method on drug and biopharmaceutical products may cause undesired contamination due to a sensor being in direct contact with the sample. The chamber pressure is measured by the pressure sensor or vacuum gauge. The measured chamber pressure value cannot directly indicate the physical variation of frozen product real-time during the freeze-drying process [20,22]. In this paper, an idea for diagnosing the physical property variations of a product in the frozen bottle to provide additional information for a machine system is presented, as shown in Figure 1.

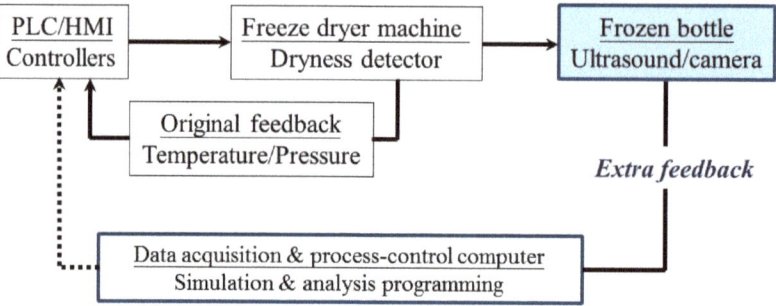

Figure 1. Controlling structure of the freeze-drying process by using the innovative frozen bottle method.

In order to circumvent the limitations of the mentioned sensors and provide real-time process diagnosis, the ultrasonic technique is one of the most widely known non-destructive and non-intrusive methods [23]. The fundamental properties of ultrasonic signals include reflection and transmission coefficients, velocity, attenuation and scatter signals given off from the materials. These signatures have specific relationships with properties of materials [24], process variations [25] and sample quality [26]. Ultrasonic echoes can also detect the temperature [27,28] and pressure [29] of a material. Passot et al. utilized ultrasonic technology to decrease the sublimation time during the primary drying stage by controlling the nucleation temperature [30]. The authors presented an innovatively designed frozen bottle with an ultrasonic transducer and ultrasonic technology to be utilized for the diagnosis of the freezing point during the frozen process in our previous paper [22]. In this study, the ultrasonic technique will be applied for real-time diagnosis of the freeze-drying process. An ultrasonic transducer (UT) is integrated into the bottom of a frozen bottle. During the freeze-drying process at various amounts of water, the process is analyzed by using ultrasonic technology to evaluate the freezing/drying points and the drying period to optimize the process and save energy.

2. Design of Frozen Bottle with Ultrasonic Transducer

In order to solve the aforementioned limitations regarding the transmission of signals in a vacuum, as well as the limitations of being unable to view the process using the steel freezing bottle apparatus or the opaque samples, an innovative freezing bottle apparatus was designed as shown in Figure 2. The innovatively designed container used for freezing is comprised of two parts: a steel plate as the bottom holder and a transparent polymethylmethacrylate (PMMA) tube as the side wall of the container. Figure 2a shows the side view of the frozen bottle. A steel plate with a diameter of 25 mm and a height of 6mm was designed for better heat transmission with the heating/cooling shelf. A transparent PMMA tube measuring 31 mm high, with a diameter of 21 mm and a thickness of 2 mm was designed for better visibility. The steel plate and PMMA tube were then glued together. Figure 2b shows the top view of the apparatus. Figure 2c shows the bottom view of the apparatus. The UT is comprised of $Pb(Zr_xTi_{1-x})O_3$ (PZT) material and integrated into the cavity of steel plate using the sol-gel spray technique, as described in previous publications [22,24,31]. The operational temperature range of the UT is between −100 and 400 °C.

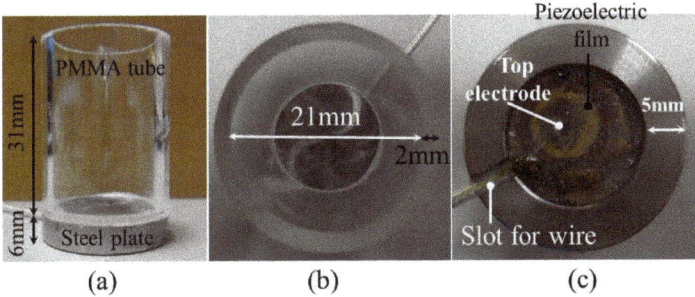

Figure 2. Photographs of the designed apparatus from (**a**) the side, (**b**) the top and (**c**) the bottom view.

To understand the transmission paths of the ultrasonic signals within the apparatus, a schematic view of the apparatus placed upright is displayed in Figure 3. A temperature sensor (Type T thermocouple) was installed in the middle of the apparatus for measuring the water/ice temperature. The temperature can be recorded and then compared with the ultrasonic signals received during the freeze-drying process. As shown in Figure 3a, an ultrasonic transducer (UT) was integrated onto the bottom of the steel plate. When electric pulses are applied to the piezoelectric film, the ultrasonic signals are transmitted into the steel plate. L^n (n = 1, 2, ...) denotes the nth round trip of longitudinal ultrasonic echoes reflected from the interface of the steel plate/water or ice and L_w is the echo reflected from the water/air interface. The L^3 and L_w echoes are used to monitor the freeze-drying process and water/ice state. The height of the water level and the thickness of the bottom of the apparatus are denoted as H and D, respectively.

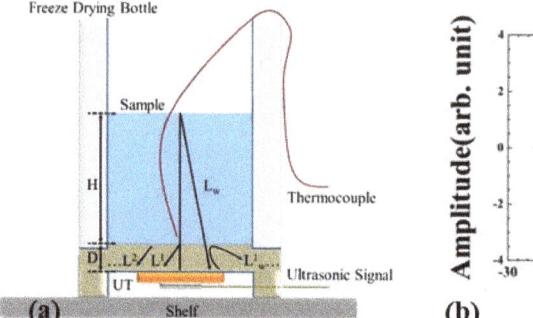 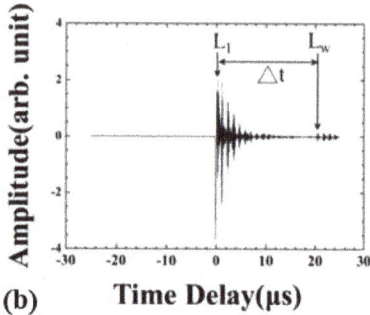

Figure 3. (a) Schematic drawing of cross-section of the designed apparatus with an ultrasonic transducer (UT) and thermocouple and (b) shows the typical ultrasonic echoes acquired by the UT.

When the water was fed into the apparatus, the typical ultrasonic signals acquired by the UT are presented in Figure 3b. It was observed that the L^1 echo, which reflected from the steel plate/water or ice interface, appeared as 0.81 µs and remained so during the entire process. The operating frequency of the ultrasonic transducer was 9.13 MHz and the −6 dB bandwidth was 8.0 MHz. The signal to noise ratio (SNR) for the ultrasonic L^1 echo was 38.5 dB. The L_w echo, which propagated in the water and reflected from the water/air interface, was observed at 20.39 µs. Following the L^1 and L_w echoes, there were several echoes reflected from the steel plate/water or ice interface. The desired ultrasonic echoes were detected by the acquiring windows of the acquiring system. The time delay difference between the L^1 and L_w echoes is denoted as Δt.

3. Experiments

In order to confirm the hypothesis that the apparatus was able to perform improved diagnostics of the freeze-drying process in a vacuum, a 4 L air-cooled split-type shelf freeze-drying machine, equipped with vacuum chamber, shelf, control units, refrigeration system and vacuum pump, was used to carry out all experiments, as shown in Figure 4a. The vacuum chamber was used for freezing and drying the samples under temperature conditions ranging from +50 to −40 °C and pressures from 760 to 0.05 Torr. The apparatus was set on the shelf of vacuum chamber for the freeze-drying process. The control unit was composed of a programmable logic controller (PLC) and a human-machine interface (HMI) for controlling the freeze-drying process. The measured temperature data from the thermocouple were recorded by the PLC control unit every second during the freeze-drying process.

The ultrasonic signal is triggered and received by the pulse/receive unit (5072PR, Olympus, Japanese), as shown in Figure 4b. It is equipped with a broadband negative spike pulser and receiver, which can be operated in reflection or transmission mode. All the experiments in this study were carried out using the ultrasonic pulse/receiver unit's pulse-echo mode. The ultrasonic signals were acquired every second during the entire cycle. The ultrasonic signals received during the freeze-drying process were monitored and recorded with the digital storage USB oscilloscope (DSO-U2400, Perytech, Taipei, Taiwan), as shown in Figure 4c.

To compare the temperatures and ultrasonic signals, a charge-coupled device (CCD) camera, digital video recorder (DVR) and PC were used to capture dynamic images of the apparatus during the freeze-drying process, as shown in Figure 4d–e. The CCD camera (CV-M10BX, JAI, Japan) is a progressive scan camera with a standard interlaced video output at a resolution compatible with VGA or SVGA formats. The camera was equipped with a 35 mm lens for better resolution. The DVR (08KD, Kingnet, New Taipei City, Taiwan) is an 8-channel recorder that uses the H.264 image compression format.

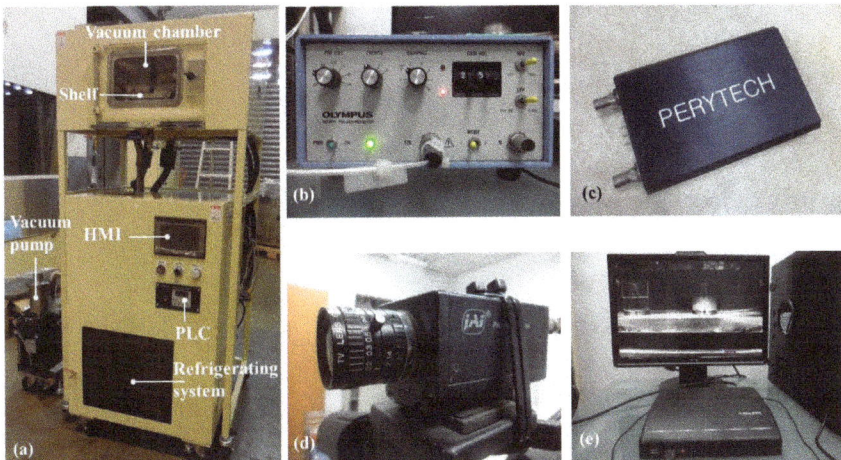

Figure 4. Photographs of (**a**) the split-type freeze-drying machine with vacuum chamber, controller unit, refrigeration unit and vacuum pump, (**b**) the ultrasonic pulse/receive unit, (**c**) a PC-based digital oscilloscope, (**d**) Charge-coupled device (CCD) camera, (**e**) digital video recorder (DVR) and PC.

In this study, different amounts of water at levels of 5, 10 and 15 mm were put into the apparatus for the duration of the freeze-drying process. The settings of a typical freeze-drying procedure are outlined in Table 1. The shelf temperature was set at −30 °C to freeze the water and the freezing period was 120 min. During the freezing period, the water in the apparatus/container was cooled down and it began to freeze. After enough cooling and freezing, the chamber was exhausted and the chamber pressure was reduced from 760 to 0.17 Torr within several seconds. Then the primary and secondary drying processes were started and the ice sublimated into gas continuously. The shelf temperature was incrementally increased by 10 degree intervals from −30 to 10 °C at heating and pausing periods of 60 and 120 min, respectively. The pausing period of 0 °C is 300 min to sublimate ice further. During the sublimation of the ice, the chamber pressure was reduced further to 0.036 Torr.

Table 1. The typical procedure used for the freeze-drying process.

Freezing Temp. (°C)	Freezing Period (min)	Chamber Pressure (Torr)	Drying Temp. (°C)	Heating Period (min)	Pausing Period (min)
−30	120	760 | 0.17	−20	60	120
			−10	60	120
			0	60	300
			10	60	120

4. Results and Discussions

For exploring the physical phenomena that occur during the freeze-drying process and searching for the indicators of freezing or drying completeness, the measured temperature and pressure in the vacuum chamber, ultrasonic echoes and velocity of the water/ice and visual observation during the freeze-drying process are analyzed in the following sections.

4.1. Temperature and Pressure Variations in the Vacuum Chamber during the Freeze-Drying Process

Temperature and pressure variations during the freeze-drying process have a close relationship with the state of the water/ice and both affect the size of the ice crystal nuclei and the freeze-dried product's quality. In the experiment, the measured room, shelf, cold trap, water/ice temperatures and pressures for the freezing and drying periods of the freeze-drying procedure of the water at a level of

15 mm are shown in Figure 5a,b, respectively. In Figure 5a, the room temperature was 32 °C during the entire process and the chamber pressure was kept at 760 Torr for the experiment's duration of 120 min. The shelf and cold trap temperatures were set at −30 and −40 °C and reached the desired values at 40 and 15 min, respectively. The water temperature was 32.1 °C in the beginning and was cooled down to a supercooled state of −0.4 °C, 16.1 min into the experiment. After the latent heat of the water was released completely, the phase change from water to ice was completed at 43.7 min and the ice temperature was −16.5 °C. Therefore, the cooling (ΔP_{CT}) and freezing (ΔP_{FT}) period determined by the sample temperature could be defined as the period from the start of cooling to the supercooled state and from the supercooled state to the phase change end, respectively.

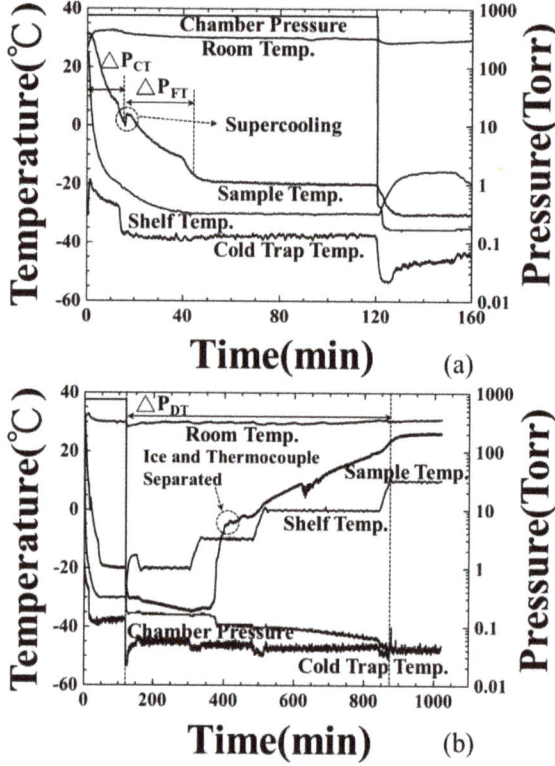

Figure 5. Temperature (room, shelf, sample, cold trap) and pressure (chamber) variations during freeze-drying process at (**a**) the freezing stage and (**b**) drying stage.

At the 120-min mark of the experiment, the drying procedure started and the chamber pressure was reduced from 760 to 0.17 Torr. The shelf temperature was increased in 10 °C increments, from −30 to 10 °C, according to the designed schedule in Table 1. The cold trap temperature was cooled down further to condense the vapor exhausted from the chamber. The chamber pressure was reduced further to 0.036 Torr during the drying process. At the 370-min mark, the ice temperature increased suddenly from −36 to −10 °C due to the sublimation of the ice and gradually increased to 23 °C by the end of the experiment. This abnormal situation was due to the detachment of the thermocouple from the ice and will be confirmed in the following section. Consequently, the ice temperatures during the second drying period is not available. The drying end could not be estimated from the temperature information and the typical experiment procedure would end at the 1020-min mark of the experiment.

4.2. Amplitude Variation of Ultrasonic Signals during the Freeze-Drying Process

The application of freeze-drying technology can guarantee the product's quality. However, the process diagnosis and control will benefit by guaranteeing product quality and energy conservation. The conventional tools for monitoring this process may be limited to visual observation and temperature measurement. According to the authors' knowledge, utilizing ultrasonic technology to monitor the freeze-drying process is rare. In order to investigate the correlation between the observed ultrasonic signals and the freeze-drying process of water, the amplitudes of the L^3 and L_w echoes in Figure 3b with respect to the process time were acquired and presented in Figure 6. The details of freeze-drying process measured by the amplitude variations of ultrasonic L^3 and L_w echoes were demonstrated as follows:

(1) At the 3.0-min mark: Water was poured into the container. At this moment, the amplitude of ultrasonic L^3 echo decreased and L_w echo appeared, due to the fact that a part of the ultrasonic energy transmitted into the water through the steel plate/water interface. Water started to cool down.

(2) At the 19.2-min mark: The amplitudes of the L^3 and L_w echoes decreased further, due to the variation of water acoustic impedance. In this supercooling period, this ultrasonic phenomenon may indicate the appearance of ice crystals. The alteration of water property will cause electrical impedance changing [32]. Hence, the amplitude decreasing point of the L^3 and L_w echoes is defined as the freezing point of water and the cooling period (ΔP_{CUT}) determined by the amplitude of the L^3 echo is defined as the period from water-in to the freezing point.

(3) At the 43.0-min mark: The amplitude of the L^3 echo increased to the relative maximum value and that of the L_w echo increased to a stable level, due to the stable ultrasonic impedance. This ultrasonic phenomenon may indicate phase change end of water and a flat ice surface. Hence, the freezing period (ΔP_{FUT}) determined by the ultrasonic L^3 echo is defined as the period from the freezing point to the phase change end.

(4) At the 125.0-min mark: The chamber was exhausted and the ice started to sublimate at the 120-min mark. The amplitude of L^3 echo decreased a little and that of L_w echo disappeared gradually, due to the sublimation of ice and the rough ice surface.

(5) At the 370.0-min mark: The amplitude of the L^3 echo increased suddenly from the bottom line and reach a stable value, due to the fact that the electromechanical coupling factor varied and the ultrasonic energy transmitting into the ice through the steel/ice interface reduced [33]. This ultrasonic phenomenon indicated the reduction of contact surface between ice and steel plate in the drying stage. At this moment, the sublimation of the ice reached a certain level and the thermocouple detached from the ice, as shown in Figure 5b.

(6) At the 885.6-min mark: The amplitude of the L^3 echo increased suddenly from the reducing tendency and reached a stable value at this moment, due to the fact that the electromechanical coupling factor varied and ultrasonic energy transmitting into the ice through the steel/ice interface reduced further. This ultrasonic phenomenon indicated the complete sublimation of the ice and only few minerals remained on the steel plate surface. This moment is defined as the drying end point. The drying period (ΔP_{DUT}) determined by the ultrasonic L^3 echo is defined as the period from the exhaust of chamber to the complete sublimation of the ice, that is, the drying end point. Compared with the typical experiment settings, the indication of drying ends for water by the amplitude variations of ultrasonic L^3 echo could reduce the processing period of the 134.4 min/cycle and save 13% of consumed electricity.

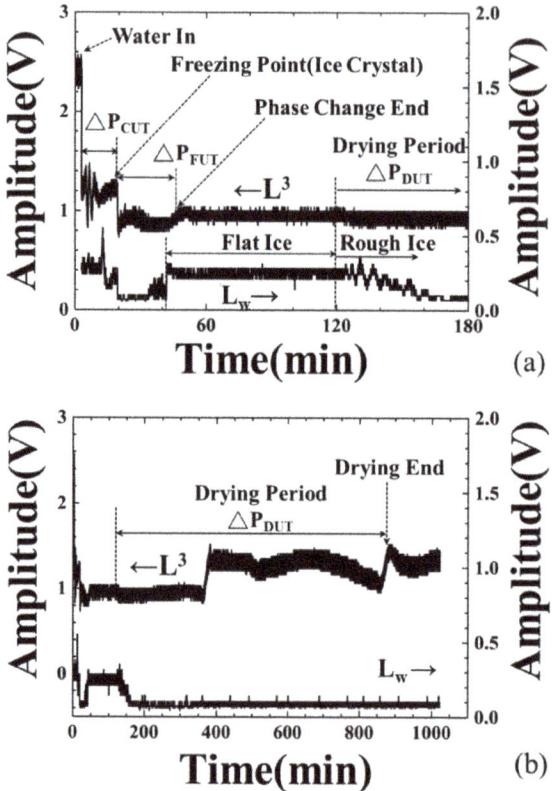

Figure 6. Amplitude variations of ultrasonic L^3 and L_w echoes during freeze-drying process at (**a**) freezing stage and (**b**) drying stage, indicating the water-in, freezing point (ice crystal), phase change end and drying end.

4.3. Variation of Ultrasonic Velocity during the Freeze-Drying Process

During the freezing process, the slow cooling speed caused bigger ice crystal nuclei than did the fast cooling time. Ultrasonic velocity has a close relationship with temperature and it is one of the indicators that shows the cooling speed of water, the phase change end and the start of sublimation. The ultrasonic velocity in the water is calculated according to the following equation:

$$v_w = 2H/\Delta t \qquad (1)$$

where H is the height of the water level in Figure 3a and Δt is the time delay between the ultrasonic L^1 and L_w echoes in Figure 3b. The result is shown in Figure 7.

In Figure 7, when the water was poured into the container, the ultrasonic velocity was first measured at the 3.0-min mark under the water temperature of 27.2 °C. At that moment, the ultrasonic velocity in the water was 1588.5 m/s. From the 3.0 to the 17.4-min mark, when the water temperature decreased to 1.0 °C, the ultrasonic velocity decreased to 1434.5 m/s, indicating that the temperature of the water had decreased during the cooling process. Comparing Figure 7 with Figure 5a, the relationship between ultrasonic velocity and water temperature is expressed by the following equation.

$$v_w(T_w) = 1.42418 \times 10^3 + 1.03190 \times 10 \times T_w - 0.15453 \times T_w^2 \qquad (2)$$

where v_w (m/s) is the ultrasonic velocity in water and T_w is the water temperature. The ultrasonic velocity decreased with the reducing water temperature. The ultrasonic velocity presented by authors in this study is higher than that presented by Bilanuik and Wang [34], due to some minerals contained in the water. Therefore, the cooling speed of water can be determined by measuring the ultrasonic velocity in the water. From the 17.4 to 38.3-min mark of the experiment, the ultrasonic velocity disappeared and the amplitude of the L_w echo also decreased in noise level due to the formation of ice crystal nuclei. After the 38.3-min mark, the ultrasonic L_w echo appeared again and the ultrasonic velocity gradually increased from 3586.4 to 3597.2 m/s, due to the phase change end of the water and the completion of the freezing process. From 43.0 to 120.0-min mark, the ultrasonic velocity maintained a steady level of 3597.2 m/s for the pausing period. Therefore, ultrasonic velocity can be used to indicate the cooling speed, the phase change end and completion of the freezing process of water.

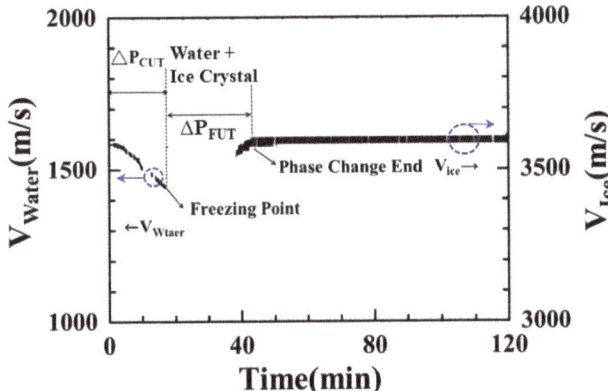

Figure 7. Ultrasonic velocity variation in the water during the freeze-drying process, which indicates the water-in, the freezing point (ice crystallization) and the phase change end.

4.4. Visual Observation of the Container during the Freeze-Drying Process

The dynamic physical phenomena that can be observed in the container shows the phase change from water to ice and the sublimation detail of the ice during the freeze-drying process in great detail. These phenomena could also be evidence of other measured parameters. Figure 8a–l are the photographs of the container with water/ice during the freeze-drying process at the 0–888-min mark, respectively. The freeze-drying process observed in Figure 8a–l is described as follows:

(1) Figure 8a: Water was poured into the container that had a thermocouple in the middle and then the container was set on the shelf of the freeze-drying machine.

(2) Figure 8b–d: Most of the water was still in a liquid state. However, the water close to the bottom of the container started to become opaque and some ice crystals began to form beside the thermocouple. The ice crystals increased gradually from the bottom to the top of the container. This was an indication that the water started to freeze.

(3) Figure 8e: Half of the water became very slushy and the visibility became worse. Some air bubbles also appeared within the slush.

(4) Figure 8f: Most of the water had become frozen into ice. Some air bubbles were pushed up from the bottom to the top. The ice surface was flat, which indicated that the stress on the ice during the freezing process was reduced.

(5) Figure 8g–h: The chamber was exhausted and a vacuum was created. The ice sublimated from the top progressing downwards towards the bottom of the container and from the exterior inwards towards the interior of the container. The ice surface was composed of porous structure and appeared rough.

(6) Figure 8i–j: The ice shrank more due to further sublimation. The contact surface between the ice and the steel plate was also reduced. The thermocouple seemed to detach from the ice at the 480-min mark of the process. This could explain why the sample temperature increased suddenly from −36 to −10 °C at 390.5-min mark. At the 540-min mark, the contact surface between the ice and the steel plate was reduced even further, which may diminish the conduction of heat being transferred from the shelf to the ice.

(7) Figure 8k–l: There was tiny amount of residual ice that remained on the surface of the steel plate at the 840-min mark. Finally, at the 888-min mark, only a little of ice crystals remained on the surface of the steel plate.

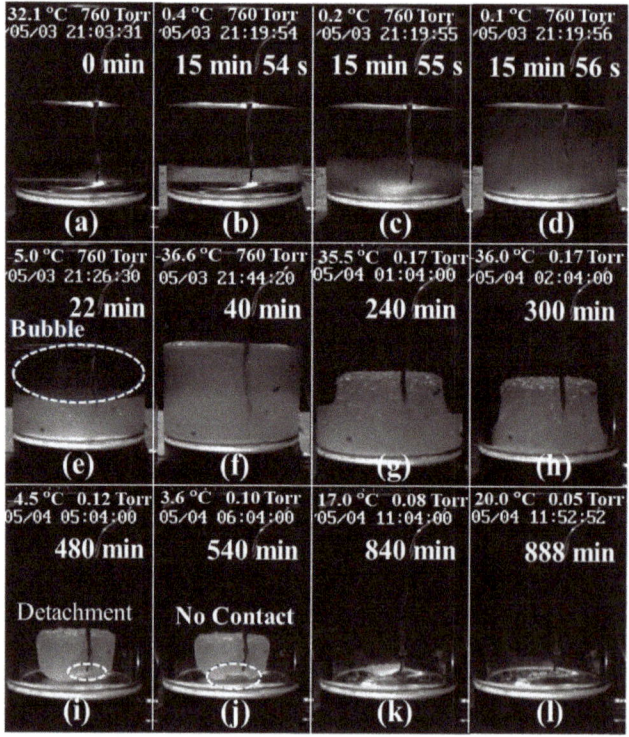

Figure 8. Photographs of the water/ice in the container during the freeze-drying process along with the corresponding process times of (**a**) 0 min, (**b–d**) 15 min 54~56 s, (**e**) 22 min, (**f**) 40 min, (**g**) 240 min, (**h**) 300 min, (**i**) 480 min, (**j**) 540 min, (**k**) 840 min and (**l**) 888 min.

The freeze-drying process can be observed clearly in Figure 8. These results can provide additional evidence for the previous measuring methods. However, this observing method is only suitable for the transparent container and plain solution. For the unobservable container and the opaque material, such as metal container and colored samples, the temperature and ultrasonic techniques are more capable to diagnose the mentioned phenomena. From the measured results, the thermocouple detached from the ice and further temperature information of the ice was not available. In this viewpoint, the ultrasonic technique is a more reliable and capable method for diagnosing the freeze-drying process.

4.5. Freeze-Drying Processes of Various Water Levels

Before applying the ultrasonic technology to the freeze-drying process, the linear relationship between ultrasonic signatures and the water level is the fundamental requirement. To verify this relationship, the containers, which were filled with water in the height of 5, 10, 15 mm, passed through the freeze-drying process under the shelf temperature setting of −30 °C. In this research, the variation during the drying period will be evaluated. During the drying stage, the thermocouple cannot measure the samples' temperature profiles due to the detachment of thermocouple from the sample. The ultrasonic signals are able to diagnose the drying phenomena. The experimental results of the amplitude of the ultrasonic L^3 echo are shown in Figure 9. Figure 9a–c are the amplitude variations of ultrasonic L^3 echo in water levels of 5, 10, 15 mm, respectively, during the drying stage. The timings of the drying end indicated by ultrasonic L^3 echo are 329.8, 526.7 and 885.6-min mark for the water levels of 5, 10 and 15 mm, respectively. The drying periods of various water levels are also indicated. The corresponding timings of the drying end indicated by ultrasonic L^3 echo are illustrated in Table 2. It seems that there is a linear relationship between the timings of drying end and the water level.

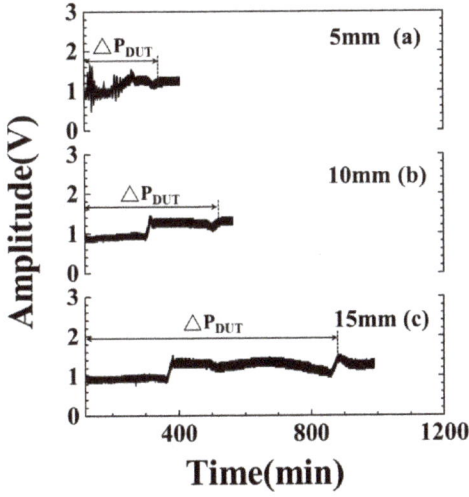

Figure 9. Amplitude variations of ultrasonic L^3 echo with respect to various water levels at drying stage. Water level: (**a**) 5, (**b**) 10 and (**c**) 15 mm.

Table 2. Timings and temperatures of water in, freezing point, phase change end and drying end indicated by temperature and ultrasonic L^3 echo for water levels of 5, 10, 15 mm.

Water Level (mm)	Items	Watering Point	Freezing Point	Phase Change End	Drying End
5	Time$_{Temp}$ (min)	3.0	16.3	24.4	N/A
	Time$_{UT}$ (min)	3.0	15.8	20.6	329.8
	Temperature (°C)	29.8	−2.8	−10.8	N/A
10	Time$_{Temp}$ (min)	3.0	14.0	28.9	N/A
	Time$_{UT}$ (min)	3.0	13.7	27.1	526.7
	Temperature (°C)	30.4	−2.4	−14.9	N/A
15	Time$_{Temp}$ (min)	3.0	19.2	46.7	N/A
	Time$_{UT}$ (min)	3.0	19.5	45.8	885.6
	Temperature (°C)	32.1	−0.4	−16.5	N/A

The measured drying periods are compared with the water level for evaluating the mentioned linear relationship. The results are shown in Figure 10. The estimated error of drying period under the experimental conditions is less than 5%. In the water level range from 5 to 15 mm, the average drying periods indicated by ultrasonic L^3 echo increase from 185.1 to 742.8 min. The slope of the fitting line is 55.8 min/mm. The drying period is expressed as:

$$\Delta P_{DUT} = -112.8 + 55.8 * H \tag{3}$$

where ΔP_{DUT} is the drying period in Figure 9 and H is the water level in Figure 3. Even though there are no temperature and visual information, one can still estimate the required drying period according to the filled water level based on Equation (3). In the future, this ultrasonic sensor and evaluating technology would be installed into freeze drying machine to detect the phase change and feedback the information to machine control system, as shown in Figure 1. Comparing with the typical experiment settings of 1020 min, the ultrasonic technique can clearly indicate the dynamic phenomena and completion of cooling/freezing/drying stages at each water level before the end of the experiment for reducing the freeze-drying process and saving energy.

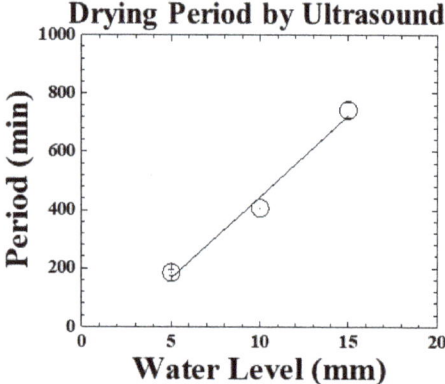

Figure 10. Drying periods measured by ultrasonic signal with respect to various water levels. Water level: 5, 10, 15 mm.

5. Conclusions

In this paper, an ultrasonic transducer (UT) is integrated onto the bottom of a specially designed container to analyze the freeze-drying process of water at various water amounts. The measured ultrasonic signatures are compared with the temperature and visual records. Among these three measured methods, ultrasonic and visual records are able to document the entire progression of the freeze-drying process, including the water-in, freezing/drying points, the phase change end of the water and the cooling/freezing/drying periods. The ultrasonic velocity in the water also indicates the cooling tendency of the water. The drying period increases with the water level. The increase rate, which is evaluated by the amplitude of the ultrasonic L^3 echo, is 55.8 min/mm. During the drying stage, the thermocouple cannot measure the entire temperature profile of the ice due to the detachment of the thermocouple from the ice. Only the ultrasonic signals and visual records are utilized for diagnosing the drying phenomena. However, the observing method is only suitable for the transparent container and plain solution. Comparing with the typical experiment settings, the indication of drying ends for water at level of 15 mm by the amplitude variations of ultrasonic L^3 echo could reduce the process period of 134.4 min/cycle and save 13% of consumed electricity. Therefore, this study demonstrates the use of a specially designed container, integrated ultrasonic sensor and technology for analyzing and optimizing the freeze-drying process of water for saving the process cost and energy.

Author Contributions: C.-C. Cheng initiated the idea, provided the draft and completed the revising of the paper. Y.-H. Tseng prepared the integrated sensor and practiced the experiments of freeze-drying process. S.-C. Huang analyzed the data and assisted to revise paper.

Funding: This work is funded by Ministry of Science and Technology of Taiwan, ROC under Contracts No: MOST 104-2221-E-027-121, MOST-105-2221-E-027-103 and MOST 107-3113-E-008-003.

Acknowledgments: The authors would like to provide special thanks to Hong-Ping Cheng (Department of Energy and Refrigerating Air-Conditioning Engineering, National Taipei University of Technology) for providing the useful discussions, Rei-Zhu Hsueh (Tai Yiaeh Enterprise Co., Ltd., New Taipei City, Taiwan) for providing useful maintaining assistance of freeze drying machine.

Conflicts of Interest: The authors declare no conflict of interest.

References

1. Zhao, J.H.; Liu, F.; Wen, X.; Xiao, H.W.; Ni, Y.Y. State diagram for freeze-dried mango: Freezing curve, glass transition line and maximal-freeze-concentration condition. *J. Food Eng.* **2015**, *157*, 49–56. [CrossRef]
2. Chizoba Ekezie, F.G.; Sun, D.W.; Han, Z.; Cheng, J.H. Microwave-assisted food processing technologies for enhancing product quality and process efficiency: A review of recent developments. *Trends Food Sci. Technol.* **2017**, *67*, 58–69. [CrossRef]
3. Ishwarya, S.P.; Anandharamakrishnan, C. Spray-Freeze-Drying approach for soluble coffee processing and its effect on quality characteristics. *J. Food Eng.* **2015**, *149*, 171–180. [CrossRef]
4. Monteiro, R.L.; Link, J.V.; Tribuzi, G.; Carciofi, B.A.M.; Laurindo, J.B. Microwave vacuum drying and multi-flash drying of pumpkin slices. *J. Food Eng.* **2018**, *232*, 1–10. [CrossRef]
5. Senadeera, W.; Alves-Filho, O.; Eikevik, T. Influence of drying conditions on the moisture diffusion and fluidization quality during multi-stage fluidized bed drying of bovine intestine for pet food. *Food Bioprod. Process.* **2013**, *91*, 549–557. [CrossRef]
6. Maisnam, D.; Rasane, P.; Dey, A.; Kaur, S.; Sarma, C. Recent advances in conventional drying of foods. *J. Food Technol. Preserv.* **2016**, *1*, 25–34.
7. Ahrens, D.C.; Villela, F.A.; Doni Filho, L. Physiological and industrial quality of white-oat (Avena sativa L.) seeds in intermittent drying. *Rev. Bras. Sementes* **2000**, *22*, 12–20. [CrossRef]
8. Toshniwal, U.; Karal, S. A review paper on Solar Dryer. *Int. J. Eng. Res. Appl.* **2013**, *3*, 896–902.
9. Kasper, J.C.; Friess, W. The freezing step in lyophilization: Physico-chemical fundamentals, freezing methods and consequences on process performance and quality attributes of biopharmaceuticals. *Eur. J. Pharm. Biopharm.* **2011**, *78*, 248–263. [CrossRef]
10. Siew, A. Freeze-drying process optimization. *Pharm. Technol.* **2018**, *42*, 18–23.
11. Ratti, C. Hot air and freeze-drying of high-value foods: A review. *J. Food Eng.* **2001**, *49*, 311–319. [CrossRef]
12. Rahman, M.S.; Guizani, N.; Al-Khaseibi, M.; Al-Hinai, S.A.; Al-Maskri, S.S.; Al-Hamhami, K. Analysis of cooling curve to determine the end point of freezing. *Food Hydrocoll.* **2002**, *16*, 653–659. [CrossRef]
13. Mellor, J.D.; Bell, G.A. Freeze-Drying—The Basic Process. In *Encyclopedia of Food Sciences and Nutrition*; Elsevier: Amsterdam, The Netherlands, 2003; pp. 2697–2701. ISBN 9780122270550.
14. Wang, D.Q.; Hey, J.M.; Nail, S.L. Effect of Collapse on the Stability of Freeze-Dried Recombinant Factor VIII and α-Amylase. *J. Pharm. Sci.* **2004**, *93*, 1253–1263. [CrossRef]
15. Tang, X.; Nail, S.L.; Pikal, M.J. Freeze-drying process design by manometric temperature measurement: Design of a smart freeze-dryer. *Pharm. Res.* **2005**, *22*, 685–700. [CrossRef]
16. Chang, B.S.; Patro, S.Y. Freeze-drying Process Development for Protein Pharmaceuticals. In *Lyophilization of Biopharmaceuticals*; American Association of Pharmaceutical Scientists: Arlington, VA, USA, 2004; pp. 113–138. ISBN 978-0-9711767-6-8.
17. Pikal, M.J.; Shah, S.; Roy, M.L.; Putman, R. The secondary drying stage of freeze drying: Drying kinetics as a function of temperature and chamber pressure. *Int. J. Pharm.* **1990**, *60*, 203–207. [CrossRef]
18. Breen, E.D.; Curley, J.G.; Overcashier, D.E.; Hsu, C.C.; Shire, S.J. Effect of moisture on the stability of a lyophilized humanized monoclonal antibody formulation. *Pharm. Res.* **2001**, *18*, 1345–1353. [CrossRef]
19. Rahman, M.S. *Handbook of Food Preservation*, 2nd ed.; Rahman, M.S., Ed.; Taylor & Francis Group: Milton Park, UK, 2007; ISBN 9781845697587.

20. Nail, S.; Tchessalov, S.; Shalaev, E.; Ganguly, A.; Renzi, E.; Dimarco, F.; Wegiel, L.; Ferris, S.; Kessler, W.; Pikal, M.; et al. Recommended Best Practices for Process Monitoring Instrumentation in Pharmaceutical Freeze Drying—2017. *AAPS PharmSciTech* **2017**, *18*, 2379–2393. [CrossRef]
21. Malik, N.; Gouseti, O.; Bakalis, S. Effect of freezing with temperature fluctuations on microstructure and dissolution behavior of freeze-dried high solid systems. *Energy Procedia* **2017**, *123*, 2–9. [CrossRef]
22. Tseng, Y.H.; Cheng, C.C.; Cheng, H.P.; Lee, D. Novel Real-Time Diagnosis of the Freezing Process Using an Ultrasonic Transducer. *Sensors* **2015**, *15*, 10332–10349. [CrossRef]
23. Hellier, C.J. *Handbook of Nondestructive Evaluation*; McGraw-Hill Companies, Inc.: New York, NY, USA, 2003; ISBN 007139947X.
24. Cheng, C.C.; Yang, S.Y.; Lee, D. Novel real-time temperature diagnosis of conventional hot-embossing process using an ultrasonic transducer. *Sens. Switz.* **2014**, *14*, 19493–19506. [CrossRef] [PubMed]
25. Wu, Y.L.; Cheng, C.C.; Kobayashi, M.; Yang, C.H. Novel design of extension nozzle and its application on real-time injection molding process diagnosed by ultrasound. *Sens. Actuators A Phys.* **2017**, *263*, 430–438. [CrossRef]
26. Cheng, C.-C.; Wu, C.-L.; Wu, K.-T.; Yang, S.-Y. Real-time diagnosis of gas-assisted hot embossing process by ultrasound. *Polym. Eng. Sci.* **2013**, *53*, 2175–2182. [CrossRef]
27. Kažys, R.; Voleišis, A.; Voleišienė, B. High temperature ultrasonic transducers: Review. *Ultrasound* **2008**, *63*, 7–17.
28. Takahashi, M.; Ihara, I. Ultrasonic monitoring of internal temperature distribution in a heated material. *Jpn. J. Appl. Phys.* **2008**, *47*, 3894–3898. [CrossRef]
29. Cheng, C.C. Micromolding of polymer nanocomposites diagnosed by ultrasound. *J. Polym. Eng.* **2010**, *30*, 95–108. [CrossRef]
30. Passot, S.; Tréléa, I.C.; Marin, M.; Galan, M.; Morris, G.J.; Fonseca, F. Effect of controlled ice nucleation on primary drying stage and protein recovery in vials cooled in a modified freeze-dryer. *J. Biomech. Eng.* **2009**, *131*, 074511. [CrossRef] [PubMed]
31. Cheng, C.-C.; Young, S.-L.; Chen, H.-Z.; Yang, S.-Y. Substrate Effect on Characteristics of PbZrxTi1−xO3 (PZT) Film. *Integr. Ferroelectr.* **2014**, *150*, 51–58. [CrossRef]
32. Shuyu, L. Load characteristics of high power sandwich piezoelectric ultrasonic transducers. *Ultrasonics* **2005**, *43*, 365–373. [CrossRef]
33. Arnold, F.J.; Roger, L.L.B.; Gonçalves, M.S.; Mühlen, S.S. Electrical Impedance of Piezoelectric Ceramics under Acoustic Loads. *Ecti Trans. Electr. Eng. Electron. Commun.* **2014**, *12*, 48–54.
34. Bilaniuk, N.; Wong, G.S.K. Speed of sound in pure water as a function of temperature. *J. Acoust. Soc. Am.* **1993**, *93*, 1609–1612. [CrossRef]

© 2019 by the authors. Licensee MDPI, Basel, Switzerland. This article is an open access article distributed under the terms and conditions of the Creative Commons Attribution (CC BY) license (http://creativecommons.org/licenses/by/4.0/).

Article

Temperature and Strain Correlation of Bridge Parallel Structure Based on Vibrating Wire Strain Sensor

Lu Peng [1,*], Genqiang Jing [1], Zhu Luo [1], Xin Yuan [2], Yixu Wang [1] and Bing Zhang [1]

[1] National Center of Metrization for Equipments of Roads and Bridges, Research Institute of Highway Ministry of Transport, Beijing 100088, China; gq.jing@rioh.cn (G.J.); ZhuLuo@ncmerb.com (Z.L.); wyx@ncmerb.com (Y.W.); b.zhang@rioh.cn (B.Z.)
[2] School of Electrical and Information Engineering, Wuhan Institute of Technology, Wuhan 430205, China; qcmhssk@163.com
* Correspondence: lu.peng@rioh.cn

Received: 29 October 2019; Accepted: 21 January 2020; Published: 24 January 2020

Abstract: Deformation is a ubiquitous phenomenon in nature. This process usually refers to the change in shape, size, and position of an object in the time and spatial domain under various loads. Under normal circumstances, during engineering construction, technicians are generally required to monitor the safe operation of structural facilities in the transportation field and the health of bridge, because monitoring in the engineering process plays an important role in construction safety. Considering the reliability risk of sensors after a long-time work period, such as signal drift, accurate measurement of strain gauges is inseparable from the value traceability system of high-precision strain gauges. In this study, two vibrating wire strain gauges with the same working principle were measured using the parallel method at similar positions. First, based on the principle of time series, the experiment used high-frequency dynamic acquisition to measure the thermometer strain of two vibrating wire strain gauges. Second, this experiment analyzed the correlation between strain and temperature measured separately. Under the condition of different prestress, this experiment studied the influencing relationship of temperature corresponding variable. In this experiment, the measurement repetitiveness was analyzed using the meteorology knowledge of single sensor data, focused on researching the influence of temperature and prestress effect on sensors by analyzing differences of their measurement results in a specified situation. Then, the reliability and stability of dynamic vibrating wire strain gauge were verified in the experiment. The final conclusion of the experiment is the actual engineering in the later stage. Onsite online meteorology in the application provides support.

Keywords: parallel position; bridge structure; temperature; vibrating wire strain sensor

1. Introduction

Bridge health monitoring and diagnostic discriminant models have always been a key challenge for the transportation sector worldwide. In a previous study, on a regular and irregular basis, construction workers used different monitoring instruments to test some components of structure, analyzed the data, and finally evaluated the performance of structure. In 2018, Mao by monitoring dynamic characteristics of Sutong cable-stayed bridge (SCB), including acceleration and strain responses, as well as modal frequencies, are investigated through one-year continuous monitoring data under operating conditions by the structural health monitoring system. One-year continuous modal frequencies of SCB are identified using the Hilbert–Huang transform method. Variability analysis of the structural modal frequencies due to environmental temperature and operational traffic is then conducted. Results show that temperature is the most important environmental factor for vertical and torsional modal frequencies. The traffic load is the second critical factor especially for the fundamental vertical frequency of SCB [1]. In 2019, Wang reported that the real-time monitoring data collected from a long-span cable-stayed bridge is utilized to demonstrate the feasibility of the improved BDLM-based method. In particular, the present BDLM-based method allows for probabilistic forecasts, offering substantial information about the target TIS response, such as mean and confidence interval. Results show that the improved BDLM is capable of capturing the relationship between temperature and TIS. Compared to the AR model, multiple linear regression (MLR) model, and BDLM without the AR component, the improved BDLM shows better forecasting performance in modeling and forecasting the TIS of a long-span bridge [2]. Such measurement methods have the drawback of large errors and discontinuities. The ideal health monitoring system should accurately reflect the change in grassroot structure under the influence of factors such as temperature, humidity, and other environmental factors, installation and deployment methods, and the sensor's own error, and establish an effective health assessment and prediction model. For example, Mao proposed according to one-year continuous monitoring of strain data recorded by the structural health monitoring system of the Sutong cable-stayed bridge, the lifetime fatigue reliability of three welded details of the orthotropic steel deck was investigated, detailed analysis of the separated components of the raw strain data was first conducted, and included the slow-varying trend and the dynamic component. The strain dynamic component was mainly induced by the local vehicle axle loads. Rainflow counting was used to obtain the stress range histograms, which were then used to calculate the equivalent stress range according to the lognormal-fitting method. Finally, a time-dependent fatigue reliability evaluation of the described three welded details was conducted using one-year monitoring strain data. Results showed that the fatigue performance of the two welded details, RTDD and DTD, remained satisfactory after 100 years of operation because the failure probabilities were both lower than 10^{-5}. The designed cutout of the diaphragm was applied to the RTD weld at the welded connection between the U rib and diaphragm. This cutout was validated as a means to help achieve better fatigue resistance for the RTD weld [3]. For bridge strain monitoring sensors, most experimental procedures use resistive strain gauges, vibrating wire strain gauges, fiber grating strain gauges, etc. Although the development of fiber grating technology in recent years has led to a new round of equipment upgrades for structural monitoring methods, vibrating wire strain gauges are most widely used in stage bridge health monitoring systems.

In recent years, studies on the heat of vibrating wire sensors at home and abroad have continued unabated, and domestic and foreign experts and scholars have conducted a lot of research to improve the sensor performance. The research on sensors is gradually extended towards high precision, large scale, small volume, and multiple applications. During the improvement in the performance of the sensor, higher requirements are also imposed on the measurement accuracy. To improve accuracy, in 2010, He proposed a low-voltage excitation with the feedback method. By pre-excitations, the vibration frequency of the sensor can be used as the output of the driver. The feedback signal frequency is very close to the real frequency of the sensor, so the wire can reach resonance state quickly. The optimal excitation strategy was verified by the new designed detection circuit. The data perform as short-time excitations with large resonance amplitude, therefore the anti-interference ability got enhanced at lower cost of signal processing circuit, finally increasing the measurement time and improving the accuracy of measured frequency [4]. In 2010, Wen proposed a method for the frequency measurement of vibrating wire sensors with LM3S6965 as the control core. An equal precision measurement was used to effectively improve the measurement accuracy of the system [5]. In 2016, Tian et al. proposed a design method of nonlinear compensation that is provided for the nonlinear relation between the measured force and the frequency of vibrating wire sensor. Data density is increased by interpolation based on the principle of cubic spline interpolation. The F-f curve of vibrating wire sensor is fitted and revised by MATLAB based on the least-square method, more accurately the smooth curve according to engineering practice is obtained and a reasonable conclusion is obtained. Experiments show that the method achieves the accurate fitting to the F-f curve of vibrating wire sensor with the limited data. The F-f function of vibrating wire sensor which is long buried can be rapidly determined by this procedure [6]. In 2017, Chen et al. proposed an adaptive vibration measurement method based on fast Fourier transform, which uses digital Fourier filtering to automatically filter noise interference and then uses Quinn algorithm for high-precision frequency calculation. Based on the STM32 processor platform, the method was used in vibrating a wire sensor frequency measurement system. The test results show that the relative error of system frequency measurement is less than 0.01% in the case of no noise interference; in the case of severe white noise interference (signal to noise ratio is −20 dB), the system's frequency measurement relative error is less than 0.3%. Compared to other frequency measurement methods, it was found that the method has better noise suppression and frequency measurement accuracy [7]. In terms of static metrology and calibration, the experiment proved that a vibrating wire strain gauge can be widely used in the field of geotechnical engineering, and the accuracy requirement for variable measurement is relatively low. However, in the application of traffic engineering, the measurement accuracy is correspondingly improved, accompanied by the needs of metrology and calibration. Many scholars used a strain sensor calibration device consisting of a calibration frame and digital display to calibrate certain metering characteristics of strain gauge. For example, in 2009, Xu invented a vibrating wire strain sensor calibration device that uses a dial gauge as a gauge for deformation, with a deformation range of up to 3 mm and a strain measurement resolution of about 0.1%. The results show that the strain in the middle of the sensitive grid of strain gauge is the largest and gradually decreases to zero at both ends when measuring. The force has no effect on the sensitive gate base layer and bonding layer at both ends and there is a strain transition zone between the base layer and the sensitive gate, the bonding layer, and the base layer. The longer the horizontal width of bonding layer is the thinner the thickness and the larger the shear modulus will be, and the strain transfer will be more efficient if the active zone of the sensitive grid is longer [8]. In 2011, Zhang et al. designed a vibrating wire strain sensor calibration device. They used a grating scale as the measuring standard for deformation. The maximum calibration distance was 300 mm, and the measurement resolution was 2 μm. The degree is about 0.2% [9]. In 2016, Mai et al. invented a vibrating wire strain gauge calibration device. A dial gauge was used as the measurement standard for deformation amount, so that the resolution was 0.01 mm [10]. During engineering application, many in-depth studies have been conducted on the dynamic calibration of strain gauges. In 2016, Bai et al. affixed FBG sensors on standard beams. Loads were applied to the standard beams to deform them, and the beams were measured. The surface strain was used to calibrate the sensitivity coefficient of FBG strain gauge, and the measured strain range could be analyzed up to 2000 με [11]. In 2017, Zhang proposed a method for the indirect calibration of fiber grating strain sensors using a fiber grating temperature sensor. The lossless calibration of strain sensors was achieved [12]. During the detection, monitoring, and metering of vibrating wire strain gauges, the temperature deformation caused by temperature changes has been the research direction of many scientists. Chen et al. monitored data in engineering applications based on the working principle of vibrating wire strain gauges. The relationship between temperature and strain was analyzed using the relevant data of bridge strain monitoring, and the relationship between temperature influence and strain of the string itself and the temperature field of structural section was evaluated [13]. Bai et al. analyzed that when the temperature changes significantly, a variety of mathematical models were used to fit and calibrate the monitoring data, eliminating the temperature drift of strain gauge, and the experimental results reflect the original characteristics of deformation [14]. Agostiono investigation of the sensing features of the long-period fiber gratings (LPGs) fabricated in hollow core photonic crystal fibers (HC-PCFs) by the pressure assisted electric arc discharge (EAD) technique. In particular, the characterization of the LPG in terms of shift in resonant wavelengths and changes in attenuation band depth to the environmental parameters: Strain, temperature, curvature, refractive index, and pressure is presented. Results show that LPGs in HC-PCFs represent a novel high performance sensing platform for measurements of different physical parameters including strain, temperature and, especially, for measurements of environmental pressure. The pressure sensitivity enhancement is about four times greater if comparing LPGs in HC and standard fibers. Moreover, differently from LPGs in standard fibers, these LPGs realized in innovative fibers, i.e., the HC-PCFs, are not sensitive to the surrounding refractive index. During the online calibration study of strain sensors [15], offline removal and reinstallation of sensing elements pose a risk to the continuity and consistency of monitoring data. At present, the load test method is mainly used, but the load test method cannot eliminate the instability of structure such as the stiffness and strength of the structure itself [16,17]. Therefore, a more effective method at this stage is to install a traceable high-frequency dynamic monitoring sensor in parallel to the sensor to perform parallel measurement, thus achieving the online calibration of a long-term monitoring sensor. At present stage, the measurement calibration of the strain sensor system in various industries only adopts static calibration before installation or carries out a static and dynamic load test for calibration, and does not carry out online calibration during use. The static calibration cannot determine the influence of the error and noise, temperature change, and prestress effect on the strain value. Static and dynamic load tests cost a lot of manpower and materials, and cannot provide reference for sensor calibration under complicated passive excitation. In the online calibration research of strain sensors, because the offline disassembly and reinstallation of the sensing elements pose a risk to the continuity and consistency of the monitoring data, the load test method is mainly used at this stage, but the load test cannot eliminate the rigidity and strength of the structure itself.

In the research process, we carried out a number of relevant tests. For example, the calibration of fixed excitation by a simply supported beam model is shown in Figure 1. The strain correlation test of a small simply supported beam at constant temperature in a high and low temperature box is shown in Figure 2. Fatigue test verification is shown in Figure 3. In order to further verify the test, a test was carried out on Jiujiang Bridge to verify Figure 4. Most of the above studies adopted common source excitation schemes, mainly to verify the calibration of the strain monitoring system under common source excitation by parallel measurement methods, which provided important verification support for the coupling relationship between temperature, prestress, and strain output in the online calibration of dynamic strain measurement in this study.

Figure 1. Schematic diagram of simply supported beam test model.

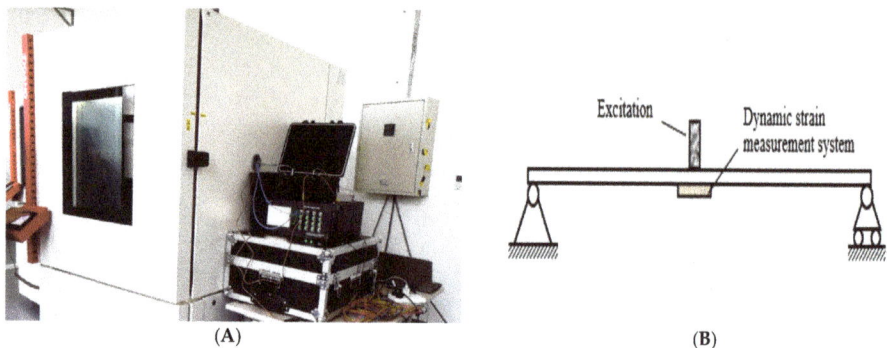

Figure 2. High and low temperature dynamic strain test. (**A**)Test equipment, (**B**) Assembly drawing of experiment beam.

Figure 3. Verified by fatigue tests on bridge model.

(A)   **(B)**

Figure 4. Test verification of Jiujiang Bridge. (**A**) Jiujiang Bridge, (**B**) Test process.

During this study, two vibrating wire strain gauges with the same working principle were measured using the parallel method at similar positions. The feasibility of a parallel measurement scheme was verified [18]. Under the condition of different prestresses, this experiment studied the influencing relationship of temperature corresponding variable [19,20]. We focused on researching the influence of temperature and prestress effect on sensors by analyzing differences of their measurement results in a specified situation. The measurement repetitiveness was analyzed using the meteorology knowledge of single sensor data, and then the reliability and stability of a dynamic vibrating wire strain gauge were verified in the experiment. The relevant measurement performance of a dynamic vibrating wire strain gauge was demonstrated.

2. Working Principle of Vibrating Wire Strain Gauge

A vibrating wire sensor is tested by steel string vibration. An experiment is conducted to characterize the force according to the variation in vibration frequency. In an actual output frequency signal, there is no strain gauge; it must be field calibration, signal drift, long distance transmission, and long time. The problem of poor durability was used, and the robustness was good [21]. This experiment solves the shortcoming of unstable strain gauges for a long time, and this product can be widely used in bridge monitoring.

A vibrating wire sensor has good measurement characteristics; it can achieve nonlinear characteristics of less than 0.1%, sensitivity of 0.05%, and less than 0.1%/10 °C temperature error.

After the strain gauge is manufactured, its steel string has a certain initial tensile force T_0 and thus has an initial frequency F_0. When the strain gauge is installed, the tensile force of vibrating wire changes with deformation, and the strain can be measured by the tensile force change of vibrating wire. Set the tension of the vibrating wire to T and the natural frequency to f. The relationship between tension and frequency can be expressed as Equation (1):

$$T = Kf^2 \qquad (1)$$

where K is related to the length of string, and the mass per unit length can be expressed using Equation (2):

$$\Delta T = T - T_0 = K(f^2 - f_0^2) \qquad (2)$$

Assuming that the strain increment of strain gauge can be set to the strain increment of vibrating wire, Equation (3) can be derived as follows:

$$\varepsilon_h = \varepsilon_g = \frac{\Delta K}{EA} \qquad (3)$$

When EA is the axial stiffness of steel string, it can be derived from Equation (4) as follows:

$$\varepsilon_h = \frac{K}{EA}(f^2 - f_0^2) = k_h(f^2 - f_0^2) \qquad (4)$$

A mathematical model of the vibrating wire sensor can be expressed using Equations (5) and (6) as follows:

$$F = K(f^2 - f_0^2) \qquad (5)$$

$$F = A(f^2 - f_0^2) + B(f - f_0) \qquad (6)$$

When the length is such that the fine string of mass m is subjected to tension F (Figure 5), the natural frequency f can be expressed as Equations (7)–(9) as follows:

$$f = \frac{1}{2}\sqrt{\frac{F}{ml}} \qquad (7)$$

$$f = \frac{1}{2l}\sqrt{\frac{ES\Delta l}{\rho l}} = \frac{1}{2l}\sqrt{\frac{E\Delta l}{\rho_v l}} \qquad (8)$$

$$f = \varphi(F) \qquad (9)$$

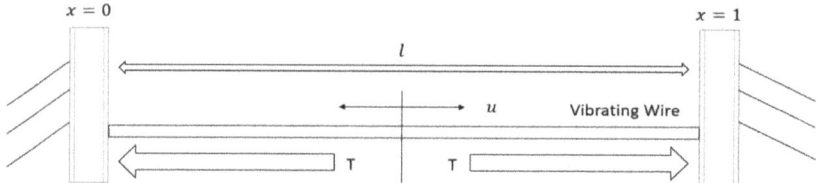

Figure 5. Working principle diagram of vibrating wire strain gauge.

Sensitivity can be derived using Equation (10) as follows:

$$f^2 = \frac{1}{4l^2}\frac{E\Delta l}{\rho_v l} = K\varepsilon \qquad (10)$$

After differentiation, Equations (11) and (12) can be deduced as follows:

$$2f df = K d\varepsilon \qquad (11)$$

$$k = \frac{df}{d\varepsilon} = \frac{K}{2f} \qquad (12)$$

The material coefficient can be calculated using Equation (13):

$$\begin{cases} K = \frac{1}{4l^2}\frac{E}{\rho_v} = \frac{ES}{4l^2\rho} \\ \varepsilon = \frac{\Delta l}{l} \end{cases} \quad (13)$$

The above formula shows that sensitivity k is directly proportional to material coefficient K and inversely proportional to the vibration frequency of string.

After many experiments, the initial frequency is f_0 when the measured tension is F_0, and the vibration frequency is f_1 when the measured tension is $F_1 = f_0 + f$. The nonlinear error of vibrating wire strain gauge can be expressed using Equations (14) and (15) as follows:

$$f_1 = \frac{1}{2}\sqrt{\frac{F_0 + \Delta F}{ml}} = \frac{1}{2}\sqrt{\frac{F_0}{ml}}\sqrt{1 + \frac{\Delta F}{F_0}} = f_0\sqrt{1 + \frac{\Delta F}{F_0}} = f_0\sqrt{1 + \varepsilon_F} = f_0(1 + \varepsilon_F)^{\frac{1}{2}} \quad (14)$$

$$f_1 = f_0(1 + \frac{1}{2}\varepsilon_F - \frac{1}{8}\varepsilon_F^2 + \frac{1}{16}\varepsilon_F^3 - \cdots) \quad (15)$$

At that time $F_2 = F_0 - \Delta F$, leading to Equation (16):

$$f_2 = f_0(1 + \frac{1}{2}\varepsilon_F - \frac{1}{8}\varepsilon_F^2 + \frac{1}{16}\varepsilon_F^3 - \cdots) \quad (16)$$

Its quadratic nonlinearity error can be expressed as Equation (17):

$$\frac{\left|\frac{1}{8}f_0\varepsilon_F^2\right|}{\frac{1}{2}f_0\varepsilon_F} = \frac{1}{4}\varepsilon_F \quad (17)$$

The above formula shows that the larger ε_F, the larger δ_m. At the same time, the ambient temperature mainly influences the frequency stability. The bulk density ρv and Δl caused by F do not change with the ambient temperature, so the frequency stability can be expressed using Equation (18):

$$\gamma_f = \frac{df}{f} = \frac{dE}{2}E - \frac{3}{2}\frac{dl}{l} \quad (18)$$

When the ambient temperature changes, the vibrating wire strain gauge material and the measured structural material have different linear expansion coefficients, and the sensor is subjected to additional stretching or compression. The additional strain can be expressed as Equation (19):

$$\varepsilon T = (\alpha - \beta) \bullet \Delta T \quad (19)$$

where:

εT: Additional strain caused by temperature effect;
α: Linear expansion coefficient (°C^{-1}) of the structural material to be tested;
B: The coefficient of linear expansion (°C^{-1}) of a steel string of a vibrating wire strain gauge;
ΔT: Temperature change amount.

In actual application, the vibrating wire strain sensing system generally adopts software compensation. After the thermistor is set to collect the working environment temperature in the electromagnetic coil, the optimized temperature compensation algorithm is combined with the software to compensate in the demodulation instrument. The temperature strain compensation of the vibrating wire strain gauge mostly utilizes the two-dimensional regression method, polynomial fitting method, and neural network method. After the test results of the vibrating wire strain gauge are linearly fitted, the strain calculation method can be expressed as Equation (20):

$$\varepsilon = a \times (f_i^4 - f_0^4) + b(f_i^2 - f_0^2) + k_T(T_i - T_0) \qquad (20)$$

where:

ε: The dependent variable of the current time relative to the initial position (10^{-6});
k: The steel string strain gauge minimum reading $10^{-6}/(\text{kHz}^2)$;
f_i^2: The Steel string strain gauge current output modulus kHz^2;
f_0^2: The Steel string strain gauge initial output frequency modulus kHz^2;
k_T: The steel string strain gauge temperature correction factor $10^{-6}\ °C$;
T_i: The steel string strain gauge current time temperature value (°C);
T_0: Temperature value when measuring f_0 (°C).

When using polynomial fitting, the coefficients a and b were calculated using the least squares fitting method, and the strain calculation method for integrating temperature changes is shown in Equation (21):

$$\varepsilon = a \times (f_i^4 - f_0^4) + b(f_i^2 - f_0^2) + k_T(T_i - T_0) \qquad (21)$$

The abovementioned various theoretical calculation methods and vibrating wire strain gauge measurement characteristics can better measure the strain, eliminate the strain generated by environmental influence, more accurately understand the mechanical strain of structure, and analyze the stress state of structural facility.

3. Test Plan

Two sensors were installed in parallel on the 45th steel tooling for free acquisition at the same frequency. The device (Figure 6) shows the material properties of No. 45 steel (Table 1). The acquisition device is equipped with a high-frequency dynamic acquisition device, which is a dynamic vibrating wire automatic acquisition system. This system uses a nonsweeping technology scheme to prevent the steel string vibration from attenuating, and an embedded mirror oscillation circuit to ensure excitation frequency. By matching the phase and true motion of steel cord and simultaneously detecting the resonant frequency of steel string through several waveform periods, noise immunity and resolution of measurement have largely improved compared with the flat-domain periodic averaging method.

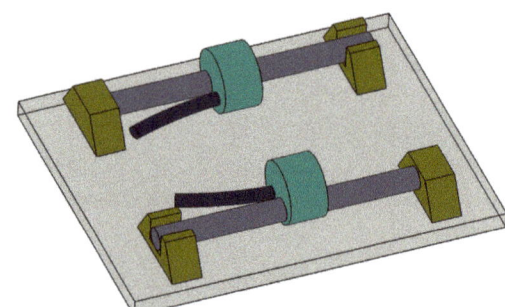

Figure 6. Caption for figure schematic diagram of the test fixture.

Table 1. Forty-five steel material characteristics.

Content	Value and Unit
Density	7.85 g/cm^3
Modulus of elasticity	210 GPa
Poisson ratio	0.269
Tensile strength	600 MPa
Yield strength	355 MPa
Elongation	16%
Section shrinkage	40%
Impact work	39 J

The measurement module used the patented 8,671,758 products, and the dynamic measurement rate is 20 to 333 Hz. At the same time of dynamic measurement, the module also performs auxiliary measurement, performs static measurement at 1 Hz, provides finer measurement resolution, and better anti-interference to external noise source performance. The thermistor input signal of each vibrating string acquisition channel was measured at a high resolution of 24 bits at 1 Hz. The performance of thermistor parameters is shown in Table 2. The excitation module has a resolution of 26 mV and a dynamic measurement rate of 20, 50, 100, 200, and 333.33 bHz. The range of sensor resonance frequency is shown in Table 3. The measurement frequency accuracy is ± (0.005% reading + measurement resolution). The noise level corresponding to the measurement resolution b at different sampling rates is shown in Tables 2–4.

Table 2. Thermistor performance parameters.

Content	Value and Unit
Half bridge arm	0.1% accurate resistance is 4.99 KΩ
Excitation voltage	1.5 V
Resolution	0.002 Ω RMS @ 5 KΩ thermistor
Accuracy (−55–85 °C)	0.15% of reading
Measurement rate	1 Hz

Table 3. Range of sensor resonance frequencies.

Sample Rate (Hz)	Minimum Sensor Frequency (Hz)	Maximum Sensor Frequency (Hz)
20	290	6000
50	290	6000
100	580	6000
200b	1150	6000
333b	2300	6000

Table 4. Measurement resolution b (typical value for a 2.5 kHz resonant frequency sensor).

Sample Rate (Hz)	Noise Level (Hz RMS)
1	0.005
20	0.008
50	0.015
100	0.035
200C	0.11
333C	0.45

In the measurement, both steel plate and strain gauge will produce different deformations due to temperature changes. To minimize the acquisition error, a correlation curve between temperature and strain was analyzed. During the test, welding was used. To solder the two sensor holders on the tooling and for better verification, the effect of temperature response, and correlation, when collecting the zero point, one sensor was set to the free state, one sensor passed the fixture. The force was applied such that the sensor zero point acquisition and the first sensor's zero point acquisition strain difference is about 140 $\mu\varepsilon$.

4. Analysis of Test Data

The relationship between temperature and strain measurement shows that the strain changes correspondingly when the temperature changes. The strain reversely sways as the temperature changes. We got a good corresponding relationship. The temperature and strain data collected using the two sensors A and B were compared, and the correlation between the two data was analyzed. During the analysis, there are usually two ways of correlation: Autocorrelation and cross-correlation. The autocorrelation function is known as the autocorrelation equation and used to describe the correlation between the correlation functions of related data at different times as shown in Equation (22):

$$R_f(\tau) = f(\tau) * f^*(-\tau) = \int_{-\infty}^{\infty} f(t+\tau)f^*(t)dt = \int_{-\infty}^{\infty} f(t)f^*(t-\tau)dt \quad (22)$$

At the same time, cross-correlation or cross-covariance were also used to represent a measure of similarity between two signals. Cross-correlation mainly analyzes the degree of correlation between two time series. Cross-correlation is essentially similar to the convolution of two functions. For discrete functions f_i and g_i, it can be defined as Equation (23):

$$(f * g)_i \equiv \sum_j f_j^* g_{i+j} \quad (23)$$

If the continuous signal is set to two sets of $f(x)$ and $g(x)$, then the cross-correlation is defined as Equation (24):

$$(f * g)(x) \equiv \int f^*(t)g(x+t)dt \quad (24)$$

During this test analysis, the data collected using the two sensors A and B were compared, and the correlation between the two sets of data was analyzed. Their similarities were derived to verify the accuracy and reliability of dynamic acquisition of data obtained using sensors A and B. The correlation coefficient between the temperature data of sensors A and B was calculated to be 0.9983, and the correlation degree is highly correlated. The correlation coefficient between the strain data of sensors A and B is 0.9895, and the correlation degree is highly correlated. Cross-calculation can be obtained. The correlation between temperature and strain of sensor A is −0.6683, and the correlation between temperature and strain of sensor B is −0.5573. The experimental results show that the values are significantly correlated (Figure 7).

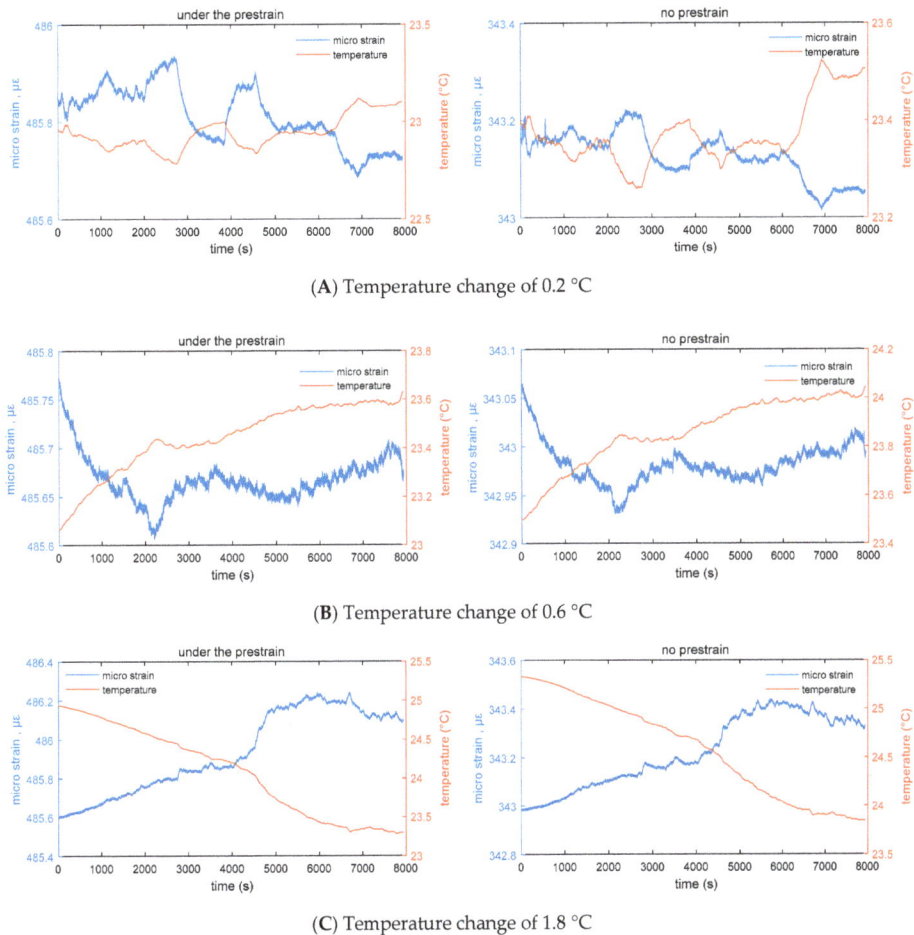

(A) Temperature change of 0.2 °C

(B) Temperature change of 0.6 °C

(C) Temperature change of 1.8 °C

Figure 7. Free state strain values of strain sensors A and B in a temperature change of 0.2, 0.6, and 1.8 °C.

During the test, to better use the knowledge of metrology to analyze the dynamic RMS change output, verify the reliability of linear relationship between temperature and strain monitoring, and repeat the calculation of temperature measurement and strain measurement, the experiment measured the comprehensive reflection of various random influencing factors, including the instability of tooling materials used, random error caused by the laying process, instability of sensing instrument, environmental conditions, and other factors, as well as the actual measured randomness. The measured object also affects the dispersion of measured values, especially when the random variation in the measured object during the bridge monitoring process is large. Therefore, the dispersion of experimentally measured values is typically slightly greater than the dispersion introduced by the sensor static calibration standard itself. To be less affected by outliers, all the deformations and the corresponding temperature were measured. During the calculation of repeatability of the experiment, the arithmetic mean value of measured value should be calculated first, and the calculation formula can be expressed as Equation (25):

$$\overline{F} = \frac{\sum_{i=1}^{n} F_i}{n} \qquad (25)$$

where:

F_i—The strain measurement indication of the i-th measurement, $\mu\varepsilon$;
n—The number of measurements.

The measurement repeatability can be quantitatively expressed using the experimental standard deviation $Sr(y)$, and the calculation formula can be expressed as Equation (26):

$$Sr\,(y) = \sqrt{\frac{\sum_{i=1}^{n}(F_i - \overline{F})^2}{n-1}} \qquad (26)$$

where:

F_i—The measured strain measured at the i-th measurement, $\mu\varepsilon$;
$\overline{F}$—The arithmetic mean of indications of strain measurement, $\mu\varepsilon$;
n—number of measurements.

Using the above formula, under the synchronous acquisition test conditions, sensors A and B were repetitively used for the temperature and strain measurements, respectively. The repeatability of sensor A strain indication is 0.16071, and the temperature measurement repeatability is 0.09971. The repeatability of strain measurement of sensor B is 0.11743, and the repeatability of temperature measurement is 0.08209. The repeatability shows that the stability of sensor B is better, and the experiment can be performed under the condition that the initial state of sensor is small. The measured values are more stable and more clearly characterized by the corresponding relationship between temperature and strain.

At the same time, the relationship between strain and temperature was analyzed during a temperature variation of 0.2, 0.6, and 1.8 °C. Figures 4–6 show that the correlation did not change with the change in temperature. A stable negative correlation curve relationship was maintained. Among them, sensor A has a relatively weak induction of temperature and deformation under the prestrain of 140 $\mu\varepsilon$, which can be characterized as shown in the figure. In the absence of external excitation, the change in experimental numerical temperature is less than the perception of sensor B. At the same time, the corresponding strain produces a small change in the output. Sensor A temperature and the strain relationship map shows that in the free state of acquisition device, the experimental temperature will also have a corresponding change during small changes. However, both sensors A and B can consistently respond to the corresponding output temperature and strain signals (Figure 8).

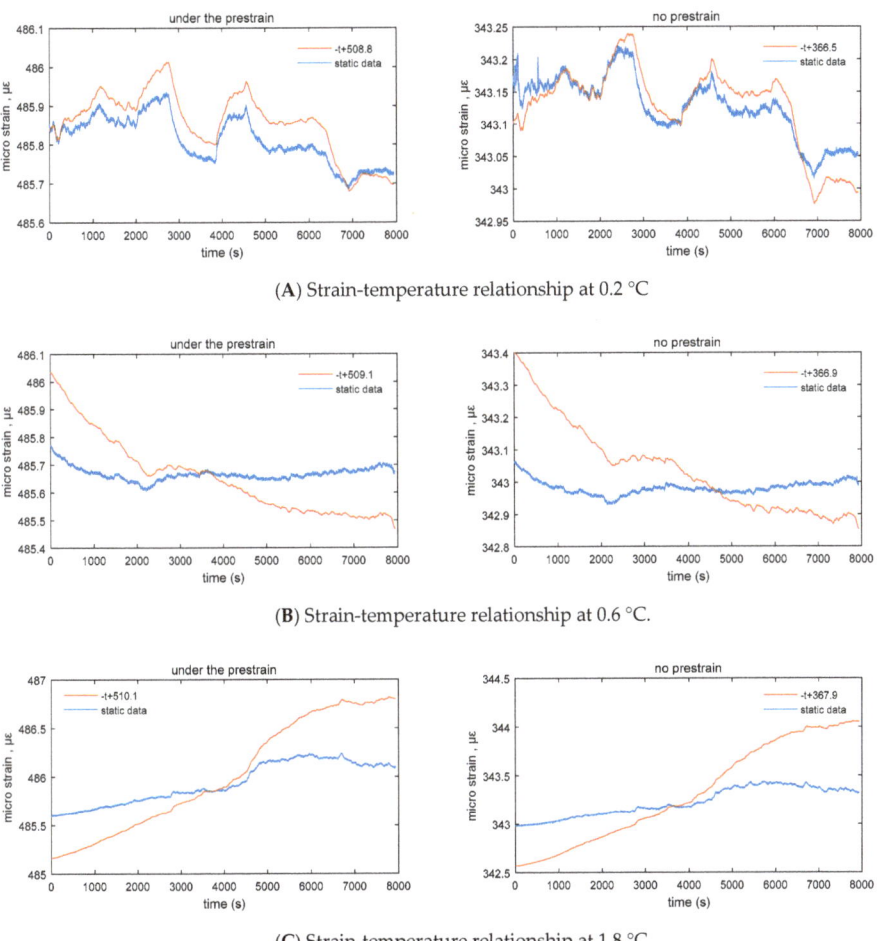

(A) Strain-temperature relationship at 0.2 °C

(B) Strain-temperature relationship at 0.6 °C.

(C) Strain-temperature relationship at 1.8 °C.

Figure 8. Strain-temperature relationship between 0.2, 0.6, and 1.8 °C intervals for strain gauges A and B.

After calculating different temperature changes, the experimental values show the correlation between strain and temperature. By analyzing the strain-temperature correlation function, it can be found that the temperature change is less than that of sensor B when there is no external excitation. At the same time, the change in output corresponding to the strain is small. In different temperature variation ranges, the experimental data were used to analyze the relationship between the temperature and strain of sensor A. It was observed that the acquisition device is in a free state, and the temperature is small. The effect of corresponding change will be produced. According to statistical analysis, the strain-temperature correlation can be expressed as $\Delta \mu \varepsilon = \Delta T + 510$, and the strain measurement result has a negative correlation with the temperature change. From the relationship between the temperature and corresponding strain curves of sensor B in the free state, it can be observed that the strain-temperature correlation can be expressed as $\Delta \mu \varepsilon = \Delta T$ in the state where sensor B is more free under the same conditions as sensor A. $\Delta \mu \varepsilon = \Delta T + 367$; the strain measurement results have a stable negative correlation with temperature change. During data acquisition, it was found that strain gauge A is larger at the same temperature due to prestressing, the strain output is larger than the output of sensor B, and the correlation between temperature, peak, and valley is more prominent. It was assumed that 0.2 is a one-fold change in temperature. During the three-fold change and nine-fold change, the steady-state strain curve is stable, and the correlation function is not fluctuating. However, in the state where the prestrain of sensor A is 140 $\mu\varepsilon$, the induction of temperature and deformation is relatively weak, and the difference is about 3–6 $\mu\varepsilon$, which is about 2–5% of the prestress, as shown in Figures 7 and 8. In the later test, more theoretical and experimental verifications are needed. Combined with the artificial intelligence model training method, the expansion effect caused by the material itself after the vibration of vibrating string was analyzed, and the environmental impact error was maximized. The tooling surface was optimized for temperature conductivity. Since this test is the calibration test for the maximum fitting, instrumentation error, etc., the welding process was used, so that the above effects can be neglected. At the same time, referring to Table 4, the noise of the device is 9.775×10^{-9} $\mu\varepsilon$, which can be ignored. Through comprehensive evaluation, it was found that the method of synchronously collecting sensors at similar positions can achieve the online comparison of the corresponding variable measuring sensors. The later test can be calibrated by high-frequency dynamic acquisition to calibrate the low-frequency acquisition and combined with the relevant optimization algorithm for calculation. The maximum optimization restores the original deformation magnitude, and at the same time reduces the instability of constitutive performance parameters such as the stiffness and strength of structure. The experimental results show the calibration of the corresponding variable monitoring sensor.

5. Conclusions

In engineering, the strain test method is usually used to determine the actual stress state of structure and monitor the structural deformation increment, and the structural strain test is an important method to solve the structural strength problem. During strain detection and monitoring, the calibration of strain gauge plays a significant role in the accuracy of data. To ensure the accuracy of traceability of magnitude, we should study the corresponding relationship between the temperature and strain of strain gauge and the online calibration test method, which is important for the online metrological traceability system of strain monitoring sensor. We focused on researching the influence of temperature and prestress effect on sensors by analyzing differences of their measurement results in specified situation. In this study, a dynamic vibrating wire measurement module was used to dynamically collect the temperature and strain of two vibrating wire sensors in the free state, and the correlation between temperature and strain was analyzed. The correlation coefficient between the temperature data of sensors A and B were statistically analyzed. The degree of relevance 0.9983 is highly correlated. The experimental results show that the correlation coefficient between the strain data of sensors A and B is 0.9895, and the correlation degree is highly correlated. After cross-calculation, a significant correlation was observed between temperature and strain indications. An experiment was carried out to analyze the measurement repeatability by using the knowledge of metrology. After analysis, it was found that the repeatability of temperature and strain values is within 0.2, and the consistency between the measured results of the same measurement is good. At the same time, the relationship between strain and temperature in different temperature variation intervals was analyzed. It was found that the temperature change has a small effect on the strain in the prestressed state, but the relationship between temperature and strain is relatively stable. Finally, the experiment verified the feasibility of parallel measurement scheme corresponding to the online calibration of variable monitoring sensor and analyzed the effect of temperature corresponding to the variable measurement under different prestress conditions, providing theoretical and experimental verification support for establishing an online metrological verification standard. This study can also better confirm the accuracy of bridge health monitoring data, effectively reduce the monitoring error caused by temperature drift, effectively eliminate the impact of environmental load, and better feedback the essential deformation of bridge structure. Considering the reliability risk of sensors after a long-time work period, such as signal drift, the approach proposed in this paper can be used as a substitute to such strain monitoring system in order to determine whether these sensors should be replaced. We expect our approach to be expandable to various applications. It can provide important support for the effective monitoring data extraction of the bridge health monitoring system.

Author Contributions: Conceptualization, L.P., G.J., Z.L., Y.W., and B.Z.; Data curation, L.P., X.Y., and Y.W.; Formal analysis, Y.W.; Funding acquisition, L.P.; Writing of original draft, L.P.; Writing—review and editing, L.P. All authors have read and agree to the published version of the manuscript.

Funding: This research was funded by the Project of National Key R & D Program of China grant number 2017YFF0206305 and The Key Projects of Public Scientific Research Institutes (2018-9031).

Acknowledgments: The authors would like to acknowledge the support of National Key R & D Program of China under grant number 2017YFF0206305 and the Key Projects of Public Scientific Research Institutes (2018-9031).

Conflicts of Interest: The authors declare no conflict of interest.

References

1. Mao, J.X.; Wang, H.; Feng, D.M.; Tao, T.Y.; Zheng, W.Z. Investigation of Dynamic Properties of Long-Span Cable-Stayed Bridges Based on One-Year Monitoring Data Under Normal Operating Condition. *Struct. Control Health Monit.* **2018**, *25*, e2146. [CrossRef]
2. Wang, H.; Zhang, Y.M.; Mao, J.X.; Wan, H.P.; Tao, T.Y.; Zhu, Q.X. Modeling and Forecasting of Temperature-induced Strain of a Long-span Bridge using an Improved Bayesian Dynamic Linear Model. *Eng. Struct.* **2019**, *192*, 220–232. [CrossRef]
3. Mao, J.X.; Wang, H.; Li, J. Fatigue Reliability Assessment of a Long-Span Cable-Stayed Bridge Based on One-Year Monitoring Strain Data. *J. Bridge Eng.* **2019**, *24*, 05018015. [CrossRef]
4. He, H.; Wang, W.; Tian, D.; Sun, J.; Xiong, C. Optimization of Vibrating Wire Sensor Excitation Strategy. *Chin. J. Sens. Actuators* **2010**, *23*, 74–77. [CrossRef]
5. Wen, Z.; Xia, Z.; Li, F. Research on Frequency Measurement Technology of String Vibration Sensor. *J. Xi'an Poly. Univ.* **2012**, *1*, 75–79.
6. Tian, Z.; Huang, Z. Nonlinear Compensation of Vibrating Wire Sensor Based on Spline Interpolation and Least-Squares Method. *Meas. Control Technol.* **2016**, *35*, 7–10.
7. Chen, N.; Li, H.; He, H.; Xie, K. An Self-Adaptive Vibrating Pick-up Method for Vibrating Wire Sensor. *J. Guangxi Univ. (Nat. Sci. Ed.)* **2017**, *42*, 1145–1150.
8. Xu, Y.; Yang, X.; Wei, T.; Yao, J. Analysis of Strain Transfer Influence Factors of Resistance Strain Sensor. *China Meas. Test* **2018**, *1*, 136–142.
9. Zhang, Y.; Liang, P.; Zhang, Y.; Chen, W.; Huang, M. Design of Calibration Device for Vibrating Wire strain Transducer. *Ind. Instrum. Autom.* **2011**, *4*, 52–55.
10. Guangzhou Guangcai Testing Instrument Co., Ltd. *Vibrating Wire Strain Gauge Calibration Instrument: China*; Guangzhou Guangcai Testing Instrument Co., Ltd.: Guangzhou, China, 2016.
11. Bai, S.; Xiao, Y.; Huang, B.; Liu, G. Research on Strain Calibration Method of Fiber Bragg Grating Sensor. *J. Vib. Mea. Diagn.* **2016**, *36*, 321–324.
12. Zhang, H.; Nie, F.; Fan, D. Calibration Method, Device and System of Fiber Bragg Grating Sensor. CN106871810A, 20 June 2017.
13. Chen, C.; Yan, D.; Chen, Z.; Tu, G.; Tian, Z. Technique Research of Vibrational Chord Strain Gauge to Concrete. *China J. Highw. Transp.* **2004**, *17*, 29–33.
14. Bai, T.; Deng, T.; Xie, J.; Hu, F.P. Accurate Mathematical Model of Vibrating Wire Sensor and Its Application. *Chin. J. Rock Mech. Eng.* **2005**, *24*, 5965–5969.
15. Iadicicco, A.; Campopiano, S. Sensing Features of Long Period Gratings in Hollow Core Fibers. *Sensors* **2015**, *15*, 8009–8019. [CrossRef] [PubMed]
16. Wang, Y.B.; Zhao, R.D.; Chen, L.; Xu, Y.; Xie, H.Q. Temperature Correction Test of Vibrating Wire Strain Sensor. *J. Archit. Civ. Eng.* **2017**, *34*, 68–75.
17. Lee, H.M.; Park, H.S. Measurement of Maximum Strain of Steel Beam Structures Based on Average; Strains from Vibrating Wire Strain Gages. *Exp. Technol.* **2013**, *37*, 23–29. [CrossRef]
18. Choi, S.W.; Kwon, E.; Kim, Y.; Hong, K.; Park, H.S. A Practical Data Recovery Technique for Long-Term Strain Monitoring of Mega Columns during Construction. *Sensors* **2013**, *13*, 10931–10943. [CrossRef] [PubMed]

19. Park, H.; Lee, H.; Choi, S.; Kim, Y. A Practical Monitoring System for the Structural Safety of Mega-Trusses Using Wireless Vibrating Wire Strain Gauges. *Sensors* **2013**, *13*, 17346–17361. [CrossRef] [PubMed]
20. Barot, D.; Wang, G.; Duan, L. High-Resolution Dynamic Strain Sensor Using a Polarization-Maintaining Fiber Bragg Grating. *IEEE Photonics Technol. Lett.* **2019**, *31*, 709–712. [CrossRef]
21. Jin, X.; Sun, C.; Duan, S.; Liu, W.; Li, G.; Zhang, S.; Chen, X.; Zhao, L.; Lu, C.; Yang, X.; et al. High Strain Sensitivity Temperature Sensor Based on a Secondary Modulated Tapered Long Period Fiber Grating. *IEEE Photonics J.* **2019**, *11*, 1–8. [CrossRef]

© 2020 by the authors. Licensee MDPI, Basel, Switzerland. This article is an open access article distributed under the terms and conditions of the Creative Commons Attribution (CC BY) license (http://creativecommons.org/licenses/by/4.0/).

Article

Train Hunting Related Fast Degradation of a Railway Crossing—Condition Monitoring and Numerical Verification

Xiangming Liu * and Valéri L. Markine *

Department of Engineering Structures, Delft University of Technology, 2628 CN Delft, The Netherlands
* Correspondence: Xiangming.Liu@tudelft.nl (X.L.); V.L.Markine@tudelft.nl (V.L.M.)

Received: 4 March 2020; Accepted: 15 April 2020; Published: 17 April 2020

Abstract: This paper presents the investigation of the root causes of the fast degradation of a railway crossing. The dynamic performance of the crossing was assessed using the sensor-based crossing instrumentation, and the measurement results were verified using the multi-body system (MBS) vehicle-crossing model. Together with the field inspections, the measurement and simulation results indicate that the fast crossing degradation was caused by the high wheel-rail impact forces related to the hunting motion of the passing trains. Additionally, it was shown that the train hunting was activated by the track geometry misalignment in front of the crossing. The obtained results have not only explained the extreme values in the measured responses, but also shown that crossing degradation is not always caused by the problems in the crossing itself, but can also be caused by problems in the adjacent track structures. The findings of this study were implemented in the condition monitoring system for railway crossings, using which timely and correctly aimed maintenance actions can be performed.

Keywords: railway crossing; wheel-rail impact; train hunting; numerical verification; railway track maintenance

1. Introduction

In the railway track system, turnouts (switches and crossings) are essential components that allow trains to pass from one track to another. A standard railway turnout is composed of three main parts: switch panel, closure panel, and crossing panel, as shown in Figure 1. In a railway turnout, the crossing panel is featured to provide the flexibility for trains to pass in different routes.

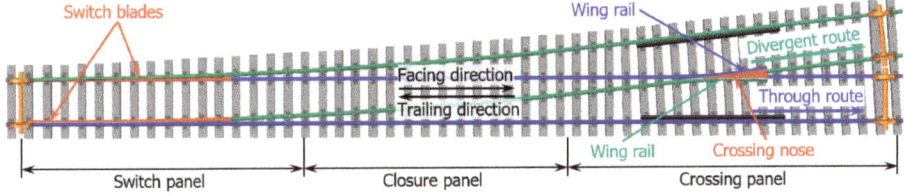

Figure 1. Standard left-hand railway turnout and the definition of the passing routes.

For rigid crossings that are commonly used in conventional railway lines, the gap between the wing rail and the nose rail usually results in high wheel-rail impacts in the transition region where the wheel load transits from the wing rail to the nose rail (vice versa, Figure 2), which makes the crossing a vulnerable spot in the railway track. In the case of crossings that are mainly used for the through route traffic (e.g., crossings in the crossover), there is no specific speed limit [1] and trains can pass through the crossings with a high velocity of up to 140 km/h. The high train velocity makes the wheel-rail impact more serious. In the Dutch railway system, around 100 crossings are urgently replaced every year [2] due to unexpected fatal defects, which not only result in substantial maintenance efforts, but also lead to traffic disruption and can even affect traffic safety.

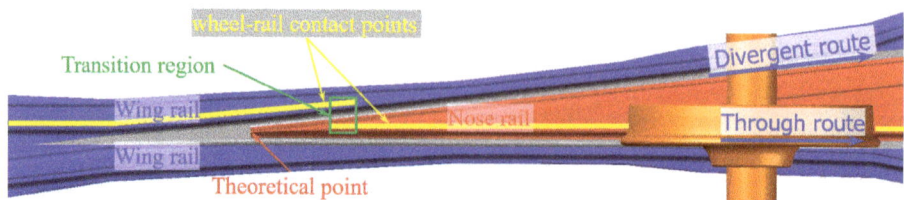

Figure 2. Wheel-rail interaction in the railway crossing for through route traffic.

In contrast to a switch panel, wherein sensors are instrumented for condition monitoring [3,4] and remaining useful life prediction [5], monitoring in a crossing panel is usually absent. As a result, the real-time information on the condition of railway crossings is limited. The present maintenance activities are mainly reactive and based on the experience of the contractors. In this case, the root causes of the crossing degradation are not always resolved by the maintenance actions, and the crossings are likely to be operated in a degraded condition. To improve this situation, necessary guidance for maintenance actions is highly required.

Proper crossing maintenance usually relies on condition assessment and degradation detection, which can be realized through field monitoring. In recent years, condition monitoring techniques have been frequently applied in the railway industry. Aside from the above-mentioned instrumentation on the turnout switches, vehicle-based monitoring systems have been applied in track stiffness measurement [6] and estimation [7], track alignment estimation [8], hanging sleepers detection [9], and track fault detection [10], etc. Compared with the normal track, the current studies on railway crossings are mainly based on numerical simulation. Typical contributions include wheel-rail interaction analysis [11–21], damage analysis [16,17,22,23], and prediction [18,24,25] as well as crossing geometry and track stiffness optimization for better dynamic performance [16,26]. Field measurements are mainly used for the validation of numerical models. The monitoring of railway crossings for condition assessment and degraded component detection is still limited.

In the previous study, key indicators for the crossing condition assessment based on the field measurement were proposed [27,28]. Additionally, a numerical vehicle-crossing model was developed using a multi-body system (MBS) method to provide the fundamental basis for the condition indicators [29]. In this study, the condition indicators, as well as the MBS model, were applied in the condition monitoring of a fast degraded railway crossing. The main goals of this study were to investigate the root causes of the crossing degradation as well as to assess the effectiveness of the current maintenance actions.

Based on the objectives, this paper is presented in the following order. The experimental and numerical tools, including the crossing condition indicators, are briefly introduced in Section 2. The measurement results and the crossing degradation analysis as well as the effectiveness of the current maintenance actions are presented in Sections 3 and 4. Based on the measurement results and field inspections, the root causes for the fast crossing degradation were investigated with the assistance of the MBS model, as presented in Section 5. In Section 6, the verification of the effectiveness of the maintenance actions is given. Finally, in Section 7, major conclusions are provided.

2. Methodology

In this section, the experimental tools for the crossing condition monitoring, as well as the indicators for the crossing condition assessment, are briefly introduced. The MBS vehicle-crossing model for the verification of the experimental findings is also presented.

2.1. Experimental Tools

The experimental tools mainly consisted of the in-site instrumentation system modified from ESAH-M (Elektronische System Analyse Herzstückbereich-Mobil) and the video gauge system (VGS) for wayside monitoring, as briefly described below. Both tools have already been introduced and actively applied in previous studies. Detailed information regarding the installation and data processing can be found in [27,30].

2.1.1. Crossing Instrumentation

The main components of the crossing instrumentation are an accelerometer attached to the crossing nose rail for 3-D acceleration measurement, a pair of inductive sensors attached in the closure panel for train detection as well as train velocity calculation, and the main unit for data collection. An overview of the instrumented crossing is shown in Figure 3.

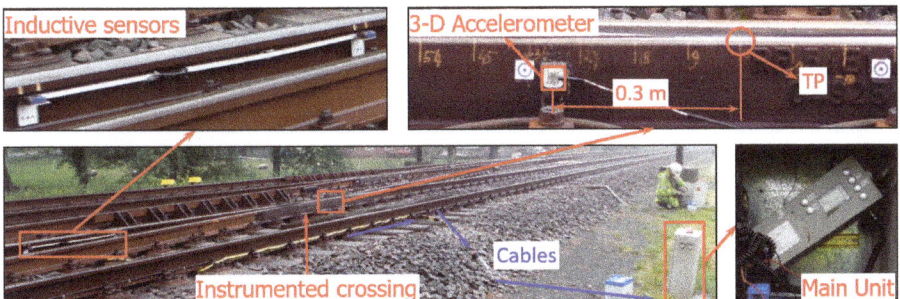

Figure 3. Crossing instrumentation based on ESAH-M.

The main outputs of the crossing instrumentation were the dynamic responses of the crossing nose, including the wheel-rail impact accelerations and locations, etc. All these responses were calculated within the transition region, which can be obtained through field inspection [29]. Based on these measured responses and the correlation analysis between the responses [28], two critical condition indicators related to the wheel impact and fatigue area, respectively, were proposed.

The wheel impact is reflected by the vertical accelerations, which were obtained from the crossing and processed through statistical analysis. This indicator is mainly based on the magnitude of the impacts due to each passing wheel (Figure 4a), and the changes in time indicate the different condition stages of the crossing (Figure 4b).

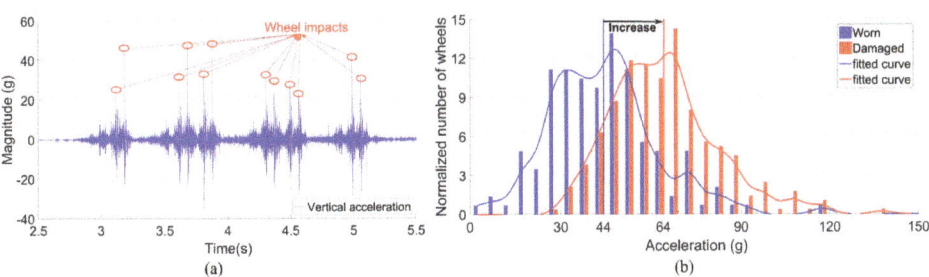

Figure 4. Indicator for the wheel impact. (**a**) Procedure for the obtainment of wheel impacts. (**b**) Example of the variation of the wheel impacts in different condition stages.

The fatigue area is defined as the region where the majority of wheel impacts are located on the crossing, and where ultimately the crack initiates (Figure 5a). In practice, the fatigue area can be simplified as the confidence interval of $[a - \sigma, a + \sigma]$, where a is the mean value of the wheel-rail impact locations, and σ is the standard deviation. The location and size of the fatigue area are critical values for the assessment of crossing wear and plastic deformation. A wide fatigue area usually represents well-maintained rail geometry. As demonstrated in Figure 5b, when the crossing condition was degraded from "Worn" to "Damaged", the fatigue area was dramatically narrowed and shifted further from the theoretical point (TP) of the crossing. More information about the fatigue area can be found in the previous study [27].

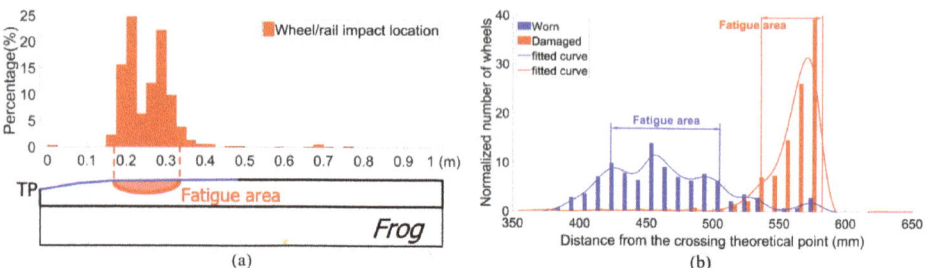

Figure 5. Demonstration of the crossing fatigue area detection. (**a**) Definition of the fatigue area. (**b**) Example of the fatigue area changes in different crossing condition stages.

2.1.2. Wayside Monitoring System

The VGS for wayside monitoring is a remote measurement device based on digital image correlation (DIC). It uses high-speed digital cameras to measure the dynamic movements of the selected targets in the track. The system, set up together with the targets installed on the crossing rail next to the instrumented accelerometer, is shown in Figure 6a, and the demo of the displacement measurement is shown in Figure 6b. The main outputs are the vertical displacements of the tracked targets with a stable sampling frequency of up to 200 Hz.

Figure 6. Wayside monitoring. (**a**) System setup. (**b**) The screen of displacement measurements.

Due to the limitation of the experimental conditions, the wayside monitoring system is usually set up close by the side of the track, which will introduce extra noise in the measured displacement results. To improve the accuracy of the measurement, the noise part needs to be eliminated. The noise mainly comes from the ground-activated camera vibration, which can be manually activated by hammering the ground near the camera. The measured camera vibrations in both the time and frequency domains are given in Figure 7.

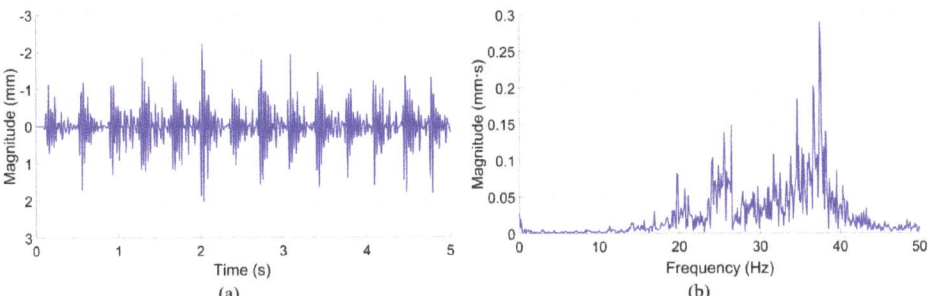

Figure 7. Ground activated camera vibration. (**a**) Time domain signal. (**b**) Frequency domain responses.

Despite the differences in the displacement responses in the two monitored crossings, the main resonance of the camera vibration was around 15–45 Hz. In the previous study [30], the main components in the displacement signal were elaborated. The train-track components related to displacement responses are mainly distributed below 10 Hz, which do not overlap with the camera vibration introduced noise. The noise part due to camera vibration can then be reduced through low-pass filtering, as shown in Figure 8.

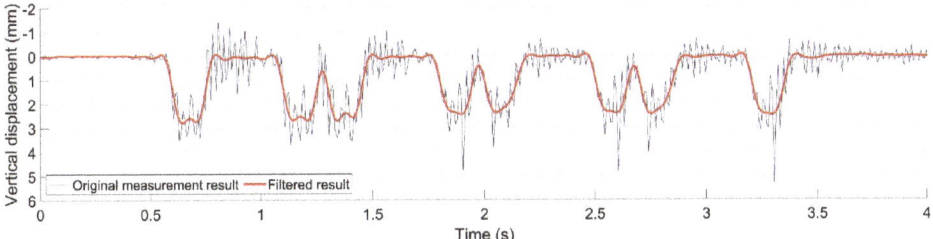

Figure 8. Examples of the measured rail vertical displacement.

The magnitude of the dynamic vertical displacement of the rail directly reflects the intensity of the track movement due to the passing trains. By comparing the measured rail displacement with the reference level, which can be obtained from numerical simulation using the parameters in the designed condition, the ballast settlement level of the monitored location can be estimated. The MBS model for the crossing performance analysis is described later in this section.

2.2. Multi-Body System (MBS) Vehicle-Crossing Model

The numerical model for the crossing performance analysis was developed using the MBS method VI-Rail (Figure 9a). The rail pads, clips, and ballast were simulated as spring and damping elements (rail busing and base busing, Figure 9b). In the vehicle model, the car body, bogie frames and the wheelsets were modeled as rigid bodies with both the primary suspension and secondary suspension taken into account (Figure 9b). The track model was a straight line with the crossing panel (Figure 9c) situated in the middle of the track. The rail element for the acceleration and displacement extraction was the lumped rail mass located 0.3 m from the TP of the crossing (Figure 9d), which is consistent with the setup of the field measurements (Figures 3 and 6a).

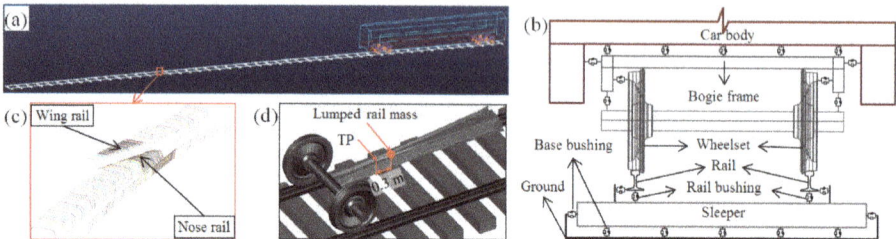

Figure 9. Multi-body system (MBS) model. (**a**) Vehicle-track model. (**b**) Flexible connections in the model. (**c**) Crossing profiles. (**d**) Rail element for acceleration extraction.

The detailed model development, experimental validation, and numerical verification can be found in the previous study [29]. Corresponding to the condition indicators, the main outputs of the MBS model are the wheel impact acceleration, transition region and wheel-rail contact forces. Using the MBS model, the condition of the monitored crossing, as well as the detected track degradations, can be verified.

3. Field Measurements and Analysis

The monitored crossing was a cast manganese crossing with an angle of 1:9. As part of a crossover, trains mainly pass the crossing in the facing through route (Figure 2) with a velocity of around 140 km/h. The on-site view of the crossing is shown in Figure 10a. According to the maintenance record, this crossing was suffering from fast degradation with the service life of only around three years (18 years on average [2]). At the beginning of the condition monitoring, the damaged crossing was completely renovated.

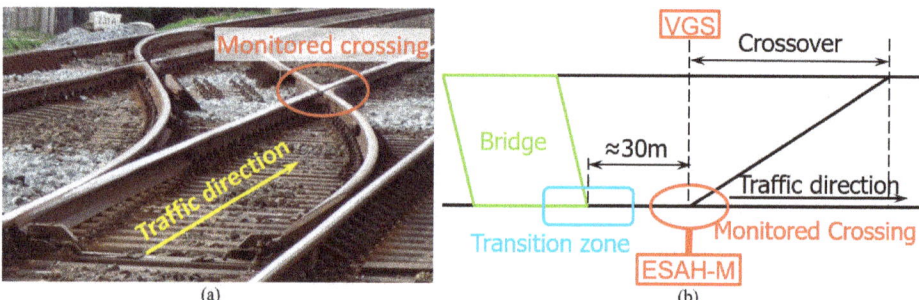

Figure 10. Overview of the monitored crossing. (**a**) On-site view. (**b**) Sketch view.

Figure 10b gives a sketch view of the crossing, including the setup of the monitoring devices and the layout of the adjacent structures, especially the small bridge in front of the crossing. Considering that the bridge is located quite close to the monitored crossing, the performance of the crossing might be affected by the bridge, which will be discussed later.

The measurement results from the crossing instrumentation were based on multiple train passages in one monitoring day. For the wayside monitoring, one sufficient train passage is enough to estimate the ballast condition. To maximally reduce the influence of the vehicle-related variables, the selected results were restricted to the commonly operated VIRM trains with velocities of around 140 km/h.

3.1. Wheel Impacts

Based on the estimated transition regions, the wheel impact accelerations were calculated. The distribution of the wheel impacts due to multiple wheel passages is shown in Figure 11a.

The sample size, in this case, was 78 passing wheels. It can be seen that the wheel impacts presented a bimodal distribution. Around 80% of the wheel impacts were below 50 g, while the remaining 20% of the wheel impacts were extremely high with a mean value of around 350 g. Such a polarized distribution of impacts indicates the highly unstable wheel-rail interaction in this crossing. It was demonstrated in a previous study [29] that for this type of railway crossing, the average level of the wheel impact is around 50 g, meaning that the 20% of high impacts of the monitored crossing are already more than seven times higher than the average impact level. It can be imagined that such high impacts will dramatically accelerate the degradation procedure of the crossing.

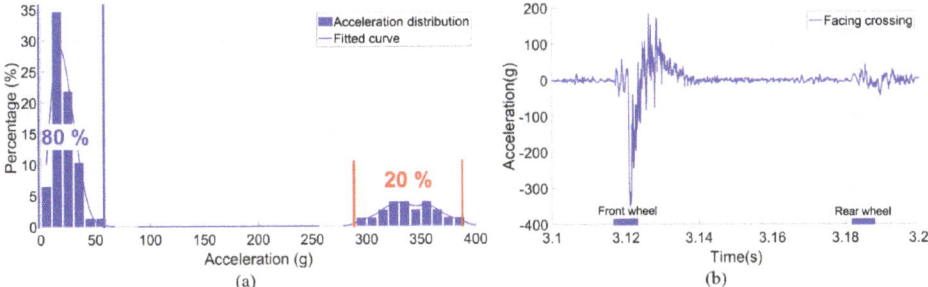

Figure 11. Vertical acceleration responses of the monitored crossings. (**a**) Distribution based on multiple train passages in one day. (**b**) Example of impacts due to one bogie.

An example of the impact acceleration response in the time-domain due to the first bogie of a VIRM train is shown in Figure 11b. It can be seen that for the two passing wheels from the same bogie, the impacts can be quite different. The impact due to the front wheel was up to 350 g, while the rear wheel activated vertical acceleration was only 20 g. It has to be noted that the high impacts were not always introduced by the front wheel, but appeared to have random occurrences. Such results further confirmed the instability of wheel-rail interaction at this crossing.

3.2. Fatigue Area

The measured fatigue area of the monitored crossing is presented in Figure 12. It can be seen that the wheel impacts were widely distributed at 0.22–0.38 m from the TP with the fatigue area size of 0.16 m. According to the previous study [28], the transition region (Figure 2) for this type of crossing is around 0.15–0.4 m. The fatigue area widely covered 64% of the transition region, which can be considered to be in line with the expectation of a new crossing profile. Such results further confirmed that the crossing rail was not worn or deformed.

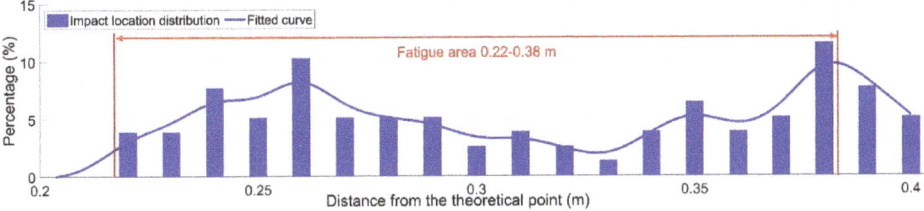

Figure 12. Measured fatigue area of the monitored crossing.

It has to be noted that the fatigue area does not conform to the normal distribution (referring to the "Worn" stage demonstrated in Figure 5b). Combined with the results of the wheel impacts such a fatigue area further confirmed the instability of the wheel-rail contact in the monitored crossing.

In a previous study [27], it was found that the crossing degradation was accompanied by the increase of wheel-rail impacts and the reduction in the fatigue area. The large number of extremely high wheel-rail impacts and relatively wide fatigue area clearly indicate the abnormal performance of the monitored crossing. Finding the root causes of such abnormality is the key to improving the dynamic performance of the crossing.

3.3. Ballast Settlement

The measured vertical displacement of the crossing rail is presented in Figure 13. It can be seen that the vertical rail displacement was around 4 mm. The measured displacement result can be considered to have two main parts: the elastic deformation and the gap between the sleeper and ballast. Considering that the ballast settlement is the accumulated effect due to multiple wheel passages, the plastic deformation caused by each passing train can be neglected. Due to the high impacts in the crossing panel, the ballast is usually settled unevenly, which results in hanging sleepers. Using the validated MBS model, it was calculated that the rail displacement in the reference condition was 1.4 mm (Figure 13), which only consisted of the elastic deformation part. By comparing these two results, it could be calculated that the gap between the sleeper and ballast was 2.6 mm, which can be estimated as the settlement of ballast. It was observed that the rail displacement obtained from the MBS simulation was much higher than that in a normal track (less than 1 mm [27,31]), which indicates the vulnerability of the ballast in the railway crossings.

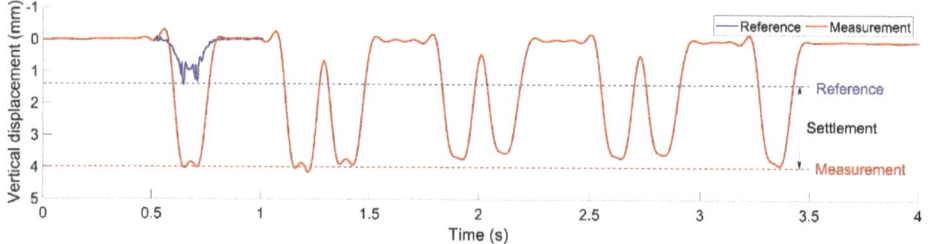

Figure 13. Ballast settlement in the monitored crossing.

In a previous study [27], it was found that track irregularities such as rail joints and turnout crossings can lead to the fast deterioration of the ballast, and the ballast settlement will in turn accelerate the degradation procedure of other related track components. In this study, the 2.6 mm ballast settlement was already higher than those in the previously monitored welded joints (≈1.5 mm) and movable crossings (≈2 mm), which revealed the seriously deteriorated ballast condition.

It can be concluded that the monitored crossing was suffering from rapidly occurring, extremely high wheel-rail impacts and severe ballast settlement. For a recently renovated crossing, such performance is quite abnormal.

4. Effectiveness Analysis of the Maintenance Actions

The constantly occurring extremely high wheel-rail impacts as well as serious ballast settlement clearly indicate the degraded condition of the crossing. In order to improve such a situation, various maintenance actions were implemented in this location including ballast tamping, fastening system renovation, etc. In this section, the effectiveness of the maintenance actions are briefly discussed, as presented below.

4.1. Ballast Tamping

Considering that the crossing rail was lately renovated with limited wear or plastic deformation, the severe ballast settlement was suspected to be the main cause for the high wheel-rail impacts.

Therefore, ballast tamping actions were frequently performed in this location by the local contractor. However, due to the lack of maintenance facilities, the tamping actions were mainly performed using the squeezing machine (Figure 14a) without track geometry correction. It can be imagined that the settled ballast cannot be fully recovered with such tamping action. As shown in Figure 14b, after tamping, the rail displacement was not dramatically reduced.

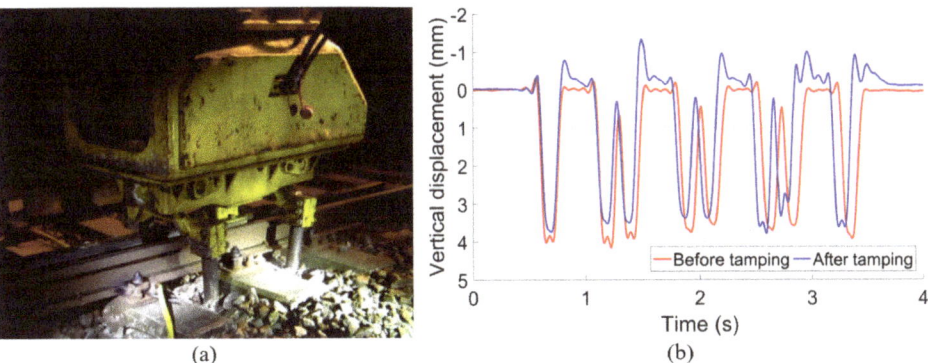

Figure 14. (**a**) Squeezing machine used for ballast tamping in the monitored crossing. (**b**) Measured rail displacement before and after ballast tamping.

The development of the wheel-rail impacts before and after tamping are presented in Figure 15. In this figure, each point represents the mean value of the impact accelerations based on multiple wheel passages in one monitoring day. It was discussed in a previous study [28] that the fluctuation of the wheel impacts was highly affected by external disturbances such as the weather. Still, it can be seen that the regression values before and after tamping were both around 100 g.

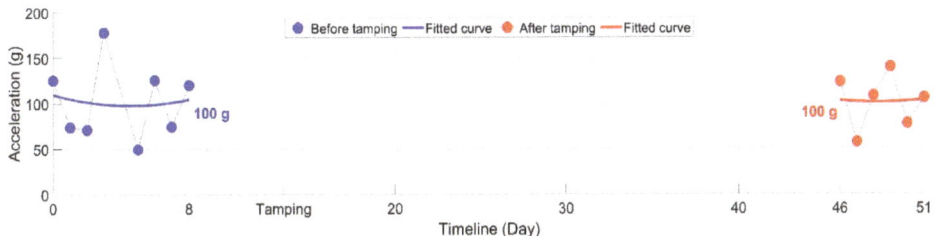

Figure 15. Development of the wheel-rail impacts before and after ballast tamping.

It can be concluded that such frequently implemented ballast tamping had no improvement in either the ballast condition or the dynamic performance of the monitored crossing. Without figuring out the root causes for the fast crossing degradation, such ineffective ballast tamping should be suspended.

4.2. Fastening System Renovation

During the monitoring period, the fastening system was found to be degraded with some broken bolts. Such degradation affected the lateral stability of the track. Therefore, the fastening system, mainly the bolts in the guard rails and the clips, was renovated, as shown in Figure 16.

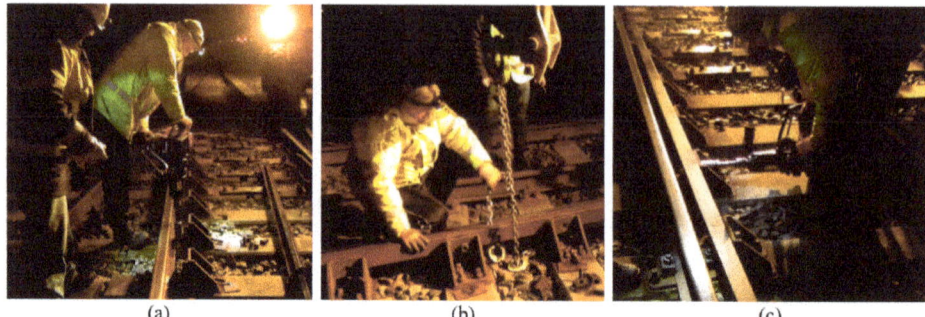

Figure 16. Fastening system renovation. (**a**) Remove the broken bolts. (**b**) Reposition the guard rail. (**c**) Install new bolts.

The development of the wheel-rail impacts before and after renovation is shown in Figure 17. The upper figure is the development of the mean value, and the lower figure gives the ratio of different impact levels in each monitoring day, corresponding to the value in the upper figure.

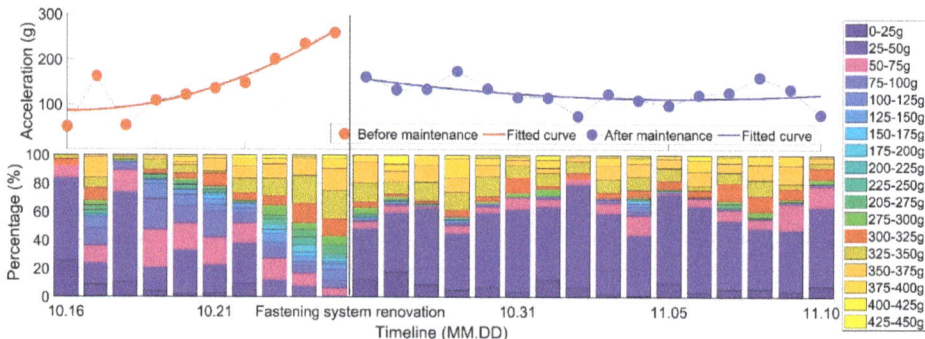

Figure 17. Effect of fastening system renovation on the dynamic performance of the crossing.

It can be seen from Figure 17 that before the renovation, the wheel-rail impact showed a clear increasing trend with the impact values widely distributed from 0 to 450 g. Such a degradation trend indicates that maintenance is urgently required due to the defects of the fastening system. After the renovation, the wheel-rail impacts were dramatically reduced in terms of the mean value and separated into two distribution modes, which is similar to those shown in Figure 11a. Such improvement is due to the enhancement in the track integrity. However, the wheel-rail impacts above 300 g were still a large proportion after maintenance, which means that the sources for such high wheel-rail impacts were not found.

In practice, ballast tamping is currently one of the few options for contractors to maintain the track. However, the unimproved crossing performance clearly indicates the ineffectiveness of tamping. The fastening system renovation was a forced action to repair damaged components. Although the crossing performance was improved, the extremely high wheel-rail impacts were not reduced, thus the sources for the fast crossing degradation were not eliminated. To figure out the root causes for the crossing damage, the track inspection was extended to the bridge in front of the crossing (Figure 10b). The results for the track inspection, as well as the numerical verification using the MBS model, are presented in the next section.

5. Damage Sources Investigation

In this section, the track inspection, including the whole turnout and the adjacent bridge, is presented. The inspected degradations will be input into the MBS model to verify the influence on the crossing performance. As a reference, the dynamic responses in the designed condition with no track degradations were also simulated and compared with those in degraded conditions. The verification results, followed by the analysis, are also presented.

5.1. Track Inspection

In the field inspection, it was found that the bridge was not well aligned in the track, but deviated around 15 cm, as shown in Figure 18a. Such deviation introduced a curve into the track, which was likely to be out of design since no elevation was set up in the outer rail. It can be imagined that the passing trains could not pass the track along the central line but tended to have wheel flange contact with the outer rail, which eventually leads to the severe wear in the switch blade (Figure 18b).

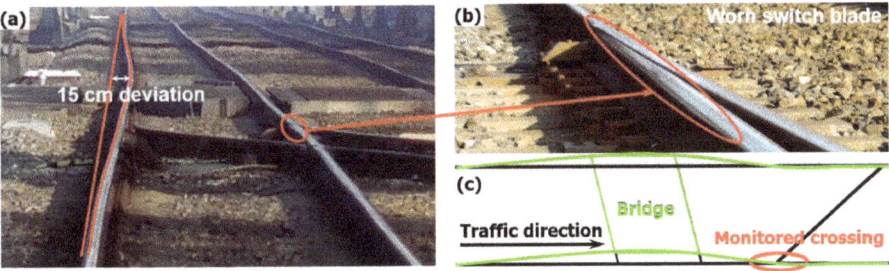

Figure 18. Track deviation in front of the crossing. (**a**) Inspected curve introduced by the bridge. (**b**) Worn switch rail. (**c**) Demonstration of the bridge deviation.

The accumulated effect of the track deviation was also reflected in the variated track gauge. It was shown in the measurement results that the gauge variations along the whole turnout were up to 3 mm, as presented in Table 1. Considering that the monitored crossing is located quite close to the bridge (Figure 18c), such track misalignment, including the track deviation in the bridge and track gauge variation along the turnout, may affect the wheel-rail interaction in the crossing.

Table 1. Track gauge measurement results in the critical sections along the turnout.

Location	A	B	C	D	E	F	G
Deviation (mm)	+2	+3	−2	−2	+2	+3	0

5.2. Numerical Verification and Analysis

In order to verify the effect of the track lateral misalignment on the performance of the crossing, both the bridge-introduced curve and the track gauge variation were input into the MBS vehicle-crossing model (Figure 9). The equivalent track lateral irregularities as the model input are shown in Figure 19.

In the MBS model, the crossing type is the same as the monitored 1:9 crossing with the rail type of UIC54 E1. The vehicle model is consistent with the recorded VIRM train with the wheel profile of S1002. The initial track parameters of Dutch railways [32] applied in the model are given in Table 2.

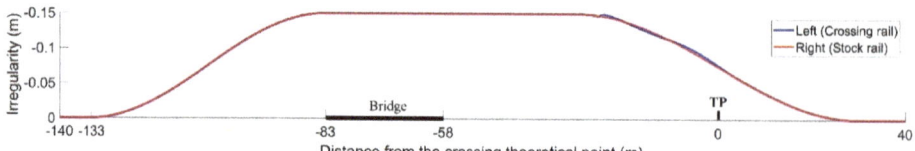

Figure 19. Equivalent lateral irregularities in the track.

Table 2. Track parameters.

Track Components		Stiffness, MN/m	Damping, kN·s/m
Rail pad/Clips	Vertical	1300	45
	Lateral	280	580
	Roll	360	390
Ballast	Vertical & lateral	45	32

With the track misalignment taken into account, the crossing condition was considered as degraded. The simulation results of both wheels in the bogie, including the wheel impact accelerations and transition regions, were compared with the results in the designed condition [29], as shown in Figure 20.

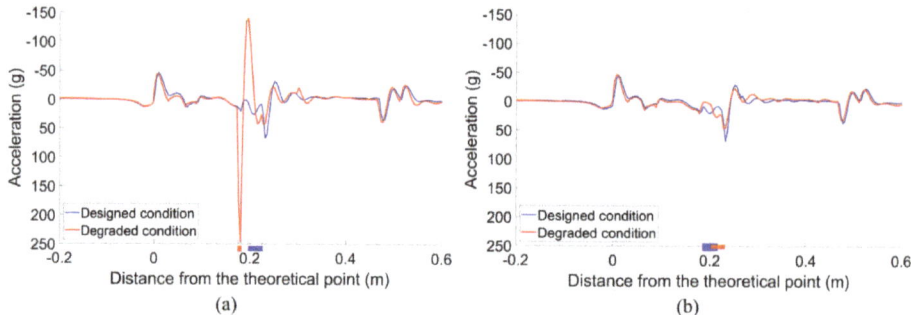

Figure 20. Vertical impact acceleration responses and transition regions. (**a**) Front wheel. (**b**) Rear wheel.

It can be seen from Figure 20a that with the lateral irregularity taken into account, the impact of the front wheel was dramatically increased to 247 g, which was 4 times higher than the reference value (around 62 g) in the designed condition. While for the rear wheel from the same bogie, the impact was 48 g, which was even lower than the reference value. Despite the slight difference in the absolute values, the simulation results were consistent with the measurement results (Figure 11). Meanwhile, the transition region of the front wheel was 0.176–0.182 m from the TP with a size of only 0.006 m. Compared with the reference level (0.196–0.217 m with a size of 0.031 m, [29]), it was much narrower and closer to the TP, indicating earlier wheel impact and much sharper wheel load transition in the crossing. For the rear wheel, although the transition region was located farther from the TP, the size was almost the same as the reference value.

Such results clearly show that the curve and lateral track misalignment in front of the crossing can lead to unstable wheel-rail contact in the crossing and sometimes result in extremely high impacts. Additionally, the front and rear wheels pass through the crossing quite differently, which indicates that the performance of the rear wheel is not independent, but is affected by the front wheel.

For the wheel-rail contact forces, the tendency was similar to the acceleration responses, as shown in Figure 21. With the degraded track condition, the maximum contact force of the front wheel in the degraded condition was 468 kN, which was twice as high as that in the designed condition (235 kN). While for the rear wheel, the difference between the degraded condition and the designed condition was limited.

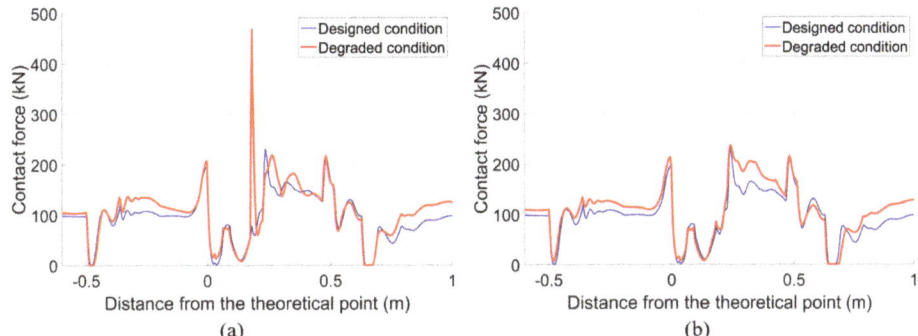

Figure 21. Vertical wheel-rail contact responses of the facing crossing. (**a**) Front wheel. (**b**) Rear wheel.

To understand how the track misalignment affects the wheel-rail interaction in the crossing, the relationship between the wheel lateral displacements and wheel-rail contact forces were analyzed. Before that, the wheel lateral displacement in the designed condition is presented in Figure 22. When the train enters the crossing panel, the variated rail geometry will lead to the lateral movement of the wheel. The maximum lateral displacement was around 0.7 mm.

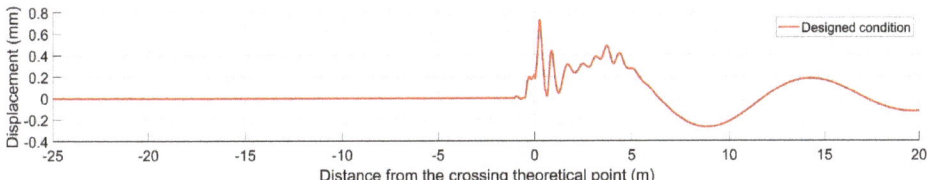

Figure 22. Wheel lateral displacement in the designed condition.

In the degraded condition with track lateral irregularities, the lateral displacements of the wheels were dramatically changed, as shown in Figure 23. It can be seen that both the front wheel and the rear wheel showed activated hunting oscillation before and after passing through the crossing, but the trajectories were quite different. For the front wheel, the lateral movement was more intense and ran toward the crossing nose rail near the TP. The maximum lateral displacement corresponding to the position with the highest contact force was 2.3 mm, which means that compared with that in the designed condition, the wheel flange was around 1.6 mm closer to the nose rail. Comparatively speaking, such displacement of the rear wheel was only 0.3 mm. Such results indicate that the wheel-rail impact was profoundly affected by the movement of the wheel. When the wheel approaches closer to the crossing nose, the wheel-rail impact is likely to be increased. It can be concluded that the train hunting activated by the lateral track misalignment in front of the crossing is the main cause of the extremely high wheel-rail impacts.

The train hunting effect also explains the unstable wheel-rail impacts. For the rear wheel, the lateral movement was affected not only by the track misalignment but also by the front wheel from the same bogie. As a result, these two wheels led to quite different wheel trajectories. It can be imagined that in the real-life situation, there are much more factors that may affect the hunting motion of each passing wheelset such as the initial position of the wheel when entering the misaligned track section, the mutual interaction between the adjacent wheelsets, the lateral resistance of the track, and even the weather condition [28], etc. The combined effect of all these factors ultimately resulted in the polarized distribution of the impact acceleration responses (Figure 11a).

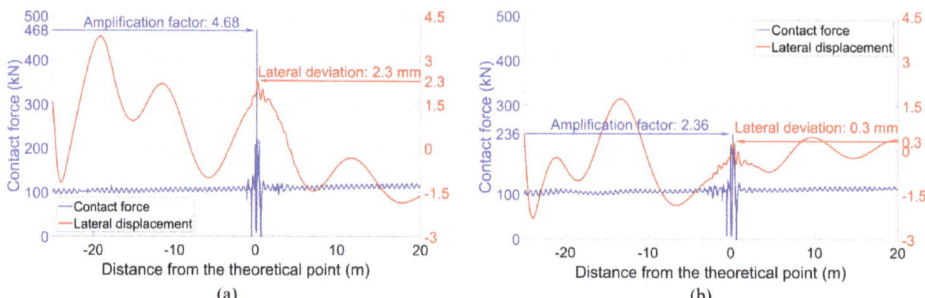

Figure 23. Wheel-rail contact forces and wheel lateral displacements. (**a**) Front wheel. (**b**) Rear wheel.

5.3. Respective Effect of Lateral Curve or Track Gauge Deviation

It can be noticed that in the previous analysis, the input track misalignment consisted of two parts: the lateral curve introduced by the bridge and the track gauge deviation. In order to understand the effect of each part in the wheel-rail interaction, these two parts were further analyzed, and the results are presented below.

Considering the bridge-introduced lateral curve, the wheel-rail contact forces and the lateral wheel displacements were calculated, as presented in Figure 24. It can be seen that in the front wheel, the bridge-introduced curve mainly resulted in the lateral shift of the wheel trajectory due to the centripetal force. Such a shift was only 0.5 mm near the crossing nose when compared with the designed condition, and the effect on the wheel impact was limited. For the rear wheel, the combined effect of the curve and the motion of the front wheel resulted in the lateral deviation of 0.9 mm, which was quite close to that in the designed condition and had no significant influence on the wheel-rail impact.

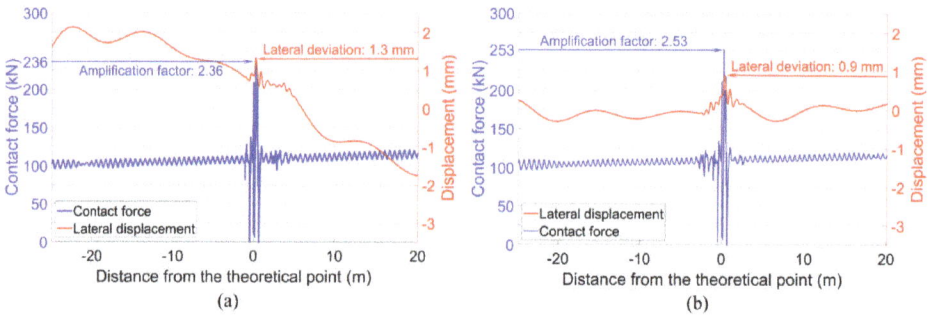

Figure 24. Wheel-rail contact forces and lateral wheel displacements. (**a**) Front wheel. (**b**) Rear wheel.

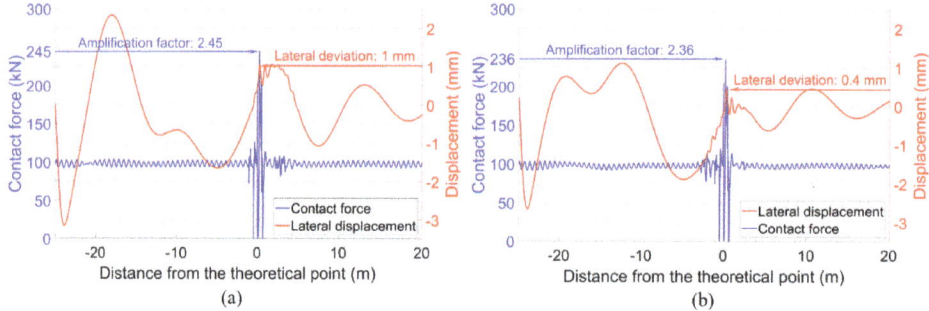

Figure 25. Wheel-rail contact forces and wheel lateral displacements. (**a**) Front wheel. (**b**) Rear wheel.

The effect of the track gauge deviation on the wheel-rail interaction is demonstrated in Figure 25. Different from the effect of the bridge-introduced curve, the deviated track gauge activated the hunting motion of the passing wheels. Still, the resulted lateral wheel displacements were not large enough to amplify the wheel-rail impact. The maximum displacements corresponding to the wheel impacts were 1 mm in the front wheel and 0.4 mm in the rear wheel, respectively.

5.4. Summary

Based on the above analysis, it can be concluded that the extremely high wheel-rail impacts in the monitored crossing were caused by the hunting oscillation of the passing trains. Such train hunting was the combined effect of the bridge-introduced curve in front of the crossing and the deviated track gauge along the turnout. When the maximum wheel lateral displacement reaches a certain level (e.g., 2.3 mm), the wheel-rail impact will be dramatically amplified.

It has to be noted that although the curve in front of the crossing did not directly activate train hunting, the activated lateral shift of the passing wheels resulted in the wear in the switch blade (Figure 18b) and contributed to the track gauge deviation. Therefore, such a curve can be considered as the root cause of the fast degradation of the monitored crossing. To improve the performance of the crossing, this curve has to be first eliminated.

In the previous study [28], it was proven that high rail temperature due to the long duration of sunshine would amplify the existing track geometry deviation in turnout and lead to the increase in the wheel-rail impacts. The train hunting activated by the track gauge deviation in this study further confirmed these results.

6. Effect of Maintenance-Related Degradation

According to the measurement results, the monitored crossing also suffered from ballast settlement and broken clips. In order to better simulate the real-life situation, these track defects were respectively added to the degraded MBS model developed in Section 5.2. The combined effects were simulated and analyzed, as presented below.

6.1. Effect of Ballast Settlement

It is shown in Figure 13 that the detected ballast settlement was around 2.6 mm. To simplify the problem, a vertical irregularity was introduced in the MBS model to simulate the ballast settlement, as shown in Figure 26. In this irregularity function, the amplitude was 1.3 mm, and the wavelength was 10 m. The trough of the wave was located 0.3 m from the TP of the crossing, which was consistent with the instrumented accelerometer and the installed displacement target.

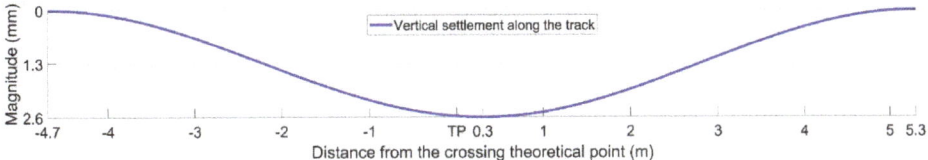

Figure 26. Ballast settlement introduced in the MBS model.

With the ballast settlement taken into account in the MBS model, the dynamic performance of the crossing was simulated. The representative results are shown in Figure 27. It can be seen that the simulation results were almost the same as those without ballast settlement (Figure 23), despite the slightly increased impact force of the front wheel (from 468 kN to 487 kN). It can be concluded that the existence of ballast settlement had a limited influence on the dynamic performance of the crossing. From another point of view, the ballast settlement was more likely to be the accumulated effect of the high wheel-rail impacts. Such results further explain the ineffectiveness of the frequently performed ballast tamping since ballast settlement is not the main cause of the extremely high wheel-rail impacts.

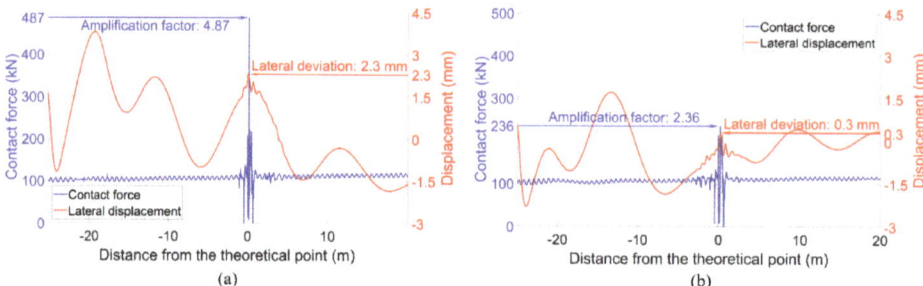

Figure 27. Wheel-rail contact forces and lateral wheel displacements. (**a**) Front wheel. (**b**) Rear wheel.

6.2. Influence of Reduced Lateral Support

It is shown in Figure 16 that the defects of the fastening system can increase the instability of the wheel-rail impact in the crossing. Combined with the maintenance action and the simulation results in Section 5, it can be inferred that this effect was caused by the reduced lateral track resistance. To verify this inference in the degraded model (Section 5.2), the input lateral stiffness of the clips in the crossing panel was reduced from 280 MN/m (Table 2) to 2.8 N/m, and the corresponded damping was reduced from 580 kN·s/m to 5.8 N·s/m. Based on these inputs, the wheel-rail contact forces and the lateral wheel displacements were calculated, as presented in Figure 28.

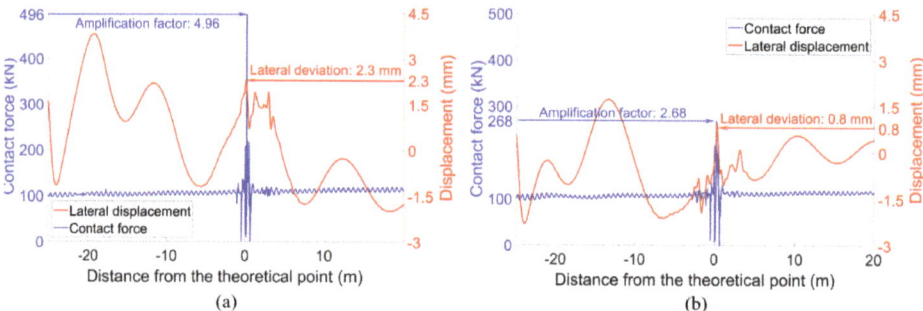

Figure 28. Wheel-rail contact forces and lateral wheel displacements. (**a**) Front wheel. (**b**) Rear wheel. Note: Ballast settlement was not taken into account.

It can be seen from Figure 28 that with the reduced lateral stiffness and damping of the clips, the impacts of both the front wheel and the rear wheel were slightly increased (compared with the results in Figure 23). Moreover, the hunting motion of wheels in the crossing panel was more intense. As a result, the lateral deviation of the rear wheel increased from 0.3 mm to 0.8 mm. It can be imagined that with the impacts of the passing trains, the track alignment will continuously be changing due to the reduced structural integrity. The changed track alignment will, in return, act on the wheel-rail interaction and eventually lead to more unstable wheel impacts in the crossing (Figure 17). From this point of view, renovating the defected fastening system is necessary for a monitored crossing. Enough track lateral resistance can help to maintain better crossing performance.

7. Conclusions

In this study, the root cause of the fast degradation of a 1:9 crossing in the Dutch railway system was investigated. The effectiveness of some typical track maintenance actions was also assessed and verified. Based on the measurement and simulation results, the following conclusions can be drawn:

- The fast crossing degradation was directly caused by the extremely high wheel-rail impacts, and the root cause for such high impacts was the hunting of the passing trains that were activated by the track lateral misalignment in front of the crossing. When the lateral deviation of the passing wheel exceeds a certain extent (e.g., 2.3 mm), the wheel-rail contact situation will change and the wheel impacts will be dramatically increased. To improve the current situation, such track misalignment needs to be eliminated;
- Ballast settlement is likely to be the accumulated effect of the high wheel-rail impacts. The influence on the crossing performance is somewhat limited. Ballast tamping, especially with only the squeezing machine, cannot improve the dynamic performance of the crossing. In the case of not knowing the sources of damage, it is better to take no action, rather than implement ballast tamping;
- Fastening system renovation helped improved the crossing performance by providing better lateral support in the track but was not targeted to the fundamental problem. Therefore, such damage repair action is useful, but not enough for an improvement in the crossing performance.

This study further verified the effectiveness of the previously proposed condition indicators in the investigation of the damage sources of the crossing. Since the root causes for the fast degradation were the deviated track in front of the crossing, this means that the degradation detection is not only restricted to the crossing itself but can also take the adjacent structures into account.

The activated train hunting reasonably explained the instability of wheel-rail interaction in the crossing, which pointed out a possible direction to maintain the problematic crossings in the Dutch railway network. As part of the Structural Health Monitoring System for railway crossings developed in TU Delft, the findings in this study will help improve the current maintenance philosophy from "failure reactive" to "failure proactive", and eventually lead to sustainable railway crossings.

Author Contributions: This article is written by X.L. and supervised by V.L.M. All authors have read and agreed to the published version of the manuscript.

Funding: The field measurements in this research were funded by ProRail and performed within the framework of the joint ProRail and TU Delft project "Long-Term Monitoring of Railway Turnouts".

Acknowledgments: The authors would like to thank the support from I.Y. Shevtsov from ProRail and technical support from G.H. Lambert and other colleagues from ID^2. Furthermore, the fruitful cooperation with Strukton Rail is highly appreciated.

Conflicts of Interest: The authors declare no conflicts of interest.

References

1. *ProRail, Ontwerpvoorschrift (Baan en Bovenbouw) Deel 6.1: Wissels en Kruisingen Alignement, Tussenafstanden, Overgangsvrije Zones en Bedieningsapparatuur, V006*; ProRail: Utrecht, The Netherlands, 2015; p. 12.
2. Shevtsov, I. Rolling Contact Fatigue Problems at Railway Turnouts. Available online: http://www.sim-flanders.be/sites/default/files/events/Meeting_Materials_Nov2013/mm_13112013_ivan_shevtsov_prorail.pdf (accessed on 13 November 2013).
3. Cornish, A.T. Life-Time Monitoring of in Service Switches and Crossings through Field Experimentation. Ph.D. Thesis, Imperial College London, London, UK, May 2014.
4. Molina, L.F.; Resendiz, E.; Edwards, J.R.; Hart, J.M.; Barkan, C.P.L.; Ahuja, N. Condition Monitoring of Railway Turnouts and Other Track Components Using Machine Vision. In Proceedings of the Transportation Research Board 90th Annual Meeting, Washington, DC, USA, 23–27 January 2011.
5. Chen, C.; Xu, T.; Wang, G.; Li, B. Railway turnout system RUL prediction based on feature fusion and genetic programming. *Measurement* **2020**, *151*, 107162. [CrossRef]

6. Wang, P.; Wang, L.; Chen, R.; Xu, J.; Gao, M. Overview and outlook on railway track stiffness measurement. *J. Mod. Transp.* **2016**, *24*, 89–102. [CrossRef]
7. Yeo, G.J. Monitoring Railway Track Condition Using Inertial Sensors on an in-Service Vehicle. Ph.D. Thesis, University of Birmingham, Birmingham, UK, June 2017.
8. De Rosa, A.; Alfi, S.; Bruni, S. Estimation of lateral and cross alignment in a railway track based on vehicle dynamics measurements. *Mech. Syst. Signal Process.* **2019**, *116*, 606–623. [CrossRef]
9. Balouchi, F.; Bevan, A.; Formston, R. Detecting Railway Under-Track Voids Using Multi-Train in-Service Vehicle Accelerometer. In Proceedings of the 7th IET Conference on Railway Condition Monitoring, Birmingham, UK, 27–28 September 2016.
10. Tsunashima, H. Condition Monitoring of Railway Tracks from Car-Body Vibration Using a Machine Learning Technique. *Appl. Sci.* **2019**, *9*, 2734. [CrossRef]
11. Kassa, E.; Nielsen, J.C.O. Stochastic analysis of dynamic interaction between train and railway turnout. *Veh. Syst. Dyn.* **2008**, *46*, 429–449. [CrossRef]
12. Alfi, S.; Bruni, S. Mathematical modelling of train-turnout interaction. *Veh. Syst. Dyn.* **2009**, *47*, 551–574. [CrossRef]
13. Ren, Z.; Sun, S.; Xie, G. A method to determine the two-point contact zone and transfer of wheel rail forces in a turnout. *Veh. Syst. Dyn.* **2010**, *48*, 1115–1133. [CrossRef]
14. Pletz, M.; Daves, W.; Ossberger, H. A wheel set/crossing model regarding impact, sliding and deformation explicit finite element approach. *Wear* **2012**, *294*, 446–456. [CrossRef]
15. Anyakwo, A.; Pislaru, C.; Ball, A. A New Method for Modelling and Simulation of the Dynamic Behaviour of the Wheel-rail contact. *Int. J. Autom. Comput.* **2012**, *9*, 237–247. [CrossRef]
16. Pålsson, B.A. Optimisation of Railway Switches and Crossings. Ph.D. Thesis, Chalmers University of Technology, Göteborg, Sweden, February 2014.
17. Wei, Z.; Chen, S.; Li, Z.; Dollevoet, R. Wheel-Rail Impact at Crossings-Relating Dynamic Frictional Contact to Degradation. *J. Comput. Nonlinear Dyn.* **2017**, *12*, 1–11. [CrossRef]
18. Xin, L. Long-Term Behaviour of Railway Crossings Wheel-Rail Interaction and Rail Fatigue Life Prediction. Ph.D. Thesis, Delft University of Technology, Delft, The Netherlands, June 2017.
19. Chiou, S.; Yen, J. Modeling of railway turnout geometry in the frog area with the vehicle wheel trajectory. *Proc. Inst. Mech. Eng. Part F J. Rail Rapid Transit* **2018**, *232*, 1598–1614. [CrossRef]
20. Skrypnyk, R.; Nielsen, J.C.O.; Ekh, M.; Pålsson, B.A. Metamodelling of wheel-rail normal contact in railway crossings with elasto-plastic material behavior. *Eng. Comput.* **2019**, *35*, 139–155. [CrossRef]
21. Torstensson, P.T.; Squicciarini, G.; Krüger, M.; Pålsson, B.A.; Nielsen, J.C.O.; Thompson, D.J. Wheel-rail impact loads and noise generated at railway crossings influence of vehicle speed and crossing dip angle. *J. Sound Vib.* **2019**, *456*, 119–136. [CrossRef]
22. Wiest, M.; Daves, W.; Fischer, F.D.; Ossberger, H. Deformation and damage of a crossing nose due to wheel passages. *Wear* **2008**, *265*, 1431–1438. [CrossRef]
23. Johansson, A.; Pålsson, B.; Ekh, M.; Nielsen, J.; Ander, M.; Brouzoulis, J.; Kassa, E. Simulation of wheel-rail contact and damage in switches & crossings. *Wear* **2010**, *271*, 472–481.
24. Nielsen, J.C.O.; Li, X. Railway track geometry degradation due to differential settlement of ballast/subgrade Numerical prediction by an iterative procedure. *J. Sound Vib.* **2018**, *412*, 441–456. [CrossRef]
25. Skrypnyk, R.; Ekh, M.; Nielsen, J.C.O.; Pålsson, B.A. Prediction of plastic deformation and wear in railway crossings—Comparing the performance of two rail steel grades. *Wear* **2019**, *428–429*, 302–314. [CrossRef]
26. Wan, C. Optimisation of Vehicle-Track Interaction at Railway Crossings. Ph.D. Thesis, Delft University of Technology, Delft, The Netherlands, September 2016.
27. Liu, X.; Markine, V.L.; Wang, H.; Shevtsov, I.Y. Experimental tools for railway crossing condition monitoring. *Measurement* **2018**, *129*, 424–435. [CrossRef]
28. Liu, X.; Markine, V.L. Correlation Analysis and Application in the Railway Crossing Condition Monitoring. *Sensors* **2019**, *19*, 4175. [CrossRef]
29. Liu, X.; Markine, V.L. Validation and Verification of the MBS Models for the Dynamic Performance Study of Railway Crossings. *Eng. Struct.* **2019**, submitted.
30. Wang, H.; Markine, V.L.; Liu, X. Experimental analysis of railway track settlement in transition zones. *Proc. Inst. Mech. Eng. Part F J. Rail Rapid Transit* **2018**, *232*, 1774–1789. [CrossRef] [PubMed]

31. Xu, L.; Yu, Z.; Shi, C. A matrix coupled model for vehicle-slab track-subgrade interactions at 3-D space. *Soil Dyn. Earthq. Eng.* **2020**, *128*, 105894. [CrossRef]
32. Hiensch, M.; Nielsen, J.C.O.; Verheijen, E. Rail corrugation in The Netherlands measurements and simulations. *Wear* **2002**, *253*, 140–149. [CrossRef]

© 2020 by the authors. Licensee MDPI, Basel, Switzerland. This article is an open access article distributed under the terms and conditions of the Creative Commons Attribution (CC BY) license (http://creativecommons.org/licenses/by/4.0/).

Article

Correlation Analysis and Verification of Railway Crossing Condition Monitoring

X. Liu * and V. L. Markine *

Section of Railway Engineering, Delft University of Technology, 2628 CN Delft, The Netherlands
* Correspondence: Xiangming.Liu@tudelft.nl (X.L.); V.L.Markine@tudelft.nl (V.L.M.)

Received: 25 August 2019; Accepted: 23 September 2019; Published: 26 September 2019

Abstract: This paper presents a correlation analysis of the structural dynamic responses and weather conditions of a railway crossing. Prior to that, the condition monitoring of the crossing as well as the indicators for crossing condition assessment are briefly introduced. In the correlation analysis, strong correlations are found between acceleration responses with irregular contact ratios and the fatigue area. The correlation results between the dynamic responses and weather variables indicate the influence of weather on the performance of the crossing, which is verified using a numerical vehicle-crossing model developed using the multi-body system (MBS) method. The combined correlation and simulation results also indicate degraded track conditions of the monitored crossing. In the condition monitoring of railway crossings, the findings of this study can be applied to data measurement simplification and regression, as well as to assessing the conditions of railway crossings.

Keywords: railway crossing; condition monitoring; condition indicator; correlation analysis; weather impact; numerical verification

1. Introduction

Railway turnouts are essential components of railway infrastructure and provide the ability for trains to transfer from one track to the other. In the meantime, a gap between the wing rail and nose rail in the crossing panel (Figure 1b) introduces a discontinuity in the rail. As a result of trains passing through, the high wheel–rail impact due to the high train velocity causes this type of crossing to suffer from severe damage such as cracks (Figure 1c) and spalling (Figure 1d), and the service lives of some railway turnouts are only 2–3 years. In order to improve the maintenance of the crossing and prolong service life, it is better to perform maintenance in a predictive way by developing a structural health monitoring (SHM) system for railway crossings [1].

In order to obtain information on damage detection, localization and condition assessment in SHM systems, it is important to get insight into the performance of the structures. In recent years, SHM has drawn increasingly more attention in the railway industry. D. Barke and W.K. Chiu reviewed the major contributions of condition monitoring in regards to wheels and bearings [2]. Based on digital image correlations, D. Bowness et al. measured railway track displacement using a high speed camera [3]. The axle box acceleration (ABA) system has been widely applied in the condition monitoring [4] and damage detection [5,6] of railway tracks. However, most of the contributions of SHM are based mainly on normal tracks. Z. Wei et al. have applied the ABA system in railway crossing damage detection [7]. However, as a special and vulnerable component in the railway track system, the study on crossings in terms of condition monitoring are still limited.

In the existing studies, the performance analysis of crossing has been based mainly on numerical approaches. For instance, finite element (FE) wheel-crossing models have been applied to calculate plastic deformation and frictional work [8], to simulate the distribution of stresses in the crossing nose [9] and to predict the fatigue life of a crossing [10]. Also, multi-body system (MBS) vehicle-crossing models have been used for general train–track interaction analysis [11], track elasticity analysis [12], crossing geometry optimization [13–15] and so on. Due to restricted track access, high costs and time consumption, field measurements have mainly been used for numerical model validation [9,16]. The numerical models are usually developed according to a certain hypothesis with a focus on specific problems. However, for damage detection and assessments of crossing conditions, the numerical approach alone is not enough, and monitoring the conditions of in-service railway crossings is highly necessary.

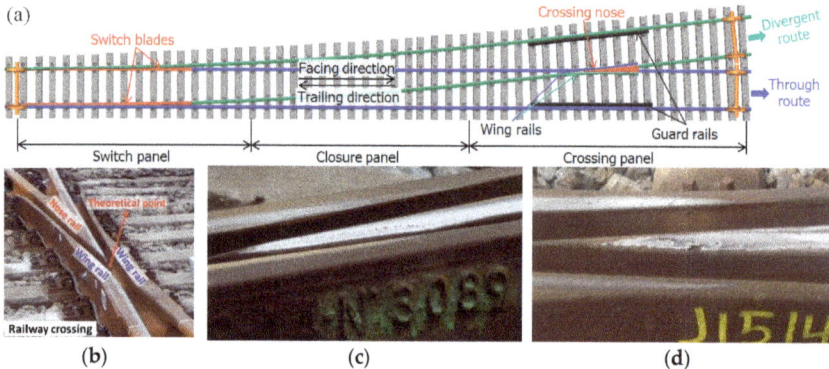

Figure 1. (**a**) Standard left-hand railway turnout with a 1:9 crossing (drawn by X. Liu); (**b**) crossing panel on site (shot by V.L. Markine); (**c**) plastic deformation with cracks (shot by X. Liu); (**d**) spalling (shot by X. Liu).

In real life, the wheel–rail contact in a crossing can be affected by many factors. Some factors are related to the train track system, such as train type [17], velocity [18], axle load [18,19], wheel–rail friction [18], crossing geometry [18,19], track alignment [19], track elasticity [12] and so on. Some factors are related to the crossing environment, such as the contaminants on the rail [19–21] and rail temperature variation [22,23]. All these factors, especially those introduced by the environment, make the measurement data noisy and the crossing condition cannot be clearly shown [24]. In order to properly analyze the measurement data for monitoring the crossing condition, the first step is to figure out the influence of the above mentioned factors on the performance of the crossing.

In this study, the influence of train track system-related factors is minimized through data selection and a filtering process. Specifically, train type, velocity and the bogie number are restricted to a certain scope. In order to estimate the influence level of the external factors (such as the weather condition), a correlation analysis using Pearson's correlation coefficient, which is usually applied to quantitatively evaluate the correlation strength between two variables, is performed. The correlation analysis results are verified using a vehicle-crossing model developed using the multi-body system (MBS) method. In this model, the weather changes are modelled according to changes in the properties of the affected track elements. The correlation between the weather condition and the dynamic responses of the crossing provides the foundation for long-term measurement data regression, which will be applied in the crossing degradation assessment procedure. In addition to weather factors, the correlation strengths between the dynamic responses of the crossing are also analyzed, which can be applied to provide guidance for the selection and post-processing of the measurement data and to improve the efficiency of analyzing a large amount of data.

The paper is organized as follows. The condition-monitoring procedure of a railway crossing, including the crossing instrumentation, is presented in Section 2. The indicators applied for the crossing condition assessment are briefly introduced in Section 3. The correlation analysis, including the dynamic responses and weather variables, are given in Section 4. In Section 5, the mechanisms of the weather effects are analyzed and verified through numerical simulation. Finally, in Section 6, the conclusions based on the correlation analysis are provided and further applications for the degradation procedure description of the monitored crossing are discussed.

2. Railway Crossing Condition Monitoring

In this section, monitoring the condition of a railway crossing is discussed. The crossing instrumentation and a brief procedure for processing the measurement data are described.

2.1. Crossing Instrumentation

The monitored crossing in this study is a cast manganese steel crossing with an angle of 1:9, which is the most commonly used crossing for Dutch railway tracks (more than 60% [25]). As part of a double crossover, the crossing is mainly used for through-facing routes (Figure 1a). This railway line is mainly used for passenger transportation with a velocity of passing trains up to 140 km/h. The crossing is instrumental for using the system that has been introduced, and has been actively used in previous studies [1,17,19,26]. An overview of the crossing instrumentation is given in Figure 2.

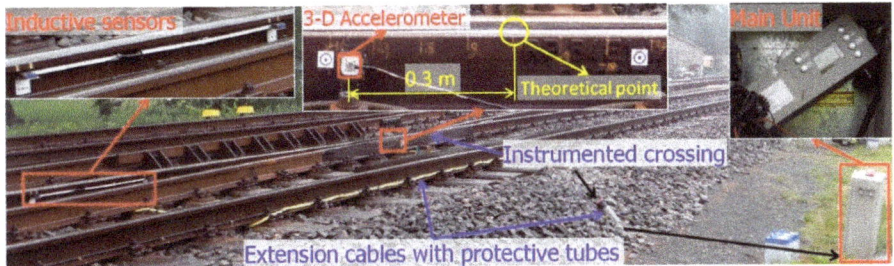

Figure 2. Overview of the crossing instrumentation.

The main components of this device are a 3-D accelerometer attached to the crossing rail, a pair of inductive sensors attached to the rails in the closure panel and the data logger (main unit) installed on the outside of the track. The inductive sensors are used for train detection and the initiation of the measurements, as well as for train velocity determination. All of the sensors are connected to the data logger for data storage and basic analysis of the data. The measurement range and sampling frequency of the acceleration sensor are 500 g and 10 kHz, respectively. The main measured data are the 3-D acceleration responses (i.e., a_x, a_y and a_z)) of the crossing due to the passing trains.

An example of the vertical acceleration response in a time domain due to one passing train with 12 wheelsets is shown in Figure 3a. It can be seen from this figure that the time and location of each wheel's impact on the crossing can easily be obtained from the acceleration responses. The region where most of the wheel impact is located is defined as the fatigue area (Figure 3b), which can be used for assessing crossing conditions based on a large amount of data.

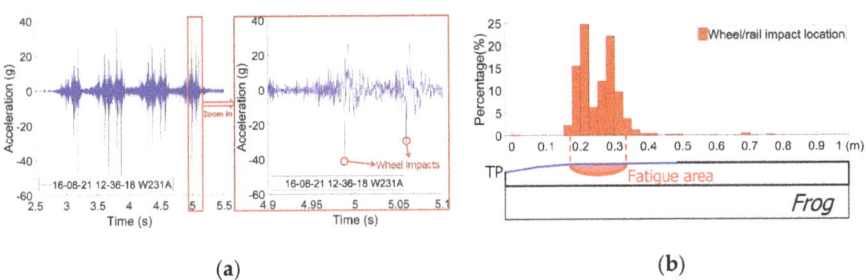

(a)   (b)

Figure 3. Examples of the output of crossing instrumentation. (a) Vertical acceleration response due to one train's passage; (b) wheel impact location distribution.

2.2. Measurement Data Selection and Processing

The crossing monitored in this study was in a new state at the beginning of the observations. In order to reduce the influence of vehicle variations, the measurement results considered here were restricted to one type of train, namely the VIRM (double-deck) trains that pass with a velocity of around 140 km/h. Moreover, the accelerations caused only by the first bogie were considered. Thus, the remaining uncertainties in the measured data mainly coming from the environment (e.g., the weather). Depending on the amount of monitoring data, the measurement results will be analyzed on three different levels, namely,

- the dynamic response due to the passage of a single wheel;
- the results of multiple-wheel passages from one monitoring day; and
- the statistical results from multiple monitoring days.

An example of vertical acceleration responses in different levels is shown in Figure 4.

The response due to single wheel passages was directly obtained from the measured time domain signal (Figure 4a). The distribution of the maximum impact acceleration from each passing wheel constituted the results of multiple wheel passages (Figure 4b). For the statistical results from multiple monitoring days, each point represented the average value of the impact vertical accelerations of the recorded passing wheels from one monitoring day (Figure 4c). It can be seen that each wheel passed the railway crossing differently. Based on a single wheel's passage it is difficult to assess the performance of the crossing. Yet, some conclusions on wheel–rail interaction can still be drawn based on these data. The statistical analysis based on multiple passing wheels was more applicable for assessing the condition of the railway crossing.

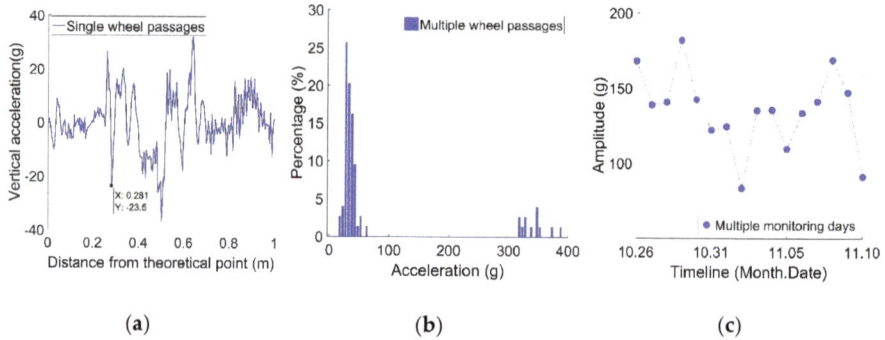

(a)   (b)   (c)

Figure 4. Example of measured vertical acceleration responses: (**a**) from single-wheel passages; (**b**) from multiple-wheel passages from one monitoring day; (**c**) from multiple monitoring days.

3. Condition Indicators

In this section, the indicators for assessing a crossing's condition are briefly described. These indicators are calculated based on the transition region and consist of the irregular contact ratio, 3-D acceleration responses and the fatigue area. To demonstrate the condition analysis procedure, some typical examples of the measurement results from the monitored crossing are presented.

3.1. Transition Region

The transition region of a crossing is the location where the wheel load is transferred from the wing rail to the nose rail (or vice versa, depending on the traveling direction). In practice, the wheel–rail contact points in the crossing can be recognized by looking at the shining band on the rail surface. An example of such a band on the monitored crossing is given in Figure 5 and denoted by the red triangle areas.

Using these bands, the transition region can be then estimated by the overlapping area of the shining bands on the wing rail and nose rail. Based on this image, the transition region of the monitored crossing is located around 0.15–0.40 m as measured from the crossing's theoretical point (TP).

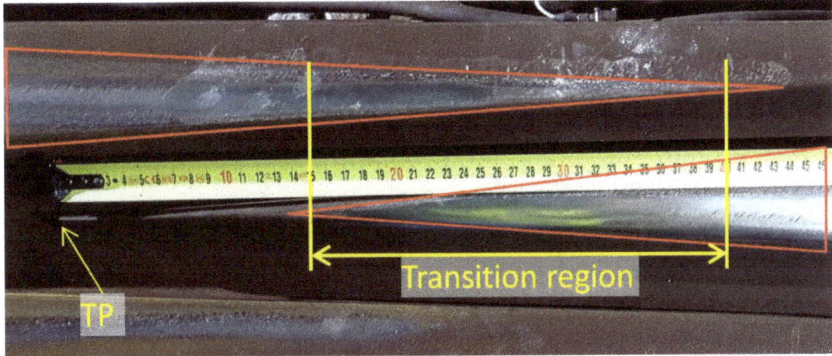

Figure 5. Transition region of the monitored crossing.

From a performance point of view, the transition region is the most vulnerable part of the crossing, since the rail is thinner and the wheel forces are higher than in the other parts of the turnout. Therefore, to analyze the dynamic performance of the crossing, only the accelerations located within the transition region are taken into account.

3.2. Wheel–Rail Impact Status

In an ideal situation, the wheel will pass through the transition region smoothly without flange contact (Figure 6a). In such a case, the vertical acceleration (a_y) will dominate the 3-D acceleration responses. However, in real life, due to disturbances existing in the track, each wheel passes the crossing at a different angle, which results in different impact accelerations in all the three directions. Referring to the measurement results, the impact angle can be defined by the factor of $k = a_z/a_y$. It has been found [1] that when an impact factor exceeds a certain level ($|k| \geq 1$), there is a large chance that the wheel flange will hit the nose rail or wing rail of the crossing (depending on the direction). Such flange contact is recognized as irregular positive (Figure 6b) or negative (Figure 6c) contact.

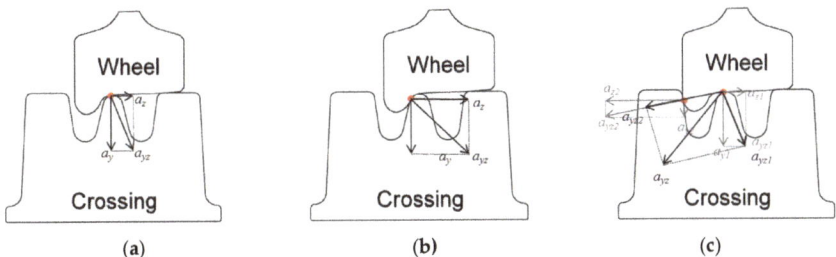

Figure 6. Wheel–rail contact situations: (a) regular contact; (b) irregular positive contact when wheel flange hits the crossing nose; (c) irregular negative contact when wheel flange hits the wing rail.

The irregular contact ratio is usually at a low level (below 3%) for well-maintained crossings, but might dramatically increase when damage occurs to the crossing (above 20%) [1]. Thus, the irregular contact ratio can be applied as a key indicator in assessing the conditions of railway crossings.

3.3. 3-D Acceleration Responses

For the monitored crossing, the regular and irregular contact wheels showed dramatic differences in the 3-D impact acceleration responses (a). For regular passing wheels, the impact vertical acceleration was usually below 50 g, while such impact could be above 300 g for irregular passing wheels. Examples of the 3-D acceleration responses from typical regular and irregular passing wheels are shown in Figures 7 and 8, respectively. In order to better understand the wheel–rail contact, the transition region obtained from field observation (Figure 5) is marked in the figures as a green line on the horizontal axis.

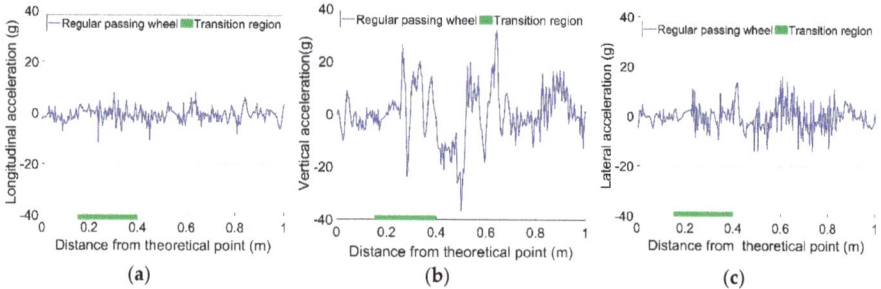

Figure 7. Examples of regular impact acceleration responses due to passing wheels. (**a**) Longitudinal acceleration; (**b**) Vertical acceleration; (**c**) Lateral acceleration.

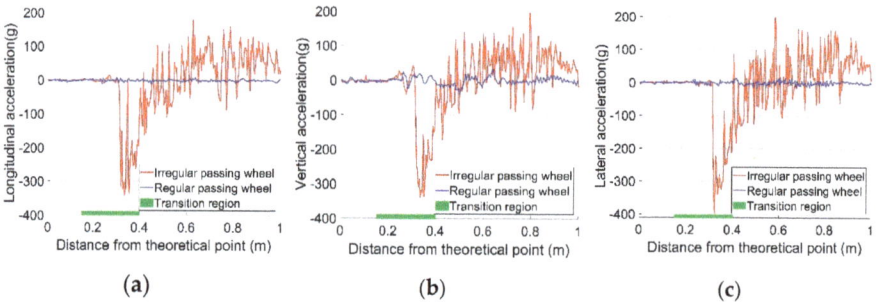

Figure 8. Examples of regular (same as Figure 7) and irregular impact acceleration responses due to passing wheels. (**a**) Longitudinal acceleration; (**b**) Vertical acceleration; (**c**) Lateral acceleration.

It can be seen from Figure 7 that a_y is much higher than a_x for a regular passing wheel, while a_z, meaning that the impact factor (a_z/a_y), is relatively small. It is also indicated that the wheel has two impacts on the crossing, with the first one (22 g) in the transition region and the second one (34 g) after the wheel load transit to the crossing nose rail. Even though the second impact has a higher amplitude, the first one is more damaging, since in the first impact location the nose rail is much thinner than in the second one.

For the irregular passing wheel presented in Figure 8, it can be seen that the impact was located in the transition region, and the accelerations in all three directions were very close to each other (in contrast to the regular passing wheel). Such a strong correlation of the acceleration responses reflects the intense wheel impact on the crossing nose rail and the rough transition of the wheel load from wing rail to the crossing nose rail. The big difference between the two typical wheel–rail impacts gives an example of the violent fluctuation of dynamic response results that can be observed in such crossings.

3.4. Impact Location and Fatigue Area

The impact location is defined as the point where the maximum wheel–rail impact occurs. As described previously, the impact location is restricted within the transition region. For the example given in Figure 4a, the impact location was 0.281 m from the TP.

The fatigue area is defined as the region where most of the wheel impacts are located and is calculated based on multiple wheel passages. In monitoring the conditions of railway crossings, the location and size of the fatigue area reflect the wheel load distribution along the crossing nose. In general, farther impact locations from the TP and wider fatigue areas indicate a better crossing condition.

In practice, to simplify the calculation procedure, the distribution of the wheel impacts due to multiple wheel passages is assumed to be normal distribution, the mean value a is the impact location and the confidence interval $[a - \sigma, a + \sigma]$ is recognized as the fatigue area. An example of the fatigue area of the monitored crossing during a single day is given in Figure 9.

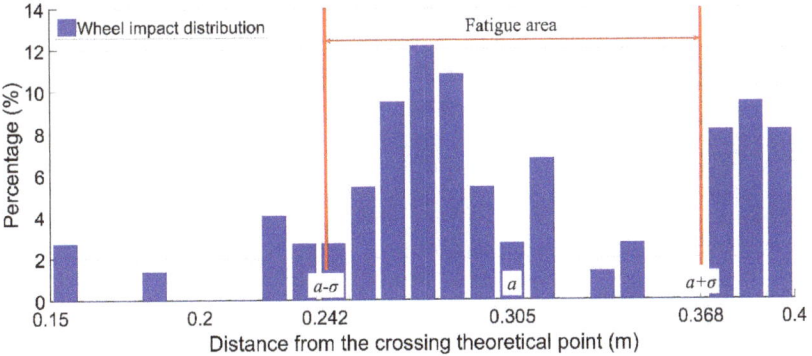

Figure 9. Example of fatigue area calculation.

In this example, the wheel impact location was $a = 0.305$ m, and the standard deviation of the simplified normal distribution was $\sigma = 0.063$ m. Therefore, the fatigue area for the crossing during this monitoring day was between 0.242 and 0.368 m, with a size of 0.126 m. It can be noticed that the calculated fatigue area is not accurate, yet for condition monitoring in the long term, such simplification can provide reasonably acceptable results and highly improve the efficiency of data analysis.

3.5. Results from Multiple Monitoring Days

In order to describe the development of the crossing's condition, the indicators are mainly used as statistical results over multiple monitoring days. An example of the development of vertical crossing acceleration responses as well as an irregular contact ratio is given for a span of 16 days in Figure 10. In this period, no track activities (e.g., maintenance) were performed, and the time frame was relatively too short for the condition of the crossing to degrade; therefore, the crossing condition can be assumed to be stable.

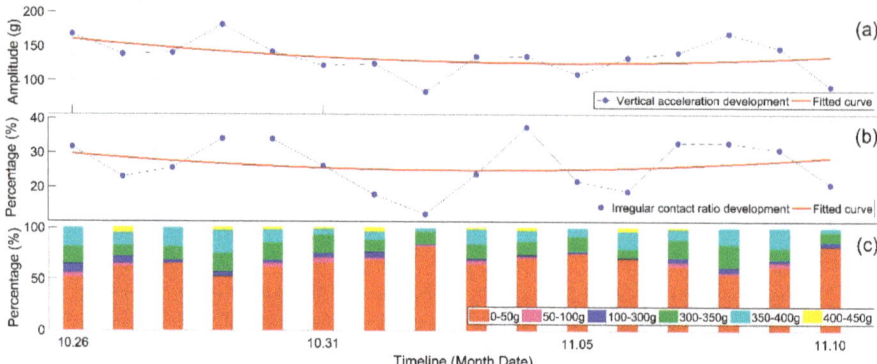

Figure 10. Development of the vertical acceleration responses in the monitored period. (**a**) Mean value of the vertical acceleration; (**b**) irregular contact ratio; (**c**) distribution of the acceleration responses for each day.

From Figure 10a it can be seen that the overall trend of the mean value of the accelerations is relatively stable, while the fluctuations of the responses are quite significant. The vertical accelerations have a minimum value of 84 g and a maximum value of 182 g. Such fluctuations resemble the fluctuations of the irregular contact ratio (Figure 10b). This resemblance will be further studied in the correlation analysis. It should be noted that the irregular contact ratio in the monitored period was above 10%, and for some days even it was higher than 30%, which is much higher than the previously studied 1:15 crossing [1] and reflects the abnormal condition of the monitored 1:9 crossing.

To summarize, the analyzed results have shown the following interesting features:

- a large difference in the dynamic responses from one passing wheel to another;
- a high irregular contact ratio due to multiple wheel passages during a single monitoring day; and
- highly fluctuating acceleration responses, as well as an irregular contact ratio during the short monitoring period.

All these features of the monitored 1:9 crossing indicate quite different performances from the previously studied 1:15 crossing. Investigating the sources of the fluctuation is necessary for a proper assessment of the crossing condition. Also, some condition indicators such as impact acceleration and the irregular contact ratio show possible correlations with each other. Figuring out the relationships between these indicators can help to reduce the amount of the required data, which will improve the efficiency of the post processing of the measurement results. These two questions can be investigated using correlation analysis, which will be presented in the next section.

4. Correlation Analysis

As discussed in the previous section, a high fluctuation was observed in the vertical acceleration responses to the monitored crossing over a short period of time, and was unlikely to be related to structural changes. Considering that the interference factors from the train were minimized by data selection, one possible cause of the fluctuating dynamic responses might have been the continuously changing weather conditions.

4.1. Influence of the Weather

It was discovered in the previous study [24] that temperature variation shows a good correlation with the acceleration fluctuation. In that study, the temperature fluctuation was considered to be the result of the duration of sunshine or precipitation. In order to assess the impact of the weather more accurately, the influences of weather conditions—including mean value of the daily temperature, daily sunshine and precipitation duration—will be analyzed. Figure 11 shows the fluctuation of crossing vertical acceleration responses with varying weather conditions.

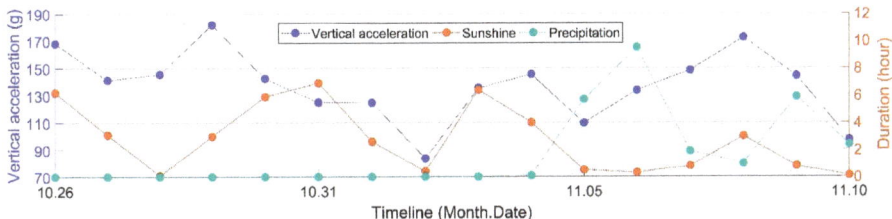

Figure 11. Development of vertical acceleration together with the durations of sunshine and precipitation.

From Figure 11 it can be seen that the fluctuating durations of sunshine showed a similar pattern to the crossing's vertical acceleration responses. There seems to be connection between these two variables. For durations of precipitation, the connection with the vertical acceleration responses was lower. In order to quantitatively assess the impact of the weather, the correlations between the weather variables and condition indicators must be analyzed.

The weather data are obtained from the Royal Dutch Meteorological Institute (KNMI) [27] in days, and mainly consist of the following items:

- sunshine duration per day (D_s); and
- precipitation duration per day (D_p).

The crossing condition indicators were obtained from the crossing instrumentation, and the statistical results based on multiple monitoring days have been applied. The analyzed indicators include the following parts:

- longitudinal, vertical and lateral acceleration responses (a: a_x, a_y and a_z, respectively);
- an irregular contact ratio (I_r); and
- wheel impact location (L_o) and size of the fatigue area (F_a).

4.2. Pearson's Correlation Coefficient

In statistics, the linear correlation between two variables is normally measured using Pearson's correlation coefficient r. For two variables X and Y with the same sample size of n, r can be obtained using the following formula:

$$r_{X,Y} = \frac{\text{cov}(X,Y)}{\sigma_X \sigma_Y} = \frac{E[(X-\mu_X)(Y-\mu_Y)]}{\sigma_X \sigma_Y} = \frac{1}{\sigma_X \sigma_Y} \cdot \frac{1}{n}\sum_{i=1}^{n}[(x_i-\mu_X)(y_i-\mu_Y)] \quad (1)$$

$$X = X(x_1, x_2, \ldots x_n), \ Y = Y(y_1, y_2, \ldots y_n) \quad (2)$$

where

- $\text{cov}(X,Y)$ is the covariance of X and Y;
- σ_X & σ_Y are the standard deviations of X & Y, respectively;
- μ_X & μ_Y are the mean values of X & Y, respectively; and
- $E[\ldots]$ is the expectation of the given variables

When X is in direct/inverse proportion to Y, then the correlation coefficient is

$$r_{X,Y} = \frac{E[(X-\mu_X)(Y-\mu_Y)]}{\sigma_X \sigma_Y} = \pm\frac{\sigma_X \sigma_Y}{\sigma_X \sigma_Y} = \pm 1 \quad (3)$$

If X and Y are independent, then the variable of $(x_i - \mu_X)(y_i - \mu_Y)$ (1) could be a random positive or negative value. In case of a large amount of data ($n \to \infty$),

$$\lim \frac{1}{n} \sum_{i=1}^{n} [(x_i - \mu_X)(y_i - \mu_Y)] = 0 \qquad (4)$$

Therefore, the value range of the correlation coefficient is $r_{X,Y} = [-1, 1]$. $r_{X,Y} = \pm 1$ means that the two variables X and Y are perfectly correlated, and $r_{X,Y} = 0$ means that X and Y have no correlation with each other. Otherwise, X and Y are considered partly correlated.

In different research fields, the gradation of the correlation index may have notable distinctions [28]. In some domains such as medicine and psychology, the requirement of the correlation coefficient—that a strong correlation is defined as $|r| \geq 0.7$—is relatively strict, while in other domains such as politics, $|r| \geq 0.4$ can already be considered a strong correlation. In this study, the structural responses and weather were indirectly associated. The three-level guideline modified from [29] is applied for the correlation strength analysis, as shown in Table 1.

Table 1. The three-level correlation strength guideline.

r	Correlation Strength
$|r| < 0.3$	Weak
$0.3 \leq |r| < 0.5$	Moderate
$0.5 \leq |r| < 1$	Strong

4.3. Correlation Analysis

In the analysis presented here, the correlations between the dynamic responses of the crossing (a, I_r, L_o and F_a) and the weather-related variables (T_m, D_s and D_p) are studied. The data used for the correlation analysis are from 16 monitoring days (the same as in Figure 10, n = 16 in Equation (2)). The correlation within each group of parameters, as well as the cross-correlation between these two groups of parameters, will be analyzed.

The results are presented in Table 2. Nomenclature in the table is presented earlier in Section 4.1. The strong, moderate and weak correlation coefficients are marked with red, blue and black colors, respectively. The correlation results will be analyzed in the different categories presented below.

Table 2. Correlation coefficients for dynamic responses and weather variables.

r	a_x	a_y	a_z	I_r	L_o	F_a	D_s	D_p
a_x	1	0.98	0.99	0.84	−0.30	−0.56	0.43	−0.23
a_y		1	0.99	0.79	−0.37	−0.51	0.36	−0.17
a_z			1	0.85	−0.32	−0.53	0.42	−0.22
I_r				1	−0.09	−0.42	0.40	−0.22
L_o					1	0.36	−0.39	0.14
F_a						1	−0.63	0.38
D_s							1	−0.54
D_p								1

4.3.1. Correlation of the Dynamic Responses

It can be seen from Table 2 it can be seen that the 3-D acceleration responses (a_x, a_y and a_z) are very strongly correlated to each other. The irregular contact ratio (I_r) and the size of fatigue area (F_a) are also strongly correlated with $a(a_x, a_y$ and $a_z)$. It can be noted that the correlations between F_a and a are negative, meaning that the increase of a is usually accompanied with the reduction of F_a.

The correlations of the impact location (L_o) with other dynamic responses are not strong, meaning that L_o is relatively independent from the other dynamic response. Some typical correlation results of the dynamic responses (framed in Table 2) are further discussed below.

The very strong correlations of a_x, a_y and a_z ($r \approx 1$) indicate that the 3-D accelerations are synchronously developed. The correlation between a_y and a_z is demonstrated in Figure 12a. Therefore in practice, it is possible to use the accelerations only in one direction (e.g., a_y) to analyze the crossing behavior, which can help improve the efficiency in processing the measurement data.

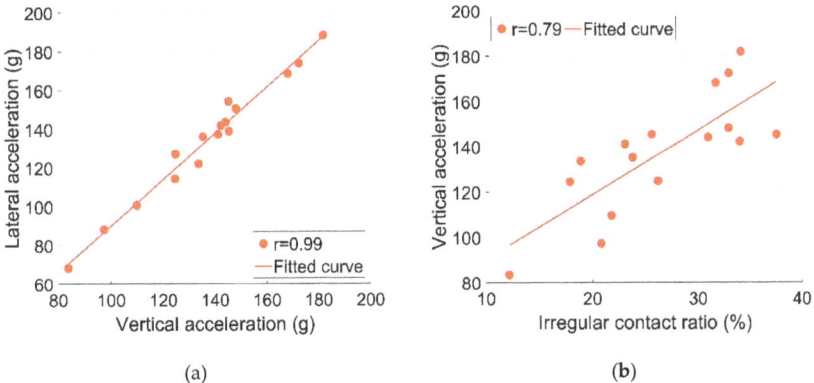

Figure 12. Correlations of the dynamic responses. (**a**) a_y-a_z; (**b**) I_r-a_y.

The strong correlations between I_r and a (Figure 12b) clearly indicate that the high acceleration responses are to a great extent contributed by the high ratio of irregular contact. This phenomenon could have been caused by temporary (not residual) rail displacements due to varying temperature forces in the rail. It has to be noted that all these responses (I_r and a) fluctuated violently, a phenomenon that was likely caused by instable track conditions that were possibly affected by changes in weather conditions. This assumption will be verified later using a numerical model.

Figure 13 shows the correlation between a_y and L_o. The negative result means that when a increased, there was a tendency for L_o to be shifted closer to the crossing's theoretical point, although the moderate correlation strength ($r = -0.37$) indicates that the connection between a and L_o was rather limited. This might have been because the shift of L_o was a long-term effect of rail geometry degradation [1]. However, the rail geometry was unlikely to be changed during the relatively short monitoring period (16 days), so the temporary change of a might not have directly resulted in the shift of L_o.

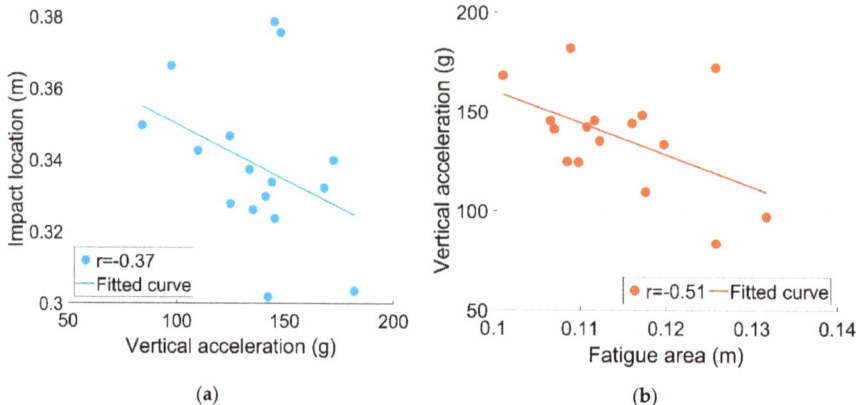

Figure 13. Correlations of the dynamic responses. (**a**) a_y-L_o; (**b**) F_a-a_y.

The correlation between F_a and a_y is shown in Figure 13b. Compared with L_o, F_a was more likely to be reduced due to the increase of a. Combined with the strong correlation between a and I_r, it can be deduced that the impact locations of the irregular contact wheels tended to be centralized, while those of regular contact wheels were decentralized. Such a result confirms that a wider fatigue area will to some extent indicate a better crossing performance.

4.3.2. Correlation of the Weather Conditions

As can be seen from Table 2, the precipitation duration (D_p) had a strongly negative correlation with the sunshine duration (D_s), as shown in Figure 14.

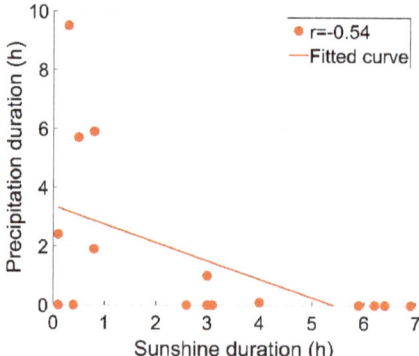

Figure 14. Correlation results between the sunshine and precipitation durations (D_s-D_p).

For the weather variables, D_s and D_p can be considered as two opposite weather conditions. From this point of view, the correlation coefficient of $r = -0.54$ is not very strong. Such results could be explained by the existence of cloudy/overcast conditions, and weather in a single day can switch among sun, rain and clouds/overcast. It can be noticed that in the monitored period, precipitation only occurred in 6 of the 16 days, which to some extent shows the complicity of the weather conditions.

4.3.3. Cross-Correlation between Dynamic Responses and Weather Conditions

According to the correlation results presented in Table 2, the cross-correlations of D_p with the dynamic responses are quite limited, except the moderate correlation with F_a. Meanwhile, D_s was strongly correlated with F_a and moderately correlated with all the other dynamic responses.

The moderate correlation between I_r and D_s is shown in Figure 15a. Such a result can be explained by the fact that an increase in rail temperature due to sunshine causes the displacements in the turnout. Due to these geometrical changes, the wheel cannot pass the crossing normally anymore and results in the increase of the irregular contact. Such a result is consistent with the moderate correlations between D_s and a.

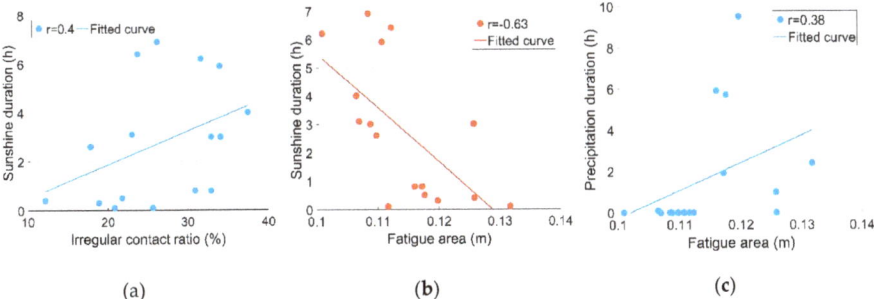

Figure 15. Cross-correlation results between the dynamic responses and weather conditions: (a) I_r-D_s; (b) F_a-D_s; (c) F_a-D_p.

The correlation of D_s with F_a was stronger than with the other dynamic responses ($r = -0.63$, Figure 15b), meaning that sunshine-initiated rail displacements were likely to occur primarily in centralized impact locations, which may have increased the likelihood of irregular contact.

An example for demonstrating the influence of sunshine on the dynamic responses of the monitored crossing is given in Figure 16. In this example, there was hardly any sunshine on one day (11.02), and a long period of sunshine on another day (11.03) (Figure 11). It can be seen that on 11.03 (with sunshine), I_r was higher (Figure 16a) and F_a was slightly narrower (Figure 16b). Such results indicate that the temporary effect of sunshine can lead to changes in the crossing performance.

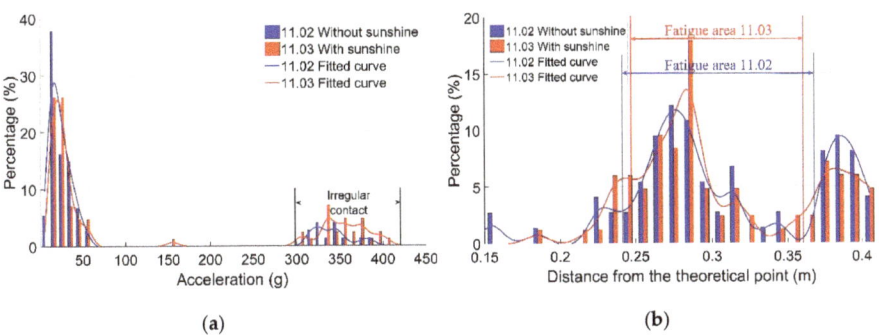

Figure 16. Influence of sunshine on the dynamic responses. (a) Vertical acceleration distribution; (b) fatigue area analysis.

The moderate correlation between D_p and F_a is shown in Figure 15c. Considering that the correlations between D_p and D_s were not very strong, the moderate correlation between the dynamic responses and weather conditions can already indicate a measure of impact. An example of the measured dynamic responses of the crossing for a day without precipitation (11.04) and a day with precipitation (11.05, Figure 11) are shown in Figure 17.

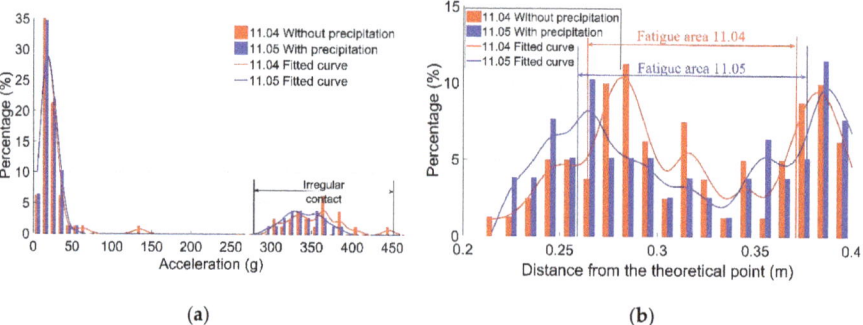

Figure 17. Influence of precipitation on the dynamic responses. (**a**) Vertical acceleration distribution; (**b**) fatigue area analysis.

It can be seen in Figure 17 that on the day with precipitation (11.05), I_r was slightly lower than on the day without precipitation (11.04), and F_a was wider. The reason for such results could be that precipitation may reduce the friction coefficient on the rail's surface and make the transition of the wheel load smoother. This assumption will be verified using a numerical model in the next section.

It should be mentioned that the subgrade of the monitored crossing was relatively soft, with canals on both sides of the track. Persistent precipitation could change the property of the subgrade and further affect the dynamic performance of the track. Therefore, the influence of precipitation can be quite complicated.

Based on the correlation analysis, the main conclusions can be drawn as follows:

- The accelerations in all three directions developed synchronously. In monitoring crossing conditions, it is sufficient to use vertical acceleration to represent the 3-D acceleration responses. Through this, the data processing procedure can be simplified.
- The strong correlation between I_r and a_y indicates that irregular contact is likely to result in high-impact accelerations. Such a result confirms that I_r can be used as an indicator for assessing crossing conditions. A high value of I_r indicates a degraded condition of the monitored crossing.
- The high (moderate/strong) correlation results between D_s and the dynamic responses of the crossing clearly indicate the influence of weather. It can be concluded that significant fluctuations in accelerations during a relatively short period are caused by changes in weather conditions. To verify this, a numerical model will be used in the next section.

5. Numerical Verification

In general, solar radiation is one of the major sources of rail thermal force. Depending on the sunshine duration, the associated rail temperature can rise to 40 °C higher than the ambient air temperature [30]. The change in rail temperature will increase the rail stress and amplify lateral displacements in the rail. The lateral displacements will then increase the uncertainty of the impact angle of a wheel in the railway crossing, eventually leading to an increase in the acceleration responses of some passing wheels, as shown in Figure 10.

Precipitation will introduce water to the rail surface that acts as a lubrication layer, which will reduce the friction coefficient in the wheel–rail interface [21]. It has been studied [31] that a low friction coefficient can be helpful in reducing hunting oscillation and, in contrast to sunshine, can reduce the impact angle of a wheel in the railway crossing.

The above-mentioned effects of temperature and friction variation corresponding to sunshine and precipitation are implemented in the multi-body system (MBS) model described below.

5.1. MBS Model Setup and Validation

In order to verify the weather effect hypotheses, a model for analyzing vehicle-crossing interaction developed according to the MBS method (implemented in VI-Rail software) will be used, as shown in Figure 18a. The track model is a straight line with a crossing panel (Figure 18b) situated in the middle of the track. The total length of the track model is 100 m, which allows enough preloading time for the vehicle before it enters into the crossing panel, as well as enough space after the vehicle passes through the crossing. The crossing geometry is defined by the control cross-sections, and the profiles between two pre-defined cross-sections are automatically interpolated using the third-order spline curve. In the track model, the rail is considered to be lumped masses on the sleepers connected with a massless beam. The flexible layers under the rail are the rail bushing that represents the rail pads and clips, and the base busing representing the ballast bed (Figure 18c).

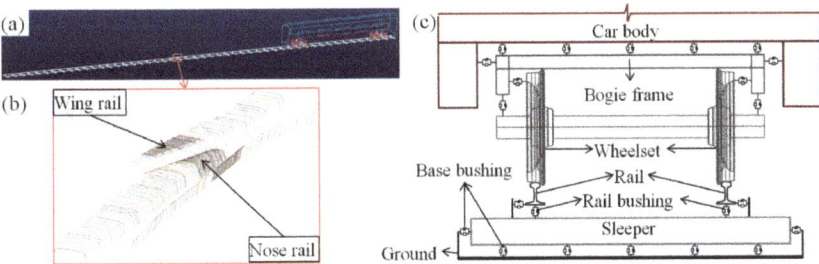

Figure 18. MBS model. (a) Vehicle-track model; (b) crossing profiles; (c) flexible connections in the model.

The crossing model is the same as the monitored 1:9 casted manganese crossing with a rail type of UIC54 E1. The track parameters of Dutch railways [32] applied in the model are shown in Table 3.

Table 3. Track parameters applied in the MBS model.

Track Components		Stiffness, MN/m	Damping, kN·s/m
Rail pad/Clips	Vertical	1300	45
	Lateral	280	580
	Roll	360	390
Ballast		45	32

The vehicle model was developed based on a VIRM locomotive with a total length of 27.5 m comprising a car body, front bogie and rear bogie. In the vehicle model, the car body and bogie frames, as well as the wheel sets, are modelled as rigid bodies with both primary and secondary suspensions taken into account (Figure 18c) [33]. The vehicle travels with a velocity of 140 km/h, the same as in the data analysis measurements. The wheels use a S1002 profile with a wheel load of 10 t. The wheel–rail contact model is defined as the general contact element and uses actual wheel and rail profiles as an input, which allows variable wheel and rail profiles.

The MBS vehicle–track model was validated using the measured acceleration responses from the crossing with the same design and stable conditions. Since the validation simulation was based on ideal track conditions, only the acceleration responses with regular wheel–rail contact were used in the comparison. The selected element for acceleration extraction was the rail with lumped mass (Figure 19a) from the same location as the instrumented accelerometer (Figure 2).

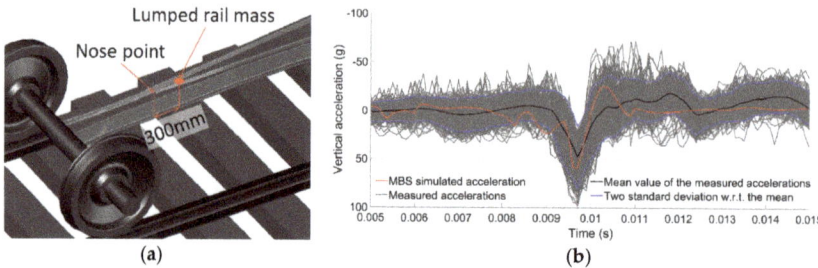

Figure 19. MBS model validation: (**a**) rail element for acceleration extraction; (**b**) comparison of MBS simulated acceleration with the measured ones in a time domain.

The validation results are shown in Figure 19b. It can be seen that the simulation results (red line) are quite comparable with the measured accelerations (black line). The magnitude of the simulated vertical acceleration during impact was around 55 g, which is comparable with the mean value of the measured acceleration responses (47 g). Although tolerable deviations of the impact signals exist, the simulation results agree reasonably with the measurements. It can be concluded that the MBS model can catch the main features of the wheel–rail impact during crossing and can be used to analyze crossing performance. Further details about the numerical model development and validation can be found in [34].

5.2. Numerical Analysis

5.2.1. Effect of Sunshine

In the previous study [35], the displacements of a turnout due to the change of the rail temperature were analyzed using a finite element (FE) model. The simulation results indicated that when the rail temperature was increased (from a stress-free temperature) by 40 °C, the turnout rails were laterally displaced up to 4 mm, as shown in Figure 20a. These results are applied in the MBS vehicle-crossing model as the sunshine-initiated lateral displacements. It should be noted that this simulation is based on ideal track conditions. In the case of a degraded track, the temperature-initiated lateral displacements could be amplified.

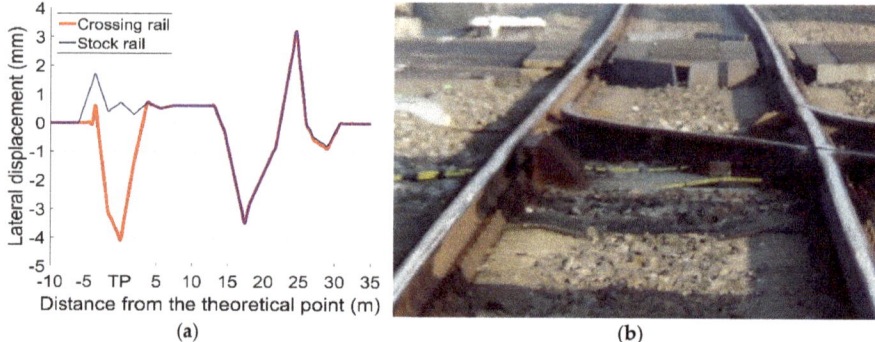

Figure 20. (**a**) Temperature-initiated rail lateral displacement in FE simulation (adapted from Figure 11.15 in [23]); (**b**) the monitored crossing.

In order to take the track degradation into account for the degraded track condition, the input lateral rail displacements in the MBS model are assumed to be twice as high as the ideal track condition (with maximum lateral rail displacements of 8 mm). The effect of precipitation is not taken into account and the friction coefficient of $f = 0.4$ is used. Based on the above assumptions, the vertical accelerations and transition regions of the rail are simulated and presented below.

The calculated transition regions under different track conditions are shown in Figure 21. In the reference condition with no lateral displacement in the track, the sizes of the transition regions for the front wheel and the rear wheel are both 0.031 m [34]. When the temperature-initiated track displacements are taken into account, the transition regions shift closer to the theoretical point and the sizes reduce dramatically to 0.015 m for the front wheel and 0.012 m for the rear wheel. For the degraded track with higher rail displacements, the size of the transition region is only 0.004 m.

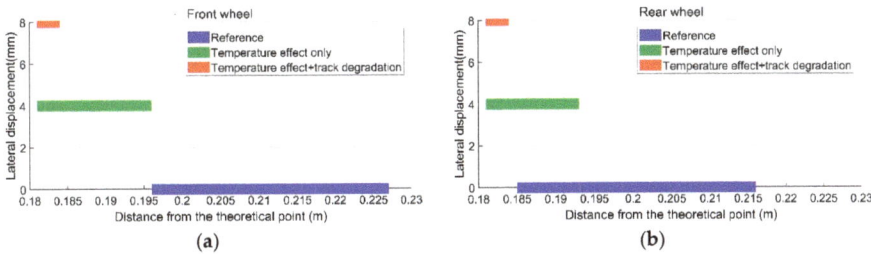

Figure 21. Transition regions of the crossing. (a) Front wheel; (b) rear wheel.

The vertical acceleration response of the rail due to passing wheels is shown in Figure 22. It can be seen that lateral displacement in the rail can result in higher acceleration responses caused by both the front and rear wheels. Combined with the results of the transition region (Figure 21), the simulation results confirm the correlation results (Figure 15a,b) that the long sunshine duration, which will result in a higher temperature in the rail, can lead to a centralized impact location and higher impact acceleration responses at the crossing.

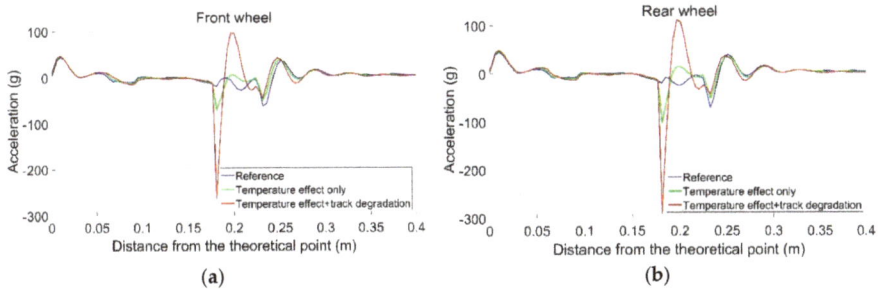

Figure 22. Vertical rail accelerations due to passing wheels. (a) Front wheel; (b) rear wheel.

It can be also seen that with the existence of rail displacement, the acceleration response caused by the rear wheel is higher than that caused by the front wheel from the same bogie. These results indicate that the performance of the rear wheel is not only affected by rail displacement, but also by the passing condition of the front wheel.

In case of a degraded track, higher rail displacements may lead to much higher acceleration responses as a result of both the front and rear wheels. Such impact accelerations (near 300 g) are close to the amplitude of the acceleration responses due to the irregular impacts in the measurements (Figures 16a and 17a). The simulation results prove that the lateral rail displacements caused by increases in rail temperature, in combination with track geometry deviations, can result in high wheel–rail impact accelerations.

5.2.2. Effect of Precipitation

With the influence of precipitation, the friction coefficient (f) in the wheel–rail interface can vary from 0.4 to 0.05 [35]. In this study, the precipitation effect is simulated by a reduction of f. The temperature-initiated rail displacements under ideal track conditions are taken into account. Calculations of rail accelerations resulting from passing wheels are shown in Figure 23.

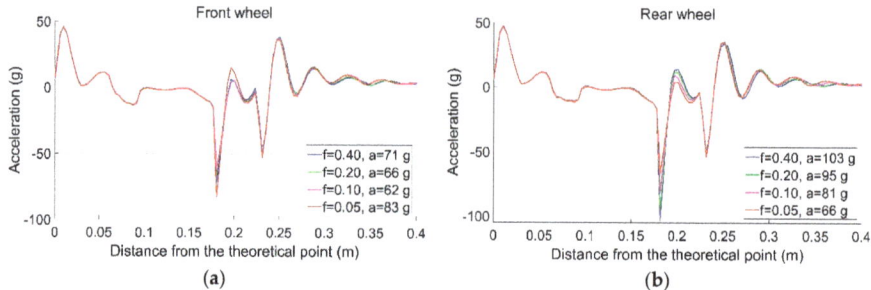

Figure 23. Vertical accelerations of the rail due to the passing wheels. (**a**) Front wheel; (**b**) rear wheel.

For the front wheel, when f is reduced from 0.4 to 0.1, the impact acceleration gradually reduces from 71 to 62 g. However, when $f = 0.05$, the maximum impact acceleration is increased to 83 g. Such results show that reducing the friction coefficient is not always helpful for the dynamic performance of the crossing. For the rear wheel, the reduction of f results in a decreased impact acceleration from 103 to 66 g. As discussed previously, the high rail acceleration responses due to the rear wheel are affected by the movement of the front wheel. In this case, the lowered f can help the wheelset return to a balanced position faster due to lower lateral restraint, which reduces the influence of the front wheelset on the rear wheelset from the same bogie.

It can be concluded that the change of f due to precipitation has an influence on the dynamic performance of the crossing, but the effect of a lower f is not always positive. Such results prove the correlation results indicating that an increase of D_p tends to result in lower acceleration responses, but the correlation strength is not high. The moderate correlation between D_p and F_a is also consistent with the simulation results that each wheel passes through the crossing more independently, which leads to less centralized impact locations.

5.3. Discussion

In this section, the MBS model for vehicle–crossing interaction analysis was briefly introduced. Using this model, the sunshine and precipitation effects were simulated as rail displacements and reduced f in the wheel–rail interface, respectively. The simulation results indicate that the rail displacements due to sunshine can lead to an increase in wheel-crossing impact acceleration. Combined with track degradation, such an effect could be highly amplified. Meanwhile, a lower f in the wheel–rail interface due to precipitation might reduce the interaction effect of two wheelsets from the same bogie, but cannot help improve track conditions. Combined with the measurement results, it can be concluded that the monitored crossing was not in the ideal condition, and possessed a certain degree of track degradation that made it more sensitive to changes in weather conditions.

6. Conclusions and Future Work

6.1. Conclusions

In this study, the conditions of a railway crossing were monitored, and the results were presented. The indicators for assessing the conditions of a crossing were briefly introduced. Inspired by the observed connection between vertical acceleration responses of the crossing and variations in the sunshine duration, correlations of the dynamic responses and weather conditions were calculated. Using the vehicle-crossing MBS model, the influence of weather on the performance of the crossing was verified. The main conclusions of this study can be drawn as follows:

- The strong correlations between the dynamic responses show that the measurement results can be simplified and the crossing conditions can be assessed by only a few indicators (e.g., vertical acceleration, irregular contact ratio and fatigue area).
- The correlation results between the dynamic responses of the crossing and sunshine duration explain the fluctuation of dynamic responses over a short period of time. Such results confirm the temporary influence of weather on the performance of a crossing.
- The correlation results between sunshine duration and precipitation duration, as well as between precipitation duration and the dynamic responses of the crossing, indicate the complexity of the effect of precipitation.
- The simulation results not only verify the impact of weather on the dynamic performance of the crossing, but also indicate that the condition of the track at the monitored crossing was degraded. In cases of track degradation, the influence of weather can be amplified.

In monitoring the conditions of railway crossings, the correlation results among dynamic responses can be used to simplify measurement data. The verified weather effects explain the fluctuation of the dynamic responses over a short time period, which provides the basis for the measurement data regression. It should be noted that although sunshine variation is a short-term effect, the interaction of sunshine with the degraded track can turn this temporary interruption into a permanent track deformation, which will further accelerate the degradation of the track. In monitoring the conditions of railway crossings, the influence of weather can be eliminated through data regression to describe the structural degradation procedure, but the reflected track problem has to draw enough attention. Ensuring good track condition will not only help prolong service life of the crossing, but will also reduce the influence of varying weather conditions.

6.2. Future Work

This study was based on monitoring the conditions of railway crossings. It can be imagined that weather variation might also have an impact on other track sections, especially vulnerable parts such as transition zones, insulated joints, sharp curves, and so on. In the future, the effects of weather on other parts can be further investigated, which will improve the universality of this study and provide broader information for railway track management.

Author Contributions: This article is written by X.L. and supervised by V.L.M.

Funding: This research received no external funding.

Acknowledgments: The first author would like to thank the China Scholarship Council (CSC) for the financial support and Rolf Dollevoet for all the support as the head of Railway Engineering section in TU Delft. Also, the technical support from Ivan Shevtsov (ProRail) and help from Greg Lambert (RailOK. B,V) in the field measurement are highly appreciated.

Conflicts of Interest: The authors declare no conflict of interest.

References

1. Liu, X.; Markine, V.L.; Wang, H.; Shevtsov, I.Y. Experimental tools for railway crossing condition monitoring. *Measurement* **2018**, *129*, 424–435. [CrossRef]
2. Barke, D.; Chiu, W.K. Structural health monitoring in the railway industry: A review. *Struct. Health. Monit.* **2005**, *4*, 81–94. [CrossRef]
3. Bowness, D.; Lock, A.C.; Powrie, W.; Priest, J.A.; Richards, D.J. Monitoring the dynamic displacements of railway track. *Proc. Inst. Mech. Eng. Part F J. Rail Rapid Transit* **2007**, *221*, 13–22. [CrossRef]
4. Chudzikiewicz, A.; Bogacz, R.; Kostrzewskim, M.; Konowrockim, R. Condition monitoring of railway track systems by using acceleration signals on wheelset axle-boxes. *Transport* **2018**, *33*, 555–566. [CrossRef]

5. Sowiński, B. Interrelation between wavelengths of track geometry irregularities and rail vehicle dynamic properties. *Arch. Transp.* **2013**, *25*, 97–108.
6. Tsunashima, H. Condition Monitoring of Railway Tracks from Car-Body Vibration Using a Machine Learning Technique. *Appl. Sci.* **2019**, *9*, 2734. [CrossRef]
7. Wei, Z.; Núñez, A.; Li, Z.; Dollevoet, R. Evaluating degradation at railway crossings using axle box acceleration measurements. *Sensors* **2017**, *17*, 2236.
8. Wei, Z.; Shen, C.; Li, Z.; Dollevoet, R. Wheel–rail impact at crossings-relating dynamic frictional contact to degradation. *J. Comput. Nonlinear Dyn.* **2017**, *12*, 041016. [CrossRef]
9. Ma, Y.; Mashal, A.A.; Markine, V.L. Modelling and experimental validation of dynamic impact in 1:9 railway crossing panel. *Tribol. Int.* **2018**, *118*, 208–226. [CrossRef]
10. Xin, L.; Markine, V.L.; Shevtsov, I.Y. Numerical procedure for fatigue life prediction for railway turnout crossings using explicit finite element approach. *Wear* **2016**, *366*, 167–179. [CrossRef]
11. Kassa, E.; Andersson, C.; Nielsen, J.C.O. Simulation of dynamic interaction between train and railway turnout. *Veh. Syst. Dyn.* **2006**, *44*, 247–258. [CrossRef]
12. Wan, C.; Markine, V.L.; Shevtsov, I.Y. Optimisation of the elastic track properties of turnout crossings. *Proc. Inst. Mech. Eng. Part F J. Rail Rapid Transit* **2016**, *230*, 360–373. [CrossRef]
13. Pålsson, B.A. Optimisation of railway crossing geometry considering a representative set of wheel profiles. *Veh. Syst. Dyn.* **2015**, *53*, 274–301. [CrossRef]
14. Wan, C.; Markine, V.L.; Shevtsov, I.Y. Improvement of vehicle–turnout interaction by optimizing the shape of crossing nose. *Veh. Syst. Dyn.* **2014**, *52*, 1517–1540. [CrossRef]
15. Wan, C.; Markine, V.L.; Dollevoet, R.P.B.J. Robust optimisation of railway crossing geometry. *Veh. Syst. Dyn.* **2016**, *54*, 617–637. [CrossRef]
16. Kassa, E.; Nielsen, J.C.O. Dynamic interaction between train and railway turnout: Full-scale field test and validation of simulation models. *Veh. Syst. Dyn.* **2008**, *46*, 521–534. [CrossRef]
17. Markine, V.L.; Shevtsov, I.Y. An experimental study on crossing nose damage of railway turnouts in the Netherlands. In Proceedings of the 14th International Conference on Civil, Structural and Environmental Engineering Computing, Cagliari, Italy, 3–6 September 2013.
18. Xin, L.; Markine, V.L.; Shevtsov, I.Y. Numerical analysis of the dynamic interaction between wheelset and turnout crossing using explicit finite element method. *Veh. Syst. Dyn.* **2016**, *54*, 301–327. [CrossRef]
19. Wan, C.; Markine, V.L. Parametric study of wheel transitions in railway crossings. *Veh. Syst. Dyn.* **2015**, *53*, 1876–1901. [CrossRef]
20. Arias-Cuevas, O. Low Adhesion in the Wheel-Rail Contact. Ph.D. Thesis, Delft University of Technology, Delft, The Netherlands, 2010.
21. Chen, H.; Ban, T.; Ishida, M.; Nakahara, T. Effect of water temperature on the adhesion between rail and wheel. *Proc. Inst. Mech. Eng. Part J J. Eng. Tribol.* **2006**, *220*, 571–579. [CrossRef]
22. Ertz, M.; Bucher, F. Improved creep force model for wheel/rail contact considering roughness and temperature. *Veh. Syst. Dyn.* **2002**, *37*, 314–325. [CrossRef]
23. Arts, T.M.H. Measuring the Neutral Temperature in Railway Track During Installation and Use. Master's Thesis, Delft University of Technology, Delft, The Netherlands, 2011.
24. Markine, V.L.; Liu, X.; Mashal, A.A.; Ma, Y. Analysis and improvement of railway crossing performance using numerical and experimental approach: Application to 1:9 double crossovers. In Proceedings of the 25th International Symposium on Dynamics of Vehicles on Roads and Tracks, Queensland, Australia, 14–18 August 2017.
25. Wegdam, J.J.J. Expert Tool for Numerical & Experimental Assessment of Turnout Crossing Geometry. Master's Thesis, Delft University of Technology, Delft, The Netherlands, December 2018.
26. Markine, V.L.; Shevtsov, I.Y. Experimental Analysis of the Dynamic Behaviour of Railway Turnouts. In Proceedings of the 11th International Conference on Computational Structures Technology, Dubrobnik, Croatia, 4–7 September 2012.
27. Koninklijk Nederlands Meteorologisch Instituut (KNMI, Royal Dutch Meteorological Institute). 2015. Available online: https://www.knmi.nl/nederland-nu/klimatologie/daggegevens (accessed on 10 November 2018).
28. Akoglu, H. User's guide to correlation coefficients. *Turk. J. Emerg. Med.* **2018**, *18*, 91–93. [CrossRef] [PubMed]
29. Cohen, J. *Statistical Power Analysis for the Behavioral Sciences*, 2nd ed.; New York University: New York, NY, USA, 1988.

30. ProRail. *Instruction of Procedures at Extreme Weather Conditions RLN00165, Version 002*; ProRail: Utrecht, The Netherlands, 2003.
31. Ohyama, T. Tribological studies on adhesion phenomena between wheel and rail at high speeds. *Wear* **1991**, *144*, 263–275. [CrossRef]
32. Hiensch, M.; Nielsen, J.C.O.; Verheijen, E. Rail corrugation in The Netherlands—Measurements and simulations. *Wear* **2002**, *253*, 140–149. [CrossRef]
33. VI-grade GmbH. *VI-Rail 16.0 Documentation*; © 2014 VI-Grade Engineering Software & Services: Marburg, Germany, 2014.
34. Liu, X.; Markine, V.L. Validation and Verification of the MBS Models for the Dynamic Performance Study of Railway Crossings. *Eng. Struct.* (under review).
35. Arts, T.M.H.; Markine, V.L.; Shevtsov, I.Y. Modelling turnout behaviour when achieving a neutral temperature. In Proceedings of the First International Conference on Railway Technology: Research, Development and Maintenance, Las Palmas de Gran Canaria, Spain, 18–20 April 2012.

© 2019 by the authors. Licensee MDPI, Basel, Switzerland. This article is an open access article distributed under the terms and conditions of the Creative Commons Attribution (CC BY) license (http://creativecommons.org/licenses/by/4.0/).

Article

A Novel Monitoring Approach for Train Tracking and Incursion Detection in Underground Structures Based on Ultra-Weak FBG Sensing Array

Qiuming Nan [1], Sheng Li [1,*], Yiqiang Yao [2], Zhengying Li [2], Honghai Wang [1], Lixing Wang [1] and Lizhi Sun [3]

1. National Engineering Laboratory for Fiber Optic Sensing Technology, Wuhan University of Technology, Wuhan 430070, China; nqm0723@whut.edu.cn (Q.N.); wanghh@whut.edu.cn (H.W.); lxwang@whut.edu.cn (L.W.)
2. School of Information Engineering, Wuhan University of Technology, Wuhan 430070, China; yqyao@whut.edu.cn (Y.Y.); zhyli@whut.edu.cn (Z.L.)
3. Department of Civil and Environmental Engineering, University of California, Irvine, CA 92697-2175, USA; lsun@uci.edu
* Correspondence: lisheng@whut.edu.cn

Received: 23 May 2019; Accepted: 12 June 2019; Published: 13 June 2019

Abstract: Tracking operating trains and identifying illegal intruders are two important and critical issues in subway safety management. One challenge is to find a reliable methodology that would enable these two needs to be addressed with high sensitivity and spatial resolution over a long-distance range. This paper proposes a novel monitoring approach based on distributed vibration, which is suitable for both train tracking and incursion detection. For an actual subway system, ultra-weak fiber Bragg grating (FBG) sensing technology was applied to collect the distributed vibration responses from moving trains and intruders. The monitoring data from the subway operation stage were directly utilized to evaluate the feasibility of the proposed method for tracking trains. Moreover, a field simulation experiment was performed to validate the possibility of detecting human intrusion. The results showed that the diagonal signal pattern in the distributed vibration response can be used to reveal the location and speed of the moving loads (e.g., train and intruders). Other train parameters, such as length and the number of compartments, can also be obtained from the vibration responses through cross-correlation and envelope processing. Experimental results in the time and frequency domains within the selected intrusion range indicated that the proposed method can distinguish designed intrusion cases in terms of strength and mode.

Keywords: underground structure safety; train tracking; incursion detection; ultra-weak FBG; distributed vibration; dynamic measurement

1. Introduction

During the last few decades, the construction of urban subways has developed rapidly worldwide and particularly in China. Aiming to ensure the operational safety of subways, a wide range of research effort has been undertaken in the fields of the subway fires [1–3], structural safety [4–7], and perimeter invasion [8–11]. Among these three fields, structural safety monitoring and perimeter intrusion detection are of more concern than fire monitoring due to the diversification of demand. For long-distance monitoring needs, especially for subway tunnels, distributed fiber-optic sensing technology has been widely recognized as the most promising means of addressing complex needs due to its advantages of large-scale monitoring, high sensitivity, and multiplexing capabilities [12]. For instance, the safety monitoring of subway structures based on Brillouin optical time domain reflectometry (BOTDR) technology [13] has been reported [14,15]. In addition to BOTDR-based static measurement, distributed fiber-optic sensors for dynamic measurement [16], especially distributed acoustic sensing (DAS) technology [17,18] have been another research hotspot. The use of DAS technology to ensure the safety of subway operations has also attracted widespread attention for both engineers and researchers.

As one of the main concerns for ensuring the operational safety of the subway, tracking e subway trains occupies an important position in the train operation control system; tracking is directly related to train safety and affects the transportation efficiency of the rail transit. In addition to tracking trains in operation, positioning illegal intruders and preventing the risk caused by such intrusion events—which usually occur during the subway outage periods—is another issue worth noting. For the former, Peng et al. [19] reviewed the shortcomings of conventional train positioning techniques and pioneered investigation of the feasibility of train positioning and speed monitoring through Φ-OTDR technology, in which the spatial resolution of the common optic fiber reaches 20 m. He et al. [20] reported a DAS-based method for condition monitoring of the running train, claiming that the train positioning error was less than 20 m. For human intrusion, Catalano et al. [21,22] reported an incursion detection system for railway security using two types of fiber Bragg grating (FBG) sensors, which is apparently only applicable to a limited protection area due to the restricted multiplexing capacity of FBG. In addition, He et al. [23] presented research on railway perimeter safety based on DAS technology, which has a spatial resolution of only 10 m in the reported application scenarios.

Obviously, the dynamic measurement techniques based on DAS provide feasible detection methods for train tracking and human intrusion. Yet, there are still few research efforts on integrated methods for both train tracking and intrusion detection. Compared with DAS technology using common optic fiber, ultra-weak FBG arrays based on the draw tower [24,25] using sensing optic fiber integrates both advantages of fiber optic point sensors and distributed sensors. This technology is a new way to achieve high-precision, fast and wide coverage distributed measurement. Previous research around this technology has focused more on monitoring strain, temperature or strain-based deformation for the object of interest [26–28]. In addition, a multi-parameter measurement system based on an ultra-weak FBG array with sensitive material was proposed in [29]. However, all this research is still limited to static indicators. In fact, ultra-weak FBG array is also adept at performing dynamic monitoring [30,31] in addition to the above positive characteristics usually witnessed in static measurement. The reports in [16,32–34] revealed that the ultra-weak FBG array can not only be used for both static and dynamic measurements, but also has a higher signal-to-noise ratio (SNR) than that of DAS sensors. It is known that higher SNR often leads to better sensing performances such as higher measurement accuracy, faster response time and simpler detection circuit. Therefore, the ultra-weak FBG array is more suitable than DAS when dealing with distributed vibration and other scenarios requiring high-speed measurement.

To eliminate the need for two separate systems, improve measurement efficiency and reduce overall cost, this paper explored the feasibility of addressing train tracking and human intrusion in subway systems using distributed vibration measurement based on the ultra-weak FBG sensing array. The experimental results of identifying running trains and intruders in an actual subway are reported. The sensing and monitoring principles make up the second part of this paper, followed by the details of the design and field arrangement used to validate the proposed method. Finally, the effectiveness on tracking and detecting the objectives of interest is discussed based on the experimental results represented by the responses of distributed vibration of the ultra-weak FBG array.

2. Sensing and Monitoring Principles

Figure 1 illustrates the distributed vibration sensing principle used to detect distributed vibration generated by moving loads, such as trains, intruders and so on. The phenomenon of light interference caused by the reflection signals of adjacent two ultra-weak FBGs is used to detect the vibration of the object of interest. Here, the ultra-weak FBG [35] is regarded as a mirror, and L represents the distance that causes light interference. In order to ensure a stable optical interference effect and overcome the occurrence of optical interference failure due to the difficulty of matching adjacent ultra-weak FBG caused by, for example, temperature variation drift, ultra-weak FBG in the array uses 3 nm wideband FBG. In addition, since the temperature changes slowly with respect to vibration, the temperature influence is ignored in the demodulation process of the vibration. The spatial resolution of the distributed vibration along the sensing optic fiber is typically determined by the parameter L. The sensitivity and the frequency response of the vibration signal measured by the strain-induced phase variation between two ultra-weak FBGs are improved by the interferometer. Here, Faraday rotating mirrors are utilized in the demodulation process of the ultra-weak FBG array to suppress the polarization effect. Moreover, the 3-by-3 coupler phase demodulation algorithm is used to reconstruct the time domain signal, and restore the phase information of the vibration signal, through which the interrogation of the vibration frequency and amplitude can be realized. Further, the optical time domain reflectometry technique is utilized to achieve vibration localization, and therefore, increasing the length of the optical cable will prolong the sampling interval of the vibration signal and reduce the response bandwidth of the system.

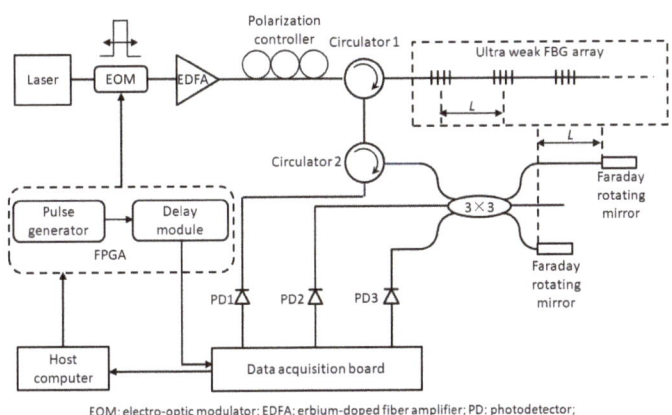

Figure 1. Sensing principle of distributed vibration detection-based on ultra-weak FBG array.

The high sensitivity of large-scale ultra-weak FBGs and the corresponding demodulation system of high speed [36] make the sensing optic fiber particularly suitable for locating structural vibration excited by moving loads occurring within a long-distance range. In addition, the previous study [37] revealed the repeatability of such a sensor represented by strain is around 3.41 nano epsilon. When dealing with train tracking and intrusion detection, either the train or intruder movement can be regarded as a vibration source. Owing to such excitation, the surface waves propagate omni-directionally on the ground. Because the surface wave couples to the track bed and rail track, distributed sensing optic fiber mounted beside the rail track along the subway can detect the vibration generated by a passing train or human footsteps (see Figure 2). The light interference region indicated by the address of ultra-weak FBG can be interrogated with the time- and wavelength-division multiplexing method [38,39], causing each known light interference region along the sensing optic fiber to have a determinate correspondence with the mileage. This also indicates that querying the interference region generated by the distributed vibration excitation is a viable way to track or detect the moving load of interest.

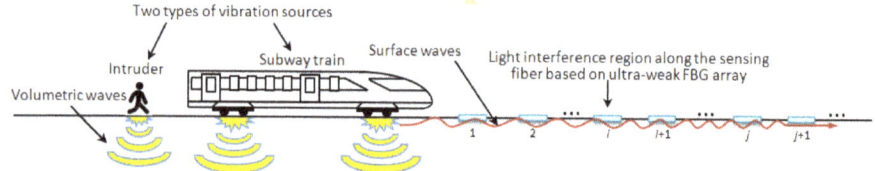

Figure 2. Monitoring principle of capturing the two types of moving loads of interest based on distributed vibration.

Moreover, the speed of the train or intruder can be determined through the τ lag time described in cross-correlation Equation (1) and the known distance between regions i and j as depicted in Figure 2.

$$R_{S_i S_j}(\tau) = \lim_{T \to \infty} \frac{1}{T} \int_0^T S_i(t) S_j(t+\tau) \qquad (1)$$

where $S_i(t)$ and $S_j(t)$ represent the vibration response at light interference regions i and j, respectively. The lag time τ is equal to the duration from the regions i to j. Further, through draw-tower grating preparation with five-meter equidistance between adjoining FBGs along the sensing optic fiber, the spatial resolution of the sensing optic fiber discussed in this paper enabled better positioning accuracy of tracking the train and intruder than that of the above-mentioned reports in actual engineering practice.

3. Design and Field Arrangement for the Experiments

3.1. Engineering Background of the Trial

An actual tunnel structure (Wuhan Metro Line 7) was used in this study. Before the operation of the subway, the ultra-weak FBG sensing optic fiber with armored protection using a layer-stranding structure with a loose tube was installed on structure surfaces of the selected tunnel segments, as in the actual layout shown in Figure 3. Here, the research on identifying the two types of moving loads was primarily based on the track bed response. To better obtain the vibration response of the track bed, three methods for fixing the sensing optic fiber to the track bed were tried to evaluate the suppression effect of the disturbance vibration—namely, fixture fixing, epoxy adhesive and shallow groove embedding. The typical vibration responses of a monitoring zone induced by passing trains in each fixing method are shown on the right side of Figure 3. It can be seen that as the coupling constraint between the sensing array and the track bed increased, the amplitude symmetry of the vibration response improved, and the peak regularity of the vibration response associated with train excitation became clear. Therefore, shallow groove embedding was adopted to affix the sensing array to secure a better signal output.

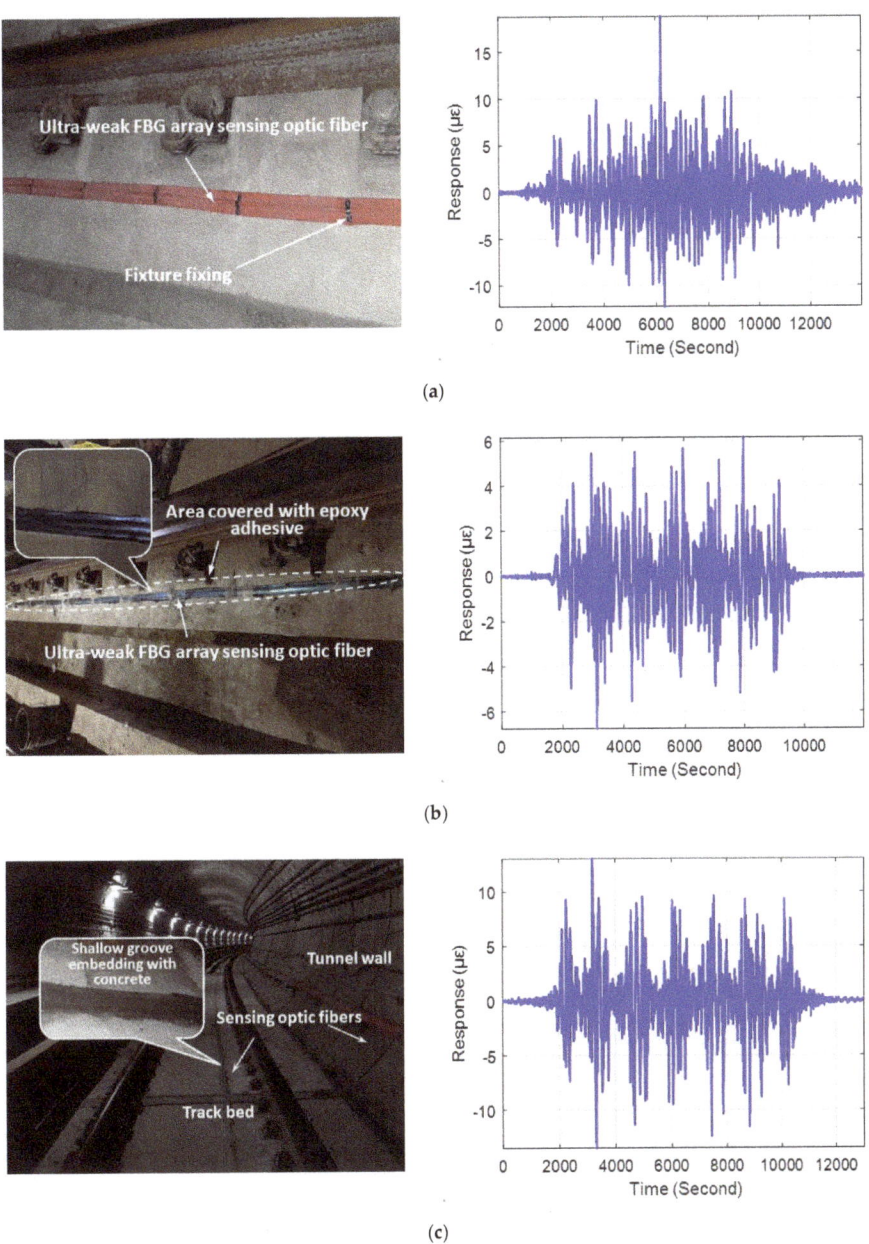

Figure 3. Methods of affixing distributed vibration sensing optic fibers and typical vibration response induced by train: (**a**) fixture fixing, (**b**) epoxy adhesive, and (**c**) shallow groove embedding.

The designed monitoring system can guarantee five kilometers array length and reach multiplexing capacity of 1000 ultra-weak FBGs. As shown in the schematic diagram in Figure 4, the experimental area covered three underground stations with a total length of nearly three kilometers. Due to the spatial resolution of the sensing optic fiber and the specific layout of the tunnel structure, more than

500 vibration regions labeled #1 to #515 along the track bed can be distinguished based on the interrogated address of the light interference. It can be seen from the right side of Figure 4 that in addition to the common track bed structure, the damping track bed was also included in the experimental area. During the trial, the real-time vibration responses with a 1 kHz sampling rate were fully transmitted back to the platform monitoring center and processed by the demodulator and servers. Since the ultra-weak FBG array was fabricated simultaneously in the optic fiber drawing, there was no additional splice in the sensing optic fiber equipped with armored protection, except for the pigtail that needed to be connected to the demodulation instrument.

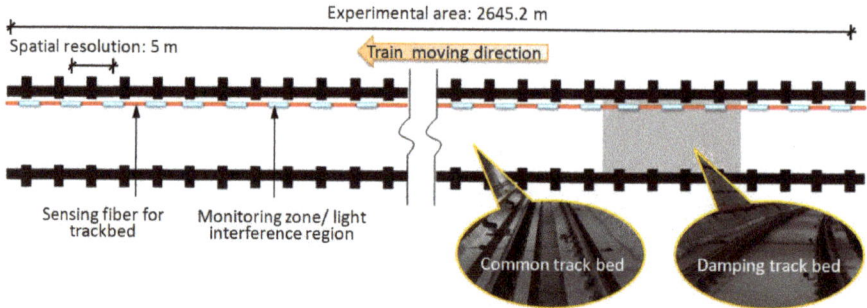

Figure 4. Schematic diagram of experimental area covering different track bed structures.

3.2. Train Tracking Trial

Because the subway line has already been in operation, the distributed vibration responses of the experimental area caused by the train were automatically collected and directly taken as the raw data for the trial. In addition to observing the response caused by the train traveling in the subway tunnel monitored by the sensing optic fiber, the test discussed the identifiability of the sensing optic fiber to the train moving in the opposite direction in the adjacent tunnel. Based on the single point response and overall distribution characteristics, the detection capabilities of the following indicators were discussed in turn: the speed and position of the train, the response difference between the common and damping tracks, and other parameters of the train.

3.3. Intruder Detection Trial

To ensure the safe operation of the subway in the following day, various manual inspections are usually carried out at the subway outage in the early hours of the morning. We conducted the incursion test at this inspection window; this is also the period in which illegal intrusion usually occurs. Based on the specific circumstances and various scheduled tasks, a range in the area of the damping track bed was approved for performing multiple sets of simulated intrusion tests. To minimize cross interference from other simultaneous inspections, the trials were primarily concentrated within a 130 m range of the selected tunnel area. As shown in Figure 5, the trials simulated single and multi-person intrusion scenarios and considered the intrusion patterns of walking and jogging. In each trial, the participant in the simulation test made a round trip within the designed intrusion area.

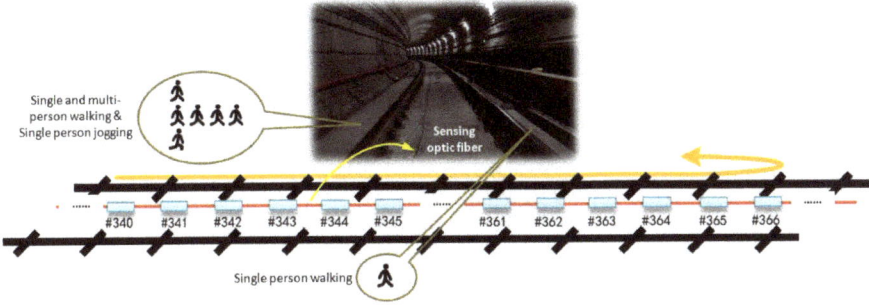

Figure 5. Simulated human intrusion scenarios along the rail track in the selected tunnel area.

4. Results and Discussion

This section reports the characteristics of distributed vibration responses under operating train and simulated incursion conditions, respectively. Feasibility, based on the proposed comprehensive approach concerning train tracking and detecting incursion, was investigated and is discussed. All the following analyses were based on the original output of the distributed vibration responses with no additional techniques adopted to improve the data quality.

4.1. Distributed Vibration Responses under Load of Passing Train

Figure 6a depicts a typical visualization relationship between the structural vibration intensity and the space and time. Here, the vibration intensity was represented by color of the figure. A waterfall diagram such as Figure 6a can be used to help analyze the train's running direction, speed change and arrival or departure interval. In the tunnel where the sensing optic fiber arrays were deployed, the train entered the experimental area from #500 monitoring zone in Figure 6a. In this case, a moving train appeared as a diagonal signal pattern where the slope depended on the speed. Here, the diagonal signal pattern highlighted the characteristics of the distributed vibration response caused by the passing train within the monitoring range. Three complete sets of such diagonal signal patterns can be clearly seen in the left part of Figure 6a. Moreover, vibrations generated by moving trains in the opposite direction in the adjacent tunnel were simultaneously detected by the sensing optic fiber, although the vibration intensity was somewhat weak. Further, the process of the train stopping at the station between the diagonal signal patterns can also be observed in the figure. For monitoring zones #250–#500, the region range of the damping track bed structure, can be clearly identified based on the height changes (along the time axis) of the diagonal signal pattern. Due to the large design distance between the tunnel where the experimental areas #250–#500 were located and the adjacent tunnel, the vibration transmitted from the adjacent tunnel becomes invisible in the right part of Figure 6a.

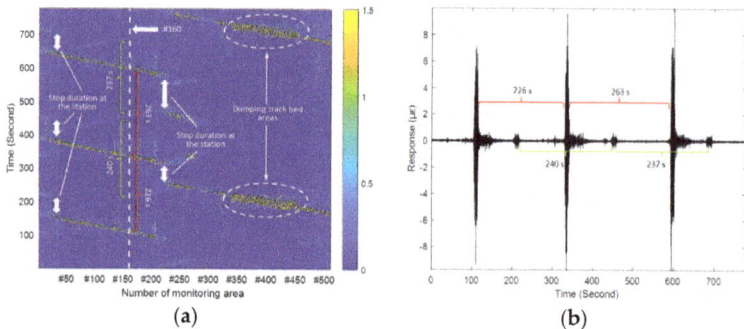

Figure 6. (**a**) Vibration intensity versus space and time under operating train; (**b**) original time series of vibrations of monitoring zone #160.

Figure 6b presents a complete vibration response of one monitoring zone in Figure 6a, which quantifies the difference in vibration intensity of the sensing optic fiber caused by the train moving in two adjacent tunnels. In addition, the time interval of the two adjacent trains was approximately four minutes as observed in Figure 6, which was consistent with the planned subway operating timetable. Moreover, compared with the report based on Φ-OTDR [18], the method using sensing optic fiber in this paper did not require the multiple averaging technique to improve the SNR of the original time series of vibrations, which was more efficient for providing location information of the detected object. Therefore, the responses of any two different monitoring zones could be used to determine the train speed. For instance, Figure 7a shows the intensity projections of the two measurement areas on the time axis of Figure 6a, the cross-correlation analysis (see Figure 7b) of the vibration sequences (see Figure 7a) of the two monitoring zones at 650 m apart indicated that the train took 37.51 s to pass through the two selected zones. In this way, the train speed of 62 km/h can be obtained. Moreover, it was found that the amplitudes of these two monitoring areas were different, although the sensing optic fiber and its fixation method were consistent. The reason for this was primarily due to uneven geological properties of the underground structure along the mileage direction of the tunnel and different design curvature along the tunnel line, and the structural stiffness of the shield segments.

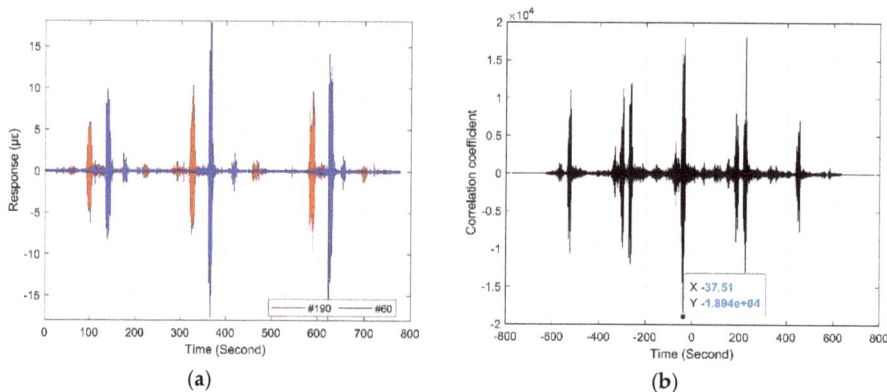

Figure 7. (**a**) Original time series of vibrations at zones #60 and #190; (**b**) time lag between zones #60 and #190.

In addition, the vibration response obtained by a particular monitoring area during the passage of the train can reflect some geometric parameters of the train, such as its length and the number of compartments. The former, length, can be estimated by the calculated speed and the known height changes of the diagonal signal pattern. The latter, number of compartments, can be revealed by the number of peaks or valleys of the envelope signal. Figure 8 shows a typical vibration response of a monitoring zone between #60 and #190 during train passage. The vibration response lasted for about 8.5 s, corresponding to the height change of the diagonal pattern shown in Figure 6a. Based on the obtained average speed of 62 km/h, the calculated train length of 146 m was close to the actual known 142 m. Also, seven envelope peaks and valleys can be recognized in Figure 8 by envelope processing. This envelope result agreed well with the axle impact of the six train compartments.

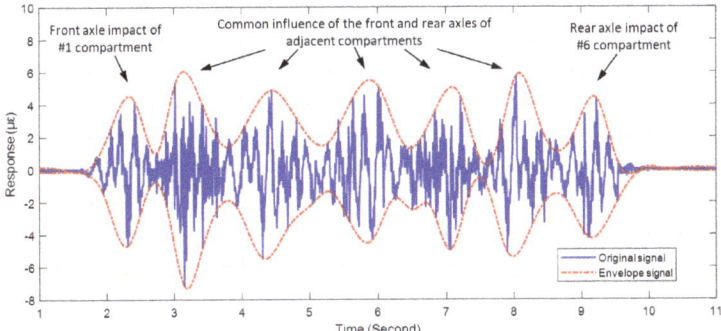

Figure 8. Original time series of vibration and corresponding envelopes during the passage of the train through one monitoring zone.

4.2. Distributed Vibration Response under Footsteps of Intruder

Figures 9 and 10 show the results of the designed human intrusion in the time and frequency domains, respectively. Moreover, the experimental results in the frequency domain for each of the designed cases are depicted two-dimensionally (left) and three-dimensionally (right) in Figure 10. Figure 9 reveals that significant distributed vibration responses generated by walking or jogging as defined in Figure 5 can be detected within the incursion range under both sides of the track. In addition, two diagonal signal patterns in the opposite direction further verified the simulated incursion process represented by round-trip walking or jogging. Furthermore, based on the different slope pattern caused by different speeds of the intruder, it was easy to distinguish the intrusion mode of jogging shown in Figure 9c from the other three intrusion modes of walking. This result was consistent with Figure 10 and was more pronounced in the frequency domain, where the intrusion caused by the jogging shown in Figure 10c led to the maximum fluctuation of the vibration intensity.

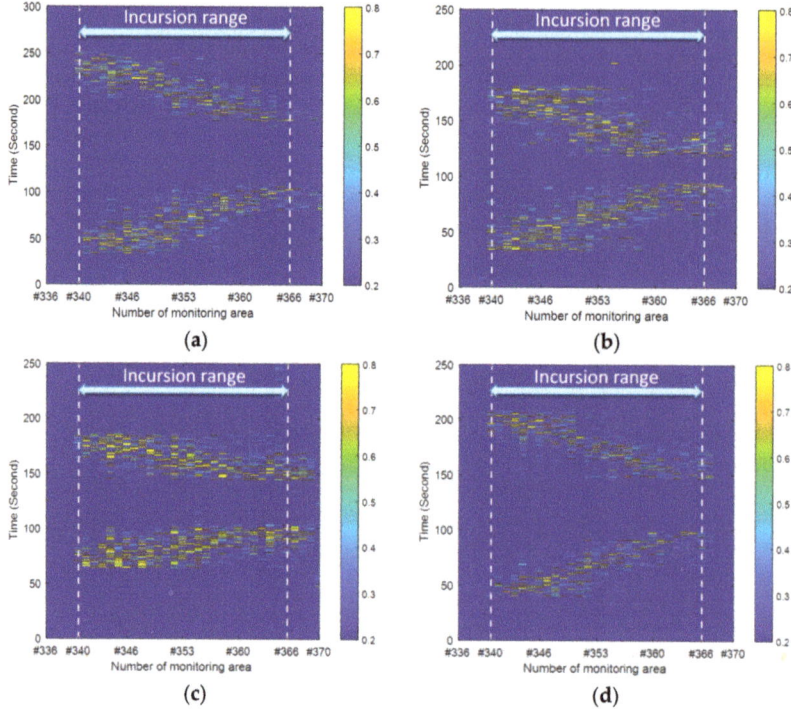

Figure 9. Vibration intensity versus space and time under different intrusion scenarios: (**a**) single person walking; (**b**) four people walking; (**c**) single person jogging; (**d**) single person walking along the other side.

To further quantify the different intrusion patterns reflected in Figure 9, the effective value represented by the root-mean-square (RMS) of the vibration response signal for each monitoring zone within the intrusion range in the whole test process was calculated and is shown in Figure 11. Here, the effects caused by personnel in the round-trip process outside the incursion range were not involved in the evaluation. As can be seen from the overall distribution of Figure 11, in addition to the significant difference between jogging and walking intrusion, the dynamic distributed vibration response can distinguish between single and multi-person walking intrusions. Moreover, subtle differences of vibration distribution caused by a single pedestrian intrusion at different distances from the sensing optic fiber can also be observed. Furthermore, Figure 12 quantifies the results represented in Figure 10 by the overall distribution of primary frequency. Here, the frequency value corresponding to the maximum energy of each column represented in Figure 10 was selected as the primary frequency for each monitoring zone. Table 1 further provides the statistical results for the four types of intrusion cases for Figure 12, where cases 1–4 represent a single person walking, four people walking, single person jogging and single person walking along the other side, respectively.

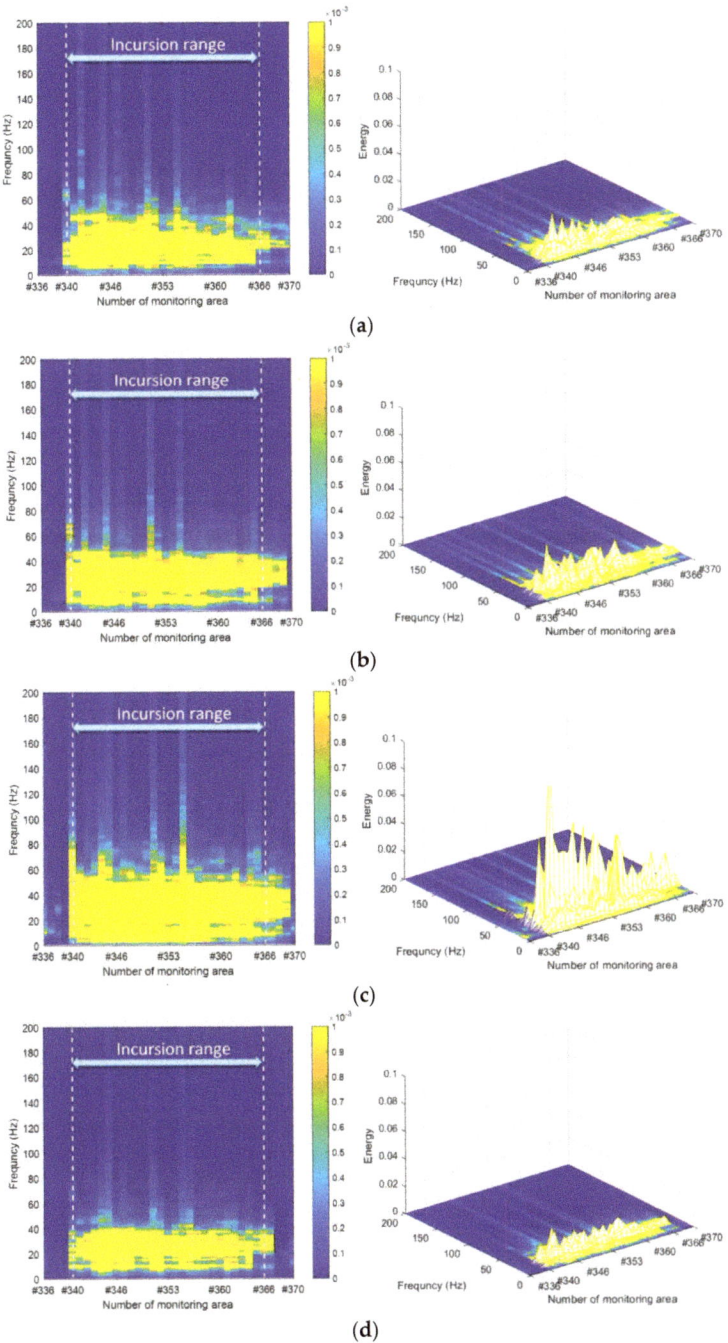

Figure 10. Vibration intensity versus space and frequency under different intrusion scenarios: (**a**) single person walking; (**b**) four people walking; (**c**) single person jogging; (**d**) single person walking along the other side.

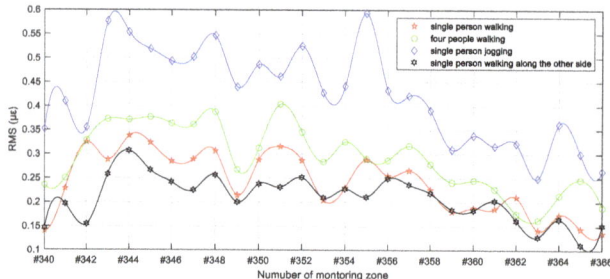

Figure 11. Fitting distribution of effective values of vibration response of incursion range under different simulated intrusion cases.

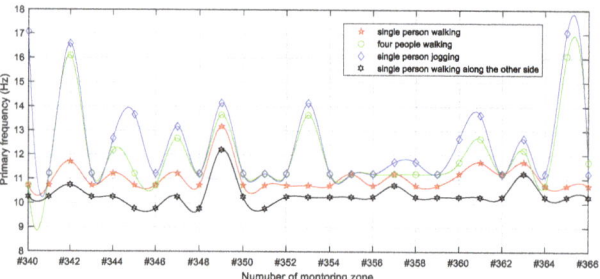

Figure 12. Fitting distribution of primary frequencies of vibration response of incursion range under different simulated intrusion cases.

Table 1. Primary frequency characteristics of the different intrusion cases within the experimental area (unit: Hz).

Comparisons	Case 1	Case 2	Case 3	Case 4
Maximum	13.18	16.11	17.09	12.21
Mode [1]	10.74	11.23	11.23	10.25

[1] The most frequent primary frequency in the monitoring zones of the incursion range.

Since the dynamic characteristics of the structure within the incursion range and the forced vibration mode related to intrusion load frequency and type were different, it can be seen from Figure 12 that the primary frequencies excited by the simulated intrusion were different in the incursion range. However, the similarity of the distribution features in the different intrusion cases shown in Figure 12 can still be observed. That is, the distribution patterns of cases 1 and 4 were closer due to single pedestrian intrusion, while cases 2 and 3 exhibited more broad frequency information under stronger and more complex excitations. The calculated result of the mode values of primary frequency under each case shown in Table 1 further verified this opinion. In addition to the distribution feature, different maximum primary frequencies shown in Table 1, and varied fluctuation strength in Figure 12, also contributed to distinguishing different simulated human intrusions based on the frequency domain results of dynamic distributed sensing of ultra-weak FBG.

5. Conclusions

This study reported an integrated monitoring technology used for ensuring the safety of subway operation, which verified that dynamic distributed measurement based on ultra-weak FBG was a feasible method, suitable for both train tracking and human intrusion detection in an actual engineering application. The analysis based on subway operation monitoring illustrated that the location, speed, length, and number of train compartments could be determined through the vibration responses and distribution on the track bed. Moreover, the results of the simulated human intrusion performed in the damping track bed area during the subway outage period demonstrated that the sensing optic fiber had the potential to distinguish the strength and pattern of intruders. In view of the available test time and experimental range, the simulated cases of human intrusion were relatively limited and the detection effectiveness in the common track bed was not taken into account; this seems to be less than complete and deserves further attention when conditions permit. However, the advantages determined by the high SNR of ultra-weak FBG, when compared to other distributed sensing technologies based on common optic fiber, make us believe that the proposed method is promising for recovering and identifying signals in more complex modes.

Author Contributions: Data curation, S.L.; Formal analysis, S.L.; Funding acquisition, S.L.; Investigation, Q.N.; Methodology, Z.L., H.W., and L.W.; Project administration, Z.L. and H.W.; Resources, Z.L.; Supervision, L.W. and L.S.; Writing—original draft, Q.N., S.L., and Y.Y.; Writing—review and editing, L.S.

Funding: This research was funded by the National Natural Science Foundation of China, grant numbers 61735013 and 61875155, and the Fundamental Research Funds for the Central Universities (WUT: 2019-III-160CG).

Acknowledgments: The research reported in this paper was supported by the National Engineering Laboratory for Fiber Optic Sensing Technology, Wuhan University of Technology, and the Smart Nanocomposites Laboratory, University of California, Irvine.

Conflicts of Interest: The authors declare no conflicts of interest.

References

1. Nordmark, A. Fire and life safety for underground facilities: Present status of fire and life safety principles related to underground facilities. *Tunn. Undergr. Space Technol* **1998**, *13*, 217–269. [CrossRef]
2. Liu, Z.; Kim, A.K. Review of recent developments in fire detection technologies. *J. Fire Prot. Eng.* **2003**, *13*, 129–151. [CrossRef]
3. Jiang, D.; Zhou, C.; Yang, M.; Li, S.; Wang, H. Research on optic fiber sensing engineering technology. In Proceedings of the 22nd International Conference on Optical Fiber Sensors, Beijing, China, 15–19 October 2012.
4. Pamukcu, S.; Cheng, L.; Pervizpour, M. Chapter 1—Introduction and overview of underground sensing for sustainable response. In *Underground Sensing. Monitoring and Hazard Detection for Environment and Infrastructure*, 1st ed.; Pamukcu, S., Cheng, L., Eds.; Academic Press: Cambridge, MA, USA, 2018; pp. 1–42.
5. Soga, K.; Kechavarzi, C.; Pelecanos, L.; Battista, N.; Williamson, M.; Gue, C.Y.; Murro, V.D.; Elshafie, M.; Monzón-Hernández, D., Sr.; Bustos, E.; et al. Chapter 6—Fiber-optic underground sensor networks. In *Underground Sensing. Monitoring and Hazard Detection for Environment and Infrastructure*, 1st ed.; Pamukcu, S., Cheng, L., Eds.; Academic Press: Cambridge, MA, USA, 2018; pp. 287–356.
6. Chen, X.; Li, X.; Zhu, H. Condition evaluation of urban metro shield tunnels in Shanghai through multiple indicators multiple causes model combined with multiple regression method. *Undergr. Space Technol.* **2019**, *85*, 170–181. [CrossRef]
7. Das, S.; Saha, P. A review of some advanced sensors used for health diagnosis of civil engineering structures. *Meas. J. Int. Meas. Confed.* **2018**, *129*, 68–90. [CrossRef]
8. Huang, J.; Zhang, W.; Huang, W.; Huang, W.; Wang, L.; Luo, Y. High-resolution fiber optic seismic sensor array for intrusion detection of subway tunnel. In Proceedings of the 2018 Asia Communications and Photonics Conference (ACP), Hangzhou, China, 26–29 October 2018.
9. Yu, Z.; Liu, F.; Yuan, Y.; Li, S.; Li, Z. Signal processing for time domain wavelengths of ultra-weak FBGs array in perimeter security monitoring based on spark streaming. *Sensors* **2018**, *18*, 2937. [CrossRef]
10. Hoentsch, J.; Scholz, S. Reliable detection of security camera misalignment in urban rail environments. In Proceedings of the 2018 Joint Rail Conference, JRC 2018, Pittsburgh, PA, USA, 19–20 April 2018.
11. Catalano, A.; Bruno, F.A.; Galliano, C.; Pisco, M.; Persiano, G.V.; Gutolo, A.; Cusano, A. An optical fiber intrusion detection system for railway security. *Sens. Actuators A Phys.* **2017**, *253*, 91–100. [CrossRef]
12. Esmailzadeh Noghani, F.; Tofighi, S.; Pishbin, N.; Bahrampour, A.R. Fast and high spatial resolution distributed optical fiber sensor. *Opt. Laser Technol.* **2019**, *115*, 277–288. [CrossRef]

13. Motil, A.; Bergman, A.; Tur, M. State of the art of Brillouin fiber-optic distributed sensing. *Opt. Laser Technol.* **2016**, *78*, 81–103. [CrossRef]
14. Hong, C.; Zhang, Y.; Li, G.; Zhang, M.; Liu, Z. Recent progress of using Brillouin distributed fiber optic sensors for geotechnical health monitoring. *Sens. Actuators A Phys.* **2017**, *258*, 131–145. [CrossRef]
15. Gong, H.; Kizil, M.S.; Chen, Z.; Amanzadeh, M.; Yang, B.; Aminossadati, S.M. Advances in fibre optic based geotechnical monitoring systems for underground excavations. *Int. J. Min. Sci. Technol.* **2017**, *29*, 229–238. [CrossRef]
16. Liu, X.; Jin, B.; Bai, Q.; Wang, Y.; Wang, D.; Wang, Y. Distributed fiber-optic sensors for vibration detection. *Sensors* **2016**, *16*, 1164. [CrossRef]
17. He, Z.; Liu, Q.; Fan, X.; Chen, D.; Wang, S.; Yang, G. A review on advances in fiber-optic distributed acoustic sensors (DAS). In Proceedings of the Conference on Lasers and Electro-Optics/Pacific Rim, CLEOPR 2018, Hong Kong, China, 29 July–3 August 2018.
18. Muanenda, Y. Recent advances in distributed acoustic sensing based on phase-sensitive optical time domain reflectometry. *J. Sens.* **2018**, *2018*, 3897873. [CrossRef]
19. Peng, F.; Duan, N.; Rao, Y.; Li, J. Real-time position and speed monitoring of trains using phase-sensitive OTDR. *IEEE Photonics Technol. Lett.* **2014**, *26*, 2055–2057. [CrossRef]
20. He, M.; Feng, L.; Zhao, D. Application of distributed acoustic sensor technology in train running condition monitoring of the heavy-haul railway. *Optik* **2019**, *181*, 343–350. [CrossRef]
21. Catalano, A.; Bruno, F.A.; Pisco, M.; Cutolo, A.; Cusano, A. An intrusion detection system based on the optical fiber technology for the protection of railway assets. In Proceedings of the 18th Conference on Sensors and Microsystems, AISEM 2015, Trento, Italy, 3–5 February 2015.
22. Catalano, A.; Bruno, F.A.; Pisco, M.; Cutolo, A.; Cusano, A. An intrusion detection system for the protection of railway assets using fiber Bragg grating sensors. *Sensors* **2014**, *14*, 18268–18285. [CrossRef]
23. He, Z.; Liu, Q.; Fan, X.; Chen, D.; Wang, S.; Yang, G. Fiber-optic distributed acoustic sensors (DAS) and applications in railway perimeter security. In Proceedings of the SPIE—The International Society for Optical Engineering, Beijing, China, 11–13 October 2018.
24. Yang, M.; Bai, W.; Guo, H.; Wen, H.; Yu, H.; Jiang, D. Huge capacity fiber-optic sensing network based on ultra-weak draw tower gratings. *Photonic Sens.* **2016**, *6*, 26–41. [CrossRef]
25. Yang, M.; Li, C.; Mei, Z.; Tang, J.; Guo, H.; Jiang, D. Thousand of fiber grating sensor array based on draw tower: A new platform for fiber-optic sensing. In Proceedings of the Optical Fiber Sensors, OFS 2018, Lausanne, Switzerland, 24–28 September 2018.
26. Bartelt, H.; Schuster, K.; Unger, S.; Chojetzki, C.; Rothhardt, M.; Latka, I. Single-pulse fiber Bragg gratings and specific coatings for use at elevated temperatures. *Appl. Opt.* **2007**, *46*, 3417–3424. [CrossRef]
27. Ecke, W.; Schmitt, M.W.; Shieh, Y.; Lindner, E.; Willsch, R. Continuous pressure and temperature monitoring in fast rotating paper machine rolls using optical FBG sensor technology. In Proceedings of the 22nd International Conference on Optical Fiber Sensors, Beijing, China, 15–19 October 2012.
28. Yao, Y.; Li, S.; Li, Z. Structural cracks detection based on distributed weak FBG. In Proceedings of the Optical Fiber Sensors, OFS 2018, Lausanne, Switzerland, 24–28 September 2018.
29. Bai, W.; Yang, M.; Hu, C.; Dai, J.; Zhong, X.; Huang, S.; Wang, G. Ultra-weak fiber Bragg grating sensing network coated with sensitive material for multi-parameter measurements. *Sensors* **2017**, *17*, 1509. [CrossRef]
30. Zhou, L.; Li, Z.; Xiang, N.; Bao, X. High-speed demodulation of weak fiber Bragg gratings based on microwave photonics and chromatic dispersion. *Opt. Lett.* **2018**, *43*, 2430–2433. [CrossRef]
31. Gan, W.; Li, S.; Li, Z.; Sun, L. Identification of ground intrusion in underground structures based on distributed structural vibration detected by ultra-weak FBG sensing technology. *Sensors* **2019**, *19*, 2160. [CrossRef]
32. Guo, H.; Qian, L.; Zhou, C.; Zheng, Z.; Yuan, Y.; Xu, R.; Jiang, D. Crosstalk and ghost gratings in a large-scale weak fiber Bragg grating array. *J. Lightwave Technol.* **2017**, *35*, 2032–2036. [CrossRef]
33. Guo, H.; Liu, F.; Yuan, Y.; Yu, H.; Yang, M. Ultra -weak FBG and its refractive index distribution in the drawing optical fiber. *Opt. Express* **2015**, *23*, 4829–4838. [CrossRef]
34. Gui, X.; Li, Z.; Wang, F.; Wang, Y.; Wang, C.; Zeng, S.; Yu, H. Distributed sensing technology of high-spatial resolution based on dense ultra-short FBG array with large multiplexing capacity. *Opt. Express* **2017**, *25*, 28112–28122.
35. Liu, S.; Ding, L.; Guo, H.; Zhou, A.; Zhou, C.; Qian, L.; Yu, H.; Jiang, D. Thermal stability of drawing-tower grating written in a single mode fiber. *J. Lightwave Technol.* **2019**, *37*, 3073–3077. [CrossRef]

36. Li, Z.; Tong, Y.; Fu, X.; Wang, J.; Guo, Q.; Yu, H.; Bao, X. Simultaneous distributed static and dynamic sensing based on ultra-short fiber Bragg gratings. *Opt. Express* **2018**, *26*, 17437–17446. [CrossRef]
37. Tong, Y.; Li, Z.; Wang, J.; Wang, H.; Yu, H. High-speed Mach-Zehnder-OTDR distributed optical fiber vibration sensor using medium-coherence laser. *Photonic Sens.* **2018**, *8*, 203–212. [CrossRef]
38. Chuang, W.H.; Tam, H.Y.; Wai, P.K.A.; Khandelwal, A. Time- and wavelength-division multiplexing of FBG sensors using a semiconductor optical amplifier in ring cavity configuration. *IEEE Photonics Technol. Lett.* **2005**, *17*, 2709–2711. [CrossRef]
39. Luo, Z.; Wen, H.; Guo, H.; Yang, M. A time- and wavelength-division multiplexing sensor network with ultra-weak fiber Bragg gratings. *Opt. Express* **2013**, *21*, 22799–22807. [CrossRef]

 © 2019 by the authors. Licensee MDPI, Basel, Switzerland. This article is an open access article distributed under the terms and conditions of the Creative Commons Attribution (CC BY) license (http://creativecommons.org/licenses/by/4.0/).

Article

Digital Approach to Rotational Speed Measurement Using an Electrostatic Sensor

Lin Li [1], Hongli Hu [1,*], Yong Qin [2] and Kaihao Tang [1]

1. State Key Laboratory of Power Equipment and Electrical Insulation, Xi'an Jiaotong University, Xi'an 710049, China; ll123xjtu.edu.cn@stu.xjtu.edu.cn (L.L.); mrerr07@stu.xjtu.edu.cn (K.T.)
2. State Key Laboratory of Rail Traffic Control and Safety, Beijing Jiaotong University, Beijing 100044, China; yqin@bjtu.edu.cn
* Correspondence: hlhu@mail.xjtu.edu.cn

Received: 24 March 2019; Accepted: 28 May 2019; Published: 4 June 2019

Abstract: In industrial production processes, rotational speed is a key parameter for equipment condition monitoring and fault diagnosis. To achieve rotational speed measurement of rotational equipment under a condition of high temperature and heavy dust, this article proposes a digital approach using an electrostatic sensor. The proposed method utilizes a strip of a predetermined material stuck on the rotational shaft which will accumulate a charge because of the relative motion with the air. Then an electrostatic sensor mounted near the strip is employed to obtain the fluctuating signal related to the rotation of the charged strip. Via a signal conversion circuit, a square wave, the frequency of which equals that of the rotation shaft can be obtained. Having the square wave, the *M/T* method and *T* method are adopted to work out the rotational speed. Experiments were conducted on a laboratory-scale test rig to compare the proposed method with the auto-correlation method. The largest relative errors of the auto-correlation method with the sampling rate of 2 ksps, 5 ksps are 3.2% and 1.3%, respectively. The relative errors using digital approaches are both within ±4‰. The linearity of the digital approach combined with the *M/T* method or *T* method is also superior to that of the auto-correlation method. The performance of the standard deviations and response speed was also compared and analyzed to show the priority of the digital approach.

Keywords: electrostatic sensor; digital approach; rotational speed; correlation algorithm

1. Introduction

In industrial applications, rotational speed measurement is a crucial part for condition monitoring, speed control, and protective supervision of rotation equipment, such as generators, steam turbines, and gas turbines. Various kinds of tachometers based on different mechanisms, such as optical, electrical, and magnetic induction, have been developed and widely used to measure the rotational speed of target objects. W.H. Yeh presented a high-resolution optical shaft encoder to monitor the rotation behavior of a motor [1] and J. N. Lygouras presented a solution for processing the pulses from an optical encoder attached to a motor shaft [2]. W. Lord and R.B. Chatto provided a homopolar tachogenerator with low inertia and noise generation, making it particularly suitable for velocity-control systems using high-performance DC motors as the power actuators [3]. C. Giebeler designed a contactless sensor based on the giant magneto-resistance (GMR) effect for position detection and speed sensing [4]. Z. Shi implemented a tachometer using a magnetoelectric composite as a magnetic field sensor which was mounted where the magnetoelectric composites had the highest sensitivity [5]. Considering the operating mode, the rotational speed measurement method can be classified into digital and analog categories. In the analog tachometer output a voltage or current signal proportional to the speed can be used to provide a feedback signal in a closed-loop speed control system [6]. Digital tachometers have been used over the years, which utilize electronic circuits to measure an average frequency of incoming pulses from an encoder mounted on a shaft [7].

In order to overcome the hash condition, such as a high temperature, heavy dust environment, the electrostatic method has been used to realize the rotational speed measurement. The electrostatic sensor is adaptable for the speed measurement in various industrial conditions for the advantages of contactless measurement, low cost, simple structure, and easy installation and maintenance. Recently, Y. Yan and L.J. Wang utilized electrostatic sensors and a correlation algorithm to calculate the period or elapsed time and successfully obtained the rotational speed of a rotational shaft [8,9]. The electrostatic method to measure rotational speed utilizes the electrode to induce the electric field generated by carried charges on the shaft. When two materials are touched or rubbed together electrical charge is usually transferred from one to the other [10]. According to the theory of tribo-electric charging, each material has its own surface work function. Then the surface electron transfer will occur in the driven Fermi energy level [11]. The material type determines the work function, which indicates the capability of a material to hold onto its free electrons. Thus, the polarity and quantity of the charge generated due to the triboelectric friction are mainly decided by the material type and surface roughness, and are also affected by its surrounding environment, like temperature and humidity. Thus, if the shaft has a greater work function than the air, the relative motion between the rotational shaft and the air will generate some charge on the surface of shaft due to triboelectric friction. If one of the materials is a good insulator, the charge persists on its surface for a long time, and the effects of the charge transfer are readily apparent [10].

The principle of rotational speed measurement using an electrostatic sensor is shown in Figure 1. According to electrostatic induction theory, when the surface of the shaft carries some charge due to triboelectric friction with the air, it will influence the electrostatic field of its surroundings, thus, the induced charge will be generated on the surface of sensing electrode when it is installed near the shaft. The fluctuation of the induced charge on the electrode generates a current which can be converted into a voltage signal via a current-to-voltage conversion circuit. Additionally, a charge amplifier circuit can be adopted to translate the charge into a voltage signal [12]. The voltage signal collected from the electrode contains a wealth of rotational information, thus processing and analyzing the output signals from the sensor will result in obtaining further information.

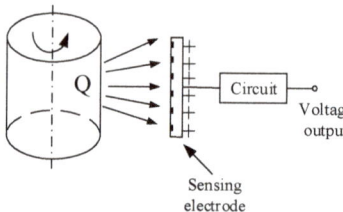

Figure 1. Measurement principle using the electrostatic method.

By now, electrostatic sensors in conjunction with correlation methods, including the cross-correlation method using dual electrostatic sensors and the auto-correlation method using a single electrostatic sensor, have been used to determine rotational speed [8,9,13]. Figure 2 describes the rotational speed measurement system which uses electrostatic sensors and the correlation method. Two or more channels of sensors and the corresponding condition units are connected to an A/D converter. Then a microprocessor system or a computer is needed to execute the correlation algorithm.

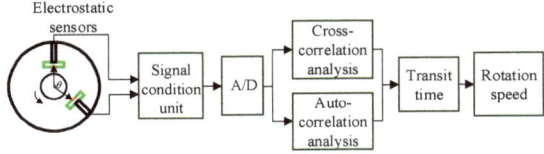

Figure 2. Principle of the rotational speed measurement using electrostatic sensors.

In the time domain, the definition of the cross-correlation function between real power signals $x(t)$ and $y(t)$ is:

$$R_{xy}(\tau) = \lim_{T \to \infty} \frac{1}{T} \int_{-T/2}^{T/2} x(t)y(t+\tau)dt \tag{1}$$

Figure 3 illustrates how to obtain the time-delay between two electrostatic signals using the cross-correlation method. The rotational speed v_c (revolutions per minute, rpm) can be calculated by the sensor angle spacing θ (degree) and the transit time τ (s):

$$v_c = \frac{1}{T} \times 60 = \frac{1}{\frac{360}{\theta} \cdot \tau} \times 60 = \frac{\theta}{6\tau} \tag{2}$$

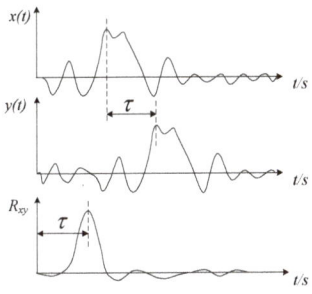

Figure 3. Illustration of the cross-correlation method.

If $x(t)$ and $y(t)$ are the same signals obtained by one electrode, Equation (1) turns out to be the auto-correlation function. With respect to the auto-correlation method, only one channel of the electrostatic signal is needed. The time-delay τ between signal $x(t)$ and signal $y(t)$ is the rotational period $T(s)$ and θ equals 360 degree. Using the auto-correlation algorithm can extract the time of rotation period. After that, the rotational speed v_a (rpm) can be obtained as follows:

$$v_a = \frac{60}{T} \tag{3}$$

Obviously, the correlation method needs to locate the coordinate of the first dominant peak in the waveform of the correlation function, which is influenced by the sampling rate to a great extent. At the same time, the waveforms collected by inducing the signal from a cylinder dielectric sleeve contain complex information and a faint sign of the periodical component. Although the correlation calculation of the waveform has good performance and successfully extracts the elapsed time, the computational accuracy of the correlation method is obviously affected by the sampling rate and signal noise [14].

For the sake of improving the performance of rotational speed measurement via the electrostatic method, this paper proposes an approach to generate a square wave from an electrostatic sensor in order to obtain the rotational speed via digital methods, thus eliminating the influence of the sampling rate and signal noise, and also simplifying the system complexity. In the following article, "square wave" refers to the output waveform from the comparison circuit which generates a pulse every rotational period. Implementation of a rotational speed measurement system based on this method is presented. Compared to the rotational speed measurement method using an electrostatic sensor in conjunction with correlation, this designation leaves out the AD converter and simplifies the computation code which is more adaptive for the implementation in a microprocessor system.

2. Measurement Principle and Finite Element Simulation

2.1. Measurement Principle

Inspired by the photoelectric method which fixes a strip of a reflection element, this experiment uses a strip of polytetrafluoroethylene (PTFE) stuck to the rotational shaft. The measurement principle is shown in Figure 4. Adopting this designation, the charge generated on the PTFE by the relative rotation with the air will pass the sensor once a revolution, which makes the waveform have a strong sign of periodicity. The electrostatic signal is firstly transformed into a voltage signal. Then, after amplifying, filtering, and comparison, the analog signal will be transformed into a square wave, which is convenient to be connected to a DSP or FPGA system to execute the following rotational speed calculation algorithm.

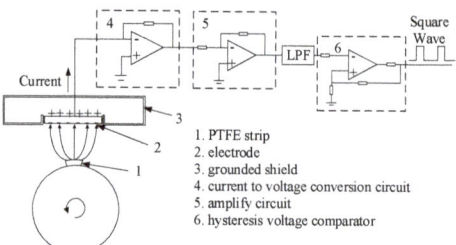

Figure 4. Measurement principle.

2.2. Rotation Speed Computation Algorithm

Usually, three methods are adopted to evaluate the speed based on these square waves: (1) Measuring the elapsed time, commonly termed as the T method, which calculates the reciprocal of the duration between consecutive pulses to obtain the frequency; (2) pulse counting, commonly termed as the M method, which counts the number of pulses generated within a prescribed period of time; and (3) constant elapsed time, commonly termed as the M/T method, is a combination of pulse counting and measuring elapsed time [15–17]. The principles of the three methods are shown in Figure 5.

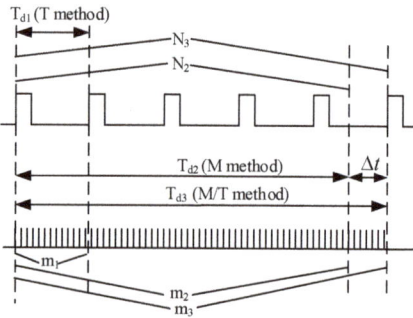

Figure 5. Principles of T, M, and M/T methods.

As seen from Figure 5, the detecting time of the T method and M method can be obtained according the Equations (4) and (5), correspondingly, where m_1 is the number of clock pulse counting during one period of square wave, m_2 is the number of pulse counting during the prescribed time, and f_s is the frequency of clock pulse used for counting and timing:

$$T_{d1} = \frac{m_1}{f_s} \qquad (4)$$

$$T_{d2} = \frac{m_2}{f_s} \qquad (5)$$

Then the rotational speed v_1 (rpm) of the T method can be calculated according to Equation (6). The rotational speed v_2 (rpm) of the M method can be calculated according to Equation (7), where N_2 is the integer number of square wave during the prescribed time:

$$v_1 = \frac{60 f_s}{m_1} \qquad (6)$$

$$v_2 = \frac{60 f_s N_2}{m_2} \qquad (7)$$

Different from the M method, which ceases the pulse counting once the prescribed time runs out, the M/T category goes on counting after the prescribed time and stops at the first pulse of rotation after the prescribed time. Thus, the detecting time of the M/T method T_{d3} (s) equals T_{d2} (s) plus Δt. Parameter N_3 is the number of square waves during the detecting time. Parameter m_3 is the number of clock pulses during the detecting time. The detecting time T_{d3} (s) can also be obtained using parameter m_2 and the frequency of clock pulse f_s used for timing, as shown in Equation (8). Thus, the rotational speed of the M/T method is calculated by Equation (9):

$$T_{d3} = \frac{m_3}{f_s} \qquad (8)$$

$$v_3 = \frac{60 f_s N_3}{m_3} \qquad (9)$$

The calculation errors of the three method can be derived according to Equations (10)–(12). Parameters m_1^0, m_2^0, and m_3^0 are the ideal pulse numbers needed to perfectly overlap the detecting time. Parameter N_2^0 equals the detecting time divided by the rotation period, which is not an integer in most cases.

$$e_1 = (\frac{60 f_s}{m_1} - \frac{60 f_s}{m_1^0}) / \frac{60 f_s}{m_1^0} = \frac{m_1^0 - m_1}{m_1} \qquad (10)$$

$$e_2 = (\frac{60 f_s N_2}{m_2} - \frac{60 f_s N_2^0}{m_2^0}) / \frac{60 f_s N_2^0}{m_2^0} = \frac{m_2^0 N_2 / N_2^0 - m_2}{m_2} \qquad (11)$$

$$e_3 = (\frac{60 f_s N_3}{m_3} - \frac{60 f_s N_3^0}{m_3^0}) / \frac{60 f_s N_3^0}{m_3^0} = \frac{m_3^0 - m_3}{m_3} \qquad (12)$$

As seen from Equation (10), it can be observed that the error of the T method is low at a high speed (m_1 decreases) and the M method resolution is not high at a low speed (N_2 is not stable). However, considering the frequency of the clock pulse used for timing in this article is 150 MHz, which is significantly greater than the frequency of rotation, the counting errors of the T method and M/T method are extremely small compared to their denominators. The T method and M/T method have absolutely accurate counting numbers of the square wave from circuits 1 and N_3, correspondingly. The calculating errors of the T method and M/T method mainly result from the counting number of the pulse clock ($m_1^0 - m_1$, $m_3^0 - m_3$). By contrast, regarding the M method, the difference between square counting N_2 and N_2^0 may result in an obvious error.

The response speed can also be determined from the principle. Among them, the T method has the fastest response speed, which enables outputting a result every period. The M/T method and M method generate a result based on the statistical average principle, which gives them a relatively slow response speed. In summary, the T method is more adaptive for the dynamic measurement of variable rotational speed and the M/T method or M method is more adaptive for constant speed measurement or mean values of a certain time. Considering the response time and accuracy simultaneously, this article utilizes the M/T method and the T method to deal with the square wave from the measurement circuits.

2.3. Finite Element Simulation

A simulation using a strip object with evenly distributed charge was conducted utilizing COMSOL software (the COMSOL Group in Stockholm, Sweden) to imitate how the rotationally charged strip influences the induced charge on the electrostatic sensor. The model is shown in Figure 6 and the simplified two-dimensional schematic of the simulation is illustrated in Figure 7. The strip object is a 7.64 degree arc with a radius of 15 mm, which is placed tightly to the surface of the metal shaft. The length of the strip is 20 mm. The radii of metal shaft and outer shielding are 15 mm and 30 mm, correspondingly. The charge is evenly distributed on the surface of the object using the Surface Charge Density setting in COMSOL. The surface charge density is set to be 0.025 C/m^2, so the total amount of charge on the strip is 1 µC. An electrode 20 mm long and 2 mm wide is placed 17 mm away from the central axis with the same z coordinate of strip. In the simulation, the strip rotates around the central axis by controlling the angle with respect to the positive axis x, beginning at −180 degree and stopping at 180 degree with a step size of one degree. The electrostatic field can be described by the Poisson equation and its corresponding boundary conditions:

$$\begin{cases} -\nabla \cdot (\varepsilon_0 \varepsilon_r \nabla \varphi) = \rho \\ V_E = 0, V_B = 0 \end{cases} \quad (13)$$

where V_E is the potential of electrode, V_B is the potential of the shielding, and ρ is the space charge density.

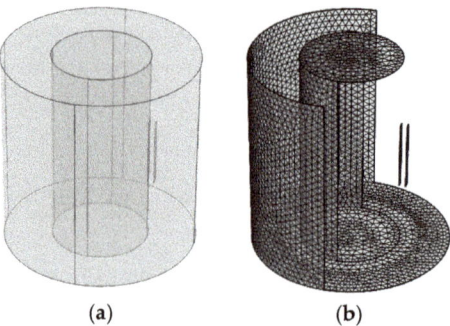

Figure 6. (a) Structure of the model, and (b) a mesh of the simulation model.

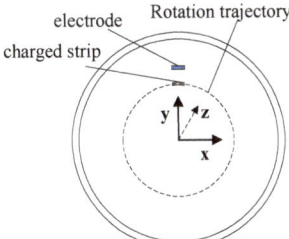

Figure 7. Simplified schematic.

Figure 8a depicts the induced charge on the electrode when the rotation of the strip begins farthest from the electrode, then passes by the electrode, and finally returns to the beginning position. The amount of induced charge on the electrode reflects the 'far-near-far' rotational process. The variation of charges on the electrode generates current i, which can be calculated by:

$$i = \frac{dq_e}{dt} \tag{14}$$

where q_e represents the amount of induced charge on the electrode. By calculating the difference of the induced charges, the current can be obtained and is shown in Figure 8c. When the strip is far from the electrode, the current is very weak and can be regarded as 0. When the strip rotates adjacent to the electrode, the current becomes larger. As seen from Figure 8c, the derivation of the induced charge contains thorns and wobbles. This can be explained due to the discretization and unavoidable computation error of the finite element simulation, the curve of induced charge is not smooth enough (shown in Figure 8b), thus leading to the thorns of the derivative curve. To acquire a more optimal result a moving average is applied to smooth the data and the result is shown in Figure 8d.

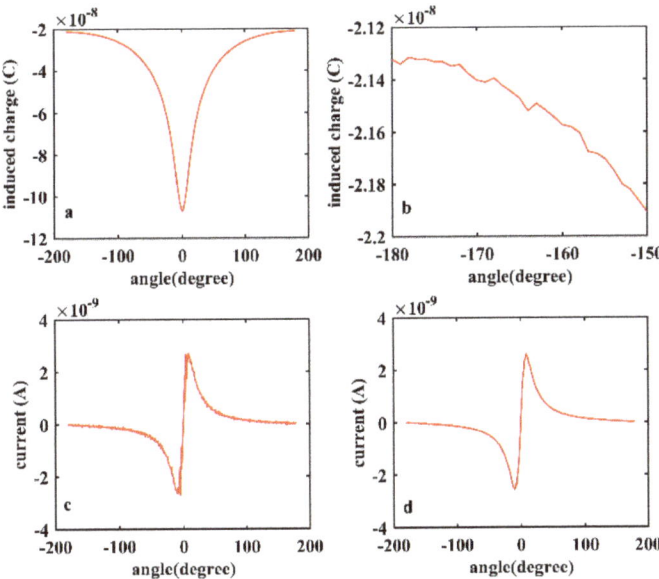

Figure 8. (**a**) Simulation result of induced charges; and (**b**) partial drawn of the induced charge. (**c**) Derivation of the induced charge; and (**d**) smoothed curve of the current.

Then, we use the rotation of the electrode to replace the rotation of shaft, so that the shaft is relatively at rest, as shown in Figure 9. According to the superposition principle of the electric field, the amount of induced charge Q on the electrode with a displacement of angle α can be calculated by Equation (15). Function $q(\theta)$ is the amount of charges on the location of angle θ. Parameter α indicates the rotated angle and is also a reflection of time. Function $f(\theta-\alpha)$ means when a unit charge is $\theta-\alpha$ degrees away from the electrode, the amount of induced charge on the electrode generated by this unit charge is $f(\theta-\alpha)$. Thus, the total amount of induced charge can be calculated by integrating over θ from $-180°$ to $180°$.

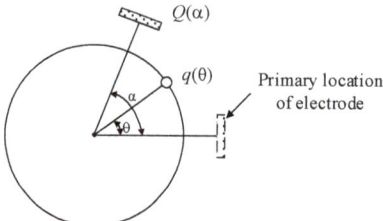

Figure 9. Schematic of electrode rotation.

$$Q(\alpha) = \int_{-180°}^{180°} q(\theta) f(\theta - \alpha) d\theta \tag{15}$$

$$Q(\alpha) = \int_{-\infty}^{+\infty} q_{circle}(\theta) f_r(\theta - \alpha) d\theta \tag{16}$$

As shown in Figure 10, Equation (15) illustrated by Figure 10a with an integral range of $[-180° \ 180°]$ can be transformed into Equation (16) illustrated by Figure 10b. In Figure 10b, the values of $f_r(\theta)$ when θ is out of range $[-180, 180]$ are zeros and the waveform of function $f_r(\theta)$ is the same as in Figure 8a, which is only different in amplitude. The induced charge on the electrode can be regarded as a weighted mean value of the contribution of the charge in a sensitive area. Meanwhile, Equation (16) is a convolution operation between the charge distribution function $q_{circle}(\theta)$ and the function $f_r(\theta)$, thus $f_r(\theta)$ can be regarded as a filter function. The low pass filter property can also be obtained from [18]. Obviously, function $f_r(\theta)$ is influenced by the rotational speed ($f_r(\theta_0 + wt)$). The cutoff frequency of $f_r(\theta)$ increases with the speed (narrow in the time domain, broad in the frequency domain). Through the analysis, the electrostatic electrode in this case of application can be regarded as a low pass filter which adaptively adjusts its cut-off frequency. Thus, the waveform of the signal mainly contains a low frequency component if the electro-magnetic interference is well shielded, which helps to explain the signal obtained in the experimental part.

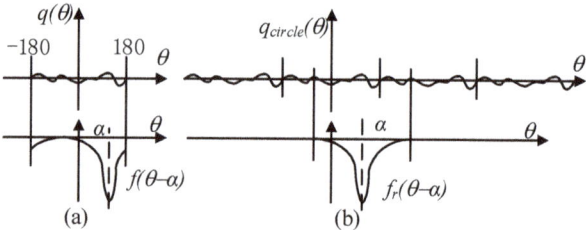

Figure 10. Explanation of Equations (15) (**a**), and Equations (16) (**b**).

3. Hardware Design

3.1. Sensor Board

The sensor board shown in Figure 11 contains an electrode and a current-to-current to voltage conversion circuit. A surface-tinned copper strip 20 mm long and 2 mm wide is utilized as the electrode. The electrode is connected to the current voltage conversion circuit, which is built on an LMP7721 (Texas Instruments in Dallas, TX, USA) amplifier with an extremely low bias current of 20 fA maximum. Figure 12 illustrates the schematic of the circuit. The feedback resistor consists of two 10^8 Ω resistors connected in series, which determines the transimpedance gain. In actual application, a feedback capacitor is needed to guarantee the stability of the circuit by inhibiting the high frequency noise.

Figure 11. Sensor board.

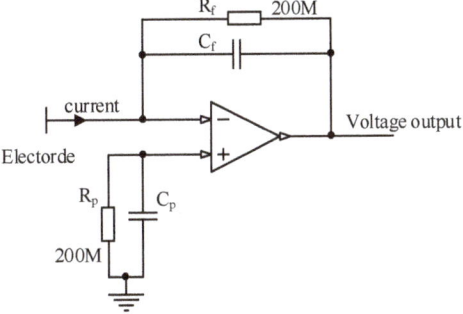

Figure 12. Schematic of current-to-voltage circuit.

The relationship between the output voltage and the input current from the electrode can be calculated according to Equation (17). Thus, the sensitivity of the circuit is 0.2 V/nA. The purpose of the balance resistor and capacitor is to make the impedance of the two inputs equal, thus, the bias current of the amplifier generates no additional offset voltage on the output.

$$U_o = R_f i_e \tag{17}$$

When the electric field near the electrode varies with the rotation of the charged strip, a small current signal will be generated and transformed into a voltage signal via the feedback resistance on the amplifier. The voltage output of the sensor board is collected by the condition unit via a shielded cable to avoid electromagnetic interference in the space.

3.2. Signal Condition Unit

The condition unit in our experiment is shown in Figure 13, which has amplifying, filtering, and comparing circuits designed to generate a square wave. Four connecters are placed on the board.

Connecter 1 is used to connect to the output of the sensor board. Connecters 2 and 3 are used to observe the result of amplifying and filtering correspondingly. The square waveform from the comparing circuit is transmitted to a DSP chip via connecter 4 or the pin headers nearby.

The amplifying circuit uses the same amplifier chip with the sensor board to meet the performance requirements. The voltage gain of the amplifier can be adjusted by a slide rheostat. Then a third-order Butterworth low pass filter with Sallen-Key topology is used for filtering and inhibiting the noise. The passband frequency is 400 Hz and the stopband frequency is 2.4 kHz. A smooth waveform improves the stability of the square wave. Finally, a hysteresis comparator is utilized to transform the waveform into a square wave which is connected to the DSP board for speed calculation.

Markers ①②③④ refer to four SMA connectors.

a. amplifying circuit
b. filter circuit
c. hysteresis comparator

Figure 13. Signal condition unit.

4. Experiment Results and Discussion

4.1. Experiment Conditions

A laboratory-scale test rig is designed and built for rotational speed measurement. Figure 14 shows the schematic of the test rig. An external power supply connects to a variable-frequency drive (VFD) via a power switch. The torque of the motor is translated to the shaft via a belt. Thus, the rotational speed of the shaft can be adjusted by the VFD. The shaft is made of steel and supported by two roller bearings with a belt pulley mounted on its side. The middle part of the shaft is surrounded by a grounded cylindrical metal shielding. As shown in Figure 15, a strip of PTFE about 2 mm wide and 20 mm long, the lengthwise direction of which is parallel to the axial direction of shaft, is glued tightly on the shaft. The sensor board is set on the inner wall of metal shielding via a copper pillar, thus the electrode is under the central axis of the shaft and the trajectory of the strip. The copper pillar is utilized to adjust the distance between the sensor and the shaft. In order to inhibit the vibration of the rig, the steel table was screwed to the ground via an expansion screw.

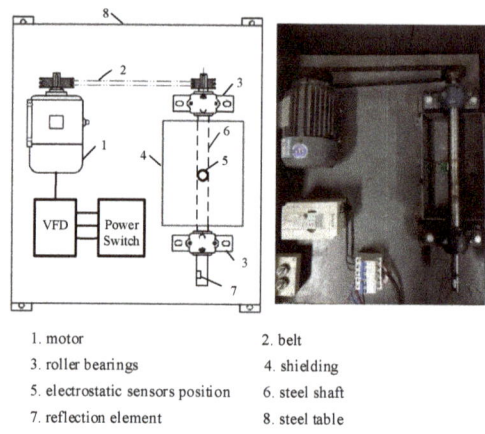

1. motor
2. belt
3. roller bearings
4. shielding
5. electrostatic sensors position
6. steel shaft
7. reflection element
8. steel table

Figure 14. Schematic and photograph of the test rig.

Figure 15. Photograph of the sensor and strip.

Experiments were conducted on the rig using the same dimension parameters as the simulation. The rotational speed of shaft was adjusted from 300 rpm to 3200 rpm with an increment of 100 rpm via the VFD. To make a comparison between the digital approach and the correlation method, each point was measured five times. Meanwhile, five values of the T method and M/T method transformed from the DSP for each point were saved for analysis. A photoelectric reflection digital tachometer with an accuracy of ±0.05% of the reading plus 1 rpm was used to provide a reference speed in our experiment. The ambient temperature was controlled between 20 °C and 24 °C and the relative humidity was kept between 55% and 65%. The square wave was connected to the external interrupt pin to contend with the square wave immediately. The code of realizing the T method and the M/T method were programed and written into the DSP board separately to test the measurement performance. The DSP transmitted the measurement results of the T method and the M/T method to computer via RS232 serial communication. In the experiment, the prescribed time of the M/T method is set to be 1 s.

Seen from the principle of the correlation method, it can be found that the auto-correlation method can be regarded as a particular case of the cross-correlation method, which leaves out the influence of the installation angle error, the distance differences of the two electrodes to the shaft, and the differences between two channels' circuits. These factors make the accuracy of cross-correlation method not as good as the auto-correlation method. Meanwhile, the cross-correlation method needs two channels of circuits, which is not consistent with the setting in this experiment. Thus, the experiment only makes a comparison between digital approaches and the auto-correlation method.

4.2. Signals

The proposed approach utilizes the electrostatic sensor to induce the charge on the strip of PTFE, which obtains a strong periodic signal. Figure 16 shows the signals before and after filtering, which contain evenly distributed waveforms similar to the simulation result (Figure 8d). The filtered waveform obviously has a higher signal-to-noise ratio. The high signal-to-noise ratio and the strong periodicity helps to improve the stability of the square wave transformed from the signal, which is very important for the rotational speed calculation based on the square wave. Figure 17 shows the square waveform generated by the hysteresis comparator. In order to illustrate the wave clearly, Figure 17 only shows 0.5 s of the signal.

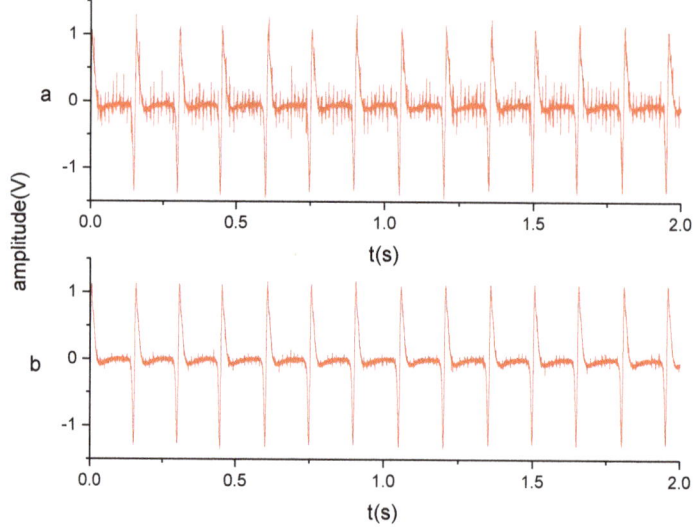

Figure 16. Input (**a**) and output (**b**) of the filtering circuit.

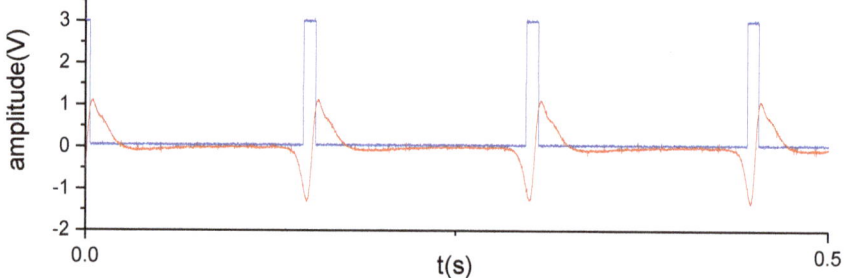

Figure 17. Analog input and digital output of the comparator circuit.

4.3. Accuracy

The mean values of the measurement results for the T method and the M/T method are plotted in Figure 18. Their relative errors are compared with the photoelectric reflection digital tachometer and are listed in Table 1. The linearity of the T method and M/T method are about 0.81‰ and 1.31‰, correspondingly. The measurement results are highly consistent with those of the photoelectric tachometer. Meanwhile, the differences between the T method and the M/T method are hardly discernible by eye.

As seen from the principle, the proposed digital method needs no sampling via an analog-digital converter, while the sampling rate is an important factor that determines the accuracy of the method based on the correlation algorithm. In order to make a comparison between these two methods, the analog signals are also collected at different sampling rates. The auto-correlation functions are calculated using the filtered analog signal. Figure 19 shows the auto-correlation of an example collected at the rotational speed of 400 rpm using a sampling rate of 2 ksps.

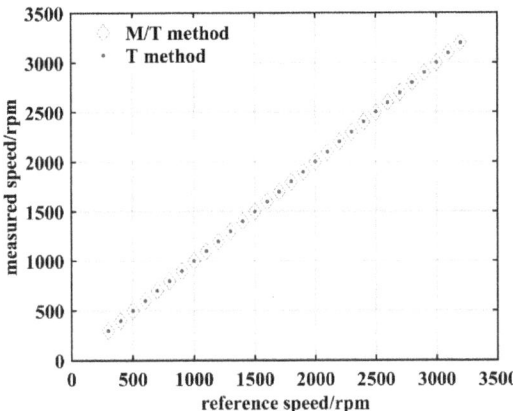

Figure 18. Measurement speed.

Table 1. The relative errors of the measurement points.

Reference Speed (rpm)	Measured Speed (rpm)		Relative Error (‰)	
	M/T	T	M/T	T
300	300.50	300.11	1.67	0.36
400	399.97	400.45	−0.08	1.12
500	501.04	501.87	2.08	3.74
600	600.63	601.08	1.05	1.80
700	701.09	701.39	1.56	1.98
800	799.12	799.26	−1.10	−0.92
900	900.10	899.38	0.11	−0.69
1000	1000.02	1001.35	0.02	1.35
1100	1099.65	1100.79	−0.32	0.72
1200	1200.06	1199.55	0.05	−0.38
1300	1300.30	1301.22	0.23	0.94
1400	1400.83	1401.27	0.59	0.91
1500	1501.81	1501.34	1.21	0.89
1600	1601.33	1601.60	0.83	0.99
1700	1700.71	1699.35	0.42	−0.38
1800	1799.75	1802.47	−0.14	1.37
1900	1900.12	1899.11	0.06	−0.47
2000	2000.34	2001.50	0.17	0.75
2100	2100.87	2101.57	0.41	0.75
2200	2201.38	2202.34	0.63	1.06
2300	2298.58	2300.50	−0.62	0.21
2400	2400.52	2404.21	0.22	1.75
2500	2497.95	2503.25	−0.82	1.30
2600	2600.32	2599.51	0.12	0.19
2700	2701.59	2698.14	0.59	−0.69
2800	2802.13	2802.10	0.76	0.75
2900	2899.71	2901.81	−0.10	0.62
3000	3000.15	3001.87	0.05	0.62
3100	3100.52	3098.77	0.17	−0.40
3200	3202.58	3202.23	0.81	0.70

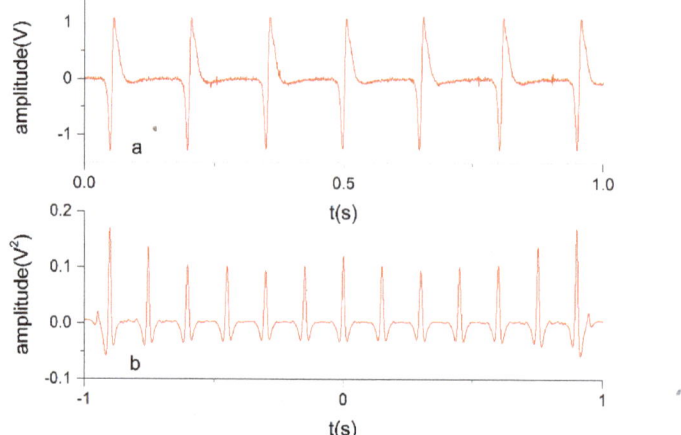

Figure 19. (a) Signal collected at the rotational speed of 400 rpm. (b) Auto-correlation of the signal.

The waveform in Figure 20 is a partial drawing of the part in Figure 19b. As shown in Figure 20, by detecting the first peak after 0 s, the period T of the rotation can be obtained. It can be observed that the waveform near the first peak after 0 s is very smooth, which benefits confirming the accurate and stable value of the period. However, due to the discretization of the data series, the obtained period T will be a time length away from the actual time of the rotation period with a significant probability despite the auto-correlation method confirming the nearest time point to the ideal time point. Moreover, when the signal contains an obvious level of noise or a weak periodicity, the waveform near the peak of the auto-correlation function will be fluctuant, which impairs the result's accuracy.

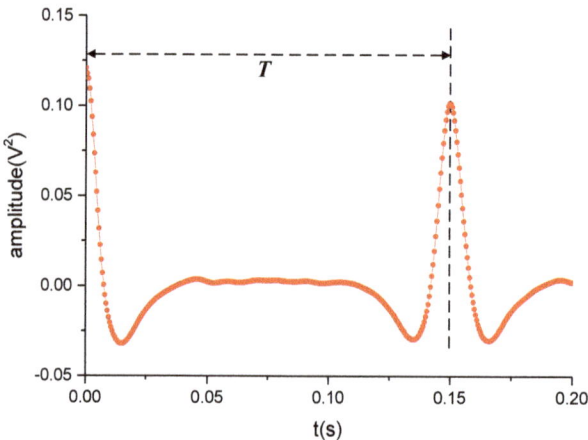

Figure 20. Partial enlarged drawing of Figure 17b.

In order to show the accuracy of this digital method, signals are collected at the sampling rate of 2 ksps and 5 ksps and analyzed using the auto-correlation method. Relative errors of the two methods are plotted and compared in Figure 21. Figure 21 shows the measurement errors of the digital approaches and the auto-correlation method at the sampling rate of 2 ksps and 5 ksps. It can be seen from Figure 21 that the digital approaches have better accuracy and the relative errors obtained using 5 ksps are smaller than those sampled at 2 ksps. The auto-correlation method is apparently influenced by the sampling rate. Meanwhile, the accuracy of auto-correlation method has the tendency to increase with the rotational speed, which has been explained in 14. The linearity of the auto-correlation method sampled at 2 ksps and 5 ksps are 3.17% and 1.33%, respectively, which are significantly greater than those of the *M/T* method and the *T* method.

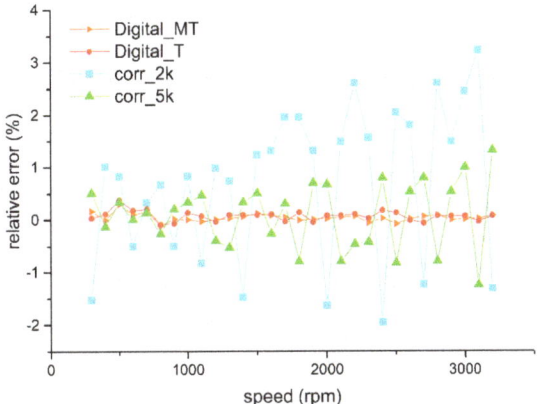

Figure 21. The absolute value of the relative error.

4.4. Standard Deviation

In order to research the robustness of the proposed method, the standard deviations of each measurement point are listed in Table 2. As seen from the table, regarding digital approaches, the *M/T* and *T* methods both have significantly small standard deviations. The standard deviations of the auto-correlation method in Table 2 contains a number of zeros and some other values, which can be easily understood from the principle. For example, using 5 ksps (Δt is 0.2 ms) to collect the signal of 1400 rpm, the period of which is about 214 times that of Δt, the calculated rotational speed by finding the first peak of the auto-correlation function will be some discrete value calculated by $60/[(214 \pm n)\Delta t]$ (n = 0, 1, 2, ...), like 1395.35, 1401.87, or 1408.45. There are two factors affecting the standard deviations of the auto-correlation method: (1) If the variation of rotation speed is not obvious enough to change the location of the first peak on the auto-correlation function, the measurement results will remain unchanged; and (2) if the locations of the first peak in the auto-correlation function differs one or two sampling intervals from each other due to signal differences, the obtained rotational speeds will show obvious fluctuations.

The standard deviations of the *M/T* and *T* methods in Table 2 are all within 1 rpm. Meanwhile, the standard deviations of the *M/T* method are much smaller than those of the *T* method. As seen from the principle, the *M/T* method can be regarded as a mean value of several consecutive *T* methods. Due to the high response speed of the *T* method, it is more sensitive to the variation of rotation, which makes its standard deviations greater than those of the *M/T* method. The minor standard deviations of the *M/T* and *T* methods mainly arise from the slight fluctuations of the actual rotation state, which is probably related to the unsteady output rotational speed of the motor and the slippage of the belt on the sheave. With respect to the digital approaches, no matter the *M/T* method or *T* method, both have very little spread in the measured speed.

Table 2. The standard deviations of the measurement points.

Reference Speed (rpm)	Standard Deviation (rpm)			
	M/T	T	2k	5k
300	0.055	0.175	1.755	2.047
400	0.013	0.292	2.641	2.250
500	0.013	0.397	4.652	1.867
600	0.032	0.274	1.651	2.161
700	0.023	0.458	3.437	1.472
800	0.097	0.142	2.367	2.592
900	0.091	0.115	3.011	1.206
1000	0.068	0.378	0	2.790
1100	0.011	0.146	0	0
1200	0.044	0.190	0	0
1300	0.015	0.284	0	2.525
1400	0.051	0.038	0	0
1500	0.010	0.027	0	0
1600	0.013	0.265	0	4.674
1700	0.017	0.190	0	0
1800	0.019	0.467	0	2.542
1900	0.010	0.065	0	0
2000	0.027	0.284	0	0
2100	0.029	0.234	0	0
2200	0.051	0.006	18.069	0
2300	0.016	0.368	0	7.878
2400	0.051	0.281	0	0
2500	0.050	0.415	0	0
2600	0.048	0.155	0	0
2700	0.018	0.364	0	0
2800	0.069	0.082	0	0
2900	0.039	0.301	0	0
3000	0.052	0.131	0	0
3100	0.012	0.327	0	0
3200	0.094	0.345	46.747	0

4.5. Response Time

The response time of each approach can be determined from their principles and data process procedures. Regarding the M/T method, the response time is decided by adding extra time to the prescribed time. Regarding the analog method, usually the sampling length should be predetermined. Thus, the time needed to acquire one measurement result is nearly confirmed. Even if the auto-correlation method self-adaptively adjusts the sampling length according to the nearest obtained rotational speed, the response speed is still not as fast as the T method for the reason that the auto-correlation method needs at least two periods of rotation to achieve the correlation calculation. Moreover, data collection and processing also consume a certain amount of time.

By contrast, the T method can output a result every rotational period for the reason that the counter in the DSP can work independently from the code and the DSP only needs to perform an easy computation of the counter number and serial communication. Experiments were conducted to test the capability of the T method to measure the variable speed. The motor was adjusted by the VFD output frequency to work at three stages: acceleration by increasing the frequency from 0 Hz to 20 Hz over 5 s, 4 s of constant speed, and deceleration by decreasing the frequency from 20 Hz to 0 Hz over 3 s. Figure 22a shows the 256 acquired rotational speeds via the T method, which successfully monitors the acceleration and deceleration processes. The rising and decline curves are not perfectly straight lines because the acceleration of the shaft is not absolutely constant. It can be observed that at a constant frequency, the measurement results are of good stability.

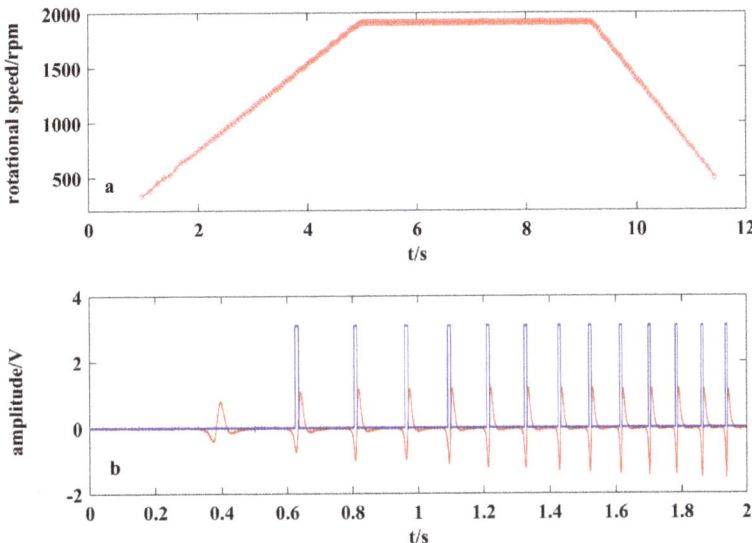

Figure 22. (a) Measured rotational speed by the T method; and (b) waveform during acceleration process.

Figure 22b shows a part of the waveform during the acceleration stage. The acceleration and deceleration process can be clearly observed through the interval variations of square waves. The waveform near 0.5 s has not been transformed into a square wave because a limited electrostatic charge is generated on the strip at low speed, thus making the signal unsuitable for the following square wave generation circuit. With the increases of rotational speed, the amount of charge rises and then becomes stable because of the dynamic balance reached between the natural discharge and recharge. The signal amplitude changes as the unbalanced charge increases or decreases at a low speed, which makes the comparison voltage appear at different positions relative to the waveform. This phenomena limits the application of this method in measuring low variable speed.

5. Conclusions

The work in this paper dedicates to find a more effective approach to cooperate with electrostatic sensors to improve the performance of rotational speed measurement. The proposed approach utilized the electrostatic sensor to induce a charge on a strip PTFE, which obtained a strong periodical signal. Simulation results also described the expected waveform when a strip of charges rotates near an electrode. By adopting a suitable signal condition unit, a square wave, the frequency of which was equal to that of the rotational speed, has been obtained. Having the square wave proportional to rotational speed, the M/T method and T method were adopted to calculate the speed in a DSP system. Experiments were conducted to compare the digital approaches with the auto-correlation method. Through experimental analysis, several conclusions can be summarized as follows:

1. Accuracy: Compared with the auto-correlation method, the M/T method and T method both have an obviously higher accuracy. The linearity of the M/T method and T method are about 0.81‰ and 1.31‰, correspondingly, which are much better than those of the auto-correlation method sampled at 2 ksps (3.17%) or 5 ksps (1.33%). Due to the signal discretization, the auto-correlation method can only obtain some discrete values. Improving the sampling rate, calculation quantity, and storage space, the hardware cost will also increase correspondingly.

2. Robustness: The auto-correlation method has a stable performance in some measurement points and also has some obvious standard deviations, which resulted from the signal discretization. However, the M/T method and T method obtained particularly small standard deviations among all

the measurement points, both within 1 rpm. The *M/T* method acquired more stable results than the *T* method due to differences of their respective principles.

3. Response speed: The proposed approach combined with the *T* method has the fastest response speed. The correlation method and *M/T* method have relatively slower response speeds. Experiments also shows that the *T* method is capable of detecting the variable speed.

Indeed, having the square wave related to the rotational period, the *M/T* method can be adopted for constant speed measurement or a mean value of rotational speed during a certain time and the *T* method can be employed for dynamic measurement of variable rotational speed. In actual programming, the *M/T* method and *T* method can be written into one piece of a DSP or FPGA simultaneously. An FPGA is more recommended to deal with the square wave for its property of parallel processing and high code execution efficiency.

There are several factors limit the application of this method working at a low speed. The amount of charge on the strip is unstable and the response time is poor at low speed. Further studies can be conducted to deal with these issues by adopting an electret material, adding adaptive numbers of strips and electrodes, and improving circuit properties.

Author Contributions: This work has been done in collaboration between all authors. Data curation, L.L.; Funding acquisition, Y.Q.; Investigation, H.H.; Methodology, L.L. and H.H.; Validation, K.T.; Writing—original draft, L.L.; Writing—review & editing, K.T.

Funding: This research was funded by the National Natural Science Foundation of China, grant number 51777151, the National Key R and D Program of China, grant number 2016YFB0901200, the Shaanxi Provincial Key Technologies R and D Programme, grant number 2016GY-001 and the Open Research Fund of State Key Laboratory of Rail Traffic Control and Safety, grant number RCS2017K006.

Conflicts of Interest: The authors declare no conflict of interest.

References

1. Yeh, W.H.; Bletscher, W.; Mansuripur, M. High resolution optical shaft encoder for motor speed control based on an optical disk pick-up. *Rev. Sci. Instrum.* **1998**, *69*, 3068–3071. [CrossRef]
2. Lygouras, J.N.; Lalakos, K.A.; Ysalides, P.G. High-performance position detection and velocity adaptive measurement for closed-loop position control. *IEEE Trans. Instrum. Meas.* **1998**, *47*, 978–985. [CrossRef]
3. Lord, W.; Chatto, R.B. Alternatives to Analog DC Tachogenerators. *IEEE Trans. Ind. Appl.* **1975**, *IA-11*, 470–478. [CrossRef]
4. Giebeler, C.; Adelerhof, D.J.; Kuiper, A.E.T.; Van Zon, J.B.A.; Oelgeschläger, D.; Schulz, G. Robust GMR sensors for angle detection and rotation speed sensing. *Sens. Actuators A Phys.* **2001**, *91*, 16–20. [CrossRef]
5. Shi, Z.; Huang, Q.; Wu, G.S.; Xu, Y.H.; Yang, M.; Liu, X.J.; Wang, C.; Ma, J. Design and Development of a Tachometer Using Magnetoelectric Composite as Magnetic Field Sensor. *IEEE Trans. Magn.* **2018**, *54*, 4000604. [CrossRef]
6. Robinson, C.E. Analog Tachometers. *IEEE Trans. Ind. Gen. Appl.* **1966**, *IGA-2*, 144–146. [CrossRef]
7. Prokin, M. Extremely wide-range speed measurement using a double-buffered method. *IEEE Trans. Ind. Electron.* **1994**, *41*, 550–559. [CrossRef]
8. Wang, L.; Yan, Y.; Hu, Y.; Qian, X. Rotational Speed Measurement through Electrostatic Sensing and Correlation Signal Processing. *IEEE Trans. Instrum. Meas.* **2014**, *63*, 1190–1199. [CrossRef]
9. Wang, L.J.; Yan, Y.; Reda, K. Comparison of single and double electrostatic sensors for rotational speed measurement. *Sens. Actuators A Phys.* **2017**, *266*, 46–55. [CrossRef]
10. Lowell, J.; Rose-Innes, A.C. Contact electrification. *Adv. Phys.* **1980**, *29*, 947–1023. [CrossRef]
11. Li, J.; Wu, G.; Xu, Z. Tribo-charging properties of waste plastic granules in process of tribo-electrostatic separation. *Waste Manag.* **2015**, *35*, 36–41. [CrossRef] [PubMed]
12. Hu, Y.H.; Zhang, S.; Yan, Y.; Wang, L.J.; Qian, X.; Yang, L. A Smart Electrostatic Sensor for Online Condition Monitoring of Power Transmission Belts. *IEEE Trans. Ind. Electron.* **2017**, *64*, 7313–7322. [CrossRef]
13. Wang, L.J.; Yan, Y.; Hu, Y.H.; Qian, X.C. Rotational Speed Measurement Using Electrostatic Sensors and Correlation Signal Processing Techniques. In Proceedings of the 2013 IEEE International Instrumentation and Measurement Technology Conference (I2MTC), Minneapolis, MN, USA, 6–9 May 2013; pp. 224–227.

14. Lin, L.; Xiaoxin, W.; Hongli, H.; Xiao, L. Use of double correlation techniques for the improvement of rotation speed measurement based on electrostatic sensors. *Meas. Sci. Technol.* **2016**, *27*, 025004.
15. Ohmae, T.; Matsuda, T.; Kamiyama, K.; Tachikawa, M. A Microprocessor-Controlled High-Accuracy Wide-Range Speed Regulator for Motor Drives. *IEEE Trans. Ind. Electron.* **1982**, *IE-29*, 207–211. [CrossRef]
16. Hachiya, K.; Ohmae, T. Digital speed control system for a motor using two speed detection methods of an incremental encoder. In Proceedings of the 2007 European Conference on Power Electronics and Applications, Aalborg, Denmark, 2–5 September 2007; pp. 1–10.
17. Hace, A. The Improved Division-Less MT-Type Velocity Estimation Algorithm for Low-Cost FPGAs. *Electronics* **2019**, *8*, 361. [CrossRef]
18. Wang, L.J.; Yan, Y. Mathematical modelling and experimental validation of electrostatic sensors for rotational speed measurement. *Meas. Sci. Technol.* **2014**, *25*, 115101. [CrossRef]

© 2019 by the authors. Licensee MDPI, Basel, Switzerland. This article is an open access article distributed under the terms and conditions of the Creative Commons Attribution (CC BY) license (http://creativecommons.org/licenses/by/4.0/).

Article

Multi-Factor Operating Condition Recognition Using 1D Convolutional Long Short-Term Network

Zhinong Jiang [1,2], Yuehua Lai [1], Jinjie Zhang [1,2], Haipeng Zhao [1] and Zhiwei Mao [1,*]

1. Key Lab of Engine Health Monitoring-Control and Networking of Ministry of Education, Beijing University of Chemical Technology, Beijing 100029, China; jiangzn@mail.buct.edu.cn (Z.J.); lyhlaibit@163.com (Y.L.); zhangjinjie@mail.buct.edu.cn (J.Z.); 18810272291@163.com (H.Z.)
2. Beijing Key Laboratory of High-End Mechanical Equipment Health Monitoring and Self-Recovery, Beijing University of Chemical Technology, Beijing 100029, China
* Correspondence: maozhiwei@mail.buct.edu.cn

Received: 18 October 2019; Accepted: 10 December 2019; Published: 12 December 2019

Abstract: For a diesel engine, operating conditions have extreme importance in fault detection and diagnosis. Limited to various special circumstances, the multi-factor operating conditions of a diesel engine are difficult to measure, and the demand of automatic condition recognition based on vibration signals is urgent. In this paper, multi-factor operating condition recognition using a one-dimensional (1D) convolutional long short-term network (1D-CLSTM) is proposed. Firstly, a deep neural network framework is proposed based on a 1D convolutional neural network (CNN) and long short-Term network (LSTM). According to the characteristics of vibration signals of a diesel engine, batch normalization is introduced to regulate the input of each convolutional layer by fixing the mean value and variance. Subsequently, adaptive dropout is proposed to improve the model sparsity and prevent overfitting in model training. Moreover, the vibration signals measured under 12 operating conditions were used to verify the performance of the trained 1D-CLSTM classifier. Lastly, the vibration signals measured from another kind of diesel engine were applied to verify the generalizability of the proposed approach. Experimental results show that the proposed method is an effective approach for multi-factor operating condition recognition. In addition, the adaptive dropout can achieve better training performance than the constant dropout ratio. Compared with some state-of-the-art methods, the trained 1D-CLSTM classifier can predict new data with higher generalization accuracy.

Keywords: diesel engine; condition recognition; CNN; LSTM; adaptive dropout

1. Introduction

A diesel engine is a kind of internal combustion engine that converts thermal energy into mechanical energy. It plays an important role in the field of national defense, in the chemical industry, in the marine industry, for nuclear power, and so on. Once a diesel engine fails, it not only causes economic losses directly or indirectly in terms of the shutdown of equipment, but it may also threaten the personal safety of users [1,2]. To enhance the availability of the diesel engine, it is imperative to monitor the engine condition and detect early faults. However, the detection of faults and the diagnosis of diesel engines [3] are not simple tasks due to the complex structure and fickle working conditions. If the operating conditions are not considered in detection and diagnosis activities, it is likely to lead to false alarms or missed detection [4,5]. With the information of operating conditions, the engineering applicability of a fault detection and diagnosis method [6–8] can be improved to avoid fatal performance degradation and huge economic losses at an early stage. Unfortunately, most fault detection methods are carried out under stable operating condition to avoiding variable operating conditions. Therefore, condition recognition is an important and urgent task in practical engineering applications.

In a diesel engine, the flywheel is attached to the crankshaft, and they rotate together. They convert the reciprocating motion of the piston into the rotational motion of the crankshaft, which outputs torque for the driving of the car and other power-driven mechanisms. Therefore, the operating conditions of a diesel engine can be determined by two parameters: load and the rotation speed of the crankshaft. The load is the output torque of the engine through the flywheel. However, the multi-factor operating conditions of a diesel engine are difficult to measure in many situations, such as for the power systems of vehicles, propulsion devices of ships, and other dynamic equipment. Therefore, the demand for automatic recognition of multi-factor operating conditions is urgent.

During the operation of a diesel engine, the corresponding status information can be obtained by using vibration analysis [9], oil analysis [10], thermal performance analysis [11], and visual inspection. Vibration is an intrinsic mechanical phenomenon, and the vibration signals contain rich information about the diesel engine's status; thus, vibration monitoring is a powerful tool for condition recognition, as well as fault detection and diagnosis. In this paper, we aim at recognizing the multi-factor operating conditions of a diesel engine based on vibration signals.

Thanks to the development of computing calculation power and powerful signal processing techniques, the recognition tasks based on vibration signals made great progress. At present, some recognition algorithms based on vibration signals exist, and most of them focus on designing various handcrafted features, fusing multiple features and training different classifiers. In Reference [12], the Hilbert spectrum entropy, which combines the Hilbert spectrum and information entropy, was proposed for the pattern recognition of diesel engine working conditions. In Reference [13], the frequency domain features of vibration signals were extracted for back propagation (BP) and radial basis function (RBF) neural network training to recognize the cylinder pressure. In Reference [14], based on the cylinder head vibration signals measured under stable operating conditions, an engine cylinder pressure identification method using a genetic algorithm with BP neural network was proposed. In Reference [15], combustion evaluation parameters were extracted using time–frequency coherence analysis and the cylinder pressure could be estimated based on the parameters and an RBF neural network. In Reference [16], the measured signal was converted into a crank angle degree signal using the rotational speed monitored by magnetic pickup sensors. Then, a real-time engine load classification algorithm was proposed based on an artificial neural network.

Most pattern recognition studies focused mainly on single-factor conditions or recognition under stable operating conditions. For single-factor conditions, the number of categories is generally no greater than five. In practical engineering applications, a single factor cannot describe complex operating conditions, and this drawback results in ambiguous boundaries among different operating conditions. As for multi-factor operating conditions, as the number of operating conditions increases, so does the complexity of condition recognition. Simultaneously, as the vibration signals are random, transient, and cyclostationary, and as the corresponding feature extraction requires rich domain knowledge, it is difficult to extract sensitive characteristics of significant importance for multi-factor operating condition recognition.

Over the last few years, with the development of deep learning, many researchers exploited deep neural networks (DNNs) as the feature extractor and classifier [17,18]. Benefiting from the powerful feature extraction ability of neural network, especially convolution neural networks (CNNs) [19], these approaches and their variations exhibit good performance in the related tasks. In Reference [20], time domain and frequency domain feature representations were selected to form a vector to act as the input parameters of a CNN. The trained CNN classifier could diagnose the fault patterns of a gearbox with outstanding performance. In Reference [21], the vibration signals of rolling bearings were analyzed using continuous wavelet transform to get time–frequency representations in grayscale. Then, all compressed time–frequency representations were taken as the input for CNN training, and the trained CNN classifier could identify the faults of rolling bearings with strong generalization ability. In Reference [22], a deep convolutional neural network of up to 38 layers, which could provide high classification accuracy, was proposed for gas classification. For CNN applications with vibration signals, there are different approaches to network input. In other words, the CNN is taken as a classifier, and the input of the CNN is mainly based on other feature extraction methods. At the same time, state-of-the-art CNN models have several parameters, which leads to problems related to storage, computation, and energy cost. In addition, recurrent neural networks (RNNs) and long short-term networks (LSTMs) [23,24] were validated in terms of their performance on one-dimensional (1D) signals. In Reference [25], a CNN and a fully connected neural network were both incorporated into a deep neural network framework to improve LSTM. The framework outperformed the original LSTM for the early diagnosis and prediction of sepsis shock. In Reference [26], an end-to-end model combining a CNN and RNN was proposed for the automatic detection of atrial fibrillation. Compared to the state-of-the-art models evaluated on standard benchmark electrocardiogram datasets, the proposed model produced better performance in detecting atrial fibrillation. The ideas in References [25,26] are very good references for multi-factor operating condition recognition based on vibration signals.

Therefore, a multi-factor operating condition recognition algorithm is proposed herein based on a 1D CNN and LSTM. In the proposed neural network framework, the 1D CNN was designed to extract local features of vibration signals through 1D convolution, and the LSTM was designed to describe the temporal relationship between local features. The contributions of this paper are summarized as follows:

1. A multi-factor operating condition recognition method is proposed using a 1D convolutional long short-term network (1D-CLSTM). As far as we know, this is the first study to combine a 1D CNN and LSTM to recognize operating conditions based on a time series of vibration signals;
2. Considering the particularity of engine vibration signals, batch normalization (BN) is introduced to regulate the input of some layers by fixing the mean value and variance of input signals in each convolutional layer;
3. Adaptive dropout is proposed for improving the model sparsity and preventing overfitting;
4. The designed 1D convolutional long short-term network (1D-CLSTM) classifier can achieve high generalization accuracy for recognizing multi-factor operating conditions.

The rest of this paper is organized as follows: Section 2 presents the test bench of a diesel engine and the experimental data acquisition. Section 3 introduces the technical background for the 1D CNN and LSTM. Section 4 describes the designed 1D-CLSTM and the flowchart of the multi-factor operating condition recognition algorithm. Section 5 shows the training performance of the designed 1D-CLSTM classifier, with generalizability verification, a performance comparison with different methods, and a training performance comparison with different dropout ratios. Finally, conclusions and future prospects are presented in Section 6.

2. Experiment and Vibration Signal

2.1. Test Bench of Diesel Engine

For data acquisition, a four-stroke diesel engine numbered TBD234 (produced by Henan Diesel Engine Industry Co. Ltd., Luoyang, China) was used and tested in different operating conditions. The parameters of the diesel engine are shown in Table 1.

As shown in Figure 1, 12 acceleration sensors were arranged on the surface of corresponding cylinder heads to monitor the status information of the diesel engine in the running state. The vibration signals formed the basis for the multi-factor operating condition recognition of the diesel engine. Moreover, an eddy current sensor was arranged on the flywheel to collect the information of rotating speed. In addition, a hydraulic dynamometer was connected with the output end of the diesel engine to adjust the load.

Table 1. Parameters of TBD234 diesel engine.

Item	Parameter
Number of cylinders	12
Shape	V-shaped 60°
Firing sequence	B1-A1-B5-A5-B3-A3-B6-A6-B2-A2-B4-A4
Rating speed	2100 rev/min
Rating power	485 kW

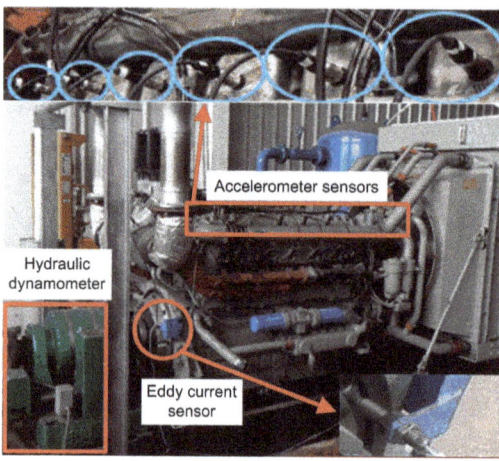

Figure 1. Test bench of the diesel engine.

All signals were measured using an online condition monitoring system (OCMS) at a sampling frequency of 51.2 kHz per channel in all tests, and the results were saved to a server through Ethernet transmission. The structure diagram of the OCMS of the diesel engine is shown in Figure 2.

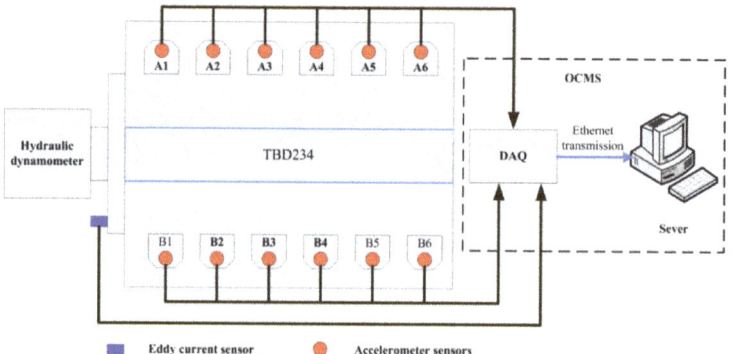

Figure 2. Structure diagram of the online condition monitoring system (OCMS) of the diesel engine.

2.2. Experimental Data Acquisition

To extract vibration data under different operating conditions, the engine was run at different levels of operating conditions. The representative operating conditions are listed in Table 2.

Table 2. Operating conditions of the diesel engine.

No.	Rev (rpm)	Load (N·m)	No.	Rev (rpm)	Load (N·m)
1	1500	700	7	1800	1600
2	1500	1000	8	2100	700
3	1500	1300	9	2100	1000
4	1800	700	10	2100	1300
5	1800	1000	11	2100	1600
6	1800	1300	12	2100	2200

Through the OCMS, vibration signals of different operating conditions could be measured. The vibration signals of 12 different operating conditions are shown in Figure 3.

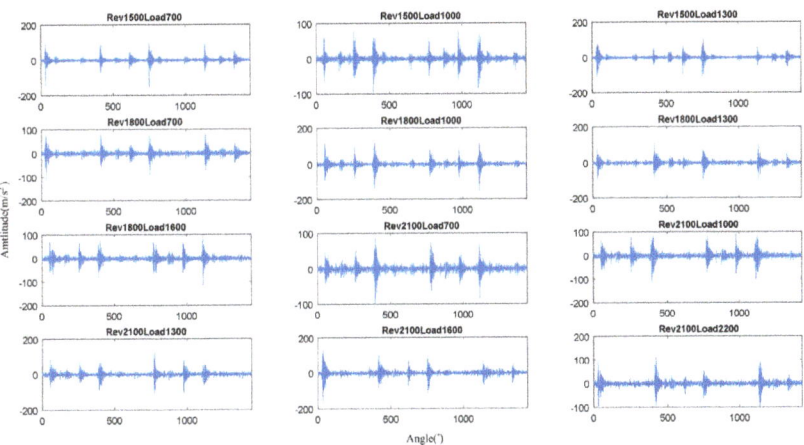

Figure 3. Vibration signals of 12 different operating conditions.

The signals in Figure 3 represent two complete periodic vibration signals, with a certain cyclic fluctuation in the angular domain. When fire combustion and closing of the intake valve and exhaust valve occur, an obvious excitation response is produced in the corresponding phase. Due to the different ignition phase points of different cylinders, the corresponding combustion excitation occurs at different positions. As the amplitude of the vibration signal features large randomness, the vibration signal of a diesel engine can be considered a non-periodic and non-stationary signal. This characteristic of the vibration signal greatly increases the difficulty of multi-factor operating condition recognition.

3. Technical Background

In this study, a deep neural network framework is proposed based on a 1D CNN and LSTM for multi-factor operating condition recognition. For the vibration signal in the form of a time series, a 1D CNN was adopted to extract local features of vibration signals through a 1D convolution kernel. Then, an LSTM was adopted to describe the temporal relationship between local features through a memory unit and gate mechanism. In this way, the combination of the 1D CNN and LSTM could perform well for the analysis of vibration signals.

3.1. 1D CNN

A typical CNN [19] contains three types of network layers: a convolutional layer, pooling layer, and fully connected layer. Some excellent variants of CNN were proposed, such as LeNet-5 [27], AlexNet [19], and VGG-16 [28]. The image recognition ability of these CNN variants is outstanding, and they achieved remarkable results. In CNNs, the receptive field, weight sharing, and pooling can greatly reduce the complexity of the network.

It was proven that a 1D CNN can be applied to the time series analysis of sensor data. In 1D CNNs, features can be extracted from segments through 1D convolution, which is a weighted sum operation between the weight matrix and the vibration data in each segment, with the addition of the overall bias. Every convolution extracts a feature from a local receptive field, and the window of the convolution kernel slides across the entire input sequence with a fixed step to achieve all features. The weight sharing exists to maintain the weights of the convolution kernel in the sliding process. As shown in Figure 4, the size of the i-th convolution kernel is shown, featuring weights (w_{i1}, w_{i2}, w_{i3}) in a 1×3 format, with the bias left out for clarity. The corresponding feature vector F ($f_{i1}, f_{i2}, f_{i3}, \dots, f_{i(n-2)}$) can be obtained from the input signal X ($x_1, x_2, x_3, \dots, x_n$) with one step of the convolution kernel.

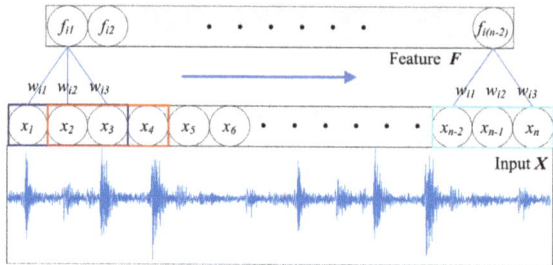

Figure 4. Temporal convolution.

Mathematically, this can be expressed as shown in Equation (1).

$$f_{ij} = \Phi\left(b_i + \sum_{k=1}^{m} w_{ik} \cdot x_{j+k-1}\right), \tag{1}$$

where m is the size of the convolution kernel, f_{ij} is the output of the j-th neuron of the i-th filter in the hidden layer, Φ is the activation function, and b_i is the overall bias of the i-th filter.

Convolution kernels of different sizes can extract features of different granularity [29]. Usually, the first convolutional layer may only extract some low-level features, and more complex features can be extracted from low-level features by stacking network layers.

As the pooling operation can maintain the variance of the translation, rotation, and scale, the pooling layer is set following each convolutional layer to retain the main features. Meanwhile, it can reduce the number of parameters to prevent overfitting and improve the generalizability of the model. In a pooling layer, the features obtained from the activation function are cut into several regions, and the maximum/average values can be taken as the new features to realize dimension reduction. By repeating operations as described above, features can be extracted continuously to improve the generalizability of the CNN.

Enough sensitive important features can be extracted by alternating convolutional and pooling layers, and the fully connected layers can map the distributed feature representation to the sample markup space. Finally, the output layer with a softmax activation function is used for classification.

3.2. LSTM

A recurrent neural network (RNN) is a kind of neural network which can be used for sequential data analysis, while the LSTM is a specific kind of RNN. Compared with a traditional RNN, a memory cell and gating mechanism are introduced to deal with the existence of gradient disappearance and gradient explosion during the training of long sequences. The gating mechanism can be used to control the transfer state, which is designed to remember the important information and forget the unimportant information. The memory cell of an LSTM is shown in Figure 5.

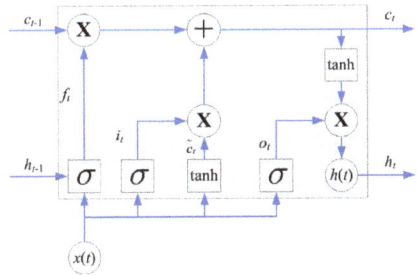

Figure 5. Memory cell of a long short-term memory network (LSTM).

As shown in Figure 5, the memory cell of an LSTM is made up of an input gate, output gate, and forget gate. The sigmoid activation function is used in the forget gate to control the weight of information that can be passed, whereas the tanh activation function is used in the input gate to deal with the input at the current sequence position, and the sigmoid activation function is used in the output gate to update the output based on the results of the input gate and forget gate. Mathematically, the parameters of the LSTM can be updated as shown in Equation (2).

$$\begin{aligned} i_t &= \sigma(W_{xi}x_t + W_{hi}h_{t-1} + b_i); \\ f_t &= \sigma(W_{xf}x_t + W_{hf}h_{t-1} + b_f); \\ o_t &= \sigma(W_{xo}x_t + W_{ho}h_{t-1} + b_o); \\ \widetilde{c_t} &= \tanh(W_{xc}x_t + W_{hc}h_{t-1} + b_c); \\ c_t &= f_t \cdot c_{t-1} + i_t \cdot \widetilde{c_t}; \\ h_t &= o_t \cdot \tanh(c_t). \end{aligned} \quad (2)$$

where x_t is the input of a sequence, c_{t-1} is the last state, and h_{t-1} is the output of the last memory cell. The state c_t and output h_t of the current memory cell can be obtained after parameter update calculation.

4. Methodologies

In this section, the 1D-CLSTM is firstly constructed for multi-factor operating condition recognition, and then adaptive dropout is proposed. Moreover, the flowchart of the multi-factor operating condition recognition method is introduced.

4.1. 1D Convolutional Long Short-Term Network

4.1.1. Overall Architecture

As described above, the features extracted by different neural networks have different characteristics. The 1D CNN can obtain the features of a receptive field through convolution, but the temporal relationship of the vibration signal is ignored as a result of the size of the convolution kernel. As for the LSTM, a temporal relationship can be described through the memory cell and gating mechanism. Therefore, the multi-factor operating condition recognition algorithm 1D-CLSTM is proposed based on a 1D CNN and LSTM. In the proposed neural network framework, the 1D CNN was designed to extract local features of vibration signals through 1D convolution, and the LSTM was designed to describe the temporal relationship between local features. The overall architecture of the 1D-CLSTM is shown in Figure 6.

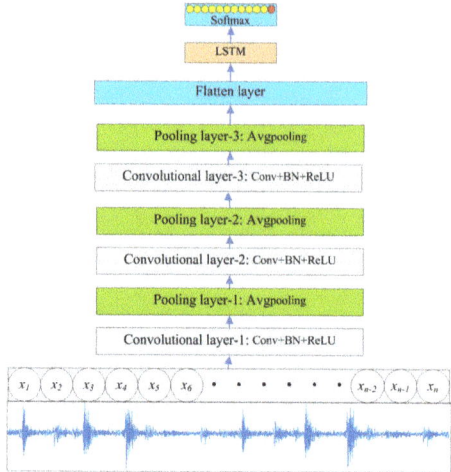

Figure 6. Overall architecture of the one-dimensional convolutional long short-term network (1D-CLSTM).

4.1.2. Architecture Design

According to the sampling frequency of the monitoring system and different operating conditions introduced in Section 2, a signal segment with a length of 4096 can be determined to contain all the information in a cycle. The crankshaft of a four-stroke diesel engine rotates 720 degrees to complete a cycle, which means complete energy conversion. Therefore, the minimum receptive field can be defined as a degree in the angular domain. Moreover, the size of the CNN filter in the first convolutional layer can be set to an odd number greater than 4096/720.

Considering the particularity of a vibration signal, which is a non-periodic and non-stationary signal, BN [30] is vital for regulating the input of some layers by fixing the mean value and variance of input signals of each convolutional layer, through which the features can maintain the same distribution in the training process of the 1D-CLSTM. Upon increasing the number of layers in a neural network, the decreasing convergence rate often leads to gradient explosion or gradient disappearance, and BN is an excellent solution. Therefore, the convolution is followed by BN in each convolutional layer. In all convolutional layers, the rectified linear unit (ReLU) activation function is adopted, and BN occurs in front of the ReLU activation function. In other words, the results of BN are the input of the ReLU activation function. The ReLU activation function makes the output of some neurons equal to 0, which results in sparsity of the network, thereby reducing the interdependence of parameters and alleviating the occurrence of the overfitting problem. The average values of features obtained from the ReLU activation function are taken as the new features to realize dimension reduction in a pooling layer. The designed 1D-CLSTM begins with a sequence input, after which the features can be extracted by alternate convolutional layers and pooling layers.

A complete periodic signal contains different sequential excitation responses; thus, the sequence length processed by the LSTM can be determined according to the degree of excitation responses in the angular domain. When the degree of an excitation response in the angular domain is 15, the number of LSTM units can be chosen to be greater than 720/15. Following the final pooling layer, there is a flattening layer to reshape the tensor as the input of the LSTM with 73 units. In order to accelerate the convergence process of 1D-CLSTM training, adaptive dropout is applied. Finally, the output layer with a softmax activation function is used for multi-class classification. The structural parameters of the 1D-CLSTM are shown in Table 3.

Table 3. Structural parameters of the one-dimensional convolutional long short-term network (1D-CLSTM).

No.	Network Layer	Size of Convolution Kernel	Stride	Output Dimension
1	Input layer	-	-	4096 × 1
2	Convolutional layer-1	11	1	4096 × 32
3	Pooling layer-1	3	2	2047 × 32
4	Convolutional layer-2	13	1	2047 × 64
5	Pooling layer-2	3	2	1023 × 64
6	Convolutional layer-3	15	1	1023 × 128
7	Pooling layer-3	3	2	511 × 128
8	Flatten layer	-	-	73 × 896
9	LSTM (two layers)	-	-	73
10	Softmax	-	-	12

4.1.3. Adaptive Dropout

Dropout is widely used for improving model sparsity and preventing overfitting in model training. The learning process of the 1D-CLSTM for multi-factor operating condition recognition is an iterative one. On account of the mutual influence among interconnected neurons, every iteration is a greedy search, whereby we find the best connections. That is, a connection may be unimportant due to the existence of some others, but it becomes important once the others are removed. Therefore, the adaptive dropout ratio is proposed to deal with this problem.

The most popular Bernoulli dropout technique [31] can be applied to neurons or weights. Assuming the input of a weight or neuron as X, the output as Y, the dropout probability as $\mathbf{P}(\alpha)$, and the weight matrix as W, each neuron is probabilistically dropped at each training step, as defined in Equation (3).

$$\mathbf{Y} = (\mathbf{X} \cdot \mathbf{P})\mathbf{W}. \quad (3)$$

Each weight in the weight matrices is probabilistically dropped at each training step, as defined in Equation (4).

$$\mathbf{Y} = \mathbf{X}(\mathbf{W} \cdot \mathbf{P}). \quad (4)$$

Usually, the dropout ratio α is constant for generating random network structures (for example, 0.5). However, the model capacity is constantly changing within the 1D-CLSTM training. Therefore, the dropout ratio needs to be adaptive to the current network. Neurons or weights are dropped temporarily during training and dropped forever after pruning to solidify the network structure. Compared with the original network structure, the parameters of the current network become sparse after pruning, and the dropout ratio should be reduced.

Assuming that the connection between the input layer and output layer is fully connected, the number of connections can be calculated as shown in Equation (5).

$$C_i = N_i N_o. \quad (5)$$

Since dropout works on neurons, taking C_{io} as the original network and C_{ic} as the current network, the dropout ratio α can be adjusted according to Equation (6).

$$\alpha_c = \frac{\alpha_o N_o}{(N_o + 1)} \sqrt{\frac{C_{ic}}{C_{io}}}, \quad (6)$$

where α_c represents the dropout rate of the current network, and α_o represents the dropout rate of the original network.

4.1.4. Implementation

The loss function, which measures the degree of difference between the predicted value and actual value, is a non-negative real value function. A smaller loss function denotes better robustness of the model. Cross-entropy is frequently used for loss calculation in neural network training, as shown in Equation (7).

$$\text{loss} = -\sum_{i=1}^{n} y_i \log(y_{i\_}), \tag{7}$$

where y_i represents the predicted value, $y_{i\_}$ represents the actual output, and n is the number of training samples.

In the training of the 1D-CLSTM designed for multi-factor operating condition recognition, the learning rate was set to 0.001. Through iterative calculation, the loss of 1D-CLSTM decreased continuously and eventually became stable. Then, the weight of 1D-CLSTM was fixed, allowing the 1D-CLSTM classifier to be used for multi-factor operating condition recognition.

To make the training of the 1D-CLSTM model more efficient and achieve better performance, the training techniques described below were introduced.

Mini-batch gradient descent. Considering the huge calculation in network training, a batch sample was adopted in the training process, and the batch size was set to 128. The batch sample strategy uses less memory and achieves a faster training speed than full batch learning. Compared with stochastic gradient descent, mini-batch gradient descent is more efficient. Compared with batch gradient descent, mini-batch gradient descent can achieve robust convergence to avoid local optimization. Therefore, mini-batch gradient descent was taken as the optimizer to minimize the loss and adjust the weights in the designed 1D-CLSTM.

Early termination. In the process of model training with the training set, the performance of the model is also evaluated with the validation set. The validation error decreases in the beginning as the training error decreases. After a certain number of training steps, the training error still decreases, but the validation error no longer decreases. Therefore, early termination can act as a regulator and effectively avoid overfitting of the model. Once the validation error stops decreasing, the early termination of model training can be enforced in the training of the 1D-CLSTM.

4.2. Multi-Factor Operating Condition Recognition

To determine the multi-factor operating condition information of a diesel engine, a condition recognition method using 1D-CLSTM is proposed. Firstly, acceleration sensors were used to monitor the status information of a diesel engine under different operating conditions. Considering the characteristics of the vibration signal, some performance improvement techniques were adopted in the 1D-CLSTM, such as BN, ReLU activation function, adaptive dropout. Moreover, mini-batch gradient descent and early termination were adopted in the training of 1D-CLSTM to achieve a fast training speed and avoid overfitting of the model. Accordingly, the 1D-CLSTM could be trained using supervised learning. After training, the trained 1D-CLSTM classifier could be used for the classification of multi-factor operating conditions. The flowchart of the multi-factor operating condition recognition method is shown in Figure 7.

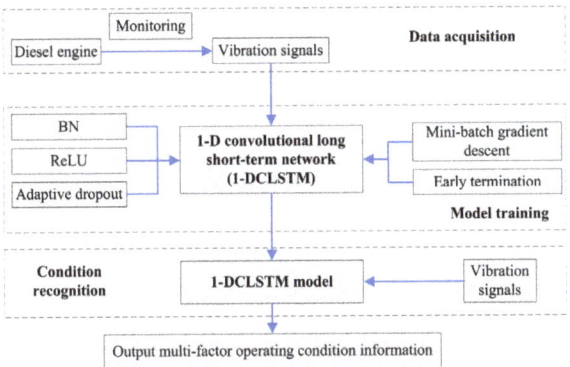

Figure 7. The flowchart of the condition recognition method.

5. Experiments

According to the flowchart shown in Figure 7, the training performance of the designed 1D-CLSTM is presented below. After training, the performance of 1D-CLSTM using vibration signals for multi-factor operating condition recognition was evaluated. Moreover, the vibration signals measured from another kind of diesel engine were applied to verify the generalizability of the proposed approach. Finally, the results of the proposed approach for multi-factor operating condition recognition were compared to other classification algorithms to verify that the designed 1D-CLSTM with strong generalizability could provide higher classification accuracy. The 1D-CLSTM model was written using Python 3.6 with TensorFlow and run on Window 10 with an NVIDIA Quadro P6000.

5.1. Training Performance of the Designed 1D-CLSTM

The vibration signals were in the form of a time series, used as the input data for training the designed network, with a total of 7200 samples. The whole dataset was randomly divided into two sets: 80% for training and 20% for validation. In other words, the training set had 5760 samples, and the validation set had 1440 samples. With the continuous iterative training of 1D-CLSTM, the losses of the training set and validation set decreased as the number of epochs increased, as depicted in Figure 8. On the contrary, the accuracies of the training set and validation set continuously improved, as depicted in Figure 9. According to the early termination, the model training stopped when the loss of the validation set stopped decreasing. The training of 1D-CLSTM stopped at the 63rd epoch when the cross-entropy of the validation set was 0.01913 and the accuracy of the training set was 0.9953. Therefore, the corresponding 1D-CLSTM classifier is a desired classification model for multi-factor operating condition recognition.

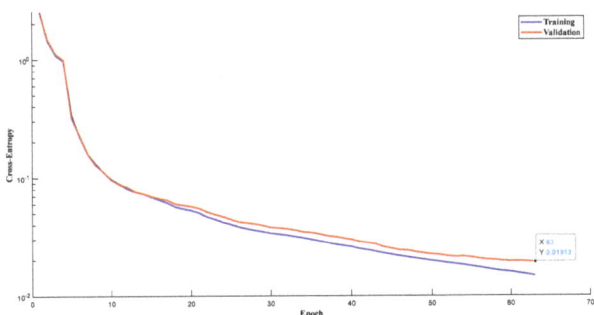

Figure 8. Losses of training set and validation set.

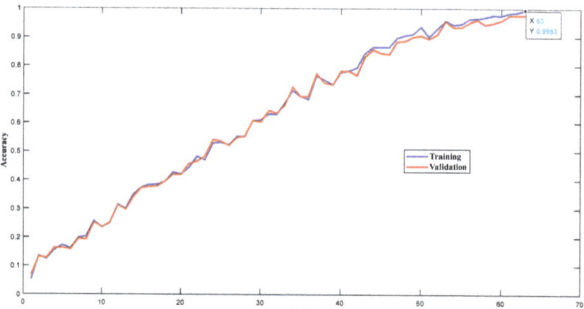

Figure 9. Accuracies of training set and validation set.

A confusion matrix, which contains information about actual and predicted classes, was used to describe the generalizability of the 1D-CLSTM classifier [32]. The testing set had a total of 1200 samples, with 100 samples for each operating condition. The confusion matrix for the testing set is shown in Figure 10.

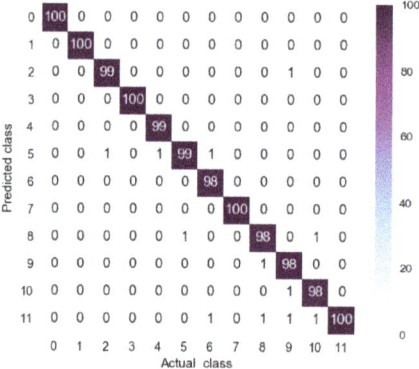

Figure 10. Confusion matrix for testing set.

The elements in row i and column j of the confusion matrix represent the number of times the j class was identified as the i class. Therefore, only the diagonal elements denote correct recognition. It can be seen from Figure 10 that only 11 samples out of 1200 were misclassified. Therefore, the designed 1D-CLSTM can classify multi-factor operating conditions with an accuracy of 99.08%.

5.2. Comparison of Training Performance with Different Dropout Ratios

The convergence process in model training is an important factor for achieving a classifier with excellent performance. Dropout serves as an effective approach to improve the model sparsity and prevent overfitting in model training. To find the best connections in the designed 1D-CLSTM, a suitable dropout ratio was very important. Adaptive dropout, due to its flexibility depending on network capacity, is able to maintain the balance between model performance and model sparsity. To check the effect of adaptive dropout, training accuracy curves of different dropout ratios were plotted, as shown in Figure 11. According to the early termination, the model training using adaptive dropout stopped at the 63rd epoch, and the comparison of training performance with different dropout ratios was conducted within 63 epochs.

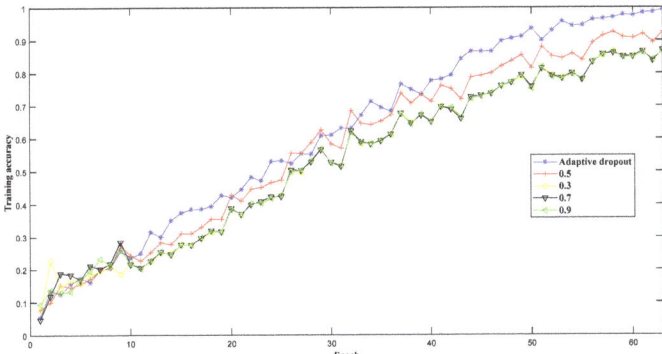

Figure 11. Training accuracy curves with different dropout ratios.

It can be seen from Figure 11 that the training performance using adaptive dropout was best; thus, adaptive dropout can improve the training performance to achieve the desired model.

5.3. Comparison Analysis

To validate the performance of the designed 1D-CLSTM, the proposed method was compared with the following baseline methods:

1. The k-nearest neighbor (kNN) algorithm, which works with a multi-domain feature set [33]. Based on the multi-domain feature set, the kNN algorithm is more suitable than other statistical learning methods.
2. The support vector machine (SVM), which works with a multi-domain feature set. SVM is a kind of generalized linear classifier that can be used for supervised learning.
3. The 1D LeNet-5, which is a convolutional network that has the same network layers as LeNet-5, i.e., two convolutional layers and two fully connected layers. The corresponding structural parameters are listed in Table 4.
4. The 1D AlexNet, which is a convolutional network that has the same network layers as AlexNet, i.e., five convolutional layers and three fully connected layers. The corresponding structural parameters are also listed in Table 4.
5. The 1D VGG-16, which is a convolutional network that has the same network layers as VGG-16, with 1D convolution kernels adopted. The corresponding structural parameters are also listed in Table 4.
6. A traditional LSTM, which has two layers and 32 LSTM units in each layer.

In Table 4, s represents the stride, and the convolution is followed by BN in each convolutional layer.

Table 4. Structural parameters of the 1D-CLSTM.

1D LeNet-5	1D AlexNet	1D VGG-16	
Conv1 [1,11] × 64, s = 1	Conv1 [1,11] × 32, s = 1	Conv1 [1,3] × 16, s = 1	Conv9 [1,3] × 128, s = 1
AveragePooling1 [1,3], s = 2	MaxPooling1 [1,3], s = 2	Conv2 [1,3] × 16, s = 1	Conv10 [1,3] × 128, s = 1
Conv2 [1,13] × 128, s = 1	Conv2 [1,5] × 64, s = 1	MaxPooling1 [1,2], s = 2	MaxPooling4 [1,2], s = 2
AveragePooling2 [1,3], s = 2	MaxPooling2 [1,3], s = 2	Conv3 [1,3] × 32, s = 1	Conv11 [1,3] × 256, s = 1
FC1 (1024)	Conv3 [1,3] × 128, s = 1	Conv4 [1,3] × 32, s = 1	Conv12 [1,3] × 256, s = 1
FC2 (512)	Conv4 [1,3] × 128, s = 1	MaxPooling2 [1,2], s = 2	Conv13 [1,3] × 256, s = 1
softmax	Conv5 [1,3] × 128, s = 1	Conv5 [1,3] × 64, s = 1	MaxPooling5 [1,2], s = 2
-	MaxPooling3 [1,3], s = 2	Conv6 [1,3] × 64, s = 1	FC1 (1024)
-	FC1 (1024)	Conv7 [1,3] × 64, s = 1	FC2 (512)
-	FC2 (512)	MaxPooling3 [1,2], s = 2	softmax
-	softmax	Conv8 [1,3] × 128, s = 1	-

For multi-factor operating condition recognition, the class domains of operating conditions are likely to overlap with each other. Our goal was to develop a multi-factor operating condition recognition method that can achieve high generalization accuracy. Therefore, the same vibration data were used for the training and testing with the above methods, and the corresponding model performance is shown in Table 5.

Table 5. Performance comparison. SVM—support vector machine.

Learning Model	Generalization Accuracy (%)
1D-CLSTM	99.08
LSTM	74.12
kNN with a multi-domain feature set	92.18
SVM with a multi-domain feature set	94.91
1D LeNet-5	94.43
1D AlexNet	97.54
1D VGG-16	98.01

It can be seen from Table 5 that the generalization accuracy of the proposed method was the best. This shows that the 1D-CLSTM learns to predict new data with higher accuracy than other machine learning models and avoids overfitting. In addition, the trained 1D-CLSTM classifier can be used as a good initializer for similar tasks of transfer learning (https://github.com/Larrylyh/Condition_Recognition).

5.4. Generalizability Verification

To verify the generalizability of the proposed approach, the designed 1D-CLSTM was applied to a diesel engine with 20 cylinders (V20DE), which is shown in Figure 12.

Figure 12. The diesel engine with 20 cylinders.

The vibration data under different operating conditions, which are listed in Table 6, were measured.

Table 6. Operating conditions of V20DE.

No.	Rev (rpm)	Load (kN·m)
1	600	0
2	1100	17.7
3	1500	22.6
4	1500	26.6
5	1500	28.3

Generally, the data measured from different engine types vary greatly, and the 1D-CLSTM classifier would need to be trained before use. The test set of V20DE contained 2101 samples, and the corresponding confusion matrix is illustrated in Figure 13. As depicted in Figure 13, 32 samples out of 2101 were misclassified, and the corresponding accuracy was 98.48%.

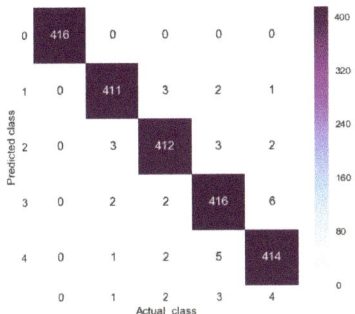

Figure 13. Confusion matrix.

6. Conclusions

In this study, an effective approach was proposed for multi-factor operating condition recognition using a 1D convolutional long short-term network. The proposed method was capable of monitoring and automatically recognizing multi-factor operating conditions based on the vibration signal measured on engine cylinder heads. Moreover, the measured vibration signals no longer needed a complex feature extraction process for condition recognition. Subsequently, adaptive dropout was proposed for improving the model sparsity and preventing overfitting in model training. The experimental results proved that the designed 1D-CLSTM classifier is indeed ideal for multi-factor operating condition recognition with high generalization accuracy. At the same time, adaptive dropout could achieve better training performance than a constant dropout ratio. In addition, this method has the potential for application in real-time scenarios because the implementation of the 1D-CLSTM classifier is simple. Last but not least, the trained 1D-CLSTM classifier can be used as a good initializer for similar tasks of transfer learning. In the future, new studies will be conducted on the transition period between the defined operating conditions to obtain a model that can identify continuous operating conditions. Moreover, continuous operating condition recognition can be the basis of fault detection or diagnosis under variable operating conditions.

Author Contributions: Data curation, Z.M.; funding acquisition, J.Z.; project administration, Z.J. and J.Z.; supervision, Z.J.; writing—original draft, Y.L.; writing—review and editing, Y.L., H.Z., and Z.M.

Funding: This work was supported by the National Key Research and Development Plan of China (Grant No. 2016YFF0203305), the Fundamental Research Funds for the Central Universities of China (Grant No. JD1815), and the Double First-Rate Construction Special Funds (Grant No. ZD1601).

Conflicts of Interest: The authors declare no conflicts of interest.

References

1. Kouremenos, D.A.; Hountalas, D.T. Diagnosis and condition monitoring of medium-speed marine diesel engines. *Lubr. Sci.* **2010**, *4*, 63–91. [CrossRef]
2. Zhiwei, M. Research on Typical Fault Diagnosis and Unstable Condition Monitoring and Evaluation for Piston Engine. Ph.D. Thesis, Beijing University of Chemical Technology, Beijing, China, 2017.
3. Porteiro, J.; Collazo, J.; Patiño, D.; Míguez, J.L. Diesel engine condition monitoring using a multi-net neural network system with nonintrusive sensors. *Appl. Therm. Eng.* **2011**, *31*, 4097–4105. [CrossRef]
4. Xu, H.F. New Intelligent Condition Monitoring and Fault Diagnosis System for Diesel Engines Using Embedded System. *Appl. Mech. Mater.* **2012**, *235*, 408–412. [CrossRef]
5. Xu, X.; Yan, X.; Sheng, C.; Yuan, C.; Xu, D.; Yang, J. A Belief Rule-Based Expert System for Fault Diagnosis of Marine Diesel Engines. *IEEE Trans. Syst. Man Cybern. Syst.* **2017**, *99*, 1–17. [CrossRef]
6. Kowalski, J.; Krawczyk, B.; Woźniak, M. Fault diagnosis of marine 4-stroke diesel engines using a one-vs-one extreme learning ensemble. *Eng. Appl. Artif. Intell.* **2017**, *57*, 134–141. [CrossRef]

7. Shen, H.; Zeng, R.; Yang, W.; Zhou, B.; Ma, W.; Zhang, L.; Projecton Equipment Surport Department, Army Military Transportation University. Diesel Engine Fault Diagnosis Based on Polar Coordinate Enhancement of Time-Frequency Diagram. *J. Vib. Meas. Diagn.* **2018**, *38*, 27–33.
8. Li, Y.; Han, M.; Han, B.; Le, X.; Kanae, S. Fault Diagnosis Method of Diesel Engine Based on Improved Structure Preserving and K-NN Algorithm. In *Advances in Neural Networks-ISNN 2018*; Springer: Cham, Switzerland, 2018.
9. Liu, Y. Research on Fault Diagnosis for Fule System and Valve Train of Diesel Engine Based on Vibration Analysis. Ph.D. Thesis, Tianjin University, Tianjin, China, 2016.
10. Wu, T.; Wu, H.; Du, Y.; Peng, Z. Progress and trend of sensor technology for on-line oil monitoring. *Sci. China Technol. Sci.* **2013**, *56*, 2914–2926. [CrossRef]
11. Tang, G.; Fu, X.; Shao, G.; Chen, N. Application of Improved Grey Model in Prediction of Thermal Parameters for Diesel Engine. *Ship Boat* **2018**, *5*, 39–46.
12. Li, H.; Zhou, P.; Ma, X. Pattern recognition on diesel engine working condition by using a novel methodology-Hilbert spectrum entropy. *J. Mar. Eng. Technol.* **2005**, *4*, 43–48. [CrossRef]
13. Ji, S.; Cheng, Y.; Wang, X. Cylinder pressure recognition based on frequency characteristic of vibration signal measured from cylinder head. *J. Vib. Shock* **2008**, *27*, 133–136.
14. Liu, J.; Li, H.; Qiao, X.; Li, X.; Shi, Y. Engine Cylinder Pressure Identification Method Based on Cylinder Head Vibration Signals. *Chin. Intern. Comhuslion Engine Eng.* **2013**, *34*, 32–37.
15. Chang, C.; Jia, J.; Zeng, R.; Mei, J.; Wang, G. Recognition of Cylinder Pressure Based on Time-frequency Coherence and RBF Network. *Veh. Engine* **2016**, *5*, 87–92.
16. Syed, M.S.; Sunghoon, K.; Sungoh, K. Real-Time Classification of Diesel Marine Engine Loads Using Machine Learning. *Sensors* **2019**, *19*, 3172. [CrossRef]
17. Yoo, Y.; Baek, J.G. A Novel Image Feature for the Remaining Useful Lifetime Prediction of Bearings Based on Continuous Wavelet Transform and Convolutional Neural Network. *Appl. Sci.* **2018**, *8*, 1102. [CrossRef]
18. Rui, Z.; Yan, R.; Chen, Z.; Mao, K.; Wang, P.; Gao, R.X. Deep learning and its applications to machine health monitoring. *Mech. Syst. Signal Process.* **2019**, *115*, 213–237.
19. Krizhevsky, A.; Sutskever, I.; Hinton, G.E. ImageNet classification with deep convolutional neural networks. In Proceedings of the 25th International Conference on Neural Information Processing Systems, Lake Tahoe, NV, USA, 3–6 December 2012.
20. Zhiqiang, C.; Chuan, L.; Sanchez, R.V. Gearbox Fault Identification and Classification with Convolutional Neural Networks. *Shock Vib.* **2015**, *2015*, 1–10.
21. Yuan, J.; Han, T.; Tang, J.; An, L. An Approach to Intelligent Fault Diagnosis of Rolling Bearing Using Wavelet Time-Frequency Representations and CNN. *Mach. Des. Res.* **2017**, *2*, 101–105.
22. Peng, P.; Zhao, X.; Pan, X.; Ye, W. Gas Classification Using Deep Convolutional Neural Networks. *Sensors* **2018**, *18*, 157. [CrossRef]
23. Wu, Y.; Yuan, M.; Dong, S.; Li, L.; Liu, Y. Remaining useful life estimation of engineered systems using vanilla LSTM neural networks. *Neurocomputing* **2018**, *275*, 167–179. [CrossRef]
24. Qing, X.; Niu, Y. Hourly day-ahead solar irradiance prediction using weather forecasts by LSTM. *Energy* **2018**, *148*, 461–468. [CrossRef]
25. Lin, C.; Yuan, Z.; Julie, I.; Muge, C.; Ryan, A.; Jeanne, M.H.; Jeanne, M.H.; Min, C. Early diagnosis and prediction of sepsis shock by combining static and dynamic information using convolutional-LSTM. In Proceedings of the 2018 IEEE International Conference on Healthcare Informatics (ICHI), New York, NY, USA, 4–7 June 2018; pp. 219–228.
26. Andersen, R.S.; Abdolrahman, P.; Sadasivan, P. A deep learning approach for real-time detection of atrial fibrillation. *Expert Syst. Appl.* **2019**, *115*, 465–473. [CrossRef]
27. Lecun, Y.; Bottou, L.; Bengio, Y.; Haffner, P. Gradient-based learning applied to document recognition. *Proc. IEEE* **1998**, *86*, 2278–2324. [CrossRef]
28. Simonyan, K.; Zisserman, A. Very Deep Convolutional Networks for Large-Scale Image Recognition. *arXiv* **2014**, arXiv:1409.1556v6.
29. Szegedy, C.; Vanhoucke, V.; Ioffe, S.; Shlen, J.; Wojna, Z. Rethinking the Inception Architecture for Computer Vision. In Proceedings of the 2016 IEEE Conference on Computer Vision and Pattern Recognition (CVPR), Las Vegas, NV, USA, 27–30 June 2016.

30. Ioffe, S.; Szegedy, C. Batch normalization: Accelerating deep network training by reducing internal covariate shift. In Proceedings of the 32nd International Conference on Machine Learning, Lille, France, 6–11 July 2015; pp. 448–456.
31. Nitish, S.; Geoffrey, H.; Alex, K.; Ilya, S.; Ruslan, S. Dropout: A simple way to prevent neural networks from overfitting. *J. Mach. Learn. Res.* **2014**, *15*, 1929–1958.
32. Visa, S.; Ramsay, B.; Ralescu, A.L.; Van Der Knaap, E. *Confusion Matrix-Based Feature Selection*; MAICS: Metro Manila, Philippines, 2011; pp. 120–127.
33. Yan, X.; Jia, M. A novel optimized SVM classification algorithm with multi-domain feature and its application to fault diagnosis of rolling bearing. *Neurocomputing* **2018**, *313*, 47–64. [CrossRef]

© 2019 by the authors. Licensee MDPI, Basel, Switzerland. This article is an open access article distributed under the terms and conditions of the Creative Commons Attribution (CC BY) license (http://creativecommons.org/licenses/by/4.0/).

Article

Comprehensive Improvement of the Sensitivity and Detectability of a Large-Aperture Electromagnetic Wear Particle Detector

Ran Jia [1], Biao Ma [1], Changsong Zheng [1,*], Xin Ba [1,2], Liyong Wang [3], Qiu Du [1] and Kai Wang [1]

1. School of Mechanical Engineering, Beijing Institute of Technology, Zhongguancun South Street No.5 Haidian District, Beijing 100081, China
2. Faculty of Engineering and Information, University of Technology Sydney, Ultimo, NSW 2007, Australia
3. The Ministry of Education Key Laboratory of Modern Measurement and Control Technology, Beijing Information Science and Technology University, Xiaoying East Street No.12, Beijing 100192, China
* Correspondence: zhengchangsong@bit.edu.cn; Tel.: +86-010-6891-8637

Received: 20 May 2019; Accepted: 16 July 2019; Published: 18 July 2019

Abstract: The electromagnetic wear particle detector has been widely studied due to its prospective applications in various fields. In order to meet the requirements of the high-precision wear particle detector, a comprehensive method of improving the sensitivity and detectability of the sensor is proposed. Based on the nature of the sensor, parallel resonant exciting coils are used to increase the impedance change of the exciting circuit caused by particles, and the serial resonant topology structure and an amorphous core are applied to the inductive coil, which improves the magnetic flux change of the inductive coil and enlarges the induced electromotive force of the sensor. Moreover, the influences of the resonance frequency on the sensitivity and effective particle detection range of the sensor are studied, which forms the basis for optimizing the frequency of the magnetic field within the sensor. For further improving the detectability of micro-particles and the real-time monitoring ability of the sensor, a simple and quick extraction method for the particle signal, based on a modified lock-in amplifier and empirical mode decomposition and reverse reconstruction (EMD-RRC), is proposed, which can effectively extract the particle signal from the raw signal with low signal-to-noise ratio (SNR). The simulation and experimental results show that the proposed methods improve the sensitivity of the sensor by more than six times.

Keywords: particle detection; sensitivity; resonance; amorphous core; signal extraction

1. Introduction

Wear is one of the major causes of failure in machine components. The excessive wear of some core parts of machineries, especially for large-scale mechanical equipment, may lead to a poor mechanical performance, which in turn causes enormous economic losses. Therefore, for online monitoring of the wear condition of machineries in order to prevent serious malfunctions, the wear particle detector has demonstrated its value [1–3]. To date, wear particle detectors with different physical principles, including optics, ultrasonics, electronics, and imaging, have been proposed, and the characteristics of the various kinds of sensors are listed in Reference [4]. Among them, electromagnetic wear particle detectors have demonstrated significant advantages in online wear condition monitoring because of their strong anti-interference ability, good temperature stability, and high reliability.

To achieve a better particle detection effect, sensors with different structures have long been objects of study. Flanagan et al. [5] proposed a wear particle detector with a single coil (inner diameter of 6 mm), which identifies particles by the fluctuation of the sensor resonance frequency. Experimental results showed that the sensor could detect iron particles with a diameter of 150 μm. Fan et al. [6]

designed a double-coil wear particle detection sensor. It estimates the size and the material properties of particles by measuring the inductance difference between the sensing coil and the reference coil of the sensor and can successfully detect 100 μm ferromagnetic particles and 500 μm non-ferromagnetic particles. To improve the consistency of the particle detection results, a sensor with planar spiral coils [7] was proposed. The simulation and experimental results showed that the uniformity of the magnetic field in the detection area was greatly improved, however, the sensor could only detect the ferromagnetic particles with a diameter of 700 μm. Further, Hong et al. [8] designed a radial inductive debris detection sensor that consisted of a C-type iron core, a drive coil, and an inductive coil. The experimental results indicated that the sensor could effectively detect a 290 μm ferromagnetic particle in a 20 mm diameter pipe. However, the magnetostatic field was adopted in this sensor, so it could not detect non-ferromagnetic particles. To improve the sensitivity of the sensor, the wear particle detector with a parallel three-coil structure was studied [9–11]. The study demonstrated that the sensor could detect approximately 100 μm ferromagnetic particles and 305 μm non-ferromagnetic particles in a 7.6 mm diameter channel. However, the sensitivity and the detectability are still the main obstacles for the development and application of the wear particle detector. Therefore, some measures have been taken to further improve the sensitivity of the sensor. The most direct and valid approach is adopting micro-channel structures [12]. The typical feature of this kind of sensor is that the diameter of the inner channel is smaller than 1 mm, which reduces the distance between target particles and sensor coils. Du et al. [13] proposed a micro-channel device based on an inductive coulter counting principle to detect metal wear particles in lubricating oil. The device could detect about 50 μm ferromagnetic particles and 125 μm non-ferromagnetic particles. Wu et al. [14] designed a microfluidic chip-based inductive wear particle detection device. For this sensor, the inner diameter of the coil was set to 200 μm, and the experimental results revealed that it could detect ferromagnetic particles with a diameter of 5–10 μm. Although the sensitivity of the sensor was greatly enhanced, the small channel diameter of the sensor greatly limits its application to large-scale machineries. Besides that, Li et al. [15] carried out a study to improve the sensitivity of a single-coil wear particle detector. They innovatively proposed that adding an external capacitor to the sensor coil and making the sensor work in a parallel resonance state could boost the sensitivity of the sensor. Recently, Zhu et al. [16] added a ferrite core to the single-coil wear debris detection sensor for the enhancement of sensor sensitivity. With this method, the sensor could detect 11 μm ferromagnetic particles in fluidic pipes with a diameter of 1 mm under a throughput of 750 mL/min.

The size of the minimum detectable particle and the real-time ability of the sensor are also limited by the noise level of the raw signal and the performance of the particle signal extraction algorithm. Fan et al. [17] presented a joint time-invariant wavelet transform and kurtosis analysis approach to extract the effective particle signal. This method depresses the background noise of a raw signal by a threshold. In this way, the wear particle detection effect is greatly influenced by the environmental noise. Li et al. [10,18,19] adopted the maximal overlap discrete wavelet transform to remove vibration interferences from the raw signal. Luo et al. [20] integrated the resonance-based signal decomposition method and fractional calculus (RSD-FC) to improve the detection accuracy of the sensor. These methods do improve the particle detection effect to a certain degree, but they are only valid when the signal-to-noise ratio (SNR) of the signal is sufficiently high, which generally means higher than 2 dB. Meanwhile, overcomplicated algorithms require a relatively high computational cost, which makes the sensor unsuitable for application to continuous real-time monitoring [21].

To meet the requirements of the high-precision wear particle detector and improve the micro-particle detection effect, a comprehensive method of improving the sensitivity and detectability of the sensor is proposed. Based on the essential features of the sensor, a parallel resonance topology and a series resonance topology are applied to the exciting coil and the inductive coil respectively, to comprehensively boost the sensitivity of the sensor. In addition, the influence of resonance frequency on the sensitivity and effective particle detection range of sensors is studied, which lays the foundation for optimizing the frequency of the magnetic field within the sensor. To further improve the induced electromotive force, an amorphous iron core is added to the inductive coil. The high permeability and the low hysteresis loss and eddy current loss of the amorphous material contribute to improving the sensitivity and keeping the performance of the sensor under a high-frequency alternating magnetic field. Additionally, to improve the real-time performance of wear monitoring, a quick extraction method of the particle signal, based on a modified lock-in amplifier and empirical mode decomposition, is proposed. This method dramatically reduces the amount of computation of the system and can quickly extract the particle signal from the raw signal with an extremely low signal-to-noise ratio (SNR).

2. Device Description and Measurement Setup

2.1. Sensor Description

The core structure of the proposed wear particle detector is shown in Figure 1. Differing from the conventional wear particle detection sensor, which only includes a coil frame, two reverse exciting coils, and an inductive coil, the proposed particle detector adopts the resonance principle and an amorphous iron core to compressively improve its sensitivity. Based on the features of the sensor, the parallel resonance topology is used for the exciting coil to boost the impedance change of the coil caused by particles. Moreover, the series resonance principle is applied to the inductive coil to improve the induced electromotive force. Therefore, the resonant capacitors C_1 and C_3 are connected to the left and right exciting coils of the sensor in parallel, and the resonant capacitor C_2 is connected to the inductive coil in series. The general working principle of the sensor has been expounded in Reference [22]. In order to achieve the flow requirements of wear monitoring for large-scale machines, the inner diameter of the sensor is set to 7 mm.

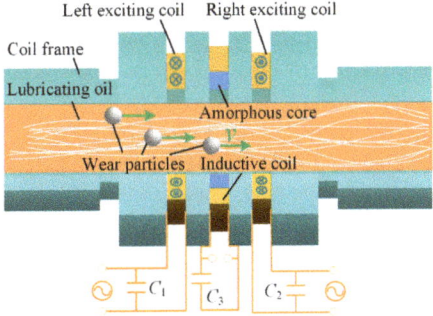

Figure 1. The structure of the proposed wear particle detector.

The metal wear particles passing through the sensor lead to magnetic perturbation of the sensor. More specifically, ferromagnetic particles enhance the local magnetic flux density, while non-ferromagnetic particles decrease the local magnetic flux density [22]. In these cases, the change of the magnetic flux through the exciting coil and the inductive coil can be expressed as (1) and (2), respectively:

$$\Delta \phi_e = \sum \int \Delta B_p(x,y) ds = \Delta(L \times I) \tag{1}$$

$$\Delta \phi_i = K(1-\lambda)(\phi_{e1} - \phi_{e2}) \tag{2}$$

where, ϕ_e is the magnetic flux through the exciting coil, ΔB_p is the change of magnetic flux density in the sensor caused by particles, L is the inductance of the exciting coil, I is the current through the exciting coil, K is the gain factor of magnetic flux through the inductive coil, λ is the magnetic flux leakage coefficient, which is closely related to the sensor structural parameters, and ϕ_{ei} is the magnetic flux through the ith exciting coil.

The induced electromotive force output by the inductive coil can be expressed as (3), where N_i is the number of turns of the inductive coil:

$$E_0 = -N_i \frac{\Delta \phi_i}{\Delta t} \approx -K N_i (1 - \lambda) \frac{\Delta (L \times I)}{\Delta t}. \tag{3}$$

From the above equation, we can see that for the sensor with certain structural parameters, the magnitude of the induced electromotive force is related to the product of the inductance of the exciting coil and current through the exciting coil, and the gain factor K. Because the change of coil inductance caused by wear particles is extremely weak, one method of improving the sensitivity of the sensor is to enlarge the current variation through the exciting coils, which is closely associated with the impedance change of the exciting circuit caused by particles. Meanwhile, this research proves that a series-resonant inductive coil and an amorphous core can boost the gain factor K. The mechanism of enhancing the sensitivity of the sensor is explained in detail in the following section.

2.2. A Sensitivity Comparison Analysis of the Sensors

To demonstrate the mechanism of sensitivity improvement by the resonant principle and the amorphous core, a sensitivity comparison analysis of the conventional and proposed wear particle detector was conducted. The circuit diagrams of the sensors are displayed in Figure 2a,b, where L_1 and L_2 are the inductances of the exciting coils, L_3 is the inductance of the inductive coil, C_1, C_2, and C_3 are the resonant capacitors for each coil, and the internal resistances of these coils are $r_1 = r_2 = 4.1\ \Omega$ and $r_3 = 4.3\ \Omega$. For the proposed sensor, as shown in Figure 2b, the resonance condition must be satisfied as Equation (4), where f_0 is the resonant frequency.

$$f_0 \approx \frac{1}{2\pi \sqrt{LC}} \tag{4}$$

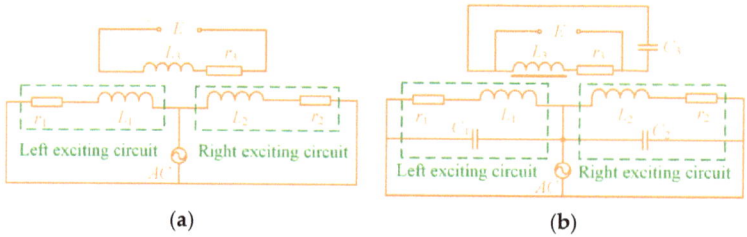

(a)   (b)

Figure 2. The circuit diagrams of the sensors. (**a**) The conventional sensor, (**b**) the proposed sensor.

The impedance change of the exciting circuit caused by particles can characterize the sensitivity of the sensor indirectly. When no particles enter the sensor, the impedance of each exciting circuit of the two sensors, as shown in Figure 2a,b, can be expressed as (5) and (6), respectively. Here, Z_a and Z_b are the impedances of the non-resonant and resonant exciting circuits respectively, $L_q = L_i - M$ is the equivalent inductance of a single exciting coil, L_i is the self-inductance of the ith exciting coil, and M is the mutual inductance between the two exciting coils. Note that, under the resonance state, $1 - \omega^2 L_q C \approx 0$ and $\omega C r \ll 1$, so it can be obtained that $Z_b \gg Z_a$.

$$Z_a = j\omega L_q + r \tag{5}$$

$$Z_b = \frac{(j\omega L_q + r)}{1 - \omega^2 L_q C + j\omega C r}. \tag{6}$$

When wear debris gets access to the sensor, the inductance of one of the two exciting coils changes, which further leads to an impedance difference between the two exciting circuits. Taking the ferromagnetic particle as an example, the inductance-change of a coil caused by a ferromagnetic particle with a radius of r_a can be expressed as (7) [23]:

$$\Delta L = \frac{(\sqrt{5}-1)\mu_0\mu_r N^2 r_a^3}{l^2}. \tag{7}$$

Here, $\mu_0 = 4\pi \times 10^{-7}$ H/m is the permeability of the vacuum, μ_r is the relative permeability, N is the number of turns of the coil, and l is the width of the coil.

The impedance differences between the exciting circuits of the two sensors, as shown in Figure 2a,b, are given by:

$$\begin{aligned} \Delta Z_a &= j\omega \Delta L \\ \Delta Z_b &= \frac{j\omega \Delta L}{(1-\omega^2 C(L+\Delta L)+iCr\omega)(1-\omega^2 CL+iCr\omega)} \end{aligned} \tag{8}$$

To characterize the sensitivity of the two sensors, the impedance differences between the exciting circuits of each sensor are calculated by MATLAB (MathWorks, USA) and shown in Figure 3. During the calculation, the equivalent inductance of the exciting coils is $L_{q1} = L_{q2} = 270.2$ μH, which is obtained from experimental measurement, the exciting frequency is set to $f_0 = 134.5$ kHz, and the corresponding resonant capacitances are $C_1 = C_2 = 5.17$ nF. It can be seen that for the sensor with a non-resonance principle, the impedance difference slowly grows with the increase of the particle diameter, and that it is merely 0.41 Ω when the diameter of the ferromagnetic particle is 750 μm. However, for the sensors with resonant exciting coils, the impedance difference rises rapidly with the increase of particle diameter, reaches a peak value (3.99 Ω) at the position of r1 (528 μm), and then decreases sharply. Therefore, the obvious impedance difference between the exciting circuits of the proposed sensor signifies that the parallel resonant exciting coil does improve the sensitivity of the sensor to a certain extent. However, the nonlinear characteristics of the impedance difference mean that different sized particles, such as the particles with the diameter of r_p and $r_p\prime$, may lead to the same impedance change, and even the impedance change, caused by the particle larger than r_2 in diameter, turns negative, which means that the large ferromagnetic particle may be recognized as a non-ferromagnetic particle. Therefore, for correctness of the particle detection result, the effective detection range of the proposed sensor is restricted to $(0, r_1)$.

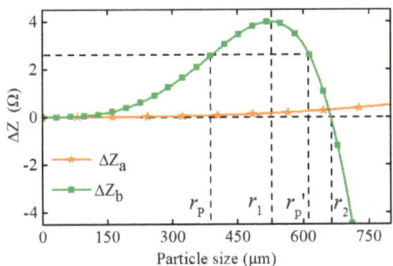

Figure 3. The impedance difference between exciting circuits of different sensors.

To effectively monitor the initial abnormal wear stage of the machinery, some measures must be taken to improve the detectability for micro particles. It is calculated that for the proposed sensor, the resonance capacitance (or resonance frequency) greatly affects the peak position of ΔZ_b. The impedance differences between the two exciting circuits with different resonance capacitors are displayed in Figure 4. It can be seen that with the decrease of the capacitance, the impedance difference curve shifts to the left, which reduces the particle detection range of the sensor to $(0, r_a\prime)$, but enhances the impedance difference between the two exciting circuits caused by micro particles. Therefore, the smaller resonance capacitance (higher resonance frequency) contributes to the detection of micro wear particles. However, that greatly increases the current through the exciting coils and makes the sensor produce more heat, which is harmful to the reliability of the sensor. Meanwhile, the excessive field frequency increases the magnetic losses in particles, which weakens the detectability for ferromagnetic particles. Considering the above factors, a real well-selecting experiment was conducted, and the results showed that a resonant capacitance of 1nF is appropriate for the detection of ferromagnetic particles. In this situation, the detection range of the sensor was restricted to (0, 300) μm.

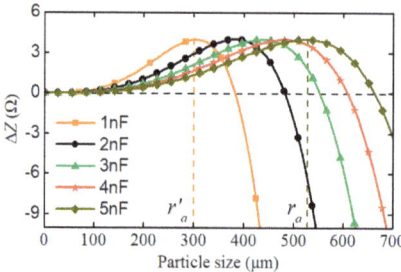

Figure 4. Impedance difference between exciting circuits of the proposed sensor with different resonant capacitors.

The impedance change of exciting coils caused by particles leads to current redistribution, which is one of the key factors of improving the sensitivity of the sensor. Under this circumstance, the current difference between exciting coils, for the sensors shown in Figure 2a,b, can be expressed as (9) and (10), respectively:

$$\Delta I_a = I_0 (\frac{1 + \Delta Z_a/Z_a}{2 + \Delta Z_a/Z_a})(1 - \frac{1}{1 + \Delta Z_a/Z_a}) \qquad (9)$$

$$\Delta I_b = I_0 \frac{Z_b}{Z_a}(\frac{1 + \Delta Z_b/Z_b}{2 + \Delta Z_b/Z_b})(1 - \frac{1}{1 + \Delta Z_a/Z_a}). \qquad (10)$$

Note here that, when the particle diameter is distributed in the range $(0, r_a')$, $Z_b > Z_a$ and $\Delta Z_b/Z_b > \Delta Z_a/Z_a$. Therefore, we obtain:

$$\Delta I_b = \frac{Z_b}{Z_a}(\frac{1 + \Delta Z_b/Z_b}{2 + \Delta Z_b/Z_b})(\frac{2 + \Delta Z_a/Z_a}{1 + \Delta Z_a/Z_a})\Delta I_a \gg \Delta I_a. \qquad (11)$$

The combination of (3) and (11) implies that the parallel resonant exciting coil can essentially improve the induced electromotive force. Meanwhile, Equations (2) and (3) indicate that increasing the magnetic flux through the inductive coil is helpful to further enhance the detectability for micro wear particles and boost the sensitivity of the sensor. Therefore, an amorphous iron core is added to the inductive coil. For the inductive coil, the difference in the magnetic flux density between the two exciting coils can be equivalent to a weak external magnetic field H_p, which produces the magnetic flux of the inductive coil. Based on the equation of $B = \mu H$, $\varphi = \sum \int B ds$, it can be obtained that a ferrite core with a high permeability can boost the external magnetic field and enhance the magnetic flux of the inductive coil. To demonstrate the enhancement effect of the magnetic flux by the amorphous core, a simulation was conducted using the software of COMSOL Multiphysics (COMSOL, Stockholm, Sweden). The simulation parameters used were obtained from the experimental system (illustrated in Section 3). The magnetic fluxes of the inductive coil caused by a 100 μm iron particle for the sensors are displayed in Figure 5. It can be seen that the magnetic flux through the inductive coil of the sensor with the amorphous core increases significantly. In this case, a larger induced electromotive force is produced by the inductive coil.

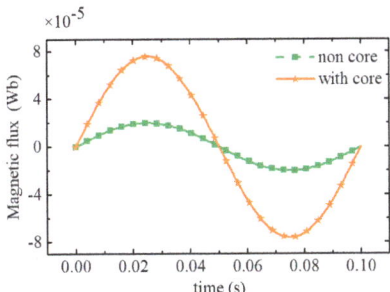

Figure 5. Magnetic flux of the inductive coil.

To further magnify the induced electromotive force caused by particles, the series resonance principle is adopted for the inductive coil and the capacitor C_3 also needs to meet the resonance condition as (4). It is noteworthy that the resonance frequency should maintain a consistent value with the exciting frequency f_0 and the inductive coil can be regarded as a power source. Under the series resonant state, the current through the coil reaches a peak as (12), and the output signal of the sensor can be expressed as (13). The result shows that the series resonant inductive coil magnifies the output signal of the sensor, and the magnification can be comprehensively described as the quality factor of the induction coil. In this situation, the stray capacitance of the coil and the equivalent series resistance of the resonant capacitor cannot be neglected, so it is difficult to directly calculate the quality factor. We measured the quality factor using a digital electric bridge tester (TH2821B) and obtained an approximate value of 3.22, which indicates that the output signal of sensor $E_s \approx 3.22 E_0$:

$$I_3 = \frac{E_0}{(r_3 + j\omega L_3 + 1/j\omega C_3)} \approx \frac{E_0}{r_3} \tag{12}$$

$$E_s = I_3(r_3 + j\omega L_3) = E_0 \sqrt{1 + (\omega L_3/r_3)^2} > E_0. \tag{13}$$

Here, I_3 is the current through the inductive coil under the resonant state, and E_0 and E_s are the induced electromotive forces output by the inductive coil and the sensor, respectively.

Consequently, adding an amorphous iron core to the inductive coil and making it work in the series resonance state are two significant methods of further improving the sensitivity of the sensor.

2.3. Particle Signal Measurement Setup

For the proposed sensor, because of the weak inhomogeneity of the magnetic field between the exciting coils, the initially induced electromotive force interference is produced when no particles pass through the sensor. By analyzing the characteristics of the sensor signal, it can be obtained that the real output signal is composed of the effective particle signal, initially induced electromotive force interference, and environmental interference. The real sensor signal can be expressed as:

$$E_s = E_0 \sqrt{1 + (\omega L_3/r_3)^2} = (E(r_a, v) \sin(\omega_1 t + \varphi_2) + E(\Delta)) \sin(\omega_0 t + \varphi_1) + N(t) \tag{14}$$

where, $E(r_a, v) \sin(\omega_1 t + \varphi_2)$ is the effective particle signal, $E(\Delta) \sin(\omega_0 t + \varphi_1)$ is the initially induced electromotive force interference, ω_0 and ω_1 are the angular frequencies of the exciting signal of the sensor and the effective particle signal respectively, and $N(t)$ is the Gaussian noise resulting from environmental interference.

A measurement system for weak signals is crucial for the detection of wear particles. For satisfying the high real-time requirements of online wear monitoring, a new signal extraction method, based on a modified lock-in amplifier (MLIA) and empirical mode decomposition (EMD), is proposed. Compared with conventional peak-detection (PD) algorithms [17,18,20], the proposed method is much simpler and faster. It can adapt to circumstances with an extremely low signal-to-noise ratio (SNR). Figure 6 shows the block diagram of the signal measurement system. The frequency synthesizer is used to adjust the frequency of the exciting signal to satisfy various monitoring situations. A capacitance matcher is applied to match suitable capacitances for sensor coils. The process of particle signal extraction includes the pre-detection process, preliminary signal extraction, and signal shaping. In the pre-detection process, the raw signal of the sensor is amplified and then filtered by a power frequency filter and an anti-aliasing filter to remove the 50 Hz interference and the high-frequency interference which is generally caused by mechanical vibration of the sensor. For preliminary signal extraction, a modified lock-in amplifier (MLIA) is proposed. In contrast to a conventional lock-in amplifier (LIA), the MLIA adopts two Bessel-type band-pass filters with a center frequency f_0 due to the essential feature of the sensor signal, and the effective particle signal is amplitude-modulated by a sinusoidal signal with a frequency of f_0. Besides that, to quickly eliminate the initially induced electromotive force interference, a Bessel high-pass filter with a cut-off frequency of 5 Hz was used. Because the extraction effect of the particle signal is relevant to the function of these filters and SNR of the raw signal, to adapt the detection requirement of the particles with different speeds, the raw signal is always under-filtered by these filters. Therefore, some unfiltered Gaussian interference still exists in the particle signal, which lowers the detection effect for particles, especially for particles with a low speed. Hence, the particle signal-shaping method based on the EMD is proposed.

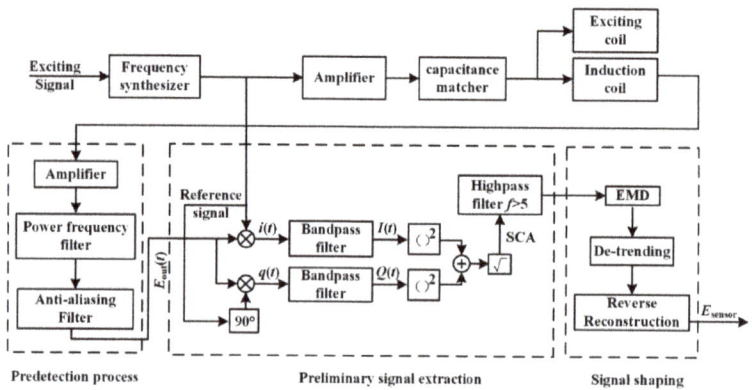

Figure 6. Block diagram of the signal measurement system.

In the procedure for preliminary signal extraction, the reference signal of MLIA is set to $A\sin(\omega_0 t + \varphi_3)$, which has the same frequency as the exciting signal. After that, the raw signal is multiplied by both the reference signal and a signal in quadrature with respect to a reference signal of $A\cos(\omega_0 t + \varphi_3)$. The signals of $i(t)$ and $q(t)$ can be obtained as (15) and (16), respectively. It can be seen that $i(t)$ and $q(t)$ consist of three parts: the amplitude component, high-frequency part (frequency is $2f_0$), and noise sector:

$$\begin{aligned}i(t) &= ((E(r_a,v)*\sin(2\pi f_1+\varphi_2)+E(\Delta))*\sin(2\pi f_0 t+\varphi_1)+N(t))A\sin(2\pi f_0 t+\varphi_3)\\ &= \tfrac{A}{2}(E(r_a,v)*\sin(2\pi f_1+\varphi_2)+E(\Delta))*\cos(\varphi_1-\varphi_3)-\tfrac{A}{2}(E(r_a,v)*\sin(2\pi f_1+\varphi_2)\\ &\quad +E(\Delta))*\cos(2*2\pi f_0 t+\varphi_1+\varphi_3)+N(t)*A\sin(2\pi f_0 t+\varphi_3)\end{aligned} \quad (15)$$

$$\begin{aligned}q(t) &= ((E(r_a,v)*\sin(2\pi f_1+\varphi_2)+E(\Delta))*\sin(2\pi f_0 t+\varphi_1)+V(t)+N(t))*A\cos(2\pi f_0 t+\varphi_2)\\ &= \tfrac{A}{2}(E(r_a,v)*\sin(2\pi f_1+\varphi_2)+E(\Delta))*\sin(\varphi_1-\varphi_3)+\tfrac{A}{2}(E(r_a,v)*\sin(2\pi f_1+\varphi_2)\\ &\quad +E(\Delta))*\sin(2*2\pi f_0 t+\varphi_1+\varphi_3)+N(t)*A\cos(2\pi f_0 t+\varphi_3)\end{aligned} \quad (16)$$

After the MLIA's band-pass filters, the high-frequency component and most of the noise interference can be removed. Therefore, the following signals are obtained:

$$I(t) = \frac{A}{2}(E(r_a,v) * \sin(2\pi f_1 + \varphi_2) + E(\Delta)) * \cos(\varphi_1 - \varphi_3) \tag{17}$$

$$Q(t) = \frac{A}{2}(E(r_a,v) * \sin(2\pi f_1 + \varphi_2) + E(\Delta)) * \sin(\varphi_1 - \varphi_3). \tag{18}$$

The estimation of the specific component amplitude (SCA) is given by (19). There are two sectors in the SCA: a sinusoidal component with a frequency of f_1, which involves the effective particle signal, and a direct component that reflects the amplitude of the initially induced electromotive force interference. Therefore, a Bessel high-pass filter with a cut-off frequency of 5 Hz is used to remove the DC interference component, and the effective particle signal is then obtained as (20):

$$\text{SCA} = \sqrt{I(t)^2 + Q(t)^2} = \frac{A}{2}(E(r_a,v) * \sin(2\pi f_1 + \varphi_2) + E(\Delta)) \tag{19}$$

$$E_{\text{sig}} = \frac{A}{2}(E(r_a,v) * \sin(2\pi f_1 + \varphi_2)). \tag{20}$$

That the cut-off frequency of the high-pass filter is 5 Hz means that the allowable minimal speed of particles passing through the sensor is $v = f_1 * l = 5 * 11 \times 10^{-3} = 5.5 \times 10^{-2}$ m/s, and the corresponding allowable minimum quantity of flow is $V = \pi v d^2/4 = 0.127$ L/min. Here, l is the outer distance between the exciting coils and d is the inner diameter of the sensor.

Although the modified lock-in amplifier can preliminarily extract the weak particle signal and greatly improve the SNR of the signal, there is still some unfiltered Gaussian interference which influences the accurate judgment of the signal amplitude. Therefore, the signal-shaping method based on the EMD-RRC (empirical mode decomposition and reverse reconstruction) is adopted. EMD is an adaptive time-frequency signal processing method used to decompose non-stationary or nonlinear data into several elementary intrinsic mode functions (IMFs), which contain the local features of the raw signal at different time scales. The detailed decomposition process is stated in [24,25]. The preliminarily extracted particle signal can be decomposed by the EMD method as:

$$E_{\text{sig}} = \sum_{i=1}^{k} c_i(t) + r(t) \tag{21}$$

where, $c_i(t)$ is the ith intrinsic mode function and $r(t)$ is the residual term.

Based on the theory of the EMD, the low-order IMFs contain the high-frequency component of the raw signal, and the high-order IMFs and the residual term represent the low-frequency trend component of the signal. Considering the preliminarily extracted particle signal, in order to eliminate the residual interference, the trend component with a low frequency should be removed first. Hence, a trend component identification method is adopted. In this method, the trend component is identified as [10]:

$$m(t) = \sum_{i=k_1}^{k} c_i(t) + r(t) \tag{22}$$

where, k_1 is the trend order of IMFs which satisfies:

$$\begin{array}{l} \prod_{i=k_1}^{k} \left(\left|\text{Mean}(c_i(t))\right| - H_T\right) > 0 \\ \prod_{i=1}^{k_1-1} \left(\left|\text{Mean}(c_i(t))\right| - H_T\right) < 0 \end{array} \tag{23}$$

where, Mean(.) denotes the mean function, and $H_T = 0.05|\text{Mean}(r(t))|$ is the threshold.

To further eliminate the high-frequency interference, a reverse reconstruction method is proposed to reconstruct the signal of the particle. This method gradually adds lower-order IMFs to the detrended highest-order IMF, which produces a series of reconstruction signals expressed as:

$$E_{\text{rsig}}^j = \sum_{i=k_1-j}^{k_1-1} c_i(t). \tag{24}$$

The best denoising effect means the maximal correlation between the particle signal and an ideal sinusoidal signal. Hence, the synthesized correlation coefficient as (25) is used to evaluate these reconstructed signals and to select the best reconstruction order:

$$\rho_{\text{rsig}}^j = \frac{\text{COV}(E_{\text{rsig}}^j, E_{\text{std}})}{\sqrt{E_{\text{rsig}}^j}\sqrt{E_{\text{std}}}}. \tag{25}$$

Here, COV(.) denotes the covariance function and E_{std} is an ideal sinusoidal signal.

The array of synthesized correlation coefficients for the different reconstruction particle signals is established as:

$$\rho_{\max} = \max(|\rho_{\text{rsig}}^1|, |\rho_{\text{rsig}}^2|, \ldots, |\rho_{\text{rsig}}^j|). \tag{26}$$

Combining Equations (24)–(26), the best reconstruction signal is expressed as:

$$E_{\text{out}} = \sum E_{\text{rsig}}^j * (\text{sgn}(|\rho_{\text{rsig}}^j| - \rho_{\max}) + 1). \tag{27}$$

The signal extraction process is simulated by MATLAB SIMULINK and the signal-to-noise ratio (SNR), as shown Equation (28), is used to evaluate the effect of the proposed signal measurement system. In addition, to illustrate the influence on the signal detection effect by the initially induced electromotive force interference, the signal-to-harmonics ratio (SHR) is defined as (29).

$$\text{SNR} = 10 \log_{10}{(P_P/P_N)} \tag{28}$$

$$\text{SHR} = \frac{E_p|_{p-p}}{E_0|_{p-p}}. \tag{29}$$

Here, P_P and P_N are the power of the effective particle signal and the noise signal respectively, E_p is the effective particle signal, E_0 is the initially induced electromotive force, and the subscript p-p means the peak-to-peak value.

The simulation is conducted on the condition that the effective particle signal is $E_0 = 5 \times 10^{-5} \sin(2\pi f_0 t)$, SHR equals 1/100, the variance of Gaussian noise is 1e-8, and the signal amplification factor is 100. In this situation, the raw signal of the sensor is demonstrated in Figure 7a, which shows that the particle signal is fully submerged in the interference, and the SNR of the raw signal is as low as −21.37 dB. The preliminarily extracted particle signal is displayed in Figure 7b. It can be seen that the interference component is greatly removed from the raw signal, however, the residual interference still influences the amplitude recognition. In the process of signal-shaping, the preliminarily extracted signal is decomposed into several IMFs and a residual component by the EMD method, as shown in Figure 7c. Based on Equations (21)–(25), the IMF5 and the residual component are regarded as low-frequency trend components and the IMF1 and IMF2 are treated as high-frequency interference. After eliminating all the interference, the reconstructed signal can be obtained, as shown in Figure 7d. It shows that the shaped particle signal has obvious sinusoidal characteristics.

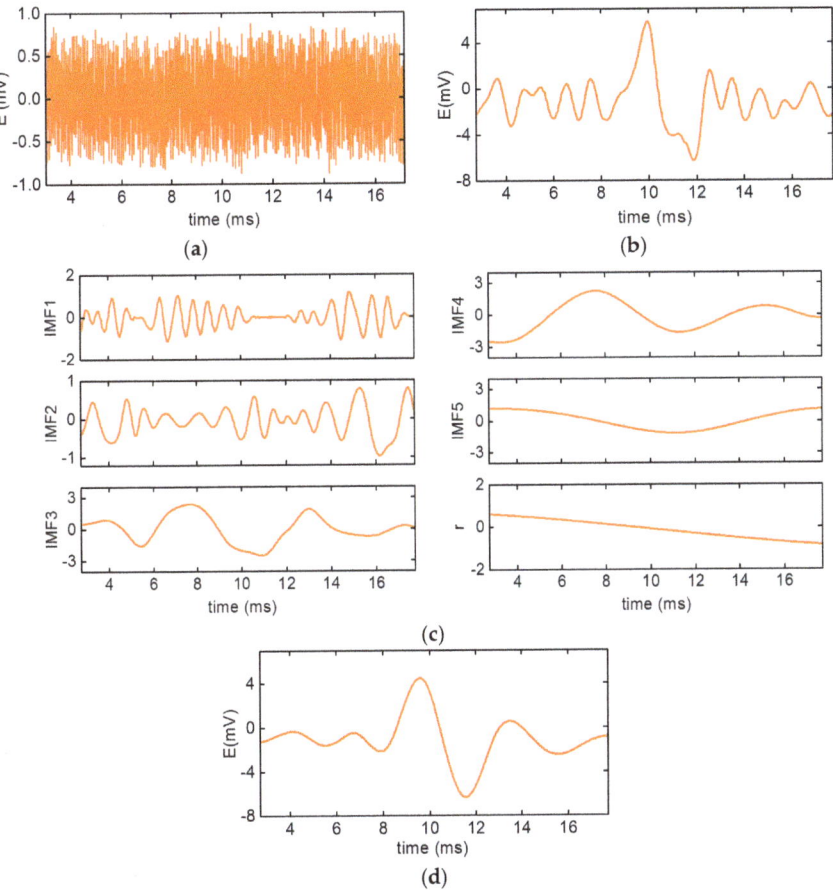

Figure 7. The simulation of the signal extraction process. (**a**) The raw signal of the sensor, (**b**) the preliminarily extracted particle signal, (**c**) the decomposed particle signal, and (**d**) the shaped particle signal.

To evaluate the validity of the proposed signal extraction and shaping method, the SNR values of the raw signal, preliminarily extracted signal, and shaped signal are calculated and presented in Table 1. The result illustrates that the SNR of the signal is greatly improved, which contributes to boosting the particle detection effect of the sensor.

Table 1. The signal-to-noise ratio (SNR) of signals.

SNR of Raw Signal	SNR of Preliminarily Extracted Signal	SNR of Shaping Signal
−21.37 dB	3.71 dB	13.181 dB

2.4. Analysis of the Computational Cost and Performance of Methods

As wear particles are monitored in real time by an electromagnetic wear particle detector, the computational efficiency of particle signal extraction algorithms and the correctness of detection results are of important concern. Therefore, in this section, a comparative analysis, involving the computational cost and extraction effect of particle signals incurred by the application of RSD-FC (resonance-based signal decomposition method and fractional calculus) [20], VMD-based method (variational mode decomposition) [26–28], and EMD-RRC (empirical mode decomposition and reverse reconstruction), is presented.

With respect to EMD and VMD, the algorithms decompose raw signals into several sub-signals (modes). However, the implementation of VMD requires first performing a Hilbert transform which involves an EMD process, so the VMD carries on a computational cost higher than the EMD. Besides that, the VMD requires a predetermined number of decomposition level k, which greatly influences its decomposition effect and computational efficiency [28]. Moreover, it's difficult to adjust the value of k for the optimal decomposition effect self-adaptively. The RSD-FC expresses a signal as the sum of a 'high-resonance' component which generally represents the interferences and a 'low-resonance' component which characterizes the particle signal. To achieve this goal, a morphology component analysis needs to be conducted, in which, an iterative optimization algorithm is utilized to update the transform coefficient matrices [20], so the method requires extensive calculations. To evaluate the computational efficiency, the preliminarily extracted particle signal with a sampling time of 1 s, extended from the data of Figure 7b, is processed using different algorithms running on a PC (Intel(R) Core(TM) i7-4720HQ CPU, 2.60 GHz, 8 GB RAM, Windows 10 operating system). For effective detection of wear particles with high speed, the sampling frequency is set to 3000 Hz. The theoretical peak-to-peak value of the particle signal output by the sensor is 10 mV. The performance of the algorithms is evaluated using the mean signal-to-noise ratio (MSNR), mean peak-to-peak value (MPPV), and mean relative amplitude error (MRAE):

$$MRAE = \frac{1}{n}\sum_{i=1}^{n}\left|\frac{T_i - M_i}{T_i}\right| \times 100\% \qquad (30)$$

where, T_i and M_i represent respectively, the theoretical and measured peak-to-peak value of particle signals, and n is the number of samples.

The extraction results of particle signals by RSD-FC, VDM-based method ($k = 7$), and the EMD-RRC are demonstrated in Figure 8a–c, which shows that the residual interferences in preliminarily extracted particle signals are removed to different degrees. The computational time and the performance of the algorithms are displayed in Table 2. It can be seen that all the methods do improve the SNR of signals to a certain degree and the MSNR of the extracted particle signals are higher than 10, which contributes to the effective detection of micro-particles. Furthermore, among these methods, the computational time of the RSD-FC is the longest and reaches to 1.9548 s, which is much larger than the sampling time (1 s). Therefore, it is difficult to guarantee real-time performance of particle detection sensors. Besides that, the correctness of the particle detection results is relatively poor. The MPPV and MRAE of particle signals extracted by the RSD-FC are 9.26 mV and 7.4%, respectively. For the VMD-based method, with the increase of the number of decomposition level k, the computational time rises accordingly. Moreover, comprehensively considering the evaluation indicators, the VMD-based method with $k = 7$ performs best (MSNR = 13.357 dB, MPPV = 9.71 mV, and MRAE = 2.9%). However, in this case, the computational time is 1.4942 s, which is also larger than the sampling time (1 s). While for the proposed EMD-RRC method, the MPPV and the MRAE of signals are 9.68 mV and 3.2%, respectively. Although, they are slightly lower than that of the VMD-based method with $k = 7$, the average computational time is only about 0.83 s which is sufficient to process the data of 1 s long with 3000 samples in real time. In summary, the proposed method is sufficiently fast for on-line application in terms of both computational efficiency and detection quality.

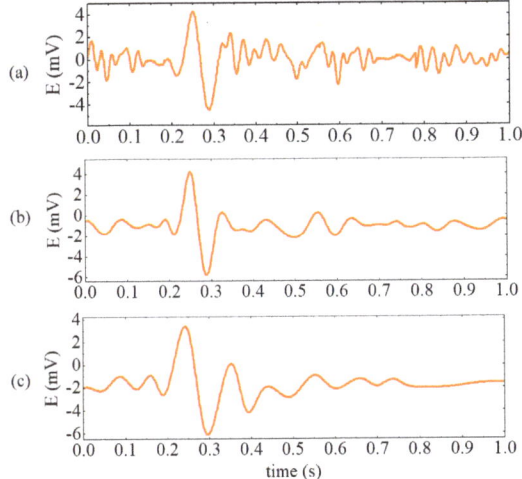

Figure 8. The extraction results of different algorithms applied on the preliminarily extracted signals. (**a**) resonance-based signal decomposition method and fractional calculus (RSD-FC), (**b**) variational mode decomposition (VMD)-based method with k = 7, (**c**) empirical mode decomposition and reverse reconstruction (EMD-RCC).

Table 2. Computational times and the performance of different algorithms.

Algorithms	MSNR (dB)	MPPV (mV)	MRAE	Mean Computational Times (s)
RSD-FC	8.854	9.26	7.4%	1.9548
VMD-Based (k = 4)	10.347	9.27	7.3%	0.9368
VMD-Based (k = 5)	11.755	9.39	6.1%	1.1256
VMD-Based (k = 6)	12.982	9.54	4.6%	1.3573
VMD-Based (k = 7)	13.357	9.69	3.1%	1.4942
VMD-Based (k = 8)	12.793	9.60	4.0%	1.5783
VMD-Based (k = 9)	12.584	9.52	4.8%	1.7137
EMD-RRC	13.181	9.68	3.2%	0.8314

3. Experiment

3.1. Experimental System

To verify the improvement of the sensitivity and the detectability of the sensor contributed by the resonance mechanism, the amorphous iron core, and the proposed signal measurement system, the detection efficiencies of the conventional and proposed sensors for wear particles were tested. The complete experimental system, as shown in Figure 9a, consists of the sensor, the excitation and detection unit, which is used to supply the exciting signal and to extract the particle signal, and the data collecting and processing software. The core parameters of the sensors adopted in the experiments are listed in Table 3. Furthermore, some sphere-like iron particles with the diameters of 75, 120, and 150 μm are selected by the scanning electron microscope as target particles, as shown in Figure 9b. The previous experimental research shows [29] that the lubricating oil does not affect the signal of the sensor, so the sensitivity analysis experiments are conducted under an oil-less condition.

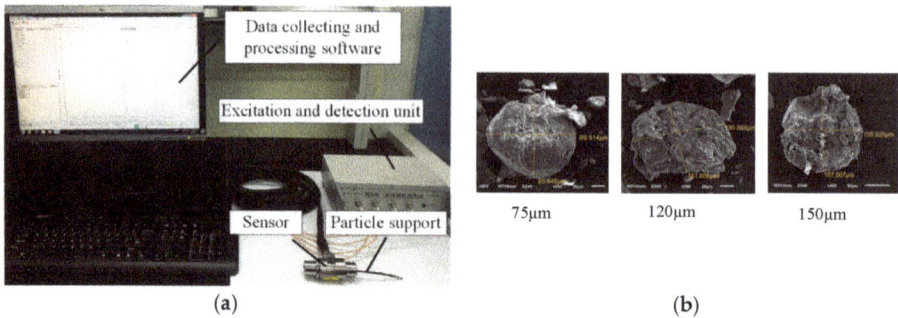

Figure 9. Experimental system. (**a**) The particle detection system, (**b**) the selected iron particles.

Table 3. The core parameters of the sensors adopted in the experiments.

Parameters	Conventional Sensor	The Proposed Sensor
Inner diameter of the sensor	7 mm	7 mm
Width of the coils	2 mm	2 mm
Inner diameter of exciting coils	9 mm	9 mm
Number of turns of exciting coils N_e	127	127
Number of turns of exciting coils N_i	110	110
Inner diameter of inductive coil	11 mm	11 mm
Inner diameter of amorphous core	-	9 mm
Outer diameter of amorphous core	-	11 mm
Resonant exciting capacitance C_1, C_2	-	1.0 nF
Resonant inductive capacitance C_3	-	0.63 nF

During the experiment, the measurement data shows that the initially induced electromotive forces of the sensors are about $E_0 = 7.3 \times 10^{-4} \sin(2\pi f_0 t)$ V and the Gaussian noise is very apparent. In this case, the particle signal is totally submerged in the inference. Taking the proposed sensor as an example, Figure 10 shows the raw signal of the sensor caused by a particle with the diameter of 120 µm. Because the particle speed may influence the signal extraction effect to a certain degree, particle detection experiments were conducted when the particle moved at the speed of 3 m/s, 5 m/s, and 8 m/s, respectively. The preliminarily extracted particle signal and the shaped particle signal are shown in Figure 11a,b, respectively. The results indicate that for the preliminarily extracted signals, a better detection is achieved at a higher particle speed. Moreover, after the signal shaping, the residual interference is further removed and the signals of the particle with different speeds can be effectively extracted. The SNR and peak-to-peak values of the particle signals are listed in Table 4, which shows that the proposed particle signal extraction method can greatly enhance the SNR of the particle signals and benefit the detection of micro wear particles. In addition, the peak-to-peak values of the signals are approximately consistent, which means that the signal measurement system has high fidelity.

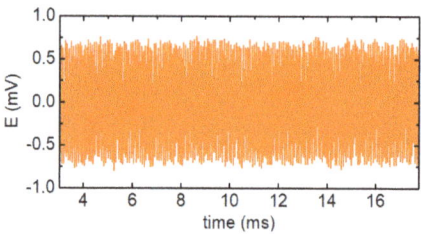

Figure 10. The raw signal of the proposed sensor.

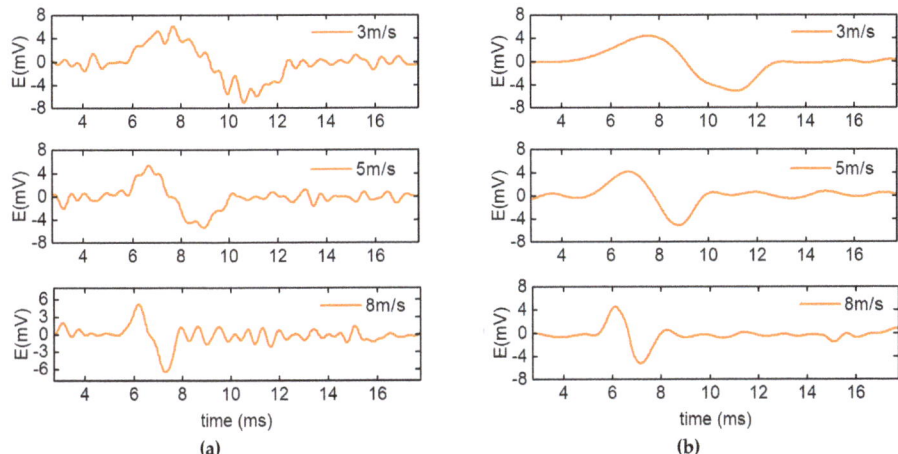

Figure 11. The signal of particles with different speeds. (**a**) The preliminarily extracted particle signals and (**b**) the shaped particle signals.

Table 4. The signal-to-noise ratio (SNR) of real signals.

Particle Speed (m/s)	SNR (dB)			Peak-To-Peak Value (mV)
	Raw Signal	Preliminarily Extracted Signal	Shaping Signal	
3	−14.318	9.341	12.933	9.62
5	−15.541	10.366	17.609	9.58
8	−19.917	11.625	15.173	9.55

3.2. Sensitivity Comparison for Ferromagnetic Particle Detection

To illustrate the sensitivity improvement by the proposed methods, both the conventional sensor, as shown in Figure 2a, and the proposed sensor, as shown in Figure 2b, were tested. Figure 12 shows the output signal of the sensors caused by the different sizes of ferromagnetic particles. In the figure, the green curve illustrates the signal output by the conventional sensor, and the orange curve represents the output signal of the proposed sensor, which adopts a resonance principle and an amorphous iron core. It can be seen that, for the conventional sensor, it is difficult to effectively detect iron particles less than 100 μm in diameter and the peak value of the induced electromotive force caused by a 100 μm iron particle is only 0.59 mV. However, for the proposed sensor, the signal amplitude of the particle with the diameter of 75 μm reaches 2.6 mV, which is much greater than that of the conventional sensor.

A comparison analysis of the detection result of the conventional sensor and the proposed sensor with various resonant capacitances is presented in Figure 13. It can be seen that the particle signal output by the proposed sensor is much larger than that of the traditional one, and with the decrease of the exciting capacitance, the sensitivity of the sensor gradually increases. The amplitude of the signal caused by a 75 μm iron particle, when the exciting capacitance equals 1 nF, is 2.6 mV, which is much greater than that under the circumstance of $C_1 = C_2 = 5$ nF (1.06 mV), and the increasing trend tends to be more evident for larger particles. However, excessive reduction of the resonant capacitance leads to a stronger eddy current effect in ferromagnetic particles and increases the current through the exciting coil rapidly, which may weaken the detectability for ferromagnetic particles and greatly reduce the reliability of the sensor. Therefore, a 1 nF resonance capacitance for the exciting coil is finally used for ferromagnetic particle detection.

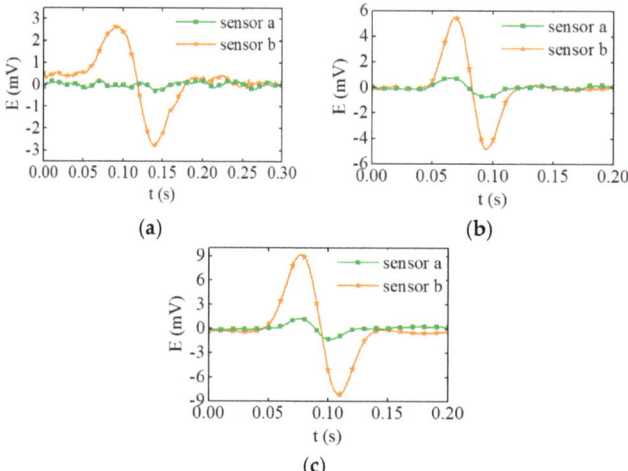

Figure 12. The particle signal output by the sensors. (**a**) 75 μm, (**b**) 120 μm, (**c**) 150 μm.

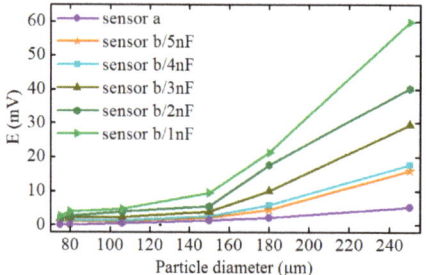

Figure 13. The comparison analysis of sensor sensitivity.

3.3. Wear Monitoring in a Real Oil Environment

To verify the detection effect of the sensor in a real oil environment, the sensor was assembled in the lubrication system with large ferromagnetic wear particles, comprised of 20 particles with a diameter of 80–100 μm, 20 particles with a diameter of 120–150 μm, and 20 particles with a diameter of 150–180 μm. These particles were added into the oil to simulate a serious wear fault of the mechanical equipment. The lubricating oil, including the wear particles, were driven by a pump and cycles through the sensor 20 times. By monitoring the wear particles using the sensor, the size distribution and the number of wear particles were estimated. The statistical result is displayed in Figure 14, which shows that the number of detected wear particles greater than 100 μm in diameter is approximately consistent with the standard value (400). However, the number of iron particles smaller than 100 μm in diameter is slightly more than the standard value. The possible reason for this phenomenon is that some parts of the larger wear particles may stick to the inner surface of the pipeline or be ground down to smaller particles by the blades of the pump during its running process. Therefore, based on the experimental result in a real oil environment, it can be concluded that the sensor can effectively monitor the quantity of the wear particles with different sizes, which helps to estimate the wear state of the mechanical equipment and to prevent mechanical failure caused by serious wear.

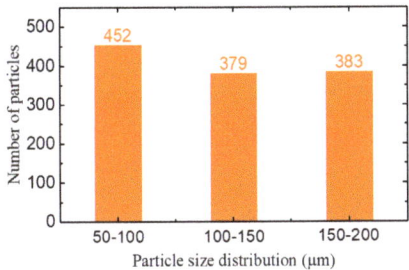

Figure 14. The statistical distribution of the wear particles.

4. Conclusions

The electromagnetic particles' detection sensor is of great importance due to its prospective application in various fields, and the sensitivity and detectability are still major obstacles in the development of wear particle detectors. Therefore, this paper has proposed that the resonance principle, an amorphous iron core, and a new signal measurement system are adopted to comprehensively improve the sensor sensitivity and detectability. Based on the work, the following conclusions are obtained:

(1) For the three-coil wear particle detector, the parallel resonant exciting coil magnifies the impedance difference between exciting circuits caused by particles. Additionally, the amorphous iron core and the series resonant inductive coil increase the magnetic flux through the coil and enhance the induced electromagnetic force of the sensor, which can comprehensively improve the particle signal more than six times compared to the conventional sensor.

(2) Under the resonance state, the nonlinear characteristics of the impedance difference between exciting circuits of the proposed sensor mean that the effective particle detection range of the sensor is restricted to $(0, r_a')$.

(3) Decreasing the resonant capacitance and increasing the exciting frequency can further improve the detection ability for micro-particles, though this reduces the effective particle detection range of sensors.

(4) By comparing different algorithms, the signal measurement system based on the MLIA and EMD-RRC guarantees the real-time ability for online particle detection and can effectively extract the particle signals from the raw signal with an extremely low SNR (≈ -20 dB). The experimental results indicate that based on the proposed method of improving the sensitivity and detectability, the large-calibre (7 mm) sensor can effectively monitor the initial abnormal wear of the heavy machines.

Author Contributions: Conceptualization, R.J., B.M., and C.Z.; methodology, R.J., L.W., and X.B.; validation, Q.D. and K.W.; writing—original draft preparation, R.J.; writing—review and editing, C.Z.; funding acquisition, C.Z. and L.W.

Funding: This work was funded by the National Natural Science Foundation of China (NSFC) (grant number: 51475044) and Beijing finance found of science and technology planning project (grant number: KZ201611232032).

Conflicts of Interest: The authors declare no conflict of interest.

References

1. Mabe, J.; Zubia, J.; Gorritxategi, E. Photonic low cost micro-sensor for in-Line wear particle detection in flowing lube oils. *Sensers* **2017**, *17*, 5863. [CrossRef] [PubMed]
2. Loutas, T.H.; Roulias, D.; Pauly, E.; Kostopoulos, V. The combined use of vibration, acoustic emission and oil debris on-line monitoring towards a more effective condition monitoring of rotating machinery. *Mech. Syst. Signal Process.* **2011**, *4*, 1339–1352. [CrossRef]

3. Guo, Z.; Yuan, C.; Yan, X.; Peng, Z. 3D surface characterizations of wear particles generated from lubricated regular concave cylinder liners. *Tribol. Lett.* **2014**, *1*, 131–142. [CrossRef]
4. Wu, T.; Wu, H.; Ying, D.; Peng, Z. Progress and trend of sensor technology for on-line oil monitoring. *Sci China Technol. Sci.* **2013**, *12*, 2914–2926. [CrossRef]
5. Flanagan, I.M.; Jordan, J.R.; Whittington, H.W. An inductive method for estimating the composition and size of metal particles. *Meas. Sci. Technol.* **1990**, *5*, 381–384. [CrossRef]
6. Fan, H.B.; Zhang, Y.T.; Ren, G.Q.; Li, Z.N. Experiment study of an on-line monitoring sensor for wear particles in oil. *Tribology* **2010**, *4*, 338–343. (In Chinese)
7. Kim, B.; Han, S.; Kim, K. Planar spiral coil design for a pulsed induction metal detector to improve the sensitivities. *IEEE Antenn. Wirel. Propag. Lett.* **2014**, *13*, 1501–1504.
8. Hong, W.; Wang, S.; Tomovic, M.; Han, L.; Shi, J. Radial inductive debris detection sensor and performance analysis. *Meas. Sci. Technol.* **2013**, *12*, 125103. [CrossRef]
9. Hong, H.; Liang, M. A fractional calculus technique for on-line detection of oil debris. *Meas. Sci. Technol.* **2008**, *5*, 55703. [CrossRef]
10. Li, C.; Liang, M. Extraction of oil debris signal using integral enhanced empirical mode decomposition and correlated reconstruction. *Meas. Sci. Technol.* **2011**, *8*, 85701. [CrossRef]
11. Zhan, H.; Song, Y.; Zhao, H.; Gu, J.; Yang, H.; Li, S. Study of the sensor for on-line lubricating oil debris monitoring. *Sens. Transducers* **2014**, *7*, 214–219.
12. Zeng, L.; Zhang, H.; Wang, Q.; Zhang, X. Monitoring of non-ferrous wear debris in hydraulic oil by detecting the equivalent resistance of inductive sensors. *Micromachines* **2018**, *3*, 117. [CrossRef] [PubMed]
13. Du, L.; Zhe, J.; Carletta, J.; Veillette, R.; Choy, F. Real-time monitoring of wear debris in lubrication oil using a microfluidic inductive Coulter counting device. *Microfluid. Nanofluid.* **2010**, *6*, 1241–1245. [CrossRef]
14. Wu, Y.; Zhang, H.; Zeng, L.; Chen, H.; Sun, Y. Determination of metal particles in oil using a microfluidic chip-based inductive sensor. *Instrum. Sci. Technol.* **2016**, *3*, 259–269. [CrossRef]
15. Du, L.; Zhu, X.; Han, Y.; Zhao, L.; Zhe, J. Improving sensitivity of an inductive pulse sensor for detection of metallic wear debris in lubricants using parallel LC resonance method. *Meas. Sci. Technol.* **2013**, *24*, 0751067. [CrossRef]
16. Zhu, X.L.; Zhong, C.; Zhe, J. A high sensitivity wear debris sensor using ferrite cores for online oil condition monitoring. *Meas. Sci. Technol.* **2017**, *7*, 075102. [CrossRef]
17. Fan, X.; Liang, M.; Yeap, T. A joint time-invariant wavelet transform and kurtosis approach to the improvement of in-line oil debris sensor capability. *Smart Mater. Struct.* **2009**, *8*, 085010. [CrossRef]
18. Li, C.; Liang, M. Enhancement of oil debris sensor capability by reliable debris signal extraction via wavelet domain target and interference signal tracking. *Measurement* **2013**, *4*, 1442–1453. [CrossRef]
19. Liang, M.; Li, C. Enhancement of the wear particle monitoring capability of oil debris sensors using a maximal overlap discrete wavelet transform with optimal decomposition depth. *Sensors* **2014**, *14*, 6207–6228.
20. Luo, J.; Yu, D.; Liang, M. Enhancement of oil particle sensor capability via resonance-based signal decomposition and fractional calculus. *Measurement* **2015**, *76*, 240–254. [CrossRef]
21. Barrios, M.L.R.; Montero, F.E.H.; Gómez Mancilla, J.C.; Marín, E.P. Application of lock-in amplifier on gear diagnosis. *Measurement* **2017**, *107*, 120–127. [CrossRef]
22. Jia, R.; Ma, B.; Zheng, C.S.; Wang, L.Y.; Ba, X.; Du, Q.; Wang, K. Magnetic properties of ferromagnetic particles under alternating magnetic fields: Focus on wear particle detection sensor applications. *Sensers* **2018**, *12*, 4144. [CrossRef] [PubMed]
23. Zeng, L.; Zhang, H.P.; Teng, H.B.; Zhang, X.M. Novel method foe detection of multi-contaminants in marine lubricants. *J. Mech. Eng.* **2018**, *54*, 125–132. [CrossRef]
24. Camarena-Martinez, D.; Valtierra-Rodriguez, M.; Perez-Ramirez, C.; Amezquita-Sanchez, J.; Rene, R.T.; Garcia-Perez, A. Novel down-sampling empirical mode decomposition approach for power quality analysis. *IEEE Trans. Ind. Electron.* **2015**, *63*, 2369–2378. [CrossRef]
25. Rai, A.; Upadhyay, S.H. Bearing performance degradation assessment based on a combination of empirical mode decomposition and k-medoids clustering. *Mech. Syst. Signal Process.* **2017**, *93*, 16–29. [CrossRef]
26. Lahmiri, S. A variational mode decomposition approach for analysis and forecasting of economic and financial time series. *Expert Syst. Appl.* **2016**, *55*, 268–273. [CrossRef]

27. Liu, Y.; Stone, J.E.; Cai, E.; Fei, J.; Lee, S.H.; Park, S.; Ha, T.; Selvin, P.R.; Schulten, K. VMD as a software for visualization and quantitative analysis of super resolution imaging and single particle tracking. *Biophys. J.* **2014**, *106*, 202a. [CrossRef]
28. Fan, J.; Zhu, Z.; Wei, L. An improved vmd with empirical mode decomposition and its application in incipient fault detection of rolling bearing. *IEEE Access* **2018**, *6*, 44483–44493.
29. Jia, R.; Ma, B.; Zheng, C.S.; Wang, L.Y.; Du, Q.; Wang, K. Sensitivity improvement method of on-line inductive wear particles monitor sensor. *J. Hunan Univ. Nat. Sci.* **2018**, *45*, 129–137.

© 2019 by the authors. Licensee MDPI, Basel, Switzerland. This article is an open access article distributed under the terms and conditions of the Creative Commons Attribution (CC BY) license (http://creativecommons.org/licenses/by/4.0/).

Article
Analysis of Satellite Compass Error's Spectrum

Andrzej Felski [1], Krzysztof Jaskólski [1,*], Karolina Zwolak [1] and Paweł Piskur [2]

1. Polish Naval Academy, Faculty of Navigation and Naval Weapons, Smidowicza St, 69, 81127 Gdynia, Poland; a.felski@amw.gdynia.pl (A.F.); k.zwolak@amw.gdynia.pl (K.Z.)
2. Polish Naval Academy, Faculty of Mechanical and Electrical Engineering, Smidowicza St, 69, 81127 Gdynia, Poland; p.piskur@amw.gdynia.pl
* Correspondence: k.jaskolski@amw.gdynia.pl

Received: 29 May 2020; Accepted: 17 July 2020; Published: 22 July 2020

Abstract: The satellite compass is one of new variants of satellite navigational devices. Is it still treated with caution on International Convention for the Safety of Life at Sea (SOLAS) vessels, but has become popular on the fishing vessels and pleasure crafts. The standard data obtained by such devices suggest accuracy of satellite compasses at a level of about 1 degree, so it seems to be as accurate as gyro or the magnetic equivalent. A changeability of heading errors, especially its frequency spectrum, is analyzed and presented in the paper. The results of comparison of an onboard standard gyrocompass, a fiber-optic gyrocompass (FOG) and a satellite compass in real shipping circumstances have been discussed based on the available literature and previous research. The similar comportment of these compasses are confirmed, however, in real circumstances it is difficult to separate heading oscillations produced by the ships yaw (or helmsman abilities) from the oscillations of the compass. Analysis of the heading oscillations has been performed based on the measurements of the heading indications of stationary compass devices and the devices mounted on the vehicles moving on the straight line (straight part of a road and tram line) to separate the impact of the vessel steering system. Results of heading changeability in the frequency domain are presented based on the Fourier transform theory.

Keywords: satellite compass; accuracy; spectrum analysis; Fourier transform

---

1. Introduction

It is impossible to navigate a vessel without any directional reference. All movements, no matter for people or for vehicles, in environments such as desserts, seas or air, require direction indicators. The contemporary ship is equipped with a magnetic compass and a gyrocompass as indispensable devices. This refers to all the open-sea ships which must be equipped in compliance with the International Convention for the Safety of Life at Sea (SOLAS). Toward the end of the 20th century, satellite compasses began to trace in a completely new way [1,2] in the form of a specific version of a multiantenna GNSS receiver with the additional option to determine a ship's heading. The most popular is the two-antenna solution, which gives an opportunity to measure two angles: heading and pitch or heading and roll, depending on how it is installed in relation to the centerline of the ship. Three-antenna solutions are also accessible. They allow measuring the full information of the ship's orientation in the space. According to Sperry Marine [3], one of the manufacturers of such devices, it has been designed as a low-cost alternative to conventional spinning-mass and fiber-optic gyrocompasses for application on workboats, commercial fishing vessels, large private yachts, naval patrol boats, and small merchant ships, which are not required to carry a gyrocompass.

The origin of this devices can be found in Very Long Base Interferometry (VLBI)—a radio-astronomical method in which space sources of electromagnetic signals (usually quasar) are collected by multiple radio telescopes distributed on the Earth [4]. On this basis, by means of correlation of random-type noise registered in the same time in different places (global network), distances between telescopes can be calculated. In the 1960s and 1970s, it was a very efficient method in geodesy, geodynamics etc. on a global scale.

Signals from satellites can be treated in the same manner. A correlation between signals received by an array of antennas, distributed in a specific way, give us an opportunity to calculate direction on the source of the signal (satellite) when the structure of this signals is known. In the case of a Global Positioning System (GPS) satellite, we are working with a 19 cm long electromagnetic wave, so the distribution of receiving antennas can be of around 1 m. By using two receiving systems and utilizing the carrier wave of GPS signals, we are in fact using (RTK) Real Time Kinematic GPS technology. In the simplest version, the two antennas, namely, base (primary) and rover (secondary), are situated along one of the axis of the ship. In the classical version of RTK, the coordinates of the primary antenna should be known; then, a spatial vector between both antennas can be calculated. As we are not interested in very accurate measurements of the antennas' positions and the base distance between both antennas is known (due to the fact that a base line between the two antennas is constant and situated in a constant position referring to the hull of the ship), the distances between each antenna and satellite enables us to determine the angles in two axes (Figure 1, axes X and Z).

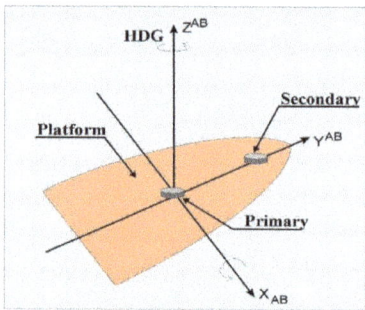

Figure 1. The idea of the use of two antennas to determine the heading of the ship [1].

Such ideas appeared at the end of the 20th century. The numerous publications of Calgary authors, including [5–8], are particularly noteworthy. Before the manufacturers proposed such devices, many researchers used them as reference systems for compass testing, for example [9,10]

Proposals for using the GPS system-derived devices to determine the angles of the spatial orientation of the object, using multiple antennas, appeared earlier. Anthony Evans is the author of significant achievements in this area by using the GPS system to determine ship orientation in the 1980s [11]. He proposed a method of measuring orientation angles through a single antenna that cyclically rotates inside the aircraft fuselage. In 1988, he began experimenting with an 18-channel receiver that used a system of three antennas spaced from 40 cm to 60 cm apart. This satellite compass precursor was tested under a marine conditions on the "USS Yorktown" to determine an accuracy during movement [12]. Parameters such as the duration of system initialization, maintaining its continuity and required accuracy in a real time in a dynamic environment, were examined at the beginning of satellite compasses development. Confirmation of the hypothesis was obtained that multi-antenna GPS receivers, in addition to the positioning ability, are able to determine reliable data regarding a ship's spatial orientation. Another example is the proposal contained in the patent of 1998 [13]. The authors proposed a compass, which determines its spatial orientation based on a construction with two antennas rotated by a stepper motor until a phase equalization of both antennas occurs.

The first devices available to wider users were introduced in the 1990s. In 1991, Ashtech launched the first multi-antenna GPS receiver: 3DF. Using this system, it was possible to determine the heading, longitudinal and transverse tilt, as well as the position, using a system consisting of four antennas, one of which served as the base antenna, and three others were supporting the base. Each antenna cooperated with a separate receiver. All of them, using signals from at least four different satellites simultaneously, by measuring the difference in phase, determined the orientation of the antenna assembly in three-dimensional space. The research was continued by scientists from the University of Calgary, who in 1994 conducted tests on the Canadian research ship, "Endeavor". Four GPS antennas were mounted on the vessel's helipad. The purpose of the tests was to compare the indications of this system with the Sperry Mark 3 Model C gyro compass available on the vessel [14,15]. Other analyses related to this subject, especially over the optimal configuration of antennas, have also been published in [5,16].

In 1994, Trimble introduced a four-antenna system called TANS Vector [17]. The system performance was based on phase measurements between one of the antennas (main, base) and each of the others, which were treated as slave antennas.

2. Materials and Methods

2.1. Background

Despite many tests confirming the usefulness of multi-antenna GNSS receivers for measurements related to spatial orientation, devices that met the criteria specified in international conventions have been developed no earlier than in 2005. Large errors appeared periodically in all the previous constructions, which were related to the changing satellite constellation parameters. An example of such measurements, taken in 2004 with the Crescent compass installed on the roof of the building is shown in Figure 2.

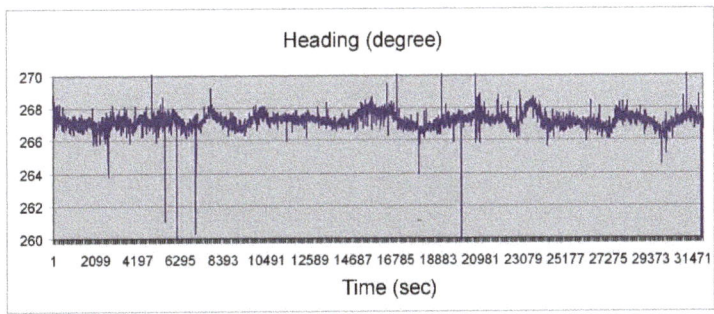

Figure 2. Heading measured with a satellite compass during stationary measurements [2].

The MX 575 compass, which as the first receiver of a certificate allowing it to be used as a heading transmitting devices, was introduced in 2005. It could be used as a backup source of heading information in IMO-compliant (International Maritime Organisation) vessels. One of the important solutions was the use of the MEMS-type (Microelectronic Mechanised System) gyroscope, which stabilizes the indications when "raw" measurements in the radio domain turn out to be temporarily inaccurate. Modern constructions often have a triad of gyroscopes and accelerometers. They are integrated systems, able to continue working for several minutes even in the event of satellite signals disappearance. The dynamic properties of such devices largely depend on the details of an algorithm used for the calculation and filtering of signals. The traditional gyro-compass has an electromechanical sensor, whose center of gravity is shifted relative to the geometric center, and thus behaves like a pendulum oscillating with a period of about 84 min (Schuler period, Schuler tuning). Maximilian Schuler made a proposal in 1923 that gyro compasses lend themselves to particularly successful tuning when the curvature of the Earth is taken into consideration. In this way, the instruments can be made insensitive to the disturbances that are caused by the result of the accelerations of the carriers along the surface of the Earth. According to this requirement, the instruments have to be tuned to an oscillation period of 84.3 min. Thus, the classical gyrocompass has its own fluctuation with a long period. There are no kinematic problems in satellite compasses, however, the results depend on changes in the satellite constellation and the properties of the measurement-processing algorithm. This is an issue addressed by the authors of this article. Manufacturers commonly describe the quality of such devices by declaring their accuracy based on an average square error or using similar methods. The assumption of white noise may not be true. Dynamic errors are caused by dynamic factors affecting the system, such as vibrations, roll, pitch or linear acceleration. According to [3] 'this error may have an amplitude and frequency related to the environmental influences and the parameters of the system itself'. However, for implementation in more complex measuring systems, when the fundamental issue is the selection of devices with different error characteristics, the question regarding the error frequency spectrum is important. The basic principle of integrating devices that perform similar functions is to vary the output error rate.

There are two main types of satellite compasses available on the market now: dual-antennas and tri-antennas. The most popular are dual-antenna constructions, which give the opportunity to measure two angles in transverse directions to the base between two antennas (pitch or roll in addition to heading). Designs with three antennas give the opportunity to measure all three angles of orientation of the carrier. Besides the number of antennas, there are devices with a constant distribution of antennas and movable antennas, so the distance between them can be changed by the owner or by the fitter. Additional sensors, commonly made in MEMS technology, are used to stabilize angular measurements. In addition to gyroscopes, these devices are often equipped with accelerometers. These are extremely useful for measuring the heave, which is thought to be important on small hydrographic units. In more extensive systems, there is also an option to include information from magnetic sensors or (and) a barometric measuring element. Possible block-diagram of a standard satellite compass was depicted in Figure 3.

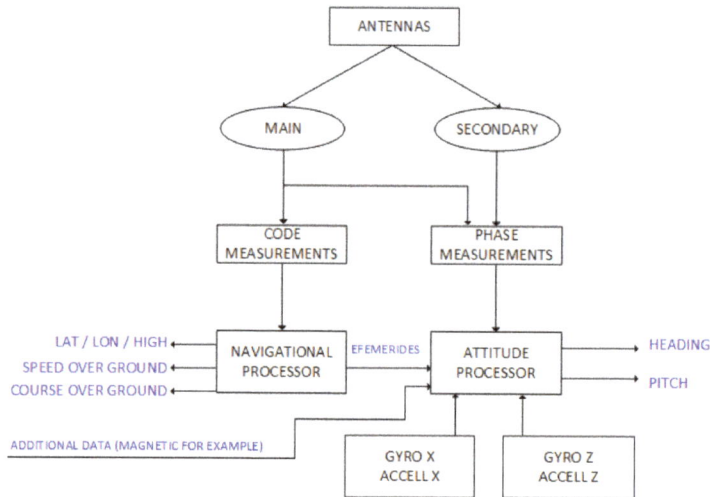

Figure 3. Possible block-diagram of a standard satellite compass. (Source: A.Felski).

Studies published in [18] proved that a satellite compass behaves similarly to a standard gyro or fiber-optic gyro (FOG) on a ship in motion. In the Figure 4 a small, systematic shift of measurements from individual devices can be noticed, however, this is due to inaccuracies in the installation of the satellite compass and FOG for the time of experiments. In general, all compasses seem to show almost the same values, and visible oscillations are probably due to imperfections in the control system and inertia of the ship. The existence of a very low frequency (Schuler tuning) characteristic of a classic gyro compass (NAVIGAT X) is noticeable. The most changes in the heading presented in the image occur due the behavior of the ship, so they are very similar to each other.

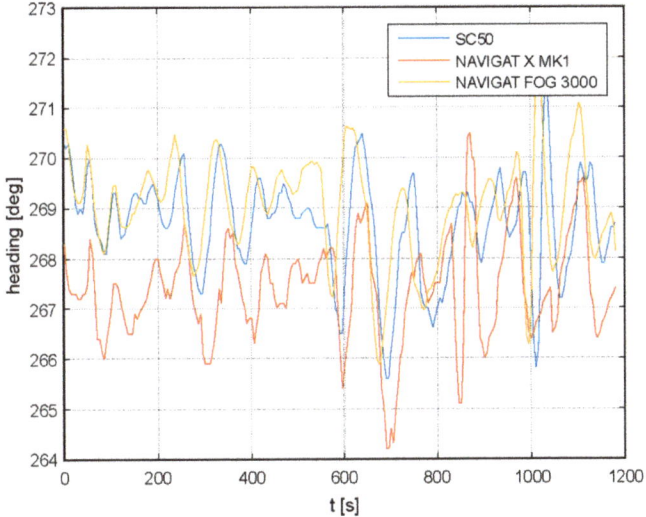

Figure 4. Example of registered headings from a classic gyrocompass (NAVIGAT X) fiber-optic gyro (NAVIGAT FOG 3000) and satellite compass FURUNO SC50 [18].

The spectral analysis of these measurements proves that low-frequency oscillations dominate, and one can also distinguish oscillations common to all three compasses, i.e., yaw resulting from the characteristics of the ships movement (0.01 Hz).

On the other hand, various frequency bands are not clearly repeated in registrations made with individual compasses. For example, significant differences occur at around 0.008 Hz for the satellite compass (as shown in Figure 5).

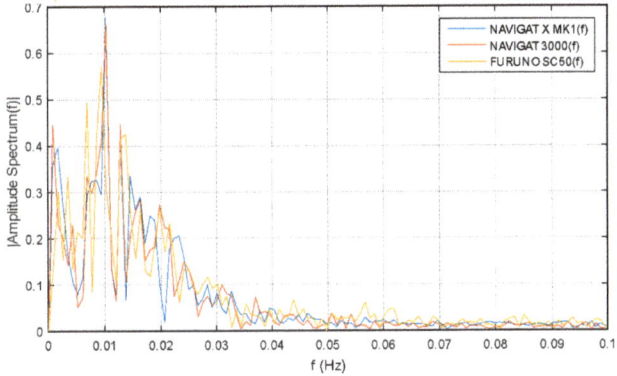

Figure 5. Single-side amplitude spectrum of oscillations presented in Figure 4 [18].

2.2. Devices

The satellite compasses of three different manufacturers were used in the experiments presented in this paper. These are Novatel PwrPak7D, the Advanced Navigation GNSS compass in a low-cost variant and Furuno SC-50. Basic parameters, based on specifications provided by manufacturers, are presented in Table 1.

Table 1. Basic technical parameters of the three compared satellite compasses used in the experiments [19–21].

No.	Feature	Novatel PwrPak7D-E1	Advanced Navigation	Furuno SC-50
1.	Position accuracy	0.6 m (DGNSS)	2.0 m or 0.6 m—(DGNSS)	3–10 m (depending on corrections)
2.	Velocity accuracy	0.03 m/s	0.05 m/s	no data
3.	Roll and pitch accuracy	no data	0.4°	no data
4.	Heading accuracy	0.1° (for 2 m base length)	0.2° (base length 620 mm)	0.5° RMS (base length 430.3 mm)
5.	Heave accuracy	no data	5% or 0.05 m (whichever is greater)	no data
6.	Output data rate	Up to 100Hz	Up to 100 Hz	Up to 40 Hz
7.	Base length	adjustable	permanent 620 mm	permanent 430.4 mm
8.	Supported navigation systems	GPS L1, L2	GPS L1 SBAS GALILEO E1 BeiDou B1	GPS L1

Where: (GPS) Global Position System, (SBAS) Satellite Based Augmentation System, L1, L2—frequency of GPS, (RMS) Root Mean Square, DGNSS - Differential Global Navigation Satellite System.

Novatel PwrPak7D-E1 is a robust GNSS receiver that combines dual antenna signal and (INS)—inertial navigation system hardware in a single enclosure to provide easy-to-deploy industry-leading position and heading data. In this experiments, two G5Ant-4AT1 models made by Antcom were used. According to the manufacturers, the device is suitable for ground vehicle, marine and air-based systems. Its software takes into account SPAN (synchronous position, attitude and navigation) technology based on GNSS+INS sensors as well as ALIGN software for angular determinations. It uses an OEM7720 receiver card and Epson G320N MEMS (IMU)—inertial measurement unit.

Advanced Navigation is a compact, low weight device, designed for marine and automotive applications, including small-size vehicles. It contains a 9 axis IMU that is integrated with a dual antenna GNSS receiver. Antennas are placed inside a 672 mm enclosure, together with all the signal processing electronic components. This seems to be around 3.25 lengths of the L1 wave of distance between the centers of the antennas. The core of this device is composed of two u-blox M8T GPS modules. The incorporation of all the processing components into the antennas enclosure makes it easy to integrate it even in restricted space conditions. Data are sent through the serial cable or via the (NMEA) National Marine Electronics Association 2000 network. The device is certified to be used on commercial vessels.

Furuno SC-50 is a popular satellite compass for commercial shipping. The large size of its processing unit qualifies it specifically for use on vessels. With its three-fold antenna, it can be useful for surveying ships. Clear display is designed for installation on the vessel's bridge, although data or data recorders can be sent to other devices using the standard marine format, NMEA 0183, or NMEA 2000 with the proper converters. The SC-303 antenna unit applied in tests consists of three antennas in one robust housing with a 650 mm diameter and a distance between the centers of the antennas of 430 mm. This is about 2.25 of the L1 wave length.

In summary, three different GPS receivers with different antennas and different distances between them, as well as different algorithms for angular calculations, were tested. The data were acquired with a data recording rate of 1 Hz.

2.3. Conducted Experiments

In this paper, two kinds of experiments are reported: stationary and dynamic. During the stationary part of experiments, all the antennas were situated on a building roof or bench in a suburban area. The second part of tests was conducted using an automotive vehicle driven directly out of the urban area; however, the horizon was partially by trees. In addition, we used the part of our previous experiments conducted in Gdańsk on tramway routes and published in [22] when the surprisingly low accuracy of the Furuno compass was observed. We intended to verify how significant the influence of the surroundings was on the observed errors in that investigation.

2.3.1. Twenty-Four-Hour Stationary Measurement Experiment

The stationary experiment results displayed the typical characteristics of this type of test. It was performed from 27 April 1300 UTC to 28 April 1200 UTC, 2020 in the suburban area of Gdynia, Poland, in the vicinity of a wall of a one-story building, partially obscuring the sky from the north side. The compasses' antennas were placed 4 m above the ground (Figure 6b). Data were recorded using the NMEA 0183 protocol through the RS-232 and RS-422 serial ports. The satellite constellation was assessed before data registration and the cut off angle of 20 degrees was applied. GPS satellite elevation during the test is presented in Figure 7 and their visibility is presented in Figure 8. The Advanced Navigation compass was set to the stationary variant of measurements. The following settings were applied for Furuno SC50: sampling frequency, 1 Hz; position smooth, 5 s; (SOG) speed over ground smooth, 5 s. A sampling frequency of 1Hz was set for the Novatel compass.

Figure 6. Experimental setup: (**a**) on a car roof for the automotive tests; (**b**) on a house roof for the stationary sets. Photo: K. Zwolak.

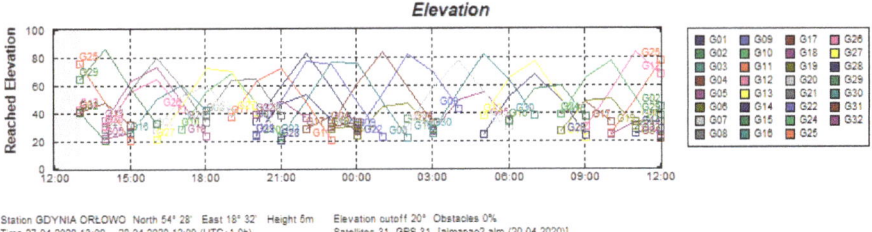

Figure 7. GPS satellite elevations during the experiment on 27–28 April 2020. (Source: Trimble Planning 2.9, 2010).

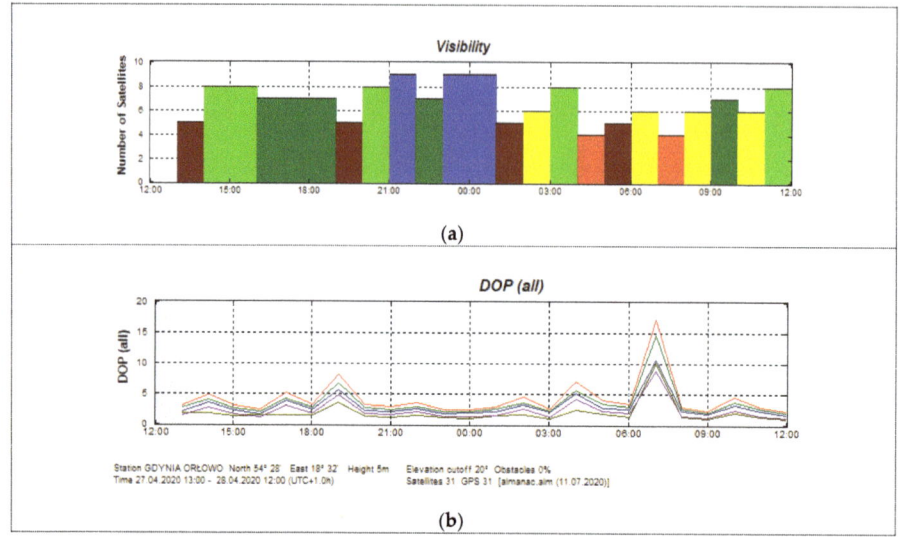

Figure 8. Visibility of satellites (**a**) and dilution of precision coefficients (**b**) in the measurement area on 27–28 April 2020. Notes: Visibility (**a**): red—4 satellites, brown—5 satellites, yellow—6 satellites, dark green—7 satellites, light green—8 satellites, blue—9 satellites. DOP (Dilution of Precision (**b**): red—geometrical DOP, green—position DOP, blue—vertical DOP, brown—horizontal DOP, magenta—time DOP. (Source: Trimble Planning 2.9, 2010).

It must be emphasized that the heading determination needs signals from at least five satellites, in contrast to the position determination, which requires four. Only four satellites have been available twice during the data registration (this occurred at about 0300 UTC and 0600 UTC on 28 April). The largest distortions between real time and average heading value in this test were observed for the FURUNO SC50 compass (Figure 9b). The maximum heading distortion for this compass is 2.8 degrees. The offset values fluctuated between −2.0 and +2.8 degrees. Similar results were observed for the Advanced Navigation compass with heading distortion values in the range of −2.2 to +2.7 degrees (Figure 9a). The lowest values of the exchange rate distortion registered for the NOVATEL PWRPAK 7D-E1 compass, however, during the tests, its antennas were 1.2 m apart, which is twice the distance of the other two cases. The heading distortion in this case varied from −0.7 to +0.8 degrees (Figure 9c).

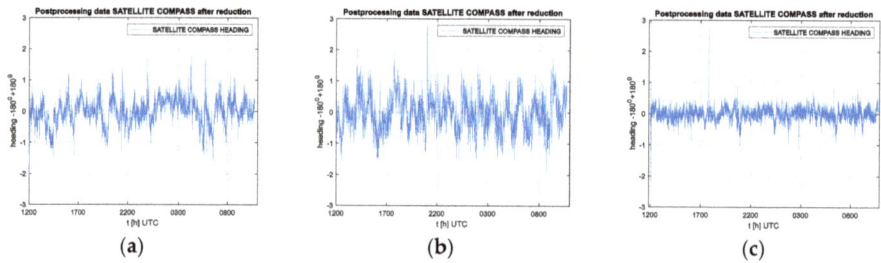

Figure 9. The example subsets of stationary heading registrations by the Advanced Navigation compass (**a**), the Furuno compass (**b**) and the Novatel compass (**c**) (27, 28 April 2020).

The root mean square of the heading distortion for the Furuno compass is 0.6 degrees, for Advance Navigation is 0.4 degrees and for Novatel is 0.2 degrees.

In order to perform the spectrum analysis of the signal in the frequency domain, the presentation of the frequency band in the range above f = 0.1 Hz was abandoned due to the negligible variability of the signal amplitude—heading distortion, which is typical for stationary measurements. Stationary registration results are very similar for all three compasses, characterized by a very low frequency of heading changes, falling in the band lower than 0.02 Hz. However, the amplitudes of these changes vary. The maximum value for the Furuno compass was 0.21 degrees, for Advanced Navigation 0.19 degrees, and for the Novatel product, it was only 0.08 degrees. Undoubtedly, this is due to the length of the base line between antennas, but it can be assumed that this is also the result of the different method of filtration or azimuth calculation. Heading registration spectrum for the three compasses are shown in the Figure 10.

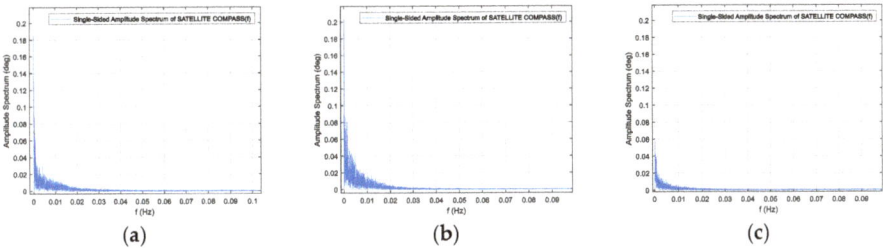

Figure 10. Heading registration spectrum for the three compasses: (**a**) Advanced Navigation, (**b**) Furuno, (**c**) Novatel. Stationary experiment, 27, 28 April 2020.

2.3.2. Automotive Experiments

The analysis of heading oscillations in compass indications in stationary conditions were performed based on calculating distortions from the average value. In this case, the direction of the compass does not matter. In dynamic conditions, the reference direction is needed. Therefore, the tests were conducted in such conditions that the direction of movement of the object was known and determined by natural conditions, i.e., on a straight sections of road or tram track. Knowing the heading of the vehicle during movement, the distortions of individual readings and the oscillations were calculated, treated as corrected measurements, and analyzed using a Fourier analysis. Matlab scripts were written to perform the analysis.

Automotive experiments, with the antennas mounted on the roof of the car, were carried out on a straight section of a rural road with a length of 1550 m and a direction of 342/162 degrees. There are single tall trees in the central part of the test section, along the road, and from the east side, which can occasionally cause interference. This is visible in the Figure 11 in the form of a break in position data registration due to incidental obstruction of the satellite signal. Such a gap is a result of a specific configuration of the satellites during this test. During other tests, similar gaps occurred in other places. Unfortunately, it was not possible to guarantee a repetitive configuration of the satellites, however, these records can be treated as examples of how important and diverse the impact of obstructions on the work of such compasses can be. The devices have options to adjust to the vehicle movement, that is, the Novatel compass has "sampling frequency: 1 Hz" and Furuno has "position smooth, 1 s; SOG smooth, 1 s; sampling frequency, 1 Hz". In addition, it is worth noting that the compasses have advanced inertial systems for the stabilization of readings, but this did not ensure the complete elimination of rapid changes at the time of appearance of another configuration of the satellites received by the device due to the appearance of obstructions.

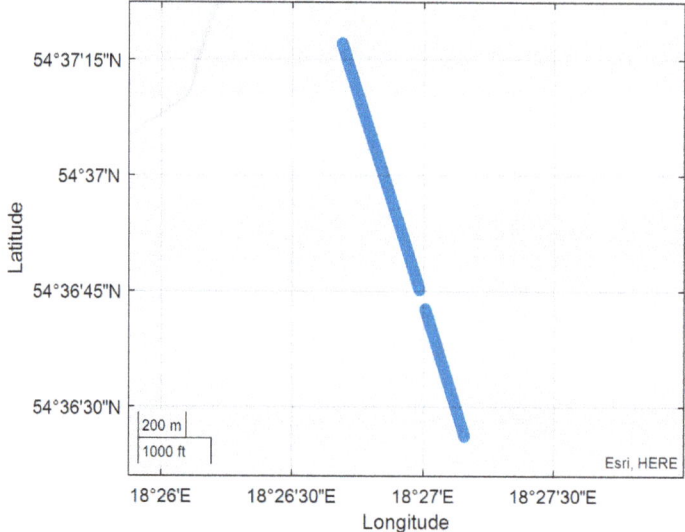

Figure 11. Positions recorded during the road test with the gap in data registration visible on the map view.

Measurements were carried out at speeds of 10, 20 and 30 km/h. Raw heading records for the compasses used in this part of the experiment are presented in Figure 12 for the Furuno compass and Figure 13 for the Novatel compass.

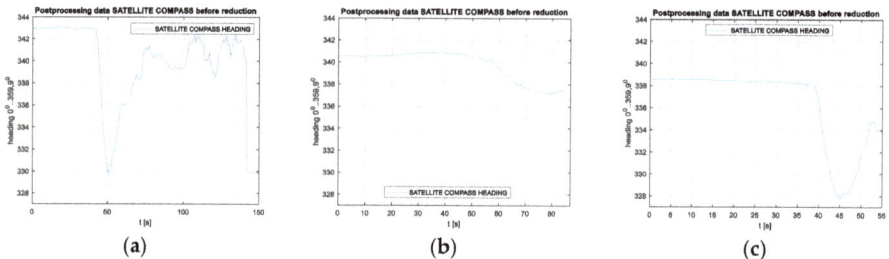

Figure 12. Raw heading records for the Furuno compass for the speeds of 10 km/h (**a**), 20 km/h (**b**) and 30 km/h (**c**).

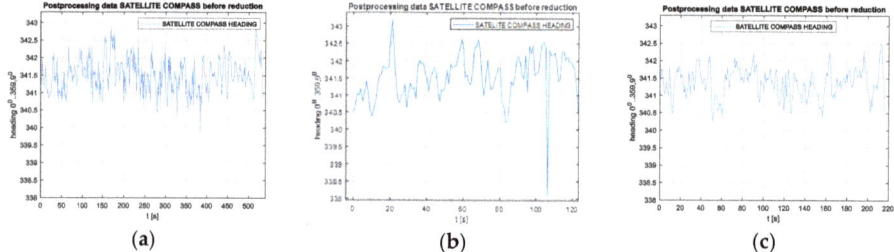

Figure 13. Raw heading records for the Novatel compass for the speeds of 10 km/h (**a**), 20 km/h (**b**) and 30 km/h (**c**).

The frequency spectrum of the signals presented above are plotted in Figures 14 and 15.

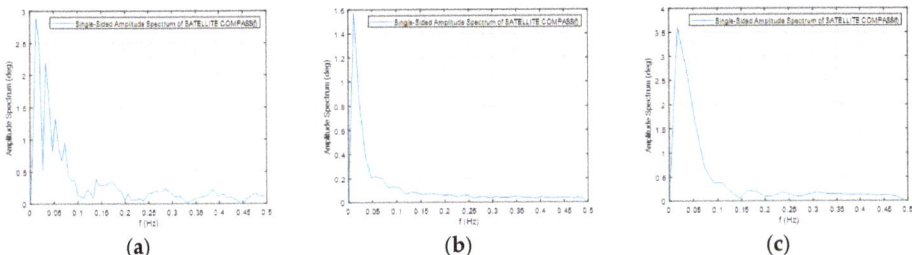

Figure 14. Frequency spectrum for the Furuno compass raw heading records in Figure 12 for the speeds of 10 km/h (**a**), 20 km/h (**b**) and 30 km/h (**c**).

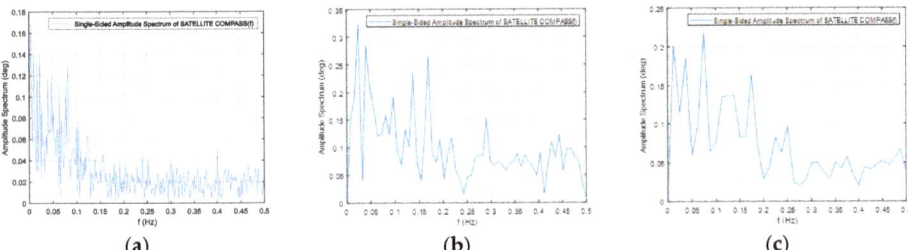

Figure 15. Frequency spectrum for the Novatel compass raw heading records in Figure 12 for the speeds of 10 km/h (**a**), 20 km/h (**b**) and 30 km/h (**c**).

In the context of a reaction to rapid changes in the satellite constellation, there is also a question regarding the impact on the stability of compass indications based on inertial sensors that are able to support the work of the radio (satellite) segment [23]. An example of the behavior of the Advanced Navigation compass in the case of complete obscuring of satellite signals (667 s after the start of registration) is shown in Figure 16. A clear drift of values is observed, which was similar in other tests, although the directions of the drift were different. After approximately 100 s, no information about the heading was reported by the device.

2.3.3. Tests on Tram Rails

The tram experiment was performed on 28 November 2018 along the route indicated in Figure 17 with the use FURUNO SC50 only. The experiment was conducted in Gdansk on a several-kilometer tram rail with variable sky visibility conditions, with the aim of assessing the performance (accuracy) of the satellite compass operation in non-standard terrain conditions. The measuring instrument used in the experiment was placed on a trolley of the DWF 300 series tram and pulled behind a tram [22] above the tram rails axis. The task of this measurements was more complex, and we now use only a small part of this registration made on the rail part, characterized by a constant direction (Figure 17).

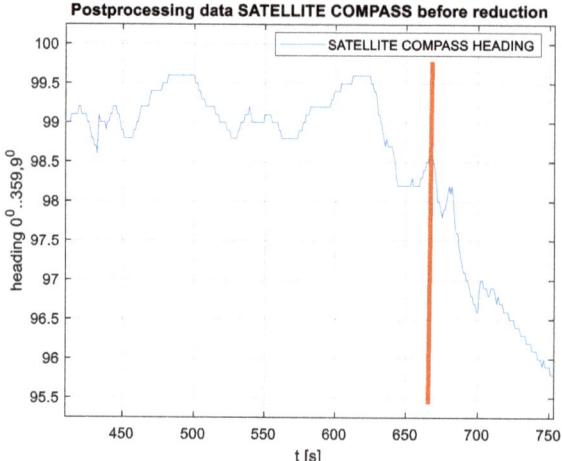

Figure 16. Raw heading value results during the stationary test with the Advance Navigation compass used as an example. The orange line denotes the satellite signals in the moment of being completely obscured.

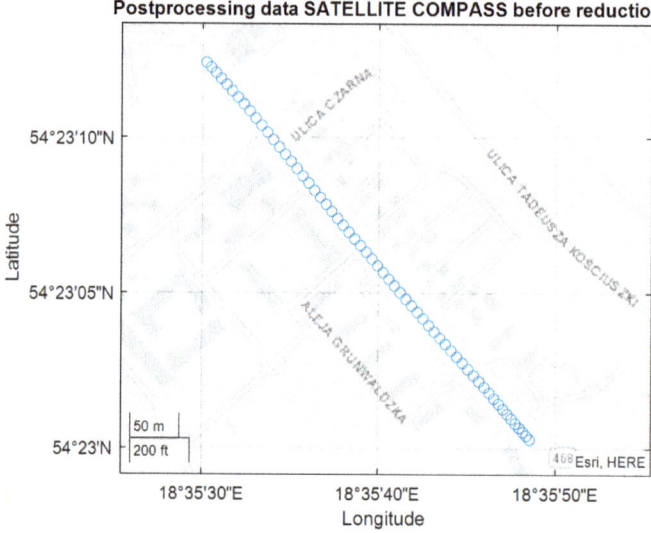

Figure 17. Position registration of the tram from 23:06:25 (UTC) - Universal Time Coordinated to 23:07:25 UTC.

For this analysis, it is important that the tram route ran through an urbanized area and some sky obstructions were observed during the registration. The GPS satellite elevations are presented in Figure 18.

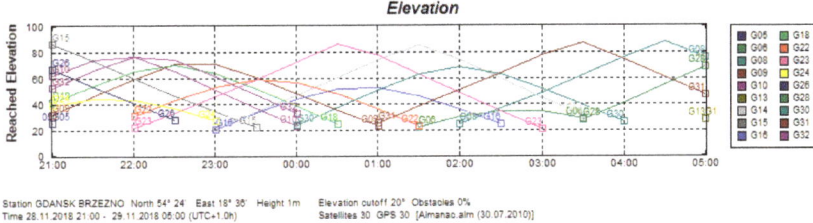

Figure 18. GPS satellite elevations on 28, 29 November 2018. (Source: Trimble Planning 2.9, 2010).

The satellite compass requires a signal from at least five satellites for each antenna to determine vehicle heading. Based on Figure 19, it can be seen that a condition of a visibility of at least five satellites to quantify the heading of the vehicle with the arbitrarily assumption of the elevation cut-off at 20 degrees has been met during the experiment.

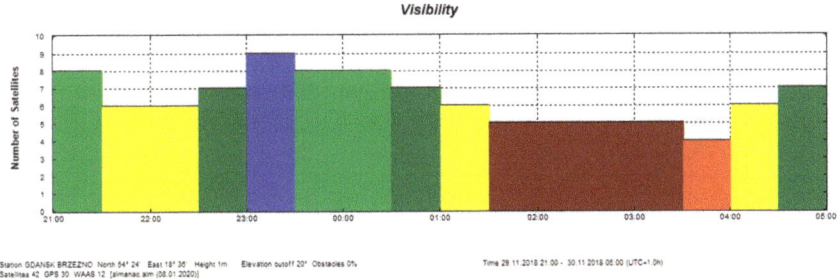

Figure 19. GPS satellite visibility on 28, 29 November 2018. (Source: Trimble Planning 2.9, 2010).

The presented test was started at 2200 UTC with the visibility of six GPS satellites (Figure 19). The problem occurred when the number of visible satellites was reduced to four (from 0230 UTC to 0300 UTC on 29 November 2018). Based on the data registered from 22:59:27 UTC on 28 November 2018 to 03:26:41 UTC on 29 November 2018, the parts of the straight tram rail section have been chosen. Data have been recorded from the 419th second to 479th second of the run, which is from 23:06:25 to 23:07:25, from the position (LAT) Latitude: 54.386780° N, (LON) Longitude: 018.591723° E to the position LAT: 54.383415° N, LON: 18.5968183° E. Heading oscillations for a vehicle on tram rails were observed in the range of −0.8. to + 0.5 degrees. The values of the heading distortions and the frequency spectrum of the heading record changes are presented in Figure 20.

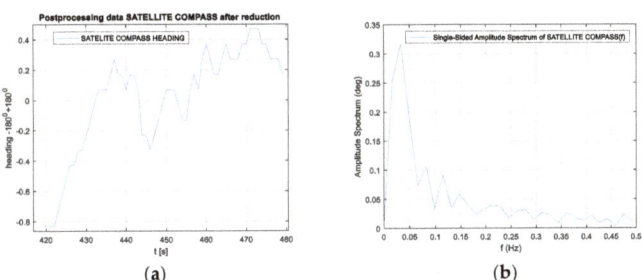

Figure 20. (a) Heading distortions. (b) Frequency spectrum of the heading records. Heading distortions and the frequency spectrum of the heading records from 23:06:25 to 23:07:25, from the position LAT: 54.386780° N, LON: 018.591723° E to the position LAT: 54.383415° N, LON: 18.5968183° E. (Source: K. Jaskólski).

The heading distortion analysis in the time domain in Figure 20a confirms the declared accuracy of the device indications in accordance with the technical specification of the device, which is 0.51 degrees (RMS)—Root Mean Square error. The spectrum analysis of the signal in the frequency domain in Figure 20b differs from that recorded during the stationary tests because oscillations appear at frequencies higher than 0.02 Hz. The maximum amplitude is slightly higher than that during the stationary record.

Another example of a registration on tram rails is shown in Figure 21 and research scores with a few course deviations are shown in Figure 22. In the heading record, there are four observed significant distortions from the track direction, which result from sky obstructions caused by high buildings in the vicinity of rails. The spectrum of this record differs significantly from others, which is undoubtedly caused by these four clearly distorted parts.

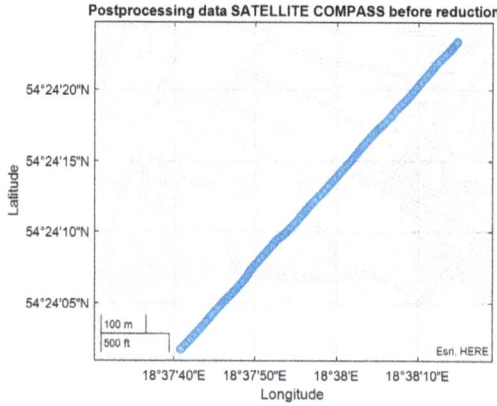

Figure 21. Position registration of the tram from 00:25:17 UTC to 00:30:37 UTC.

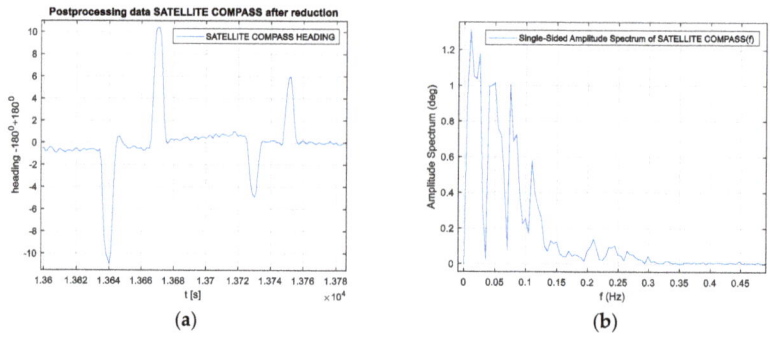

Figure 22. Heading distortions from 00:25:17 UTC to 00:30:37 UTC (**a**) and a frequency spectrum of heading changes (**b**).

3. Results

Studies have confirmed the different accuracies of the devices used in the experiment. It can be assumed that compasses with a fixed antenna system (Furuno and Advanced Navigation), built primarily for seagoing vessels, have an RMS error of approximately 0.5 degrees. The Novatel compass with adjustable distance between the antennas was observed to have higher accuracy, but this is obvious due to the fact that a longer antenna base line (over 1 meter) has been used.

It was confirmed that changes in the constellation of satellites accepted for the solution are the reason for the oscillations occurring in heading registrations. Typical frequencies appearing in the error spectrum are very low, but less than 0.02 Hz for objects in movement. These oscillations are caused by changes in the set of tracked satellites and the satellites included in the calculations. For moving objects, oscillations occur due to changes in the orientation of the vehicle, as well as unpredictable changes occurring as a result of obstructing the satellites by obstacles in the environment.

4. Discussion

The motivation to conduct the research was the authors' experience, published in [18], and especially the results described in the paper [22], in which the results of Furuno compass errors turned out to be surprisingly high. In this paper, the authors were interested in a solution that would be potentially useful on a small floating object used for hydrographic measurements. The issue of the accuracy of the position determined by such a device was purposely not analyzed here, because today it seems to be a trivial issue and comes down to the choice of support service (augmentation) of the GNSS system. This text focuses on the issue of accuracy of the heading, based on the knowledge that the accuracy of the heading measured with each type of compass has different dynamic characteristics. An assessment of a measurement of heading uncertainty on a moving object requires taking into account changes in object orientation angles, because the readings will include both this information and the inaccuracy of the compass or any other gauges. For this reason, three different compasses were tested during static and dynamic experiments. Dynamic experiments, with heading reference data included, were carried out at different speeds on straight sections of tram rails and on straight sections of a roads.

The results of stationary experiments confirm the clear relationship between the antenna base line length (distance between antennas) and heading accuracy. Analyzing the compass documentation, this relationship can be presented in a form of a curve, presented in Figure 23. The curve presented here is obtained by interpolating the sparse data regarding the Novatel compass with the quadratic function. The red square represents the manufacturer's information regarding the Furuno compass, which is consistent with data on Novatel. However, Advance Navigation data deviate from this relationship. The results of stationary tests confirm the declarations of the Novatel and Furuno compass manufacturers, although the Advance Navigation compass tests showed an error of twice the value—0.4 degrees.

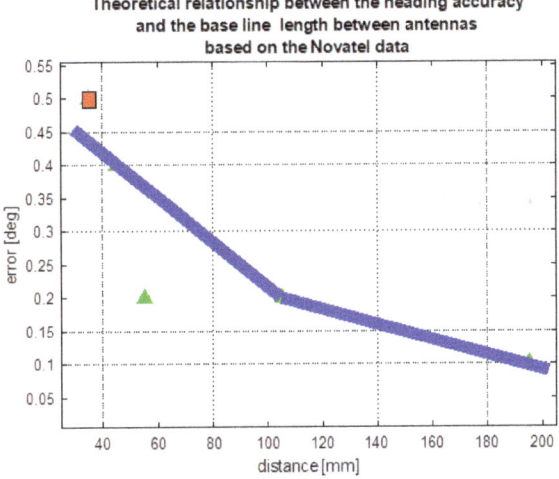

Figure 23. Theoretical relationship between the heading accuracy and the base line length between antennas, based on the Novatel data published in the device documentation (**black curve**). Declared accuracy of the Furuno compass as a function of baseline (**red square**) and of the Advanced Navigation compass (**green triangle**).

For dynamic applications, knowledge of the error spectrum is extremely important. This spectrum is characterized by the dominance of very low frequencies, which is understandable with the use of filtration and inertial sensor support. One can clearly indicate the 0.02 Hz value as the upper limit of these oscillations for all tested compasses. However, in dynamic conditions, higher frequencies appear, which result from slight changes in the orientation of the object during movement. In addition, there are also harmonics, which are in fact the results of rapid deviations of results from previous values. The reason is that despite the solutions using inertial sensors and filtration, these compasses are still sensitive to changes in the temporary constellation of satellites used for calculations. Even during the stationary tests, one can easily identify the relationship between changes in indicated heading values and changes in the satellite constellation. This statement is confirmed by the analysis of the devices' behavior at the time of signal obstructions. It is always associated with a sharp change in the value at the compass output, although the natural change in the set of satellites seems to have less effect on the results than the short-term screening of one or two satellites, which disappear after a few seconds.

The suppositions that appeared in relation to the results presented in [22], that the accuracy of the entire population of the recorded results do not reflect the whole truth regarding the compass, were also confirmed. It is necessary to take into account the effects of appearing obstructions, and the user should be aware of the effects of the nearby obstructions of the operation of the satellite compass.

The mentioned jumps in the heading measurement for a particular compass cannot be clearly determined with regards to time and amplitude. These changes, although similar and occurring almost at the same time in different compasses, are not identical. Sometimes they are slightly shifted in time for different compasses, relative to the change in constellation. This is clearly seen in Figure 9, where synchronous registrations made in the same place are shown, and sharp changes occur at different times. For example, after 1700 and just before 2200, the sharp changes can be observed in all three registrations at almost the same moments, but their amplitudes differ. There is no rule that one particular compass always shows larger changes. These changes can be considered as a typical filtration effect used in the integrated system. This kind of effect also appears in the coordinate indications referred to by the GNSS receiver, which is a separate element of the tested device. However, the code principle of a position measurement results in fundamental differences from the phase principle of angle measurement. Therefore, during measurements, especially in motion, there were several cases observed in which the heading indications were less accurate or even incorrect, while the coordinate values were correct. Such cases occur when, due to the presence of the obstructions, the configuration of the satellites has decreased to four, while setting the heading requires observation of signals from five satellites. Practitioners using such devices should pay attention to this fact. It is common to observe the values of DOP and (HDOP) – Horizontal Dilution of Precision as indicators of the quality of the receiver's work, while this refers to the position and is not true in relation to the heading (Figures 8 and 9). It seems that for applications, such as hydrographic vessels, it would be advisable to propose a new indicator to facilitate the control of heading information quality in satellite compasses.

5. Conclusions

Satellite compasses have been known for about 20 years and certainly now they can be treated as well-developed solutions. They are becoming increasingly popular in a wide range of applications, including shipping and aviation. They are particularly attractive in controlling self-propelled robots and machines. Their small size and light weight make them very attractive for small autonomous vehicles, except for underwater devices.

At the same time, it is obvious that, like any technical solution, satellite compasses have some limitations. The key is the dependence on satellite signals, which, in the case of the occurrence of terrain obstructions, becomes a significant limitation on land or in inland waters, where the near-shore objects often obstruct satellites. This aspect, however, also appeared recently in the context of potential interference with GNSS signals or spoofing. The literature and media report many cases of interference with satellite signals, even in the sea and in the air. This leads to the question of the legitimacy of the full dependency of navigation systems only on a satellite compass in terms of heading. This aspect is particularly important in the context of planned offshore autonomous vessels, where the risk of interference with satellite signals can cause many complications, not only in the context of the position, but also the heading and effective satellite communication.

When assessing the accuracy of satellite compasses, it should be noted that for several years, they have been constructed as systems supported by inertial sensors (gyroscopes and accelerometers), and now also by other sensors, such as magnetic sensors or those based on pressure. As a result, integrated systems are created, whose measurement properties largely depend on the structure of the data-processing algorithm. This, in turn, takes into account the purpose of the designed system. When assessing a modern satellite compass, one should take into account the branch of applications for which the device was constructed. There are different expectations for the fishing boat and the quadcopter. The requirements may even be different for a ship performing hydrographic measurements in river estuaries, and different for a ship performing similar measurements in the middle of the North Sea or the Gulf of Mexico.

The experiments presented here focused primarily on the spectrum of frequencies that appear in the recorded results of the heading measurements by the mean of satellite compasses. It was confirmed that the reason for the occurring oscillations is the changes in the constellation of satellites accepted for a heading solution. Typical frequencies appearing in the error spectrum are very low, below 0.02 Hz; however, the movement of the object on which they are installed, as well as obstacles causing obscuring satellites result in rapid changes in measurements values, which is evident in the registration of the occurrence of various higher frequencies in the spectrum of heading changes.

For people using such devices, importance should be placed on being aware that the quality of the satellite compass and the quality of the positioning receiver depend on various factors. Therefore, although they are in the same device and many users treat them as one device, one cannot draw conclusions about the quality of the course on the basis of indicators resulting from DOP, which characterize only the positional service. Geomagnetic storms and traveling ionospheric disturbances (TIDs) are also known as sources of GPS positioning quality deterioration. GPS scintillations lead to a range of errors in GNSS due to diffraction [24]. Deep signal fades that appear during small-scale irregularities effect the result in navigation outages [25,26]. Satellite compasses are very popular in high-latitude regions, however, the aurora borealis becomes especially visible in such regions, and the effect of the Earth's ionosphere on GNSS signal propagation (total electron content) is one of the main error sources which limits the accuracy and reliability of GNSS applications [27].

Author Contributions: Conceptualization, A.F. and K.J.; methodology, K.J.; software, K.J. and P.P.; validation, K.J., K.Z.; formal analysis, K.J.; investigation, A.F. and K.J.; data curation, K.J. and K.Z.; writing—original draft preparation, A.F., K.Z., K.J.; writing—review and editing, A.F., K.Z., K.J.; visualization, K.J., P.P.; supervision, A.F., K.Z., K.J. All authors have read and agree to the published version of the manuscript.

Funding: This research received no external funding.

Acknowledgments: The authors would like to thank the K2sea Company, for making the Advanced Navigation satellite compass available for testing.

Conflicts of Interest: The authors declare no conflict of interest.

References

1. Felski, A. Exploitative properties of different types of satellite compasses. *Annu. Navig.* **2010**, *16*, 33–40.
2. Felski, A.; Nowak, A. Practical Experiences from the Use of the Satellite Compass. In Proceedings of the ENC Conference, Parthenope University in Naples, Naples, Italy, 4–6 May 2009.
3. Satellite Compass. A Practical Guide to GNSS Transmitting Heading Devices for Marine Navigation. Northrop Grumman. 2005. Available online: https://www.sperrymarine.com/PageResources/767/NavistarGuide.pdf (accessed on 13 April 2020).

4. Bomford, G. *Geodesy*; Clarendon Press: Oxford, UK, 1980.
5. Comp, C. Optimal Antenna Configuration for GPS Based Attitude Determination. In Proceedings of the ION-GPS, Salt Lake City, UT, USA, 22–24 September 1993.
6. Lu, G. Development of a GPS Multi-Antenna System for Attitude Determination. Ph.D. Thesis, The University of Calgary, Calgary, AB, Canada, 1995.
7. Schleppe, J.B. A Real-Time Attitude System. Master's Thesis, The University of Calgary, Calgary, AB, Canada, 1996.
8. Jiun, H.K. Determining Heading and Pitch Using a Single Difference GPS/GLONASS Approach. Master's Thesis, The University of Calgary, Calgary, AB, Canada, 1999.
9. Felski, A.; Mięsikowski, M. Some Method of Determining the Characteristic Frequencies of Ship's Yawing and Errors of Ship's Compasses During the Sea-Trials. *Annu. Navig.* **2000**, *2*, 17–24.
10. Ruiz, S.; Font, J.; Griffiths, G.; Castellon, A. Estimation of heading gyrocompass error using a GPS 3DF system: Impact on ADCP measurements. *Sci. Mar.* **2002**, *66*, 347–354. [CrossRef]
11. Evans, A. Roll, Pitch and Yaw Determination Using a Global Positioning System Receiver and an Antenna Periodically Moving in a Plane. *Mar. Geod.* **1986**, *10*, 43–52. [CrossRef]
12. Kruczynski, L.R.; Li, P.C.; Evans, A.G.; Hermann, B.R. Using GPS to Determine Vehicle Attitude: USS Yorktown Test Results. In Proceedings of the 2nd International Technical Meeting of the Satellite Division of The Institute of Navigation (ION GPS 1989), Colorado Springs, CO, USA, 27–29 September 1989; pp. 163–171.
13. U.S. Patent No 5777578. Available online: http://patft.uspto.gov/netacgi/nph-Parser?Sect1=PTO1&Sect2=HITOFF&d=PALL&p=1&u=%2Fnetahtml%2FPTO%2Fsrchnum.htm&r=1&f=G&l=50&s1=5777578.PN.&OS=PN/5777578&RS=PN/5777578 (accessed on 12 April 2020).
14. Lu, G.; Lachapelle, G.; Kielland, P. Attitude determination in a survey launches using multi-antenna GPS technology. In Proceedings of the National Technical Meeting of the Institute of Navigation, San Francisco, CA, USA, 20–22 January 1993; pp. 251–259.
15. Lu, G.; Lachapelle, G.; Cannon, M.E.; Vogel, B. Performance Analysis of a Shipborne Gyrocompass with a Multi-Antenna GPS System. In Proceedings of the IEEE PLANS'94, Las Vegas, NV, USA, 11–15 April 1994; Available online: https://www.novatel.com/assets/Documents/Papers/File16.pdf (accessed on 12 April 2020).
16. Ueno, M.; Santerre, R.; Babineau, S. Impact of the Antenna Configuration on GPS Attitude Determination. In Proceedings of the 9th World Congress of the International Association of the Institutes of Navigation, Amsterdam, The Netherlands, 18–21 November 1997.
17. Tans Vector. *GPS Attitude Determination System. Specification and User's Manual*; Trimble: Sunnyvale, CA, USA, 1995.
18. Jaskolski, K.; Felski, A.; Piskur, P. The Compass Error Comparison of an Onboard Standard Gyrocompass, Fiber-Optic Gyrocompass (FOG) and Satellite Compass. *Sensors* **2019**, *19*, 1942. [CrossRef] [PubMed]
19. Novatel, PwrPak7 Family Installation and Operation User Manual. Available online: https://docs.novatel.com/OEM7/Content/PDFs/PwrPak7_Installation_Operation_Manual.pdf (accessed on 11 May 2020).
20. Advanced Navigation. Available online: https://www.advancednavigation.com/products/category/satellite-compass (accessed on 12 May 2020).
21. Furuno. Available online: http://www.furuno.se/fileadmin/files/Manuals/6_Pilot_Comp/SC-50/SC-50_OM_ENG_72510F.pdf (accessed on 9 May 2020).
22. Dąbrowski, P.S.; Specht, C.; Felski, A.; Koc, W.; Wilk, A.; Czaplewski, K.; Karwowski, K.; Jaskólski, K.; Specht, M.; Chrostowski, P.; et al. The Accuracy of a Marine Satellite Compass under Terrestrial Urban Conditions. *J. Mar. Sci. Eng.* **2020**, *8*, 18. [CrossRef]
23. Groves, P.D. *Principles of GPS, Inertial, and Multisensor Integrated Navigation Systems*; Artech House: Boston, MA, USA, 2008; pp. 58–521.
24. Gherm, V.E.; Zernov, N.; Strangeways, H. Effects of diffraction by ionospheric electron density irregularities on the range error in GNSS dual-frequency positioning and phase decorrelation. *Radio Sci.* **2011**, *46*, 1–10. [CrossRef]
25. Andrew, O.-O.A.; Doherty, P.H.; Carrano, C.S.; Valladares, C.E.; Groves, K.M. Impacts of ionospheric scintillations on GPS receivers intended for equatorial aviation applications. *Radio Sci.* **2012**, *47*, 1–11.

26. Yasyukevich, Y.; Vasilyev, R.; Ratovsky, K.; Setov, A.; Globa, M.; Syrovatskii, S.; Yasyukevich, A.; Kiselev, A.; Vesnin, A. Small-Scale Ionospheric Irregularities of Auroral Origin at Mid-latitudes during the 22 June 2015 Magnetic Storm and Their Effect on GPS Positioning. *Remote Sens.* **2020**, *12*, 1579. [CrossRef]
27. Warnant, R.; Lejeune, S.; Bavier, M. Space Weather Influence on Satellite-based Navigation and Precise Positioning. In *Space Weather. Astrophysics and Space Science Library*; Lilensten, J., Ed.; Springer: Dordrecht, The Netherlands, 2007; Volume 344. [CrossRef]

 © 2020 by the authors. Licensee MDPI, Basel, Switzerland. This article is an open access article distributed under the terms and conditions of the Creative Commons Attribution (CC BY) license (http://creativecommons.org/licenses/by/4.0/).

MDPI
St. Alban-Anlage 66
4052 Basel
Switzerland
Tel. +41 61 683 77 34
Fax +41 61 302 89 18
www.mdpi.com

Sensors Editorial Office
E-mail: sensors@mdpi.com
www.mdpi.com/journal/sensors